Advances in Environmental Microbiology

Volume 9

Series Editor
Christon J. Hurst
Cincinnati, Ohio
USA

and

Universidad del Valle
Santiago de Cali, Valle
Colombia

This book series addresses the questions of which microbes, microbial genes and gene products are present at particular places and times, as well as the environmental transport and survival capabilities of microbes. The authors define the ways in which microorganisms interact chemically as well as physically with their surroundings, including microbial actions that change our planet's geochemistry. *Advances in Environmental Microbiology* facilitates an understanding of how microbes have contributed towards coevolutionary processes and addresses microbial contributions to the successional colonization of environmental locations. The explorations of topics include a microbiological perspective of public health, animal husbandry and agricultural issues, including consideration of the fact that infectious diseases are often either acquired from environmental reservoirs or transmitted through the environment, plus an explanation of how microbial establishment either on or within a host results in transformation of the colonization site into a microbially modified environment. This series also will include both microbial pest control and microbial diversity, along with insights into industrial production processes that are connected to environmental microbiology.

More information about this series at https://link.springer.com/bookseries/11961

Christon J. Hurst
Editor

The Biological Role of a Virus

Springer

Editor
Christon J. Hurst
Cincinnati, Ohio, USA

Universidad del Valle
Santiago de Cali, Valle, Colombia

ISSN 2366-3324 ISSN 2366-3332 (electronic)
Advances in Environmental Microbiology
ISBN 978-3-030-85393-8 ISBN 978-3-030-85395-2 (eBook)
https://doi.org/10.1007/978-3-030-85395-2

Cover illustration: "Volvox reimagined" is a montage image created in 2015 and is being used with permission of the artist, Christon J. Hurst. Those images incorporated into this montage were: center image "Aspergillus flavus" (author: Hossein Mirhendi, image supplied by and used with author's permission); clockwise from upper right "Micrasterias" (author: Rogelio Moreno Gill, image supplied by and used with author's permission), "BrownGiantKelp3600ppx" (author: FASTILY, Creative Commons Attribution Share Alike 3.0 Unported license), "Stumpfungus" (author: Ecornerdropshop, public domain image), "Letharia vulpina JHollinger crop" (author: Jason Hollinger, Creative Commons Attribution-Share Alike 3.0 Unported license), "Phytoplankton - the foundation of the oceanic food chain", (author: NOAA MESA Project, public domain image), "Spirogyra cell" (author: Jasper Nance, Creative Commons Attribution-Share Alike 3.0 Unported license), "Morelasci" (author: Peter G. Werner, Creative Commons Attribution-Share Alike 3.0 Unported license).

This Springer imprint is published by the registered company Springer Nature Switzerland AG.
The registered company address is: Gewerbestrasse 11, 6330 Cham, Switzerland

“Mother”. This image is a self-portrait made by Natalia Osiatynska, 12 days into motherhood with her infant son Anker. It appears here courtesy of Natalia. Mammals rely upon usage of their endogenous Retroviridae in order to initiate the development of a placenta.

Dedication

I met Bill Benton in 1980 shortly after I had begun doing research for the United States Environmental Protection Agency in Cincinnati, Ohio. When I told Bill that he was the person whom I most wanted to meet at the EPA, he replied with surprise "Why?". My response was that his name had been on every publication from that Cincinnati group. Bill represented a kindly mixture of intelligence, pride, and humility. I wish that this world had more people like him.

Bill had a lot of good life stories to tell! He had experienced a difficult childhood and determined that he wanted to enter the US Army by attending the United States Military Academy in West Point, New York. Unfortunately, for the year when he applied, Bill ended up being the second person in line just below the cutoff level. He might have reapplied the following year, but Bill instead decided upon joining the US Army for infantry. Much to his disappointment Bill was transferred to the US Air Force where he then proudly served for 20 years as a medical corpsman. During his tours of duty in the United States, the United Kingdom, and Europe, Bill spent some of his spare time learning to farm. He cropped asparagus, which I think he enjoyed. The year that he spent raising tobacco was, from his perspective, not worth the financial gain considering the many required hours of hand labor plus the fact that harvesting and sorting tobacco leaves covers your hands with plant resin. Perhaps his most interesting adventure had been managing a British pub along with his wife Doris for a few years while Bill was stationed in England. After he retired from the US Air Force, Bill started working as a laboratory technician for the US Public Health Service. Eventually, the Public Health Service transferred him to the Environmental Protection Agency in Cincinnati where he became a cell culture technician.

A few years after I had begun working for the EPA, I was assigned to be the leader of a research group that included Bill. At that point, I had 10 years of laboratory experience compared to Bill's more than 35 years of laboratory experience. It also was interesting and a bit daunting for me to have a technician with so much more intelligence than I possess. Bill's kindness and humility never let me feel bad about my limitations. I do remember one time when I walked into his laboratory and told him that I had an idea for a new project. Bill listened, then reached into one of his desk drawers and pulled out a stack of perhaps 12 photocopied journal articles. I do not now remember what the project idea had been, but Bill handed the articles to me and said "These will get you started on it." Bill was amazingly far ahead of me on that idea, and he also was well ahead of me for many other research ideas. One of Bill's ideas was to try isolating viruses with two host cell lines grown in cocultivation and that worked very well.

The level of formal education that we achieve in life depends in large part upon the encouragement and expectations which our families demonstrate and foster inside of us during our childhood years. Bill merited a doctorate, but unfortunately had not been raised with expectations that he should try for that level of formal education. Otherwise, if allowed better expectations, then Bill more rightfully would have been my supervisor and I honorably would have reported my research to him.

I also remember one time when Bill mentioned to me that he never understood the 1960s. My answer was that he should listen to the song "Eleanor Rigby" by The Beatles. A few days later he happily told me of hearing the song over the radio in his automobile while he was driving to work. Bill was not buried along with his name as suggested by the lyrics of that song, because Bill's name will always remain recorded in the pages of scientific literature. He certainly remains recorded in the memories lodged within my mind.

Bill's laboratory equipment cart always displayed a State of Ohio temporary automobile license tag on which Bill had written the date of his eligibility for Civil Service retirement. He eventually did retire from the Civil Service, and then enjoyed spending many happy years with Doris. They watched the birds come to a feeding station behind their house and Bill worked at his golfing skills.

I feel very honored to have known Bill and to have worked with him. Appreciatively, I dedicate my work on this book to my friend and colleague William H. Benton. What might Bill's opinion be of this dedication? His humility might lead him to say that he appreciated the honor but somehow could not feel as though he ever had been that special as either a person or a scientist. Bill, you clearly were that special and I thank you for having been present in my life.

William H. Benton 1928–2000

Series Preface

The light of natural philosophy illuminates many subject areas including an understanding that microorganisms represent the foundation stone of our biosphere by having been the origin of life on Earth. Microbes, therefore, comprise the basis of our biological legacy. Comprehending the role of microbes in this world which together all species must share, studying not only the survival of microorganisms but as well their involvement in environmental processes, and defining their role in the ecology of other species, does represent for many of us the Mount Everest of science. Research in this area of biology dates to the original discovery of microorganisms by Antonie van Leeuwenhoek, when in 1675 and 1676 he used a microscope of his own creation to view what he termed "animalcula," or the "little animals," which lived and replicated in environmental samples of rainwater, well water, seawater, and water from snowmelt. van Leeuwenhoek maintained those environmental samples in his house and observed that the types and relative concentrations of organisms present in his samples changed and fluctuated with respect to time. During the intervening centuries, we have expanded our collective knowledge of these subjects which we now term to be environmental microbiology, but easily still recognize that many of the individual topics we have come to better understand and characterize initially were described by van Leeuwenhoek. van Leeuwenhoek was a draper by profession and fortunately for us his academic interests as a hobbyist went far beyond his professional challenges.

It is the goal of this series to present a broadly encompassing perspective regarding the principles of environmental microbiology and general microbial ecology. I am not sure whether Antonie van Leeuwenhoek could have foreseen where his discoveries have led, to the diversity of environmental microbiology subjects that we now study and the wealth of knowledge that we have accumulated. However, just as I always have enjoyed reading his account of environmental microorganisms, I feel that he would enjoy our efforts through this series to summarize what we have learned. I wonder, too, what the microbiologists of still future centuries would think of our efforts in comparison with those now unimaginable discoveries which they will have achieved. While we study the many wonders of microbiology, we also further our recognition that the microbes are our biological critics, and in the end they undoubtedly will have the final word regarding life on this planet.

Christon J. Hurst in Heidelberg

Indebted with gratitude, I wish to thank the numerous scientists whose collaborative efforts will be creating this series and those giants in microbiology upon whose shoulders we have stood, for we could not accomplish this goal without the advantage that those giants have afforded us. The confidence and very positive encouragement of the editorial staff at Springer DE have been appreciated tremendously and it is through their help that my colleagues and I are able to present this book series to you, our audience.

Cincinnati, OH Christon J. Hurst

Volume Preface

Long ago, a person asked me the question "What is it that viruses do? Viruses must do something because otherwise biology surely would have found a way to eliminate them." This book provides our best current answer to that question by presenting three aspects of viral ecology.

The first aspect presented is the ways in which viruses affect the population diversity and energetics of their host communities. Perhaps the most notable example for this concept would be our understanding that primary production within ecosystems often is dependent upon those viruses which serve as controllers of nutrient recycling. The result of that dependence is successful cycles that connect the aquatic and terrestrial realms on scales that can be assessed locally and globally.

The second aspect that we present is the genetic partnerships which exist between a host and many of its viruses. Interactions between viruses and their host organisms, and equally the interactions of viruses with their vectors, provide a force that drives mutual coevolution. Comprehending this partnership requires an understanding that many groups of viruses have two possible life cycles. One of those cycles is an option that we describe as being replicative, with its reliance upon the production of transmissible progeny virus particles. The second option, which is used by many virus groups, is termed to be either endogenous or lysogenous and entails the host carrying at least a partial genomic copy of the virus. Endogeny and lysogeny thus represent the forging of shared genomic fate which obligates partnership of the virus and its host. Genetic transference between a host species and its viruses results in the development of a collective genome, which has fluidity that operates on an evolutionary time scale. Collectively, the copies of viral genomes which the host carries represent a source of potential benefit and also potential peril for the host. These viral genomes can implement phenotypic changes in the host and the host often uses those changes as tools. As humans, the most notable example would be that mammals rely upon temporary activation of their endogenous viral genes in order to successfully develop a placenta. Fatal disease can result if those genes are activated at any other time during our life cycle.

The third aspect that we present is defending the health of a host and this defense relies upon activity targeted in several directions. Our presentations will help the reader to understand that hosts often use their captured viral genes to identify and subsequently direct battle against both those same viruses and against related viruses. We have borrowed this natural concept and engineered it as a means of combating cancer and as a technique for suppressing the detrimental consequences of genetic diseases. We also use this same approach to develop specifically targeted antiviral vaccines. Viral infections of a host are two-sided contests that often include collective participation by additional microbial groups. On the invading side, infecting viruses often attack in conjunction with other pathogenic microorganisms. The defending response mounted against those attacks includes symbiotic microbes that act as a shield to protect their host, and the host can play an active role in the functioning of that symbiotic microbial shield.

The information which we have presented in this book can be only a partial answer because virology still seems a few centuries away from understanding the full answer.

James Zmuda at home in Montana

In appreciation of the author Norman Maclean, and memory of my dear friend the virologist James Zmuda, both of whom liked the state of Montana, I would say that "Eventually, all things merge into one, and a virus runs through it."

Cincinnati, OH Christon J. Hurst

Contents

Part I Viral Control of Community Energetics

1 Viral Nature of the Aquatic Ecosystems 3
Daichi Morimoto, Kento Tominaga, Hiroaki Takebe, Sigitas Šulčius, and Takashi Yoshida

2 Life Continues as Viruses Close Land, Water and Atmosphere Nutrient Cycle 27
Peter Pollard

Part II Understanding the Genetic Partnership Between a Host and Its Viruses

3 Cataloging the Presence of Endogenous Viruses 47
Christon J. Hurst

4 Do the Biological Roles of Endogenous and Lysogenous Viruses Represent Faustian Bargains? 113
Christon J. Hurst

5 Einstein's Capsid: Bacteriophages Solve the Problems of Space and Time for Bacteria with Emergency Dead to Alive Horizontal Gene Transfer (EDA-HGT) 155
Leigh Owens

6 Diverse Phage-Encoded Toxins and Their Role in Bacterial Ecology 173
Sheryl L. W. Zajdowicz

7 Mycoviruses as Antivirulence Elements of Fungal Pathogens 209
Mirna Ćurković-Perica, Marin Ježić, and Daniel Rigling

Part III Defending the Health of Its Hosts

8 The Contribution of Viruses to Immune Systems 253
Felix Broecker

9 Application of Viruses for Gene Therapy and Vaccine Development 285
Kenneth Lundstrom

10 Eukaryotic Virus Interactions with Bacteria: Implications for Pathogenesis and Control 343
Melissa K. Jones, Erin A. Almand, Anand Soorneedi, and Matthew D. Moore

Part I
Viral Control of Community Energetics

Chapter 1
Viral Nature of the Aquatic Ecosystems

Daichi Morimoto, Kento Tominaga, Hiroaki Takebe, Sigitas Šulčius, and Takashi Yoshida

Abstract Viruses infecting microorganisms are ubiquitous and highly abundant in aquatic environments. They considerably affect the dynamics, diversity, and evolution of their host microorganisms. In this review, we discuss the ecological implications of viruses from the perspectives of the biogeochemical cycles, microbial diversity, and virus–host coevolutionary dynamics in aquatic environments. Generally, viruses redirect host metabolism toward reproduction through molecular host–virus interactions characterized by the compositional and stoichiometric changes in intracellular metabolites, which are eventually released into the environment when the infected host cells are lysed, thus also changing the chemical composition of the water. Therefore, the modulation of metabolite biosynthesis and promotion of their recycling are major viral functions. Viruses also maintain microbial community diversity via increased infection and lysis rates of the dominant taxa and genotypes in a frequency-dependent manner, thereby allowing the co-existence of members with various competitive abilities. Finally, viruses can expand their own genotypic diversity and that of the host through complex defense and counter-defense interactions, including loss of host fitness due to the cost of resistance and the possible need for antiviral defense-specific (e.g., intra- vs. extracellular) changes in the hosts genome diversification. Continuous interactions drive the coevolution of hosts and viruses, thereby increasing both the host and viral micro-diversity. Hence, these fundamental functions are viral "raison d'etre" and are essential for the functioning of aquatic ecosystems and its components.

D. Morimoto · K. Tominaga · H. Takebe · T. Yoshida (✉)
Graduate School of Agriculture, Kyoto University, Kyoto, Japan
e-mail: yoshida.takashi.7a@kyoto-u.ac.jp

S. Šulčius
Laboratory of Algology and Microbial Ecology, Nature Research Centre, Vilnius, Lithuania

C. J. Hurst (ed.), *The Biological Role of a Virus*, Advances in Environmental Microbiology 9, https://doi.org/10.1007/978-3-030-85395-2_1

1.1 Introduction

Viruses infecting microorganisms are ubiquitous and abundant in aquatic ecosystems (Suttle 2005, 2007). They typically are small particles (generally 20–200 nm in length) comprised of nucleic acids (single- or double-stranded DNA or RNA) and structural proteins and have no intrinsic metabolism. Thus, their reproduction depends entirely on host cellular metabolism and replication machinery. Viral reproduction can be classified as lytic or lysogenic (Guttman et al. 2005). During lytic infection, viruses inject their genomes into host microorganisms, redirect host metabolism for efficient viral genomic nucleic acid replication and protein synthesis, and are finally released through host cell lysis (Guttman et al. 2005). In contrast, in lysogenic infection, the viral genome is integrated into the host genome as a provirus (also called prophage if the virus integrates into the bacterial chromosome) and is propagated vertically within the host lineage until the induction of the lytic cycle under specific conditions (e.g., depending on host cell density or environmental conditions) (Howard-Varona et al. 2017a).

Both types of viral infections, lytic and lysogenic, have great potential to affect microbial communities in aquatic ecosystems. For example, viral-mediated cell lysis releases nutrients and organic matter from cells to the environment, thus stimulating biogeochemical cycling (Fuhrman 1999; Suttle 2005, 2007). In addition, viruses affect host microbial diversity in at least three different ways (Marston et al. 2012). First, viruses contribute to the maintenance of host microbial diversity by frequency-dependent infection, often seeming to have a greater effect upon those microbial taxa and genotypes that either are highly abundant or most metabolically active in the environment (Thingstad 2000). Second, viruses increase host genetic diversity via the reciprocal co-evolution of host resistance and viral infectivity (Buckling and Rainey 2002a). Lastly, viruses affect the genomic evolution and the fitness of microbial hosts through horizontal gene transfer (HGT) including the presence and movement of auxiliary metabolic genes (Breitbart et al. 2007; Hurwitz and U'Ren 2016), generalized transduction during lytic infection (Thierauf et al. 2009; Touchon et al. 2017), and specialized transduction during lysogenic infection (Gottesman and Yarmolinsky 1968; Fernandes et al. 1989; Campos et al. 2003; Touchon et al. 2017).

The ever-growing number of culturable viral isolates, together with recent advances in sequencing technologies and bioinformatics, provides us with a deeper understanding of the viral nature in aquatic ecosystems. This chapter summarizes current understanding of viral effects on biogeochemical cycles, microbial diversity, and virus–host evolutionary dynamics in aquatic ecosystems. The viral role in the promotion of HGT that affects host genomic evolution and fitness has been reviewed extensively elsewhere and is therefore not discussed here (e.g. Balcázar 2018; Yoshida et al. 2019).

1.2 Viral Influence on Biogeochemical Cycle

1.2.1 Viral Modulation on Patterns of Geochemical Cycling in the Ocean

In the ocean, approximately 10^{29} cells of different microorganisms form the basis of the marine food web (Whitman et al. 1998). Photosynthetic eukaryotes and prokaryotes contribute to up to 50% of the total net primary production on Earth (Field et al. 1998). Approximately half of the resultingly fixed carbon is released into the environment, re-mineralized by heterotrophic prokaryotes, and then incorporated into higher trophic levels of the aquatic food web (Azam et al. 1983). This process of recycling of photosynthetic products is called "microbial loop" and is an essential pathway of biogeochemical cycling in the ocean (Azam et al. 1983).

Viruses outnumber prokaryotes, and up to 20% of marine microorganisms are thought to be infected and lysed by viruses daily (according to a certain view in Suttle 2007). Lysis of infected cells leads to the release of organic matter and nutrients, which would otherwise be incorporated into higher trophic levels by grazing (Fuhrman 1999; Wilhelm and Suttle 1999). This pathway of carbon flux regulated by viruses is called "viral shunt" (Wilhelm and Suttle 1999) (Fig. 1.1). Calculations based on the presumed microbial biomass, its turnover rates, and the predicted daily lysis suggest that viral shunts are responsible for the release of approximately 25% of primary production in the surface ocean, which amounts to up to 3 gigatons of carbon into the oceans per year (Wilhelm and Suttle 1999; Suttle 2005). However, the quantification of virus-mediated carbon flux in natural environments remains challenging owing to methodological limitations and ecosystem complexity. Recently, more advanced nutrient–phytoplankton–zooplankton (NPZ) models, including heterotrophic bacteria and viruses, have proposed that viral shunts accelerate organic matter recycling and increase net primary productivity while reducing transfer to higher trophic levels (Weitz et al. 2015). Moreover, the effect of virus-mediated carbon flux may depend on the trophic status of the system and the limiting nutrients (Pourtois et al. 2020).

On the other hand, viruses can also contribute to carbon removal from the surface ocean (Weinbauer 2004; Sullivan et al. 2017; Laber et al. 2018). Cellular debris, including cytoplasmic material and components of the cell wall, released via viral lysis can easily aggregate and sink to the deeper layers leading to carbon sequestration (Weinbauer 2004; Laber et al. 2018). This alternative viral influence on the geochemical flux that promotes the biological carbon pump is referred to as the "viral shuttle" (Sullivan et al. 2017) (Fig. 1.1). For example, virus-induced carbon transportation has been extensively studied in *Emiliania huxleyi* (haptophyte) and its known virus (Laber et al. 2018). Several laboratory studies have reported that viral infection stimulates the production of transparent exopolymer particles (TEP), which increases the stickiness of cells, thus promoting aggregation (Rosenwasser et al. 2014; Nissimov et al. 2018). Field studies monitoring *E. huxleyi* blooms in the North Atlantic have revealed that TEP concentrations increase during the early stages of

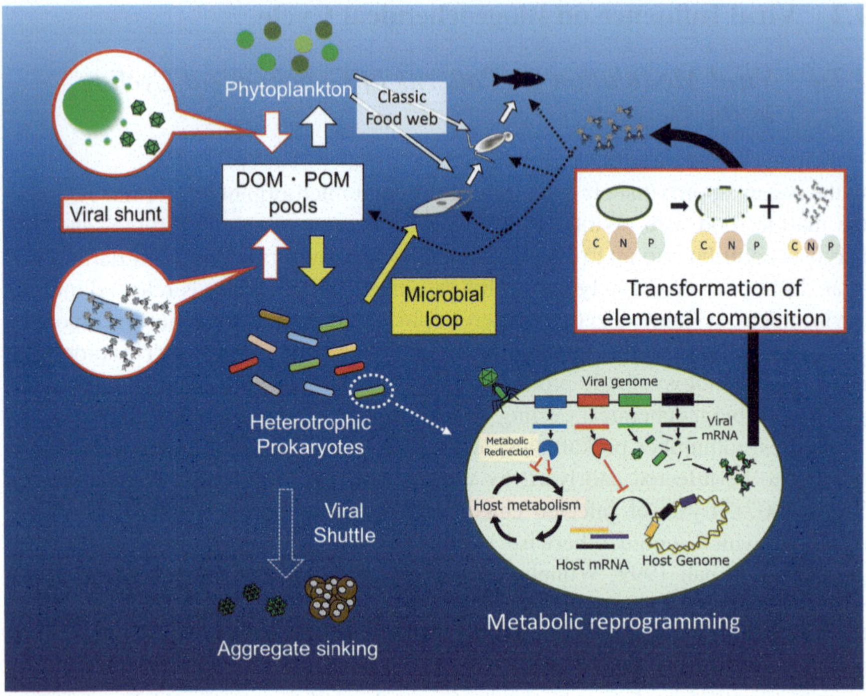

Fig. 1.1 Overview of virus-mediated biogeochemical cycling in aquatic microbial ecosystem. When viruses lyse their host, intracellular organic matter is released from host cells to the particulate organic matter (POM) and dissolved organic matter (DOM) pools, a process that has been termed a Viral shunt. This process is accompanied by compositional and stoichiometric changes in chemical properties of intracellular metabolites via viral metabolic redirection, i.e., hijacking host transcription-translation systems and expression of viral auxiliary metabolic genes (AMG). Viral particles are also a source of phosphorus-rich DOM (compared with host debris) and consumed by direct grazing. Viral infection facilitates carbon export to the deeper layer (biological pump) through particle aggregation driven by the release of lysis products and virus-induced alterations in host physiology (Viral shuttle)

infection and that infected cells are preferentially transported to the deep ocean (Sheyn et al. 2018). Considering the huge abundance of *E. huxleyi* (reaching 10^7 cells/mL during their bloom period; Silkin et al. 2020), the virus-mediated sinking of its cells would thus have a significant impact on the available carbon in the surface ocean. Moreover, laboratory experiments have reported that the virus-infected culture of *Chaetoceros tenuissimus* (Diatomea) is up to 59-fold enriched in particulate organic matter compared with uninfected controls (Yamada et al. 2018). At the global scale, a metagenomic study based on data from the Tara Ocean has suggested that the infection and lysis of the widespread and abundant cyanobacteria *Synechococcus* significantly contribute to carbon export compared with other microorganisms (Guidi et al. 2016). Further studies using quantitative methods are required for a better understanding of the effect of viral shuttles in the global ocean.

In addition, viral particles themselves contribute to the biogeochemical cycling of carbon and other nutrients (Breitbart et al. 2018). For example, approximately 0.03 Gt C per year (Bar-On and Milo 2019), and most of the viral biomass is attributed to the dissolved organic matter (DOM) fraction ($<0.45\mu m$) due to the size of the virion (e.g., bacterial viruses generally range between 20 and 200 nm) (Zsolnay 2003; Leenheer and Croué 2003; Findlay and Parr 2017). Recently, a relatively large number of marine viruses have been identified to attach to non-host organisms and particles (Yamada et al. 2020). Thus, if or when these non-host organisms or particles are predated, their attached viruses indirectly contribute to classical marine food webs even when they are not infecting any organisms. Altogether, viral particles could contribute to both DOM and the particulate organic matter (POM) pools in the ocean.

A previous study investigating the elemental composition of both virus particles and viral lysates eluted from their host debris revealed that viral lysates tend to be more depleted in phosphorus than are uninfected cells because the amount of genomic nucleic acids contained in progeny viral particles results in those viruses being relatively phosphorus-rich as compared with their amounts of carbon and nitrogen (Jover et al. 2014). Extrapolation of this model to the ecosystem scale revealed that marine viruses are estimated to constitute a comparatively high proportion ($>5\%$) of the total dissolved organic phosphorus pool in the surface ocean, suggesting viruses themselves can be regarded as an abundant phosphorus source (Jover et al. 2014).

Predation of viral particles by predators has been demonstrated in several studies using culture experiments (Suttle and Chen 1992; Bettarel et al. 2005; Lawrence et al. 2018; Welsh et al. 2020). For example, a co-cultivation study exposing *Phaeocystis globosa* and its virus to various predators has demonstrated that viruses can be effectively removed (up to 98% within 24 h) from the water column by non-host organisms, including sea anemones, polychaete larvae, sea squirts, crabs, cockles, oysters, and sponges (Welsh et al. 2020). Therefore, although the rate of viral particle predation may vary depending on both the virus and predator strains (Suttle and Chen 1992; Gonzalez and Suttle 1993; Lawrence et al. 2018; Welsh et al. 2020), the ingestion of viruses can serve as a possible source of nutrients, especially phosphorus.

These viral influences on the biogeochemical cycles fluctuate across short- and long-time scales. On a long-term scale, for instance, the microbial community exhibits seasonal compositional changes (Cram et al. 2015; Parada and Fuhrman 2017; Needham et al. 2018; Choi et al. 2020), which are followed by the seasonal dynamics of their viruses (Needham et al. 2017; Ignacio-Espinoza et al. 2020). Furthermore, diverse taxa of photosynthetic microorganisms and even some heterotrophic ones show diel activity in culture and environmental studies, which is partly explained by the viral infection cycle (Morimoto et al. 2020 and references therein). Both the seasonality and the diel cycle activity of microorganisms and their viruses suggest that host–virus interactions could generate temporal fluctuations in geochemical cycles in aquatic environments.

1.2.2 Virus–Host Interactions-Mediated Modification of Host Cell Metabolism

Viruses switch their host metabolism from cellular replication to progeny production. Compared with non-infected hosts, metabolically reprogrammed cells can be generally distinguished based on changes in the host transcription program, eventually leading to distinctiveness in the proportion of end-point products between infected and uninfected cells (Ankrah et al. 2014; Jover et al. 2014; Rosenwasser et al. 2014; Ma et al. 2018) (Fig. 1.1). For instance, disproportioning of phosphorus between infected and uninfected hosts (as discussed in the previous section) could be attributed to a viral reprograming mechanism in which viruses degrade host DNA and utilize the resultant nucleic acids for the synthesis of viral DNA (Wikner et al. 1993; Kutter et al. 2018).

Currently, cell metabolic reprogramming by viruses has been studied using both transcriptomic and metabolomic analyses and is found to be a highly regulated process. For instance, the infection strategy of T4-like viruses follows the three temporal expression classes of early, middle, and late genes, corresponding to host takeover, replication, and virion morphogenesis, respectively, and occurs in accordance with the downregulation of genes related to host replication (Roucourt and Lavigne 2009). Such transcriptional regulation by viruses has also been investigated in several lineages of marine and freshwater prokaryotic as well as eukaryotic phytoplankton (Lindell et al. 2007; Rosenwasser et al. 2014; Bachy et al. 2018; Moniruzzaman et al. 2018; Morimoto et al. 2018; Ku et al. 2020) and heterotrophic bacteria (Ankrah et al. 2014; Howard-Varona et al. 2017b, 2020). The metabolic regulation of host cells may also depend on host taxonomy and thus differ between various species (e.g., cyanoviruses with broad and narrow host range seem to have different infection strategies; *E. huxleyi* virus EhV possesses five gene expression phases during infection) (Lindell et al. 2005; Clokie et al. 2006; Doron et al. 2016; Morimoto et al. 2018; Ku et al. 2020). In addition, it may be influenced by host physiological states, which in turn depends on nutrient availability (e.g. phosphate) (Kelly et al. 2013; Lin et al. 2016; Bachy et al. 2018). Thus, viruses could affect the proportion of end-point products in the infected cells while those hosts either directly or indirectly are responding to environmental conditions.

Viruses can also possess host-derived genes (also called as auxiliary metabolic genes, AMGs) that are expressed during infection, thus altering host metabolism, and increasing the efficiency of viral reproduction (Breitbart et al. 2007; Hurwitz and U'Ren 2016). These AMGs can largely be classified into two classes (Class I and II) based on their function according to the Kyoto Encyclopedia of Genes and Genomes database (Hurwitz and U'Ren 2016). Viral-encoded AMGs not only maintain cellular functions necessary for viral DNA replication and virion production (e.g., ATP production and nucleotide synthesis) during infection (Lindell et al. 2004, 2005), but also both down and up-regulate a range of targeted metabolic pathways that can substantially alter cell stoichiometry and nutrient metabolism (De Smet et al. 2016, 2017). Most AMGs that have been identified to date are directly involved in either the utilization and uptake of limiting nutrients or energy production (Enav et al. 2014; Hurwitz and U'Ren 2016), which in turn may have (at least temporarily) a positive feedback effect on the host cell by improving its fitness during infection (Zeng and Chisholm 2012): the acquisition and metabolism of carbon (e.g., *psbA* and *psbD*; Lindell et al. 2004, 2005; Thompson et al. 2011), nitrogen (e.g., *amt*; Monier et al. 2017), and phosphorus (e.g., *pstS* and *phoA*; Zeng and Chisholm 2012). New putative AMGs are continuously being discovered in bacterial (Breitbart 2011; Crummett et al. 2016; Breitbart et al. 2018; Warwick-Dugdale et al. 2019), eukaryotic (Schvarcz and Steward 2018; Needham et al. 2019), and archaea viruses (Ahlgren et al. 2019) or the more broadly defined environmental viromes (Williamson et al. 2008; Anantharaman et al. 2014; Hurwitz et al. 2015; Moniruzzaman et al. 2020; Schulz et al. 2020; Kieft et al. 2020). Thus, considering that viral-encoded AMGs are abundant and widespread in aquatic environments (Williamson et al. 2008), AMG-mediated metabolic reprogramming can substantially contribute to major biogeochemical cycles and ecosystem functioning at the global scale (Sieradzki et al. 2019).

Indeed, changes in metabolites mediated by virus–host interactions have been observed in both eukaryotes and prokaryotes using a metabolomic approach (Rosenwasser et al. 2014; Ma et al. 2018). For example, *E. huxleyi* virus EhV downregulates host *de novo* sphingolipid genes and simultaneously promotes the induction of a viral-encoded homologous pathway, resulting in the metabolic shift toward viral sphingolipid production (Rosenwasser et al. 2014). A recent culture-based study on *Synechococcus* and its viruses revealed that the composition of chemical compounds (proteins, carbohydrates, and lipids) of organic matter differs between infected and uninfected cells (Ma et al. 2018). Such compositional changes induced by viruses have been detected during viral infection of *Sulfitobacter* (C:N ratio of host cell shifted to nitrogen-rich state compared with uninfected cells) (Ankrah et al. 2014).

Thus, from an ecological perspective, the virus-induced metabolic reprograming depends on both the virus-host pair (and therefore on the diversity and composition of microbial assemblages) and the host physiological state during infection and can modulate the generation as well as the diversity of end-point products, thus leading to altered biogeochemical cycling. It is also speculated that halting the viral reproduction at various infection stages using antiviral responses, including signal transduction, cell cycle regulation (Moniruzzaman et al. 2018), and metabolic pathway (Rosenwasser et al. 2014), might establish metabolite diversity in the infected cells via the generation of intermediate products of viral progeny (Zborowsky and Lindell 2019) or unusual proteins in the uninfected cells.

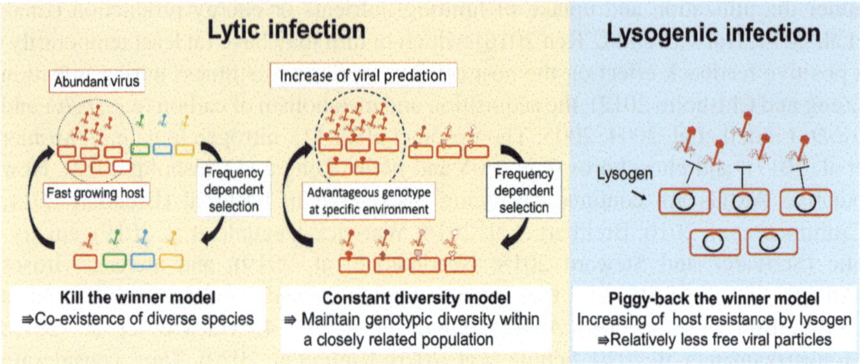

Fig. 1.2 Schematic diagram of mechanisms in virus-mediated maintenance of microbial community diversity. Preferential viral infection of abundant species enables the co-existence of diverse competing microbial species by preventing the dominance of only few species ("Kill the Winner" hypothesis). Similar viral top-down control is proposed as a mechanism to maintain high genotypic diversity within a single microbial population ("Constant Diversity dynamics" model). Lastly, the prevalence of lysogeny in dominant microbial population may be another potential mechanism that allows abundant host species to be dominant by taking advantages benefit from lysogenic conversion such as superinfection exclusion ("Piggyback-the-Winner" model)

1.3 Viral Infection Shaping Microbial Community Diversity

1.3.1 Contribution of Lytic Infection to Microbial Diversity Maintenance

In aquatic ecosystems, diverse microorganisms compete for nutrient resources but can co-exist and account for a large proportion of total aquatic biodiversity, a concept that has been known as the classical question "paradox of the plankton"[1] (Hutchinson 1961). Viruses are currently thought to contribute to the co-existence of microbial species and genotypes.

Basically, viruses are believed to infect their specific microbial hosts in a frequency-dependent manner (Fuhrman and Suttle 1993). Therefore, viral infection checkmates microbial species that become dominant through the competition among co-existing microorganisms that possess different substrate affinity, and thereby enables the co-existence of multiple competing microbial species ("Kill the Winner" hypothesis)[2] (Thingstad 2000) (Fig. 1.2). Indeed, several culture and environmental

[1] The concept arguing paradoxical situation of coexistence of various plankton species competing for identical resources in homogeneous and resource limited environment (Hutchinson 1961).

[2] A model proposing the dynamics of virus-host interactions in which an increase of host population (winner) is accompanied with increasing of its infectious viruses, and thereby viruses prevent their hosts from becoming dominant through increased mortality of the winner (Thingstad 2000).

studies have demonstrated that viral top-down control modulates microbial abundance, which is consistent with the results expected from a Kill the Winner hypothesis (Tarutani et al. 2000; Schwalbach et al. 2004; Bouvier and Del Giorgio 2007; Yoshida et al. 2008a; Rodriguez-Brito et al. 2010; Kuno et al. 2012; Parsons et al. 2012; Kimura et al. 2013; Needham et al. 2013; Cram et al. 2016).

The viral-mediated co-existence mechanism, by which viruses are expected to affect host diversity in a frequency dependent manner according to the Kill the Winner hypothesis, could provide a mechanism that explains the continuing co-existence of diverse genotypes within a single microbial species (or closely related lineage) rather than the co-existence of diverse microbial species. Conventionally, the philosophy has been that phenotypic and genotypic diversity within a microbial population was expected to become homogenized to a greater level of fitness in the environment. The microbial population (species or closely related lineage) that are genetically cohesive and ecologically distinct are called an "ecotype" (ecotype hypothesis) (Maharjan et al. 2006; Cohan and Koeppel 2008). The ecotype had been believed to be periodically replaced as fitter ecotypes emerged after profitable mutation or preferable environmental changes (Maharjan et al. 2006; Cohan and Koeppel 2008). However, metagenome sequence alignment analyses have demonstrated that several genomic regions (metagenomic islands; MGIs) are underrepresented even within a single microbial population in similar environments (Coleman and Chisholm 2007; Cuadros-Orellana et al. 2007; Kettler et al. 2007; Wilhelm et al. 2007; Frias-Lopez et al. 2008; Rodriguez-Valera et al. 2009; Rodriguez-Valera and Ussery 2012). This suggests that diverse genotypes coexist within a single microbial population. Furthermore, these MGIs include diverse accessory genes, such as extracellular structure-related genes that can be viral recognition sites (Reva and Tümmler 2008; Sharma et al. 2008; Rodriguez-Valera et al. 2009; Rodriguez-Valera and Ussery 2012) and antiviral response-related genes such as CRISPR in addition to genes that affect restriction-modification (Sorek et al. 2008; Wilmes et al. 2009). Therefore, MGIs are thought to play an important role in the ability of hosts to escape or survive from viral infection, in which host genotypic diversity is driven by viral predation pressure, thereby leading to the co-existence of multiple competing microbial genotypes ("Constant Diversity dynamics" model[3]) (Rodriguez-Valera et al. 2009) (Fig. 1.2). Indeed, multiple genotypes of *Microcystis aeruginosa* possessing different CRISPR arrays and its virus Ma-LMM01 coexist and oscillate during the massive bloom of this nuisance and toxic species (Kuno et al. 2012, 2014; Kimura et al. 2013). Hence, lytic viruses play important roles in not only the co-existence of microbial species but also the maintenance of high diversity within a single microbial population.

[3] A hypothetical model proposing that the diversity of prokaryotic populations is maintained by viral predation, because the best-adapted populations are selected by viral predation. The hypothesis assumes that each microbial population has distinct viral receptor (Rodriguez-Valera et al. 2009).

1.3.2 *Potential Contribution of Lysogenic Infection to Microbial Diversity Maintenance*

So far, we have focused on lytic viruses and described their contribution to the maintenance of microbial diversity in aquatic environments. Another key question related to viral impact on microbial diversity is whether temperate viruses also contribute to shaping microbial community diversity.

Provirus integration can occur either through repeated random transposition events or at specific integration sites (e.g., host tRNA genes; also called site-specific recombination) and is associated with the immediate transcriptional suppression (e.g. via specific virus repressors) of lytic promoters and genes associated with virion production (Casjens and Hendrix 2015). This mode of viral infection that generates lysogenic cells (Hobbs and Abedon 2016) is considered as an adaptive strategy of temperate viruses to ensure their persistence in the environment, in which novel phenotypic or metabolic advantages are sometimes conferred to host microorganisms via concomitant effects by mechanisms such as HGT (Hendrix et al. 2000; Howard-Varona et al. 2017a), the capability of up and down-regulation of host genes (Argov et al. 2017), and integration-driven gene disruption (Feiner et al. 2015).

Although one study estimated that approximately half of 100 marine bacterial isolates harbored temperate viruses in their genomes (Paul 2008), proviruses are rarely found in the marine-dominant bacterial lineages (e.g., only one provirus was recently reported in SAR11 clades; see Morris et al. 2020) possibly due to genome streamlining, in which a bacterial genome is minimized to a highly constrained gene set that confers maximum fitness (Touchon et al. 2016). Therefore, the ecological significance of lysogenic infection on microbial diversity maintenance in marine ecosystems remains under debate. Traditionally, it was believed that bacterial viruses control their host abundances in a frequency-dependent manner as described above, and thus viral abundance is typically 10-folds higher than that of prokaryotes (Wommack and Colwell 2000; Weinbauer 2004). Therefore, lysogenic infection is presumed to be the preferred viral strategy under conditions of reduced host cell number and activity (Stewart and Levin 1984; Sime-Ngando 2014; Brum et al. 2016). However, viral metagenomic and metadata approaches have revealed that viral particles are relatively less abundant at high microbial densities (Knowles et al. 2016; Wigington et al. 2016). Likewise, it was demonstrated that the virus/host genome abundance ratio was negatively correlated with the host abundance at the genus or phylum levels (Coutinho et al. 2017). Additionally, the relative abundance of hallmark genes encoded by temperate viruses increased with microbial density in a coral reef (Knowles et al. 2016). These findings suggest that lysogenic infection may become dominant at high-cell densities because proviruses can replicate quickly in a way that will keep pace with their fast-growing host and provirus-mediated superinfection resistance might become increasingly important at high cell densities (called "Piggyback-the-Winner" model[4]) (Knowles et al. 2016, 2017; Coutinho et al. 2017) (Fig. 1.2).

1.4 Evolutionary Roles of Viruses that Generate Genotype-Level Microbial Diversification

In short-term laboratory experiments, host–virus coevolution appeared to be suppressed by the emergence of a viral-resistant genotype, which the virus could not evolve to overcome (Dennehy 2012). In particular, *de novo* mutations that cause changes in those bacterial cell-surface structures which serve as viral receptor sites are one of the major factors that prevent viral attachment, and thereby confer the potential host with resistance against viral infection (Lenski and Levin 1985). Therefore, host–virus coevolution has been considered to be constrained by the asymmetry of evolutionary potential between hosts and viruses (Cannon et al. 1971; Cowlishaw and Mrsa 1975; Barnet et al. 1981; Lenski and Levin 1985; Waterbury and Valois 1993; Middelboe et al. 2001; Wei et al. 2010, 2011).

On the other hand, long-term evolutionary laboratory studies have indicated that the host and its virus undergo persistent coevolution over a prolonged period, as evidenced with the soil bacterium *Pseudomonas fluorescens* (Buckling and Rainey 2002a, b; Brockhurst et al. 2007; Hall et al. 2011a, b) and *E. coli* (Mizoguchi et al. 2003). Similar co-evolutionary dynamics have also been observed in the marine bacteria *Prochlorococcus* (Avrani et al. 2011), *Synechococcus* (Marston et al. 2012), and *Cellulophaga baltica* (Middelboe et al. 2009).

The above-mentioned co-evolution scenario, based on the one-virus-to-one-bacteria relationship, predicts that the genetic contents of both bacteria and virus would converge over repeated interactions. As described above, however, genotypic diversity is observed in MGIs even within a single microbial population (Coleman and Chisholm 2007; Cuadros-Orellana et al. 2007; Kettler et al. 2007; Wilhelm et al. 2007; Frias-Lopez et al. 2008; Rodriguez-Valera et al. 2009; Rodriguez-Valera and Ussery 2012). Additionally, under-represented genomic regions have been found in marine viruses (metaviromic islands) with a large fraction of their identified genes (e.g., 59 out of 138) associated with host recognition of viruses (Mizuno et al. 2014). Thus, the recent understanding of host–virus coevolution is based on multiple genotype (strain)-level interactions within both the microbial host and viral species which seem to represent defense and counter-defense strategies (e.g., CRISPR and mutation in protospacer) or resistance and overcome of resistance (e.g., mutation in cell surface and viral tail gene) (Fig. 1.3).

---

[4]A hypothetical model proposing that lysogeny would be favored when a bacterial host is dominated in the environment because a provirus can replicate rapidly together with host DNA replication (Knowles et al. 2016).

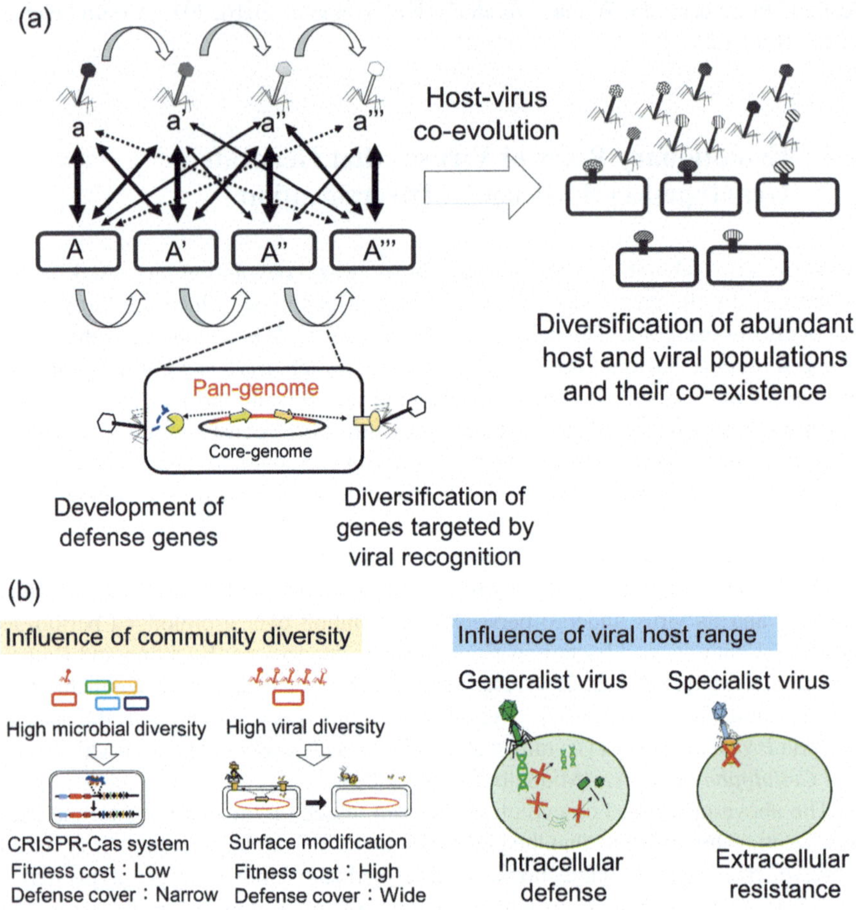

Fig. 1.3 Proposed mechanisms in host pan-genome expansion via virus–host interactions. (**a**) Under the viral top-down control toward dominant species and genotypes, "Red Queen" like host–virus co-diversification can be established in abundant host and abundant viruses. Continuous arms race via intracellular defense (e.g., CRISPR-Cas system) and extracellular resistance (e.g., viral recognition sites) plays a part in genomic diversification of both host and virus. (**b**) A complicated balance between trade-off and specificities of each defense mechanism can affect the arms race. For example, extracellular resistance (e.g., surface modification) is preferable to intercellular defense (e.g., CRISPR-Cas) under conditions of high viral genetic diversity because the former promotes resistance toward against broader viral genotypes. However, because extracellular resistance may result in a higher cost, such as the impairment of nutrient uptake ability, intercellular defense is possibly preferable under competitive situations with host competitors

In freshwater ecosystems, host-virus coevolution have been intensively studied in the bloom-forming cyanobacterium *M. aeruginosa* and its viruses (Yoshida et al. 2005, 2008a, b; Kimura et al. 2012, 2013, 2018; Kuno et al. 2012, 2014; Yoshida-Takashima et al. 2012; Morimoto et al. 2019). Interestingly, the most abundant *Microcystis* CRISPR genotype is known to coexist with that derived by novel spacer acquisitions from cyanoviruses in the environment. This finding suggested that both abundant host and viral genotypes have diversified in the bloom without a complete selective sweep (Kimura et al. 2018). Thus, the Red Queen like dynamics[5] could be established, to some extent, between the abundant host genotype and its cyanoviruses under high viral contact rate; with reciprocal adaptation via defense and counter-defense continuously occurring in multiple-to-multiple relationships (Koskella and Brockhurst 2014), thus subsequently increasing the diversity of host organisms and viruses (Fig. 1.3). Also, a recent metagenomic survey revealed the co-existence of highly host-specific (narrow host range) and broad host range *Microcystis* viruses and the high co-expression of antiviral defense and viral genes in the environment (Morimoto et al. 2019). Considering that they often induce antiviral responses, broad host range viruses might be important for host genotype diversification (Morimoto et al. 2019).

In contrast with freshwater cyanobacteria typically having the greatest overall numbers of defense genes (Makarova et al. 2011), marine prokaryotes rarely possess distinctive defense genes such as CRISPR-Cas due to their genome streamlining (Touchon et al. 2016). However, instances of co-existence between dominant marine prokaryotes and their viruses were observed, which are presumably sustained by Red Queen-like co-evolution dynamics. For example, the dominance of the SAR11 clade bacteria and their viruses was predicted to be maintained by host rapid adaptation to viruses. The rapid adaptation was achieved by high recombination rates among SAR11 in their variable genomic region that encoded genes involved in synthesis of cell surface proteins, and this hypothesis is possibly supported by their high host cell density (King of the Mountain hypothesis[6]) (Zhao et al. 2013). Furthermore, constant turnover of single-nucleotide polymorphism variants in relatively abundant marine viruses also suggests that Red Queen-like virus–host coexistence could be established by perpetually changing minor variants.

As described in this section, one of the major forces driving host and viral genotypic diversification is the reciprocal defense and counter-defense that occur through extracellular (e.g., *de novo* mutation of cell-surface structure) and intracellular resistance (e.g., acquisition of novel CRISPR spacers). Comparative antiviral resistance analyses in *Synechococcus* and *Prochlorococcus* provided new insights into host-favored antiviral defenses in narrow and broad host range viruses (Doron et al. 2016; Zborowsky and Lindell 2019). Host cyanobacteria resist against narrow host range viruses irrespective of the viral family by preventing viral entry into the cell, whereas intracellular resistance arrests the infection cycle of broad host range viruses at various infection stages (Doron et al. 2016; Zborowsky and Lindell 2019) (Fig. 1.3). These differences in antiviral responses that seemingly occur according to

[5] A hypothesis proposing co-evolutionary process between competing species (Valen 1973); in the case of virus-host interactions, this hypothesis explains continuous dynamics of resistance acquisition in microbial hosts and viral avoidance to the host resistance (Brockhurst et al. 2014).

[6] A hypothesis proposing that high recombination rate enables dominance of a competitive prokaryote in the ecosystem via horizontal transfer of genes involved in resistance to viral infection (Zhao et al. 2013).

viral host range are speculated to be associated with fitness trade-offs in extracellular and intracellular antiviral responses. In extracellular antiviral responses, mutations in cellular surface structures can impair nutrient uptake and utilization but can protect against diverse viral attacks, leading to an increase in host fitness trade-off favoring the host (Winter et al. 2010; Avrani et al. 2012) (Fig. 1.3). On the other hand, choosing to modify intracellular antiviral responses techniques may be energetically favorable, especially the CRISPR-Cas system, because the cost of a new spacer acquisition is speculated to be low, although it is possible that additional types of resistance costs may exist (Thingstad et al. 2014) (Fig. 1.3). Indeed, recent studies focusing on the differences in biotic complexity between *in vitro* and environments have revealed that coexistence among human pathogens amplified the fitness trade-offs associated with viral receptor mutations in *Pseudomonas aeruginosa* and therefore enhanced the evolution of CRISPR-based resistance (Alseth et al. 2019). Higher viral genetic diversity can also influence CRISPR-based evolution, with an example being that the majority of a *Pseudomonas* population more favorably evolved based on the mutation of viral receptors to resist a broader range of viral genotypes than CRISPR-based specific resistance (Broniewski et al. 2020). Thus, fitness trade-offs in bacterial host species and diversity of both bacterial host and viruses in the environment could be another important factor that affects host–virus coevolution. From the perspective of fitness trade-off, most recently, the "pan-immune system" concept has been proposed. This states that a single strain can access immune defense mechanisms in closely related strain via HGT, although it cannot possess all possible defense systems (Bernheim and Sorek 2020).

1.5 Conclusion

Viruses, which are highly abundant biological entities lacking their own metabolism, can reprogram host cells toward the production of virus progeny, after which the host cell is lysed, releasing new virions into the surrounding environment. This reprogramming viral strategy can change the content and composition of host metabolites and releases a large amount of organic matter, thus considerably affecting biogeochemical cycles. In addition, viral infection checkmates not only microbial species that become dominant but also abundant genotypes within a single microbial species, and thereby enables the coexistence of diverse microbial species and genotypes in the aquatic ecosystem. Thus, viral infection is one of the key factors shaping microbial community diversity and maintaining high diversity within a single microbial population. Likewise, because temperate viruses may be prevalent under specific combinations of environment and conditions, they could affect microbial diversity via superinfection exclusion during either the lysogenic cycle or induction of lytic cycle. Meanwhile, microorganisms have evolved extracellular and intracellular antiviral mechanisms with different fitness costs and specificities. Therefore, viruses with diverse host range and genetic diversity interact with abundant hosts by a complex balancing between fitness trade-offs and the specificity of extracellular- and intracellular antiviral resistance in hosts. Together, this results in continuous virus–host coevolution, leading to diversification in aquatic ecosystems. Collectively, viruses are important biological entities that sustain and generate microbial diversity and control the biogeochemical cycle in aquatic environments.

Acknowledgements This work was supported by the Grants-in Aids for Scientific Research (B) (grant number 17H03850) from the Japan Society for the Promotion of Science (JSPS); the Research Council of Lithuania (grant number S-MIP-17-28 to SŠ); the Canon Foundation (grant number 203143100025); the JSPS Scientific Research on Innovative Areas (grant number 16H06437); and the Bilateral Open Partnership Joint Research Project (Japan–Lithuania Research Cooperative Program) "Research on prediction of environmental change in Baltic Sea based on comprehensive metagenomic analysis of microbial viruses". DM, KT, HT, SŠ and TY all contributed to manuscript preparation, discussion, and revision.

References

Ahlgren NA, Fuchsman CA, Rocap G, Fuhrman JA (2019) Discovery of several novel, widespread, and ecologically distinct marine Thaumarchaeota viruses that encode amoC nitrification genes. ISME J 13:618–631. https://doi.org/10.1038/s41396-018-0289-4

Alseth EO, Pursey E, Luján AM et al (2019) Bacterial biodiversity drives the evolution of CRISPR-based phage resistance. Nature 574:549–552. https://doi.org/10.1038/s41586-019-1662-9

Anantharaman K, Duhaime MB, Breier JA et al (2014) Sulfur oxidation genes in diverse deep-sea viruses. Science (80-) 344:757–760. https://doi.org/10.1126/science.1252229

Ankrah NYD, May AL, Middleton JL et al (2014) Phage infection of an environmentally relevant marine bacterium alters host metabolism and lysate composition. ISME J 8:1089–1100. https://doi.org/10.1038/ismej.2013.216

Argov T, Azulay G, Pasechnek A et al (2017) Temperate bacteriophages as regulators of host behavior. Curr Opin Microbiol 38:81–87. https://doi.org/10.1016/j.mib.2017.05.002

Avrani S, Wurtzel O, Sharon I et al (2011) Genomic island variability facilitates Prochlorococcus-virus coexistence. Nature 474:604–608. https://doi.org/10.1038/nature10172

Avrani S, Schwartz DA, Lindell D (2012) Virus-host swinging party in the oceans. Mob Genet Elements 2:88–95. https://doi.org/10.4161/mge.20031

Azam F, Fenchel T, Field JG et al (1983) The ecological role of water-column microbes in the sea. Mar Ecol Prog Ser 10:257–263

Bachy C, Charlesworth CJ, Chan AM et al (2018) Transcriptional responses of the marine green alga Micromonas pusilla and an infecting prasinovirus under different phosphate conditions. Environ Microbiol 20:2898–2912. https://doi.org/10.1111/1462-2920.14273

Balcázar JL (2018) How do bacteriophages promote antibiotic resistance in the environment? Clin Microbiol Infect 24:447–449. https://doi.org/10.1016/j.cmi.2017.10.010

Barnet YM, Daft MJ, Stewart WDP (1981) Cyanobacteria-cyanophage interactions in continuous culture. J Appl Bacteriol 51:541–552. https://doi.org/10.1111/j.1365-2672.1981.tb01273.x

Bar-On YM, Milo R (2019) The biomass composition of the oceans: a blueprint of our blue planet. Cell 179:1451–1454. https://doi.org/10.1016/j.cell.2019.11.018

Bernheim A, Sorek R (2020) The pan-immune system of bacteria: antiviral defence as a community resource. Nat Rev Microbiol 18:113–119. https://doi.org/10.1038/s41579-019-0278-2

Bettarel Y, Sime-Ngando T, Bouvy M et al (2005) Low consumption of virus-sized particles by heterotrophic nanoflagellates in two lakes of the French Massif Central. Aquat Microb Ecol 39:205–209

Bouvier T, Del Giorgio PA (2007) Key role of selective viral-induced mortality in determining marine bacterial community composition. Environ Microbiol 9:287–297. https://doi.org/10.1111/j.1462-2920.2006.01137.x

Breitbart M (2011) Marine viruses: truth or dare. Annu Rev Mar Sci 4:425–448. https://doi.org/10.1146/annurev-marine-120709-142805

Breitbart M, Thompson L, Suttle C, Sullivan M (2007) Exploring the vast diversity of marine viruses. Oceanography 20:135–139. https://doi.org/10.5670/oceanog.2007.58

Breitbart M, Bonnain C, Malki K, Sawaya NA (2018) Phage puppet masters of the marine microbial realm. Nat Microbiol 3:754–766. https://doi.org/10.1038/s41564-018-0166-y

Brockhurst MA, Morgan AD, Fenton A, Buckling A (2007) Experimental coevolution with bacteria and phage: the Pseudomonas fluorescens—Φ2 model system. Infect Genet Evol 7:547–552. https://doi.org/10.1016/j.meegid.2007.01.005

Brockhurst MA, Chapman T, King KC et al (2014) Running with the Red Queen: the role of biotic conflicts in evolution. Proc R Soc B Biol Sci 281:20141382. https://doi.org/10.1098/rspb.2014.1382

Broniewski JM, Meaden S, Paterson S et al (2020) The effect of phage genetic diversity on bacterial resistance evolution. ISME J 14:828–836. https://doi.org/10.1038/s41396-019-0577-7

Brum JR, Hurwitz BL, Schofield O et al (2016) Seasonal time bombs: dominant temperate viruses affect Southern Ocean microbial dynamics. ISME J 10:437–449. https://doi.org/10.1038/ismej.2015.125

Buckling A, Rainey PB (2002a) Antagonistic coevolution between a bacterium and a bacteriophage. Proc R Soc B Biol Sci 269:931–936. https://doi.org/10.1098/rspb.2001.1945

Buckling A, Rainey PB (2002b) The role of parasites in sympatric and allopatric host diversification. Nature 420:496–499. https://doi.org/10.1038/nature01164

Campos J, Martínez E, Marrero K et al (2003) Novel type of specialized transduction for CTXφ or its satellite phage RS1 mediated by filamentous phageVGJφ in Vibriocholerae. J Bacteriol 185:7231–7240. https://doi.org/10.1128/JB.185.24.7231-7240.2003

Cannon RE, Shane MS, Bush VN (1971) Lysogeny of a blue-green alga, Plectonema boryanum. Virology 45:149–153. https://doi.org/10.1016/0042-6822(71)90121-8

Casjens SR, Hendrix RW (2015) Bacteriophage lambda: early pioneer and still relevant. Virology 479–480:310–330. https://doi.org/10.1016/j.virol.2015.02.010

Choi C, Jimenez V, Needham D et al (2020) Seasonal and geographical transitions in eukaryotic phytoplankton community structure in the Atlantic and Pacific Oceans. Frontiers (Boulder) 11:542372. https://doi.org/10.3389/fmicb.2020.542372

Clokie MRJ, Shan J, Bailey S et al (2006) Transcription of a "photosynthetic" T4-type phage during infection of a marine cyanobacterium. Environ Microbiol 8:827–835. https://doi.org/10.1111/j.1462-2920.2005.00969.x

Cohan FM, Koeppel AF (2008) The origins of ecological diversity in prokaryotes. Curr Biol 18:1024–1034. https://doi.org/10.1016/j.cub.2008.09.014

Coleman ML, Chisholm SW (2007) Code and context: prochlorococcus as a model for cross-scale biology. Trends Microbiol 15:398–407. https://doi.org/10.1016/j.tim.2007.07.001

Coutinho FH, Silveira CB, Gregoracci GB et al (2017) Marine viruses discovered via metagenomics shed light on viral strategies throughout the oceans. Nat Commun 8:15955. https://doi.org/10.1038/ncomms15955

Cowlishaw J, Mrsa M (1975) Co-evolution of a virus-alga system. Appl Microbiol 29:234–239

Cram JA, Chow CET, Sachdeva R et al (2015) Seasonal and interannual variability of the marine bacterioplankton community throughout the water column over ten years. ISME J 9:563–580. https://doi.org/10.1038/ismej.2014.153

Cram JA, Parada AE, Fuhrman JA (2016) Dilution reveals how viral lysis and grazing shape microbial communities. Limnol Oceanogr 61:889–905. https://doi.org/10.1002/lno.10259

Crummett LT, Puxty RJ, Weihe C et al (2016) The genomic content and context of auxiliary metabolic genes in marine cyanomyoviruses. Virology 499:219–229. https://doi.org/10.1016/j.virol.2016.09.016

Cuadros-Orellana S, Martin-Cuadrado AB, Legault B et al (2007) Genomic plasticity in prokaryotes: the case of the square haloarchaeon. ISME J 1:235–245. https://doi.org/10.1038/ismej.2007.35

De Smet J, Zimmermann M, Kogadeeva M et al (2016) High coverage metabolomics analysis reveals phage-specific alterations to Pseudomonas aeruginosa physiology during infection. ISME J 10:1823–1835. https://doi.org/10.1038/ismej.2016.3

De Smet J, Hendrix H, Blasdel BG et al (2017) Pseudomonas predators: understanding and exploiting phage–host interactions. Nat Rev Microbiol 15:517–530. https://doi.org/10.1038/nrmicro.2017.61

Dennehy JJ (2012) What can phages tell us about host-pathogen coevolution? Int J Evol Biol 2012:396165. https://doi.org/10.1155/2012/396165

Doron S, Fedida A, Hernández-Prieto MA et al (2016) Transcriptome dynamics of a broad host-range cyanophage and its hosts. ISME J 10:1437–1455. https://doi.org/10.1007/s13398-014-0173-7.2

Enav H, Mandel-Gutfreund Y, Béjà O (2014) Comparative metagenomic analyses reveal viral-induced shifts of host metabolism towards nucleotide biosynthesis. Microbiome 2:1–11. https://doi.org/10.1186/2049-2618-2-9

Feiner R, Argov T, Rabinovich L et al (2015) A new perspective on lysogeny: prophages as active regulatory switches of bacteria. Nat Rev Microbiol 13:641–650. https://doi.org/10.1038/nrmicro3527

Fernandes RM, de Lencastre H, Archer LJ (1989) Specialized transduction in Bacillus subtilis by the phages IG1, IG3, and IG4. Arch Virol 105:137–140

Field CB, Behrenfeld MJ, Randerson JT, Falkowski P (1998) Primary production of the biosphere: Integrating terrestrial and oceanic components. Science (80-) 281:237–240. https://doi.org/10.1126/science.281.5374.237

Findlay SEG, Parr TB (2017) Chapter 24 - Dissolved organic matter. In: Lamberti GA, Hauer FR (eds) BT-M in stream ecology, 3rd edn. Academic, pp 21–36

Frias-Lopez J, Shi Y, Tyson GW et al (2008) Microbial community gene expression in ocean surface waters. Proc Natl Acad Sci U S A 105:3805–3810. https://doi.org/10.1073/pnas.0708897105

Fuhrman JA (1999) Marine viruses and their biogeochemical and ecological effects. Nature 399:541–548. https://doi.org/10.1038/21119

Fuhrman J, Suttle C (1993) Viruses in marine planktonic systems. Oceanography 6:51–63. https://doi.org/10.5670/oceanog.1993.14

Gonzalez JM, Suttle CA (1993) Grazing by marine nanoflagellates on viruses and virus-sized particles: ingestion and digestion. Mar Ecol Prog Ser 94:1–10. https://doi.org/10.3354/meps094001

Gottesman ME, Yarmolinsky MB (1968) Integration-negative mutants of bacteriophage lambda. J Mol Biol 31:487–505. https://doi.org/10.1016/0022-2836(68)90423-3

Guidi L, Chaffron S, Bittner L et al (2016) Plankton networks driving carbon export in the oligotrophic ocean. Nature 532:465–470. https://doi.org/10.1038/nature16942

Guttman B, Raya R, Kutter E (2005) Basic phage biology. In: Bacteriophages: biology and applications, vol 4. CRC Press

Hall AR, Scanlan PD, Buckling A (2011a) Bacteria-phage coevolution and the emergence of generalist pathogens. Am Nat 177:44–53. https://doi.org/10.1086/657441

Hall AR, Scanlan PD, Morgan AD, Buckling A (2011b) Host–parasite coevolutionary arms races give way to fluctuating selection. Ecol Lett 14:635–642. https://doi.org/10.1111/j.1461-0248.2011.01624.x

Hendrix RW, Lawrence JG, Hatfull GF, Casjens S (2000) The origins and ongoing evolution of viruses. Trends Microbiol 8:504–508. https://doi.org/10.1016/S0966-842X(00)01863-1

Hobbs Z, Abedon ST (2016) Diversity of phage infection types and associated terminology: the problem with 'Lytic or lysogenic'. FEMS Microbiol Lett 363:fnw047. https://doi.org/10.1093/femsle/fnw047

Howard-Varona C, Hargreaves KR, Abedon ST, Sullivan MB (2017a) Lysogeny in nature: mechanisms, impact and ecology of temperate phages. ISME J 11:1511–1520. https://doi.org/10.1038/ismej.2017.16

Howard-Varona C, Roux S, Dore H et al (2017b) Regulation of infection efficiency in a globally abundant marine Bacteriodetes virus. ISME J 11:284–295. https://doi.org/10.1038/ismej.2016.81

Howard-Varona C, Lindback MM, Bastien GE et al (2020) Phage-specific metabolic reprogramming of virocells. ISME J 14:881–895. https://doi.org/10.1038/s41396-019-0580-z

Hurwitz BL, U'Ren JM (2016) Viral metabolic reprogramming in marine ecosystems. Curr Opin Microbiol 31:161–168. https://doi.org/10.1016/j.mib.2016.04.002

Hurwitz BL, Brum JR, Sullivan MB (2015) Depth-stratified functional and taxonomic niche specialization in the "core" and "flexible" Pacific Ocean Virome. ISME J 9:472–484. https://doi.org/10.1038/ismej.2014.143

Hutchinson GE (1961) The Paradox of the Plankton. Am Nat 95:137–145

Ignacio-Espinoza JC, Ahlgren NA, Fuhrman JA (2020) Long-term stability and Red Queen-like strain dynamics in marine viruses. Nat Microbiol 5:265–271. https://doi.org/10.1038/s41564-019-0628-x

Jover LF, Effler TC, Buchan A et al (2014) The elemental composition of virus particles: implications for marine biogeochemical cycles. Nat Rev Microbiol 12:519–528. https://doi.org/10.1038/nrmicro3289

Kelly L, Ding H, Huang KH et al (2013) Genetic diversity in cultured and wild marine cyanomyoviruses reveals phosphorus stress as a strong selective agent. ISME J 7:1827–1841. https://doi.org/10.1038/ismej.2013.58

Kettler GC, Martiny AC, Huang K et al (2007) Patterns and implications of gene gain and loss in the evolution of Prochlorococcus. PLoS Genet 3:2515–2528. https://doi.org/10.1371/journal.pgen.0030231

Kieft K, Zhou Z, Anderson RE, et al (2020) Ecology of inorganic sulfur auxiliary metabolism in widespread bacteriophages.bio Rxiv 2020.08.24.253096. doi:https://doi.org/10.1101/2020.08.24.253096

Kimura S, Yoshida T, Hosoda N et al (2012) Diurnal infection patterns and impact of microcystis cyanophages in a Japanese pond. Appl Environ Microbiol 78:5805–5811. https://doi.org/10.1128/AEM.00571-12

Kimura S, Sako Y, Yoshida T (2013) Rapid Microcystis cyanophage gene diversification revealed by long and short-term genetic analyses of the tail sheath gene in a natural pond. Appl Environ Microbiol 79:2789–2795. https://doi.org/10.1128/AEM.03751-12

Kimura S, Uehara M, Morimoto D et al (2018) Incomplete selective sweeps of Microcystis population detected by the leader-end CRISPR fragment analysis in a natural pond. Front Microbiol 9:article 425. https://doi.org/10.3389/fmicb.2018.00425

Knowles B, Silveira CB, Bailey BA et al (2016) Lytic to temperate switching of viral communities. Nature 531:466–470. https://doi.org/10.1038/nature17193

Knowles B, Bailey B, Boling L et al (2017) Variability and host density independence in inductions-based estimates of environmental lysogeny. Nat Microbiol 2:17064. https://doi.org/10.1038/nmicrobiol.2017.64

Koskella B, Brockhurst MA (2014) Bacteria-phage coevolution as a driver of ecological and evolutionary processes in microbial communities. FEMS Microbiol Rev 38:916–931. https://doi.org/10.1111/1574-6976.12072

Ku C, Ku C, Sheyn U et al (2020) A single-cell view on alga-virus interactions reveals sequential transcriptional programs and infection states. Sci Adv 6. https://doi.org/10.1126/sciadv.aba4137

Kuno S, Yoshida T, Kaneko T, Sako Y (2012) Intricate interactions between the bloom-forming cyanobacterium Microcystis aeruginosa and foreign genetic elements, revealed by diversified clustered regularly interspaced short palindromic repeat (CRISPR) signatures. Appl Environ Microbiol 78:5353–5360. https://doi.org/10.1128/AEM.00626-12

Kuno S, Sako Y, Yoshida T (2014) Diversification of CRISPR within coexisting genotypes in a natural population of the bloom-forming cyanobacterium Microcystis aeruginosa. Microbiology 160:903–916. https://doi.org/10.1099/mic.0.073494-0

Kutter E, Bryan D, Ray G et al (2018) From host to phage metabolism: Hot tales of phage T4's takeover of E. coli. Viruses 10:387. https://doi.org/10.3390/v10070387

Laber CP, Hunter JE, Carvalho F et al (2018) Coccolithovirus facilitation of carbon export in the North Atlantic. Nat Microbiol 3:537–547. https://doi.org/10.1038/s41564-018-0128-4

Lawrence J, Töpper J, Petelenz-Kurdziel E et al (2018) Viruses on the menu: the appendicularian Oikopleura dioica efficiently removes viruses from seawater. Limnol Oceanogr 63:S244–S253. https://doi.org/10.1002/lno.10734

Leenheer JA, Croué J-P (2003) Peer reviewed: characterizing aquatic dissolved organic matter. Environ Sci Technol 37:18A–26A. https://doi.org/10.1021/es032333c

Lenski RE, Levin BR (1985) Constraints on the coevolution of bacteria and virulent phage: a model, some experiments, and predictions for natural communities. Am Nat 125:585–602. https://doi.org/10.1086/284364

Lin X, Ding H, Zeng Q (2016) Transcriptomic response during phage infection of a marine cyanobacterium under phosphorus-limited conditions. Environ Microbiol 18:450–460. https://doi.org/10.1111/1462-2920.13104

Lindell D, Sullivan MB, Johnson ZI et al (2004) Transfer of photosynthesis genes to and from Prochlorococcus viruses. Proc Natl Acad Sci U S A 101:11013–11018. https://doi.org/10.1073/pnas.0401526101

Lindell D, Jaffe JD, Johnson ZI et al (2005) Photosynthesis genes in marine viruses yield proteins during host infection. Nature 438:86–89. https://doi.org/10.1038/nature04111

Lindell D, Jaffe JD, Coleman ML et al (2007) Genome-wide expression dynamics of a marine virus and host reveal features of co-evolution. Nature 449:83–86. https://doi.org/10.1038/nature06130

Ma X, Coleman ML, Waldbauer JR (2018) Distinct molecular signatures in dissolved organic matter produced by viral lysis of marine cyanobacteria. Environ Microbiol 20:3001–3011. https://doi.org/10.1111/1462-2920.14338

Maharjan R, Seeto S, Notley-McRobb L, Ferenci T (2006) Clonal adaptive radiation in a constant environment. Science (80-) 313:514–517. https://doi.org/10.1126/science.1129865

Makarova KS, Wolf YI, Snir S, Koonin EV (2011) Defense islands in bacterial and archaeal genomes and prediction of novel defense systems. J Bacteriol 193:6039–6056. https://doi.org/10.1128/JB.05535-11

Marston MF, Pierciey FJ, Shepard A et al (2012) Rapid diversification of coevolving marine Synechococcus and a virus. Proc Natl Acad Sci U S A 109:4544–4549. https://doi.org/10.1073/pnas.1120310109

Middelboe M, Hagström A, Blackburn N et al (2001) Effects of bacteriophages on the population dynamics of four strains of pelagic marine bacteria. Microb Ecol 42:395–406. https://doi.org/10.1007/s00248-001-0012-1

Middelboe M, Holmfeldt K, Riemann L et al (2009) Bacteriophages drive strain diversification in a marine Flavobacterium: implications for phage resistance and physiological properties. Environ Microbiol 11:1971–1982. https://doi.org/10.1111/j.1462-2920.2009.01920.x

Mizoguchi K, Morita M, Fischer CR et al (2003) Coevolution of bacteriophage PP01 and Escherichia coli O157:H7 in continuous culture. Appl Environ Microbiol 69:170–176. https://doi.org/10.1128/AEM.69.1.170-176.2003

Mizuno CM, Ghai R, Rodriguez-Valera F (2014) Evidence for metaviromic islands in marine phages. Front Microbiol 5:1–10. https://doi.org/10.3389/fmicb.2014.00027

Monier A, Chambouvet A, Milner DS et al (2017) Host-derived viral transporter protein for nitrogen uptake in infected marine phytoplankton. Proc Natl Acad Sci U S A 114:E7489–E7498. https://doi.org/10.1073/pnas.1708097114

Moniruzzaman M, Gann ER, Wilhelm SW (2018) Infection by a giant virus (AaV) induces widespread physiological reprogramming in aureococcus anophagefferens CCMP1984 – a harmful bloom algae. Front Microbiol 9:752

Moniruzzaman M, Martinez-Gutierrez CA, Weinheimer AR, Aylward FO (2020) Dynamic genome evolution and complex virocell metabolism of globally-distributed giant viruses. Nat Commun 11:1710. https://doi.org/10.1038/s41467-020-15507-2

Morimoto D, Kimura S, Sako Y, Yoshida T (2018) Transcriptome analysis of a bloom-forming cyanobacterium Microcystis aeruginosa during Ma-LMM01 phage infection. Front Microbiol 9: article 2. https://doi.org/10.3389/fmicb.2018.00002

Morimoto D, Tominaga K, Nishimura Y et al (2019) Cooccurrence of broad- and narrow-host-range viruses infecting the bloom-forming toxic cyanobacterium Microcystis aeruginosa. Appl Environ Microbiol 85:e01170–e01119

Morimoto D, Šulčius S, Tominaga K, Yoshida T (2020) Predetermined clockwork microbial worlds: current understanding of aquatic microbial diel response from model systems to complex environments. Adv Appl Microbiol 113:163–191. https://doi.org/10.1016/bs.aambs.2020.06.001

Morris RM, Cain KR, Hvorecny KL, Kollman JM (2020) Lysogenic host–virus interactions in SAR11 marine bacteria. Nat Microbiol 5:1011–1015. https://doi.org/10.1038/s41564-020-0725-x

Needham DM, Chow CET, Cram JA et al (2013) Short-term observations of marine bacterial and viral communities: patterns, connections and resilience. ISME J 7:1274–1285. https://doi.org/10.1038/ismej.2013.19

Needham DM, Sachdeva R, Fuhrman JA (2017) Ecological dynamics and co-occurrence among marine phytoplankton, bacteria and myoviruses shows microdiversity matters. ISME J 11:1614–1629. https://doi.org/10.1038/ismej.2017.29

Needham DM, Fichot EB, Wang E et al (2018) Dynamics and interactions of highly resolved marine plankton via automated high-frequency sampling. ISME J 12:2417–2432. https://doi.org/10.1038/s41396-018-0169-y

Needham DM, Yoshizawa S, Hosaka T et al (2019) A distinct lineage of giant viruses brings a rhodopsin photosystem to unicellular marine predators. Proc Natl Acad Sci U S A 116:20574–20583. https://doi.org/10.1073/pnas.1907517116

Nissimov JI, Vandzura R, Johns CT et al (2018) Dynamics of transparent exopolymer particle production and aggregation during viral infection of the coccolithophore, Emiliania huxleyi. Environ Microbiol 20:2880–2897. https://doi.org/10.1111/1462-2920.14261

Parada AE, Fuhrman JA (2017) Marine archaeal dynamics and interactions with the microbial community over 5 years from surface to seafloor. ISME J 11:2510–2525. https://doi.org/10.1038/ismej.2017.104

Parsons RJ, Breitbart M, Lomas MW, Carlson CA (2012) Ocean time-series reveals recurring seasonal patterns of virioplankton dynamics in the northwestern Sargasso Sea. ISME J 6:273–284. https://doi.org/10.1038/ismej.2011.101

Paul JH (2008) Prophages in marine bacteria: dangerous molecular time bombs or the key to survival in the seas? ISME J 2:579–589. https://doi.org/10.1038/ismej.2008.35

Pourtois J, Tarnita CE, Bonachela JA (2020) Impact of lytic phages on phosphorus- vs nitrogen-limited marine microbes. Front Microbiol 11:221. https://doi.org/10.3389/fmicb.2020.00221

Reva O, Tümmler B (2008) Think big - Giant genes in bacteria. Environ Microbiol 10:768–777. https://doi.org/10.1111/j.1462-2920.2007.01500.x

Rodriguez-Brito B, Li L, Wegley L et al (2010) Viral and microbial community dynamics in four aquatic environments. ISME J 4:739–751. https://doi.org/10.1038/ismej.2010.1

Rodriguez-Valera F, Ussery DW (2012) Is the pan-genome also a pan-selectome? F1000Research 1:1–7. https://doi.org/10.3410/f1000research.1-16.v1

Rodriguez-Valera F, Martin-Cuadrado AB, Rodriguez-Brito B et al (2009) Explaining microbial population genomics through phage predation. Nat Rev Microbiol 7:828–836. https://doi.org/10.1038/nrmicro2235

Rosenwasser S, Mausz MA, Schatz D et al (2014) Rewiring host lipid metabolism by large viruses determines the fate of Emiliania huxleyi, a bloom-forming alga in the ocean. Plant Cell 26:2689–2707. https://doi.org/10.1105/tpc.114.125641

Roucourt B, Lavigne R (2009) The role of interactions between phage and bacterial proteins within the infected cell: a diverse and puzzling interactome. Environ Microbiol 11:2789–2805. https://doi.org/10.1111/j.1462-2920.2009.02029.x

Schulz F, Roux S, Paez-Espino D et al (2020) Giant virus diversity and host interactions through global metagenomics. Nature 578:432–436. https://doi.org/10.1038/s41586-020-1957-x

Schvarcz CR, Steward GF (2018) A giant virus infecting green algae encodes key fermentation genes. Virology 518:423–433. https://doi.org/10.1016/j.virol.2018.03.010

Schwalbach MS, Hewson I, Fuhrman JA (2004) Viral effects on bacterial community composition in marine plankton microcosms. Aquat Microb Ecol 34:117–127. https://doi.org/10.3354/ame034117

Sharma RS, Mishra V, Mohmmed A, Babu CR (2008) Phage specificity and lipopolysaccarides of stem- and root-nodulating bacteria (Azorhizobium caulinodans, Sinorhizobium spp., and Rhizobium spp.) of Sesbania spp. Arch Microbiol 189:411–418. https://doi.org/10.1007/s00203-007-0322-x

Sheyn U, Rosenwasser S, Lehahn Y et al (2018) Expression profiling of host and virus during a coccolithophore bloom provides insights into the role of viral infection in promoting carbon export. ISME J 12:704–713. https://doi.org/10.1038/s41396-017-0004-x

Sieradzki ET, Ignacio-Espinoza JC, Needham DM et al (2019) Dynamic marine viral infections and major contribution to photosynthetic processes shown by spatiotemporal picoplankton metatranscriptomes. Nat Commun 10:1169. https://doi.org/10.1038/s41467-019-09106-z

Silkin V, Pautova L, Kravchishina M et al (2020) Dataset of the Emiliania huxleyi abundance and phytoplankton composition in the Barents Sea in summer 2014–2018. Data Br 32:106251. https://doi.org/10.1016/j.dib.2020.106251

Sime-Ngando T (2014) Environmental bacteriophages: viruses of microbes in aquatic ecosystems. Front Microbiol 5:1–14. https://doi.org/10.3389/fmicb.2014.00355

Sorek R, Kunin V, Hugenholtz P (2008) CRISPR - a widespread system that provides acquired resistance against phages in bacteria and archaea. Nat Rev Microbiol 6:181–186. https://doi.org/10.1038/nrmicro1793

Stewart FM, Levin BR (1984) The population biology of bacterial viruses: Why be temperate. Theor Popul Biol 26:93–117. https://doi.org/10.1016/0040-5809(84)90026-1

Sullivan MB, Weitz JS, Wilhelm S (2017) Viral ecology comes of age. Environ Microbiol Rep 9:33–35. https://doi.org/10.1111/1758-2229.12504

Suttle CA (2005) Viruses in the sea. Nature 437:356–361. https://doi.org/10.1038/nature04160

Suttle CA (2007) Marine viruses - major players in the global ecosystem. Nat Rev Microbiol 5:801–812. https://doi.org/10.1038/nrmicro1750

Suttle CA, Chen F (1992) Mechanisms and rates of decay of marine viruses in seawater. Appl Environ Microbiol 58:3721–3729. https://doi.org/10.1128/aem.58.11.3721-3729.1992

Tarutani K, Nagasaki K, Yamaguchi M (2000) Viral impacts on total abundance and clonal composition of the harmful bloom-forming phytoplankton Heterosigma akashiwo. Appl Environ Microbiol 66:4916–4920. https://doi.org/10.1128/AEM.66.11.4916-4920.2000

Thierauf A, Perez G, Maloy S (2009) Generalized transduction. Methods Mol Biol 501:267–286. https://doi.org/10.1007/978-1-60327-565-1

Thingstad TF (2000) Elements of a theory for the mechanisms controlling abundance, diversity, and biogeochemical role of lytic bacterial viruses in aquatic systems. Limnol Oceanogr 45:1320–1328. https://doi.org/10.4319/lo.2000.45.6.1320

Thingstad TF, Vage S, Storesund JE et al (2014) A theoretical analysis of how strain-specific viruses can control microbial species diversity. Proc Natl Acad Sci U S A 111:7813–7818. https://doi.org/10.1073/pnas.1400909111

Thompson LR, Zeng Q, Kelly L et al (2011) Phage auxiliary metabolic genes and the redirection of cyanobacterial host carbon metabolism. Proc Natl Acad Sci U S A 108:E757–E764. https://doi.org/10.1073/pnas.1102164108

Touchon M, Bernheim A, Rocha EPC (2016) Genetic and life-history traits associated with the distribution of prophages in bacteria. ISME J 10:2744–2754. https://doi.org/10.1038/ismej.2016.47

Touchon M, Moura de Sousa JA, Rocha EP (2017) Embracing the enemy: the diversification of microbial gene repertoires by phage-mediated horizontal gene transfer. Curr Opin Microbiol 38:66–73. https://doi.org/10.1016/j.mib.2017.04.010

Valen VL (1973) A new evolutionary law. Evol Theory 1:1–30

Warwick-Dugdale J, Buchholz HH, Allen MJ, Temperton B (2019) Host-hijacking and planktonic piracy: how phages command the microbial high seas. Virol J 16:15. https://doi.org/10.1186/s12985-019-1120-1

Waterbury JB, Valois FW (1993) Resistance to co-occurring phages enables marine Synechococcus communities to coexist with cyanophages abundant in seawater. Appl Environ Microbiol 59:3393–3399. https://doi.org/10.1128/aem.59.10.3393-3399.1993

Wei Y, Ocampo P, Levin BR (2010) An experimental study of the population and evolutionary dynamics of Vibrio cholerae O1 and the bacteriophage JSF4. Proc R Soc B Biol Sci 277:3247–3254. https://doi.org/10.1098/rspb.2010.0651

Wei Y, Kirby A, Levin BR et al (2011) The population and evolutionary dynamics of Vibrio cholerae and its bacteriophage: conditions for maintaining phage-limited communities. Am Nat 178:715–725. https://doi.org/10.1086/662677

Weinbauer MG (2004) Ecology of prokaryotic viruses. FEMS Microbiol Rev 28:127–181. https://doi.org/10.1016/j.femsre.2003.08.001

Weitz JS, Stock CA, Wilhelm SW et al (2015) A multitrophic model to quantify the effects of marine viruses on microbial food webs and ecosystem processes. ISME J 9:1352–1364. https://doi.org/10.1038/ismej.2014.220

Welsh JE, Steenhuis P, de Moraes KR et al (2020) Marine virus predation by non-host organisms. Sci Rep 10:1–9. https://doi.org/10.1038/s41598-020-61691-y

Whitman WB, Coleman DC, Wiebe WJ (1998) Prokaryotes: the unseen majority. Proc Natl Acad Sci U S A 95:6578–6583. https://doi.org/10.1073/pnas.95.12.6578

Wigington CH, Sonderegger DL, Brussaard CPD et al (2016) Re-examining of the relationship between marine virus and microbial cell abundances. Nat Microbiol 1:15024. https://doi.org/10.1038/nmicrobiol.2015.24

Wikner J, Vallino JJ, Steward GF et al (1993) Nucleic acids from the host bacterium as a major source of nucleotides for three marine bacteriophages. FEMS Microbiol Ecol 12:237–248. https://doi.org/10.1111/j.1574-6941.1993.tb00036.x

Wilhelm SW, Suttle CA (1999) Viruses and nutrient cycles in the sea: viruses play critical roles in the structure and function of aquatic food webs. Bioscience 49:781–788. https://doi.org/10.2307/1313569

Wilhelm LJ, Tripp HJ, Givan SA et al (2007) Natural variation in SAR11 marine bacterioplankton genomes inferred from metagenomic data. Biol Direct 2:1–19. https://doi.org/10.1186/1745-6150-2-27

Williamson SJ, Rusch DB, Yooseph S et al (2008) The sorcerer II global ocean sampling expedition: metagenomic characterization of viruses within aquatic microbial samples. PLoS One 3:e1456. https://doi.org/10.1371/journal.pone.0001456

Wilmes P, Simmons SL, Denef VJ, Banfield JF (2009) The dynamic genetic repertoire of microbial communities. FEMS Microbiol Rev 33:109–132. https://doi.org/10.1111/j.1574-6976.2008.00144.x

Winter C, Bouvier T, Weinbauer MG, Thingstad TF (2010) Trade-offs between competition and defense specialists among unicellular planktonic organisms: the "killing the winner" hypothesis revisited. Microbiol Mol Biol Rev 74:42–57. https://doi.org/10.1128/mmbr.00034-09

Wommack KE, Colwell RR (2000) Virioplankton: viruses in aquatic ecosystems. Microbiol Mol Biol Rev 64:69–114. https://doi.org/10.1128/mmbr.64.1.69-114.2000

Yamada Y, Tomaru Y, Fukuda H, Nagata T (2018) Aggregate formation during the viral lysis of a marine diatom. Front Mar Sci 5:1–7. https://doi.org/10.3389/fmars.2018.00167

Yamada Y, Guillemette R, Baudoux AC et al (2020) Viral attachment to biotic and abiotic surfaces in seawater. Appl Environ Microbiol 86:1–18. https://doi.org/10.1128/AEM.01687-19

Yoshida M, Yoshida T, Takashima Y et al (2005) Genetic diversity of the toxic cyanobacterium Microcystis in Lake Mikata. Environ Toxicol 20:229–234. https://doi.org/10.1002/tox.20102

Yoshida M, Yoshida T, Kashima A et al (2008a) Ecological dynamics of the toxic bloom-forming cyanobacterium Microcystis aeruginosa and its cyanophages in freshwater. Appl Environ Microbiol 74:3269–3273. https://doi.org/10.1128/AEM.02240-07

Yoshida M, Yoshida T, Satomi M et al (2008b) Intra-specific phenotypic and genotypic variation in toxic cyanobacterial Microcystis strains. J Appl Microbiol 105:407–415. https://doi.org/10.1111/j.1365-2672.2008.03754.x

Yoshida T, Morimoto D, Kimura S (2019) Bacteria-virus interactions. In: Nishida H, Oshima T (eds) DNA traffic in the environment. Springer Singapore, Singapore, pp 95–108

Yoshida-Takashima Y, Yoshida M, Ogata H et al (2012) Cyanophage infection in the bloom-forming cyanobacteria Microcystis aeruginosa in surface freshwater. Microbes Environ 27:350–355. https://doi.org/10.1264/jsme2.ME12037

Zborowsky S, Lindell D (2019) Resistance in marine cyanobacteria differs against specialist and generalist cyanophages. Proc Natl Acad Sci U S A 116:16899–16908. https://doi.org/10.1073/pnas.1906897116

Zeng Q, Chisholm SW (2012) Marine viruses exploit their host's two-component regulatory system in response to resource limitation. Curr Biol 22:124–128. https://doi.org/10.1016/j.cub.2011.11.055

Zhao Y, Temperton B, Thrash JC et al (2013) Abundant SAR11 viruses in the ocean. Nature 494:357–360. https://doi.org/10.1038/nature11921

Zsolnay Á (2003) Dissolved organic matter: artefacts, definitions, and functions. Geoderma 113:187–209. https://doi.org/10.1016/S0016-7061(02)00361-0

Chapter 2
Life Continues as Viruses Close Land, Water and Atmosphere Nutrient Cycle

Peter Pollard

Abstract

All life has green roots

—Peter Pollard

Every element on our planet that is used to sustain life is recycled. We would not exist without microbes breaking down complex organic compounds to recycle the resulting inorganic nutrients. Most people are oblivious to the fundamental regulatory role viruses play in these cycles. Indeed, most people revile viruses. However, the role of viruses in global nutrient cycling aligns with their bacterial hosts. Together, bacteria and viruses are partners in degrading organic matter to regenerate inorganic nutrients such as CO_2, nitrogen and phosphorus. Monitoring DOC concentration in freshwater ecosystems without quantifying the turnover, a common practice, tells us nothing about the inputs of freshwater DOC or how much DOC has been respired. The bacterial viral relationship is like a furnace burning DOC dissolved organic carbon and concomitantly emitting carbon dioxide through respiration. Poor bacterial growth efficiencies coupled to viral lysis of their bacterial host ensures that nutrient cycles are closed. Together, bacteria under virus regulation ensures that the organisms responsible for primary production have sufficient inorganic nutrients available to them in the Open Oceans, in freshwater and on the land so life can go on. This chapter describes the viral-bacterial relationship with dissolved organic carbon that is so important to life.

2.1 Our 'Love-Hate' Relationship with Viruses

"I love viruses", you hate them. You just saw the word "Viruses!!" Today (2021), your mind, and the daily news, would have immediately shot to COVID-19. Then possibly your mind darted to influenza or Ebola. Of course, these viruses are foremost on our minds because they can cause us disease and sometimes an

P. Pollard (✉)
School of Engineering and Built Environment, Griffith University, Southport, QLD, Australia
e-mail: p.pollard@griffith.edu.au

C. J. Hurst (ed.), *The Biological Role of a Virus*, Advances in Environmental Microbiology 9, https://doi.org/10.1007/978-3-030-85395-2_2

excruciating and painful death. Viruses ensure no one species gets out of control to disrupt the global ecological balance. We are not exempt from this natural process of viral control. However, these human pathogenic viruses represent just a tiny fraction of all the other beneficial viruses on earth. In reality, most viruses are vital to our very existence on earth. Their sheer number is astonishing.

The concentration of beneficial viruses in the oceans is around 10^7 viruses/mL (Bergh et al. 1989) while in freshwater there can be 10^9 viruses/mL (Pollard and Ducklow 2011). That equates to more than a third of the population of the USA squeezed into the bottom of a test tube in 1 mL of water. Globally the oceans contain 10^{30} viruses (Whitman et al. 1998). There are more viruses on earth than there are stars in the sky. If you lined them all up, the viruses on earth would extend for ten million light years; that's 100 times the distance across our galaxy. Collectively they would weigh as much as 75 million blue whales (Suttle 2005).

Not until 1989, using Transmission Electron Microscopy (TEM), did Bergh et al., show the exceptionally high number of viruses in aquatic environments (10^8 mL^{-1}). They also showed that these viruses were infecting their bacterial hosts at a rate of 4×10^5 d^{-1} mL^{-1} assuming a burst size of 50 viruses per cell. Daily, as much as a third of bacterial populations are being infected, lysed and recycling nutrients. Despite the inordinate number of viral numbers on the planet, their role in the global nutrient cycles has gone mostly unnoticed.

Freshwater connects the soil with the oceans and the atmosphere to complete the global nutrient cycle (Cole and Caraco 2001; del Giorgio and Williams 2005; Kirchman 2008). Aerobic freshwater bacterial respiration contributes profoundly to the global atmospheric carbon budget (Aufdenkampe et al. 2011; Ward et al. 2017). These cycles depend on the intimate and dynamic relationship between bacteria and their viral partner/nemesis (Wommack and Colwell 2000; Gómez and Buckling 2011; Zimmerman et al. 2020). This chapter aims to explain how dependent life really is on the regulatory role of viruses in nutrient cycling.

Figure 2.1 shows an epifluorescence microscope picture of viruses and bacteria sampled from a freshwater river in subtropical Australia with viruses at a concentration of 10^8 mL^{-1}. On average viruses are 0.1μm across while bacteria are most often >10 times larger. Viruses are an inanimate, an unpretentious collection of crystalline chemical units—not living. They are parasites that cannot replicate without their host.

Much like a computer virus, they carry a piece of code that dictates their function. A virus' genetic code is carried in a small piece of DNA or RNA that is protected inside a protein coat. When a virus recognises and attaches to its target host, the virus actively injects or otherwise liberates its coded nucleic material into the host to take over the metabolic machinery of its host's cells (Griffith University 2015).

Figure 2.2 captures the exact moment a virus lysed a cyanobacterial cell; the cell burst. Notice, as the cell is lysed the newly formed viral particles are released. The "g-host" bacterial cell also releases its remaining cell contents (such as Dissolved Organic Carbon) into the surrounding environment—an important source of nutrients for the surrounding intact growing bacteria (Middelboe and Lyck 2002: Weinbauer et al. 2002). Viral lysis of bacteria is an important part of carbon and

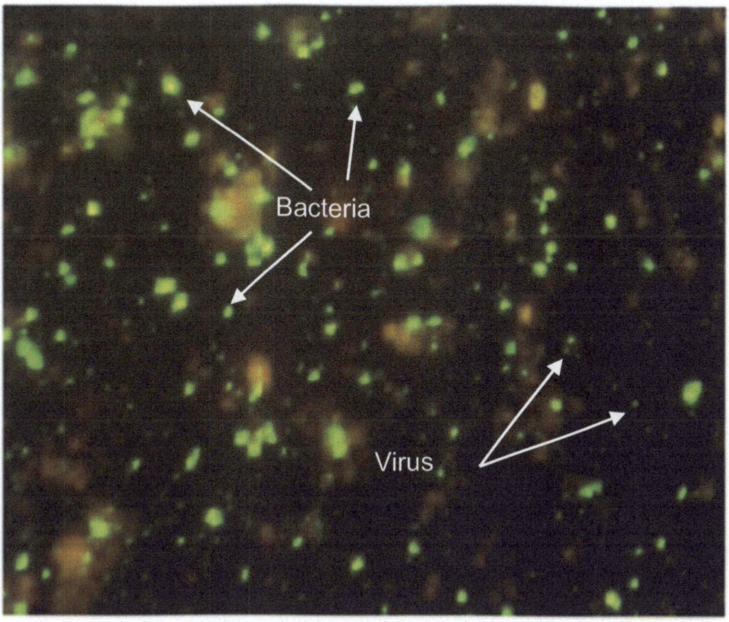

Fig. 2.1 A sample of the Bremer River (South East Queensland, Australia) freshwater where the bacteria and viruses were stained with a DNA stain. The bacteria (larger green objects, indicated by a cluster of three arrows) and viruses (smaller green objects, indicated by two joined arrows) were collected on a 0.02μm Anodisc filter. They were viewed using an epifluorescence microscope through a non-fluorescent oil. Magnification Magnification ×1000

nutrient cycling both in aquatic (Malits et al. 2014; Zhang et al. 2020b) and terrestrial environments (Daly et al. 2019; Williamson et al. 2017; Roy et al. 2020). Viral lysis forces more DOC through bacterial respiration faster and can also influence our climate as global temperatures increase rates of bacterial production and respiration (Zhang et al. 2020b).

2.2 The Open Ocean

2.2.1 Autotrophy Dominates

Globally, net primary production (eg from algae and phytoplankton) in the Open Ocean uses approximately 2.1 Pg CO_2 per year. This represents a mere 0.3% of all the DOC pool (700 Pg Carbon) in the Open Oceans (Carlson 2002). Yet, microzooplankton and mesozooplankton subsequently consume three quarters of this primary production and its chemical bond energy then feeds the higher trophic groups of the ocean's food webs (del Giorgio and Williams 2005: Armengol et al. 2019). Hence Autotrophy dominates the open oceans.

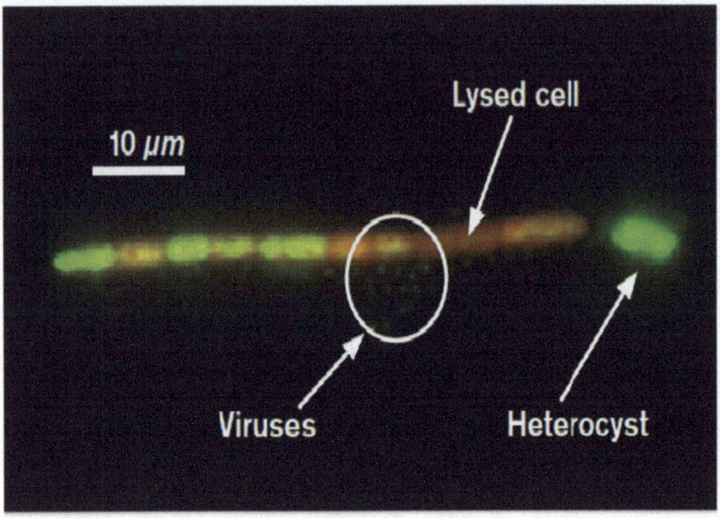

Fig. 2.2 This figure shows the moment viruses explode from a bacterial cell in a chain of *C. raciborskii* cells. All have been stained using a SYBR Green DNA stain and were viewed using an epifluorescence microscope through a non-fluorescent oil. Magnification ×1000. Reproduced from Pollard and Young (2010) with permission from *Acta Oecologia*

Here, there is little if any input of carbon from terrestrial sources nor open ocean sediments (del Giorgio and Williams 2005). Together, with the classic phytoplankton-zooplankton-nekton food web carbon moves up the food chain to the higher trophic groups. Direct use of primary production as a food source is the most efficient form of energy and carbon transfer up the food chain with very high microzooplankton herbivory (Armengol et al. 2019).

2.2.2 *Heterotrophic Bacterial Production Does Not Move Up the Food Chain*

In 1983 Azam et al. proposed a microbial loop that carried bacterial production up the food chain. Nanoplanktonic heterotrophic flagellates (nanoflagellates) control the abundance of bacterioplankton, by consuming rapidly growing bacterial which had used the dissolved organic carbon DOC being leached from living algal cells or dying cells. Microzooplankton then graze on the nano flagellates; the microzooplankton are in the same general size range as the phytoplankton and were assumed by Azam to be returning some energy and organic carbon from the 'microbial loop' to higher trophic levels in a conventional planktonic food chain. But this was a hypothesis without evidence and viruses were not considered important in the oceans until much later (Bergh et al. 1989).

Around the same time, other researchers working on the microbial ecology of planktonic marine systems concluded that marine bacterioplankton were decomposing most of the DOC in the water column straight back to CO_2 (Ducklow et al. 1986; Ducklow 1999). In other words, their work was showing that bacteria, rather than being carbon links to higher trophic groups, were in fact carbon sinks. Bacteria respired the dissolved organic carbon to carbon dioxide/bicarbonate in marine ecosystems. For that they did have evidence. To this day there are no quantitative data to show significant global transfer of bacterial production to the top of the food chain in the Open Oceans.

By the late 1990s, viruses were being introduced into the marine food web models with estimates of viral mediated DOC release in the range of 2–20 Gt DOC per year (Fuhrman 1999; Wilhelm and Suttle 1999). They described the role of viruses as a 'Shunt'—after viral infection and lysis of its microbial host, assigning an estimation that dissolved and particulate organic matter from pelagic phytoplankton and bacterial populations were released into the water column. Buchan et al. (2014) show the viral shunt as items 2 and 7 in their Fig. 2.3.

Viral lysis of phytoplankton and bacterial populations releases dissolved and particulate organic matter into the water column. This leads to internal recycling of DOC and loss of DOC through bacterial respiration generating CO_2. Even small perturbations in the input and output of DOC pool of the oceans can alter the balance between oceanic and atmospheric CO_2 (Carlson 2002).

Today, 2021, viruses are seen as having a key role in biogeochemical nutrient cycling in the oceans (Zimmerman et al. 2020). On a global scale the viral mediated release of DOC into the ocean is almost half the 50 Gt Carbon/year of phytoplankton primary production, so viruses play a major quantitative nutrient recycling role (Wilhelm and Suttle 1999; Breitbart et al. 2008). To understand that vital role of

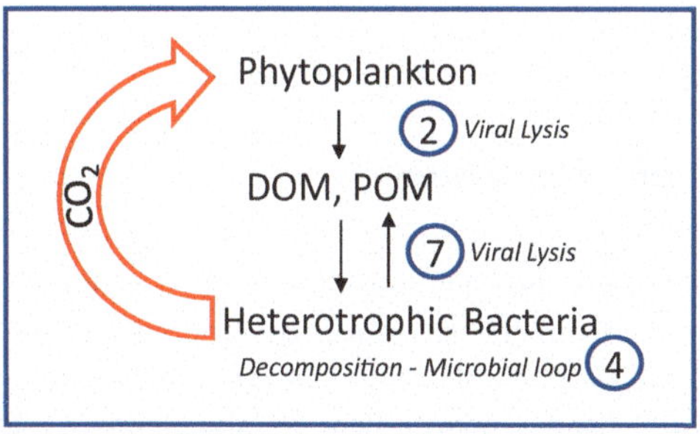

Fig. 2.3 This figure is based on the Fig. 1 of Buchan et al. 2014 who show the bacterial transformation of phytoplankton derived organic matter in water. However, here, the return of carbon to the DOM and POM through viral mediated lysis of bacterial is highlighted. (2) is the release of both dissolved organic matter (DOM) and particulate organic matter (POM) from phytoplankton. (4) Mineralization (that is the release of CO2 via respiration during the catabolism of organic matter) and recycling of organic matter through heterotrophic bacteria, (microbial loop). (7) The viral shunt describes the contributions of viral-mediated cell lysis to release dissolved and particulate matter from both the phytoplankton and bacterial pools

viruses I'm turning to a discussion of freshwater microbial ecology research. These ecosystems have an order of magnitude higher rate of bacterial production compared to the Open Oceans, making it easier to see the role of viruses.

As with the Open Ocean, freshwater higher trophic groups still depend on photosynthetic primary production despite the much higher rates of bacterial production. The major difference between the Open Ocean and freshwater is the latter has an endless supply of bacterial substrate—terrestrial dissolved organic matter. This makes it easier to see where bacterial production energy and carbon is being transferred in freshwater food webs.

2.3 Freshwater

Freshwater couples the biogeochemical cycles between land, oceans, and atmosphere. However, most of the DOC is respired before it reaches the ocean (del Giorgio and Williams 2005; Aufdenkampe et al. 2011). Freshwater bacteria and viruses per mL are 10 times greater than those of the Open Ocean. While bacterial growth rates in open marine systems vary from 0.05 to 7 d^{-1} (White et al. 1991; Ducklow 2000), they are much higher in freshwater from 0.03 to 1.8 d^{-1} (White et al. 1991; Pollard and Ducklow 2011). The much higher growth rate and concentration of bacteria in freshwater leads to much higher rates of lytic viral regulation of their bacterial hosts (Thingstad and Lignell 1997; Winget et al. 2011).

Compared to the open Oceans, the freshwater ecosystems' heterotrophic bacteria and their associated viruses out performs their marine counterparts in biogeochemical and metabolic processing an almost endless supply of dissolved organic matter.

2.3.1 *Heterotrophy Dominates*

Freshwater ecosystems are net heterotrophic with a major source of dissolved organic carbon coming from the terrestrial environment (Cole et al. 2000; Cole and Caraco 2001; Cole 2013). By definition, the rates of bacterial production exceed primary production in the same environment. Freshwater ecosystems main source of DOC is allochthonous (defined here as organic carbon from elsewhere, in either space or time). Indeed, when you consider lake and river sources of DOC from the surrounding landscape, you are hard pressed to find any system without a watershed/catchment inputting terrestrial organic carbon. Terrestrial bacteria readily mineralise terrestrially derived macromolecules, like lignin and phenolic compounds, that might be considered refractory in freshwater (Ward et al. 2013).

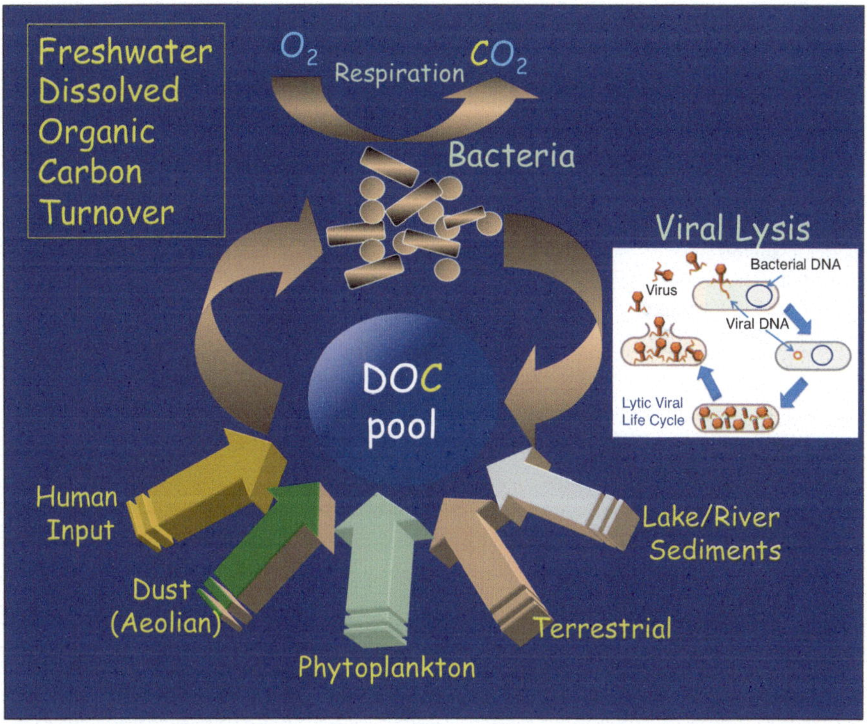

Fig. 2.4 Model of how bacteria use carbon from the DOC pool to respire the sources of DOC as CO_2 from freshwater. [DOC] in the pool remains constant; so input = output. Viral lysis of the bacteria facilitates high bacterial respiration by recycling organic carbon back through the DOC pool. This process of viral re-release of bacterial carbon and energy is an exponential decay of DOC from the DOC pool. Viruses are subverting the transfer of bacterial energy and carbon away from the higher end of the food web

2.3.2 Carbon Sources Entering the DOC Pool

Figure 2.4 shows sources of DOC entering the DOC pool of freshwater. Bacterial metabolic activity (respiration; mineralisation of organic carbon) and pCO_2 supersaturating freshwater are positively correlated (Marotta et al. 2009; Cardoso et al. 2013). A substantial amount of terrestrial organic carbon is processed within freshwater lakes and rivers through bacterial respiration (BR) (Cole et al. 2007, 2011).

There are large scale diffuse sources of organic carbon entering freshwater (Mulholland 2003). These are topping up the DOC pool as fast as bacterial and viral processes remove it. Figure 2.4 shows how viral lysis of their bacterial host contributes to this DOC pool. The bacterial respiration depends on the steady supply of DOC (del Giorgio et al. 1997; Cole et al. 2000, 2007; Mayorga et al. 2005; Pace and Prairie 2005; Pollard and Ducklow 2011; Cole 2013; Cardoso et al. 2013).

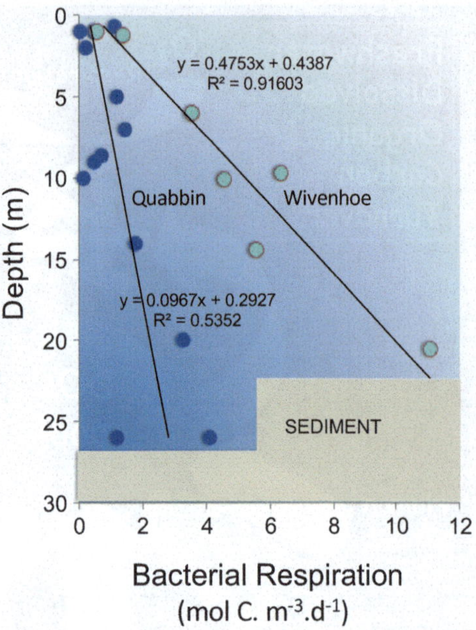

Fig. 2.5 Bacterial respiration is plotted as a function of water depth for two major reservoirs. In each case highest mineralisation rates of organic carbon were at the bottom of each reservoir. This supports bacterioplankton and respiration. The *in situ* method used to measure bacterial respiration in Fig. 2.5 is described in Pollard (2013a)

Aeolian (wind-blown) sources also contribute to this aquatic DOC pool of organic matter that has been transported from any and every corner of the globe (Swap et al. 1992), not to mention the anthropogenic inputs from agriculture, land clearing and urbanisation. Lake sediments also contribute to the DOC pool to the water column (Fig. 2.5; Zhang et al. 2020a).

Freshwater benthic processes of decomposition play a large role releasing DOC into the water column of lakes. Sometime these DOC sources can be ancient. The carbon having remained locked up in the sediments for many, many years (McCallister and Del Giorgio 2008, 2012; Zhang et al. 2020a).

Figure 2.5 is a plot of bacterial respiration as a function of depth for two lakes. Quabbin Reservoir (depth 28 m), Massachusetts, USA (Boston's drinking water supply) in the Northern Hemisphere. The other, Lake Wivenhoe (depth 22 m) South East Queensland, Australia (Brisbane's drinking water supply) is in the Southern Hemisphere. The temperate and sub-tropical freshwaters, respectively can thus be compared. Both water bodies showed a positive correlation (r2 = 0.54 and 0.92, respectively) of the rate of bacterial respiration as the depth increased.

In contrast to the Open Oceans, microbes in sediments of deep and shallow freshwater lakes and reservoirs are a major source of the DOC substrate for bacterial production and respiration in the water column. They are an important part of the global energy and nutrients cycles (Zhang et al. 2020b). Sediment sources of DOC support high rates of bacterial production in the water column (Fig. 2.4) coupled with terrestrial catchment sources of DOC.

2.3.3 Dissolved Organic Carbon [DOC] Pool

The dissolved organic carbon pool in freshwater is the source from where bacteria draw their substrate for production and respiration (Fig. 2.4). It is central to nutrient cycling in aquatic ecosystems.

Table 2.1 compares DOC concentrations in freshwater lakes and rivers across a diverse range of biomes. From low to high Latitudes of the Northern and the Southern Hemispheres the DOC concentration ranged from 1 to 13 mg.L^{-1} (Mulholland 2003; Pollard and Ducklow 2011; Oliver et al. 2017). What is most interesting is that, on a global scale, the DOC concentrations of this pool varies little across rivers and lakes and between very different biomes (Table 2.1).

Globally, DOC, freshwater bacterial numbers along with their viral partners, per mL also remain remarkably unchanged, while bacterial growth and production can vary by an order of magnitude (Church 2008) which is consistent with tight top-down viral regulation of bacteria (Zimmerman et al. 2020). Viral numbers (10^9/mL) are usually 10 times those of their bacterial hosts that are also doubling every 20 min with specific growth rates varying between 0.2 and 1.8 d^{-1} (Pollard and Ducklow 2011) with viral lysis rates of bacteria directly related to the growth rate of their bacteria hosts. Given the variable bacterial growth rates coupled with relatively stable bacterial numbers, the growth rate and death rate of bacterial populations must be equal. Viruses exert a major top-down control upon this bacterial production (Zimmerman et al. 2020).

2.3.4 [DOC] Pool Versus DOC Turnover: Flux

There is a fundamental difference between DOC concentrations and DOC turnover that is not readily appreciated. Quantifying the turnover of the DOC pool is a prerequisite to understanding the proportion of organic carbon entering and leaving

Table 2.1 DOC concentrations in freshwater lakes and rivers in different biomes from low to high Latitudes of Northern and Southern Hemispheres [Adapted from Mulholland (2003) additional data from [a]Pollard and Ducklow (2011) and [b]Oliver et al. (2017)]

Freshwater Biomes	DOC mg L^{-}1 (Mean)
Tundra	2
Boreal forests	7
Temperature	4
Temperature northern[b] rainforest	6–11
Semi-arid	1
Wet tropics	8
Dry tropics	3
Dry-subtropics[a]	5
Humid climates	4–13

the DOC pool that is supporting freshwater bacterial production and respiration. Indirectly, the bacterial associated viruses are also reliant on their bacterial host receiving a steady source of DOC substrate. The standing stock or concentration of the DOC pool is a constant in freshwater ecosystems (Table 2.1). Bacteria are using DOC substrate from the DOC pool at the same rate that it is being supplied from a whole range of sources shown in Fig. 2.4.

High bacterial respiration rates in freshwater coupled with a low and stable concentration pool of DOC (5–10 mg/L) requires the rate of heterotrophic bacterial removal of DOC from the pool to be the same as the rate of DOC input from all the external sources shown in Fig. 2.4. Monitoring DOC concentration in freshwater ecosystems without quantifying the turnover, a common practice, tells us nothing about the inputs of freshwater DOC or how much DOC has been respired. It's the DOC being drawn into a DOC pool that is the critical parameter to monitor (Fig. 2.4). We only see the high rates of allochthonous DOC removal from the pool when measuring rates of bacterial production and respiration *in situ* (Pollard and Ducklow 2011; Pollard 2013a, b). Visualising these dynamics is crucial to understand microbial processes in aquatic ecosystems (Sebastián and Gasol 2019).

2.4 Viral Lysis Is Recycling DOC Through the DOC Pool

Figure 2.4 shows bacteria respiring organic carbon from a central 'common' DOC pool to emit CO_2 into the atmosphere. The quantity of DOC in the pool remains constant; input = output. Viral lysis of bacteria has been shown to contribute to this DOC pool as a form of recycling. Their lysis releases bacterial DOC back into the DOC pool (McCarthy et al. 1998; Middelboe and Lyck 2002; Fig. 2.2) along with a release of insoluble cell remnants that ultimately enter the dissolve DOC pool. This is one of the mechanisms that regulate bacterial metabolism (Pradeep Ram et al. 2016). Other bacteria redraw DOC from the pool as substrate for use in their further respiration and growth; effectively, the DOC that had been incorporated into bacterial biomass thus becomes cannibalized (Nagata 2000). Viral lysis of the bacteria enables high rates of other bacterial respiration thereby recycling organic carbon through the DOC pool and out into the atmosphere as CO_2.

This process of viral re-release of bacterial carbon and energy is an exponential loss of DOC from the DOC pool. Viruses are subverting the transfer of bacterial energy and carbon away from the higher end of the food web. This bacterial-viral relationship is like a furnace burning the DOC and emitting CO_2 through respiration. Describing this process as a "Shunt" (Fuhrman 1999; Wilhelm and Suttle 1999), underplays the true enormity of viral role as a top-down regulator of bacteria production, Fig. 2.3. Viruses are adding significantly to the global DOC pool (2–20 Gt per year) in the oceans alone. Thus, generating CO_2 gas into the atmosphere through more bacterial respiration that would not have occurred otherwise.

2.4.1 *Low Bacterial Growth Efficiencies Coupled with Viral Lysis*

Poor bacterial growth efficiencies of 0.4–10.4% (Eiler et al. 2003) combined with high rates of viral lysis means that little of the bacterial production is going to make it higher up the food chain. Adding viral lysis to the poor bacterial growth efficiencies is like adding fuel to the fire. Viruses are enhancing bacterial growth and respiration rates that convert DOC to CO_2. There is no opportunity for bacterial production to be shunted up to higher trophic groups.

The major carbon and energy link to higher trophic groups in freshwater is primary production (Thorp and Delong 2002). Freshwater food web studies from tropical and arid environments also show autochthonous (within the system) algal production as the major source of organic carbon to metazoans (Bunn et al. 2003, 2006; Clapcott and Bunn 2003), the algae having a higher nutritional value than do bacteria (Brett et al. 2009, 2017). Increasingly, viral infection and lysis of bacteria are viewed as "sinks" of DOC by increasing the amount of DOC that bacteria process (Suttle 2005; Weitz et al. 2015), thereby preventing bacterial production from passing to higher trophic groups. While freshwater heterotrophic bacteria respire DOC rapidly and efficiently, further diminishing the amount of carbon and energy that is transferred to higher trophic groups.

In the tropical rainforests of Central America, forest production has been shown to also be lost to aquatic microbial respiration. Terrestrial primary production that is being returned to the atmosphere through bacterial respiration generating CO_2 passes through the freshwater that is surrounded by major rainforest stands. Terrestrial primary production through aquatic bacterial respiration is thus linked to the atmosphere, closing the carbon cycle (Pollard 2013b: Fig. 2.6). The same is true for

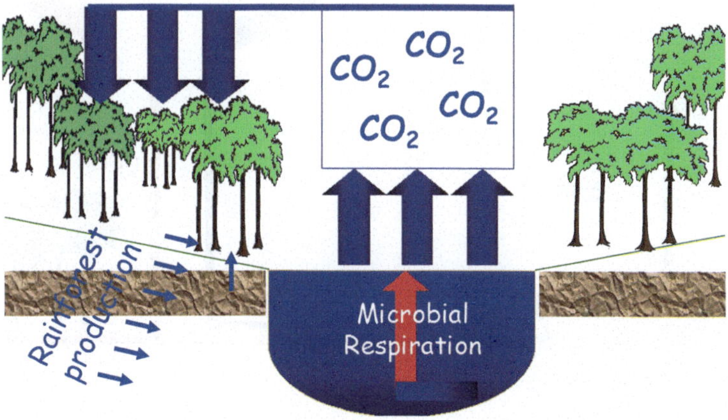

Fig. 2.6 Rainforest production, energy and carbon, is passed into the freshwater DOC pools. Here most is respired as CO_2 and returned to the atmosphere through viral enhanced bacterial respiration, leaving behind inorganic nutrients for aquatic primary production

temperate aquatic environments (Cole and Caraco 2001). While the bacterial respiration rate is enhanced through viral lysis, the respired DOC leaves behind inorganic nutrients, like N and P, that support aquatic primary production (Fig. 2.6).

2.4.2 DOC—Bacterial Respiration—Viral Lysis

Lakes and rivers are now quantitatively being seen as connecting the lithosphere to the atmosphere (Ward et al. 2017). Freshwater's critical role in the global carbon balance is being unraveled through continuing research efforts.

In a mass-balance study of a subtropical freshwater ecosystem in Eastern Australia the bacterial respiration, and its links which connect to viral lysis, are shown in a detailed flow diagram (Pollard and Ducklow 2011). Figure 2.7 shows

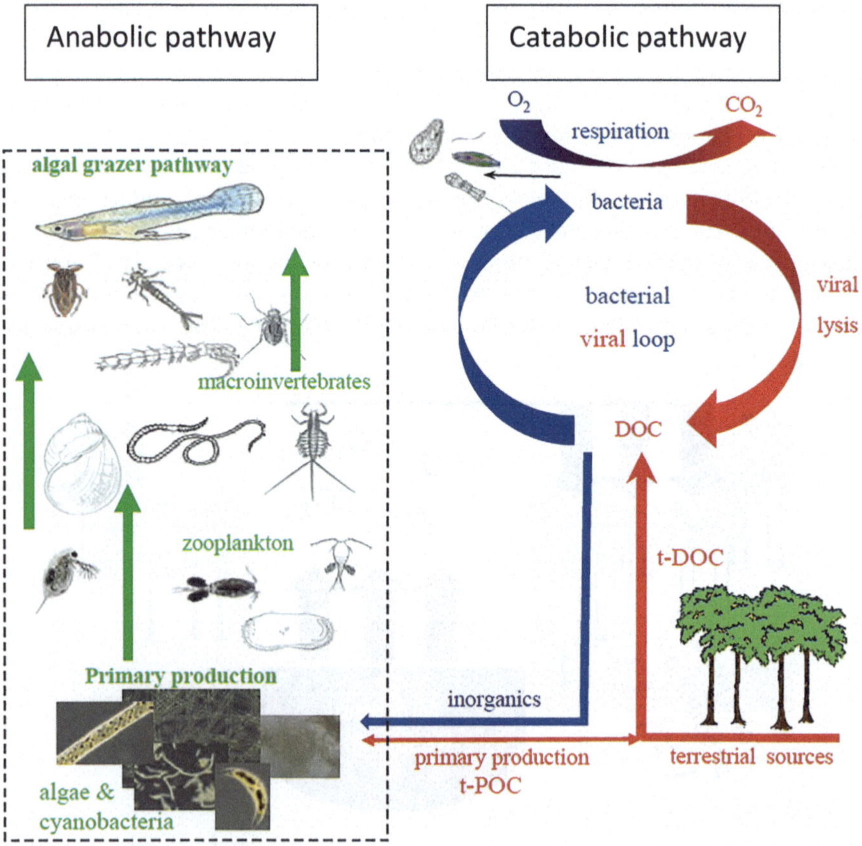

Fig. 2.7 Conceptual model showing how heterotrophic bacteria contribute to food-web structure and flow of nutrients via a bacterial-viral-vented microbial cycle. t-POC is terrestrial primary production; t-DOC is terrestrial DOC. Based on figure from Pollard and Ducklow (2011)

how a viral-bacterial loop would operate to short-circuit bacterial carbon and energy flow from reaching higher trophic groups. This semi enclosed bacterial-viral loop diverts bacterial production away from the bacterivorous protists.

On each pass of this CO_2—vented bacterial–viral loop, more and more DOC is respired as CO_2. Viral lysis acts to recycle DOC through bacterial respiration until all the DOC has been outgasses as CO_2. Thus, carbon and energy from the DOC pool never reaches higher trophic organisms such as zooplankton and fish. The net effect of viral lysis of bacterial cells, along with their poor growth efficiencies, is to regenerate the inorganic nutrients that support primary production (Haaber and Middelboe 2009). The same model would apply to Open Ocean food webs. However, in the ocean the sources of DOC essentially are limited to primary production and aeolian delivery of terrestrial DOC.

What is most striking about Fig. 2.7 is that it seems counter intuitive. We tend to think in terms of "*from little things big things grow*". Whereas, in reality, looking at Fig. 2.7, we see there are two very different pathways. One is anabolic, building more and more complex forms of life that are based on primary production. While the other is catabolic, breaking down complex forms of organic matter to their constitutive elements through bacterial respiration spurred on through viral lysis. Both pathways connected at the bottom where inorganic elements are supplied to the primary producers through viral regulated bacterial respiration. Here also is where the catabolic processes supply the essential inorganic nutrients to support both primary production and the anabolic processes (Haaber and Middelboe 2009). Thereby, through its key participation in these cycles, the bacterial viral relationship has a role in supplying essential elements for the anabolic reactions, a process that starts with primary production that eventually supports the higher trophic groups.

2.5 Viral-Bacterial Nutrient Recycling Supports Primary Production

The role of viruses in the global nutrient cycling aligns with their bacterial hosts. They are both partners as degraders of organic matter and regenerators of inorganic nutrients such as CO_2, nitrogen and phosphorus. With bacteria and viruses working together they ensure that nutrient cycles are closed. Importantly, CO_2 is recycled back into the atmosphere from the land through freshwater and to the atmosphere. Other inorganic nutrients are then, in the same process, made available for primary production both in the Open Oceans and in freshwater ecosystems and on the land. Together bacteria under viral regulation ensure that primary production has sufficient nutrients available in the oceans, freshwater and on the land, so life goes on.

2.6 Conclusion

Now when you see or hear the word virus, I hope, you acknowledge a virus' direct control over our immediate wellbeing is the same as for any other species in nature. Most importantly, that you will remember the importance of the viral-bacterial relationship that is essential to the biological recycling of inorganic nutrients on which photosynthetic organisms depend, in both terrestrial and aquatic environments. The viral bacterial relationship ensures that inorganic nutrients like C, N, P H, and O continue to be available for the 'green roots' on which all life depends.

References

Armengol L, Calbet A, Franchy G, Rodríguez-Santos A, Hernández-León S (2019) Planktonic food web structure and trophic transfer efficiency along a productivity gradient in the tropical and subtropical Atlantic Ocean. Sci Rep 9(1):2044. https://doi.org/10.1038/s41598-019-38507-9

Aufdenkampe AK et al (2011) Riverine coupling of biogeochemical cycles between land, oceans, and atmosphere. Front Ecol Environ 9:53–60

Azam F, Fenchel T, Field JG, Gray JS, Meyer-Reil LA, Thingstad F (1983) The ecological role of water-column microbes in the sea. Mar Ecol Prog Ser 10:257–263. https://doi.org/10.3354/meps010257

Bergh Ø, Borsheim KY, Bratbakt G, Heldal M (1989) High abundance of viruses foundin aquatic environments. Nature 340:467–468

Breitbart M, Middelboe M, Rohwer F (2008) Marine viruses: community dynamics, diversity and impact on microbial processes. In: Kirchman DL (ed) Microbial ecology of the oceans, 2nd edn. Wiley-Blackwell, Hoboken, NJ, pp 443–479

Brett MT, Kainz MJ, Taipale SJ, Seshan H (2009) Phytoplankton, not allochthonous carbon, sustains herbivorous zooplankton production. Proc Natl Acad Sci 106(50):21197–21201. https://doi.org/10.1073/pnas.0904129106

Brett MT, Bunn SE, Chandra S, Galloway AWE, Guo F, Kainz MJ, Kankaala P, Lau DCP, Moulton TP, Power ME, Rasmussen JB, Taipale SJ, Thorp JH, Wehr JD (2017) How important are terrestrial organic carbon inputs for secondary production in freshwater ecosystems? Freshw Biol 62(5):833–853. https://doi.org/10.1111/fwb.12909

Buchan A, LeCleir GR, Gulvik CA, González JM (2014) Master recyclers: features and functions of bacteria associated with phytoplankton blooms. Nat Rev Microbiol 12(10):686–698. https://doi.org/10.1038/nrmicro3326

Bunn S, Davies P, Winning M (2003) Sources of organic carbon supporting the food web of an arid zone floodplain river. Freshw Biol 48. https://doi.org/10.1046/j.1365-2427.2003.01031.x

Bunn SE, Balcombe SR, Davies PM, Fellows CS, McKenzie-Smith FJ (2006) Aquatic productivity and food webs of desert river ecosystems. In: Kingsford RT (ed) Ecology of desert rivers. Cambridge University Press, Cambridge, pp 76–99

Cardoso S, Vidal L, Mendonca R, Tranvik L, Sobek S, Roland F (2013) Spatial variation of sediment mineralization supports differential CO2 emissions from a tropical hydroelectric reservoir Frontiers in. Microbiology 4:115–122

Carlson CA (2002) Production and removal processes. In: Hansell DA, Carlson CA (eds) Biogeochemistry of marine dissolved organic matter. Academic, San Diego, CA, pp 91–151

Church M (2008) Resource control of bacteria in the sea. In: Kirchman DL (ed) Microbial ecology of the oceans, 2nd edn. Wiley, pp 335–382

Clapcott JE, Bunn SE (2003) Can C4 plants contribute to aquatic food webs of subtropical streams? Freshw Biol 48(6):1105–1116. https://doi.org/10.1046/j.1365-2427.2003.01077.x
Cole JJ (ed) (2013) Freshwater ecology and the carbon cycle, vol 18. Excellence in ecology. International Ecology Institute
Cole J, Caraco N (2001) Carbon in catchments: connecting terrestrial carbon losses with aquatic metabolism. Mar Freshw Res 52:101–110
Cole JJ, Carpenter SRC, Mitchell JFK (2000) Persistence of net heterotrophy in lakes during nutrient addition and food web manipulations. Limnol Oceanogr 45:1718–1730
Cole JJ, Prairie YT, Caraco NF, McDowell WH, Tranvik LJ, Striegl RG, Duarte CM, Kortelainen P, Downing JA, Middleburg J, Melack J (2007) Plumbing the global carbon cycle: integrating inland waters into the terrestrial carbon budget. Ecosystems 10:171–184
Cole JJ, Carpenter SR, Kitchell JL, Pace ML, Solomon CT, Weidel B (2011) Strong evidence for terrestrial support of zooplankton in small lakes based on stable isotopes of carbon, nitrogen, and hydrogen. Proc Natl Acad Sci U S A 108:1975–1980
Daly RA, Roux S, Borton MA, Morgan DM, Johnston MD, Booker AE, Hoyt DW, Meulia T, Wolfe RA, Hanson AJ, Mouser PJ, Moore JD, Wunch K, Sullivan MB, Wrighton KC, Wilkins MJ (2019) Viruses control dominant bacteria colonizing the terrestrial deep biosphere after hydraulic fracturing. Nat Microbiol 4(2):352–361. https://doi.org/10.1038/s41564-018-0312-6
del Giorgio PA, Williams PJ (eds) (2005) Respiration in aquatic ecosystems. Oxford University Press, New York
del Giorgio PA, Cole JJ, Cimbleris A (1997) Respiration rates in bacteria exceed plankton production in unproductive aquatic systems. Nat Geosci 385:148–151. https://doi.org/10.1038/385148a0
Ducklow HW (1999) The bacterial component of the oceanic euphotic zone. FEMS Microbiol Ecol 30:1–10. https://doi.org/10.1111/j.1574-6941.1999.tb00630.x
Ducklow HW (2000) Bacterial production and biomass in the oceans. In: Kirchman DL (ed) Microbial ecology of the oceans. Wiley-Liss, New York, pp 85–120
Ducklow HW, Purdie DA, Williams PJL, Davies JM (1986) Bacterioplankton: a sink for carbon in a coastal marine plankton community. Science 232:865–867. https://doi.org/10.1126/science.232.4752.865
Eiler A, Langenheder S, Bertilsson S, Tranvik LJ (2003) Heterotrophic bacterial growth efficiency and community structure at different natural organic carbon concentrations. Appl Environ Microbiol 69(7):3701–3709. https://doi.org/10.1128/aem.69.7.3701-3709.2003
Fuhrman J (1999) Marine viruses and their biogeochemical and ecological effects Nature. Nature 399(6736):541–548. https://doi.org/10.1038/21119
Gómez P, Buckling A (2011) Bacteria-phage antagonistic — coevolution in soil. Sci Total Environ 332:106–109. https://doi.org/10.1126/science.1198767
Griffith University (2015) Why we need viruses to survive. https://www.youtube.com/watch?v=ORHXh7iPOpE. Accessed 2 Jan 2021
Haaber J, Middelboe M (2009) Viral lysis of Phaeocystis pouchetii: implications for algal population dynamics and heterotrophic C, N and P cycling. ISME J 3(4):430–441. https://doi.org/10.1038/ismej.2008.125
Kirchman DL (ed) (2008) Microbial ecology of the oceans. Wiley, Hoboken, NJ
Malits A, Christaki U, Obernosterer I, Weinbauer MG (2014) Enhanced viral production and virus-mediated mortality of bacterioplankton in a natural iron-fertilized bloom event above the Kerguelen Plateau. Biogeosciences 11(23):6841–6853. https://doi.org/10.5194/bg-11-6841-2014
Marotta H, Duarte CM, Sobek S, Enrich-Prast A (2009) Large CO2 disequilibria in tropical lakes. Global Biogeochem Cycles 23(GB4022):1–4. https://doi.org/10.1029/2008GB003434
Mayorga E, Aufdenkampe AK, Masiello CA, Krusche AV, Hedges JI, Quay PD, Richey JE, Brown TA (2005) Young organic matter as a source of carbon dioxide outgassing from Amazonian rivers. Nature 436:538–541

McCallister SL, Del Giorgio PA (2008) Direct measurement of the delta C-13 signature of carbon respired by bacteria in lakes: linkages to potential carbon sources, ecosystem baseline metabolism, and CO2 fluxes. Limnol Oceanogr 53:1204–1216

McCallister SL, del Giorgio PA (2012) Evidence for the respiration of ancient terrestrial organic C in northern temperate lakes and streams. Proc Natl Acad Sci U S A 109:16963–16968. https://doi.org/10.1073/pnas.1207305109

McCarthy MD, Hedges JI, Benner R (1998) Major bacterial contributions to marine dissolved organic nitrogen. Science 28:231–234. https://doi.org/10.1126/science.281.5374.231

Middelboe M, Lyck PG (2002) Regeneration of dissolved organic matter by viral lysis in marine microbial communities. Aquat Microb Ecol 27:187–194. https://doi.org/10.3354/ame027187

Mulholland PJ (2003) Large scale patterns of dissolved organic carbon concentrations, flux and sources. In: Findlay SEG, Sinsabaugh RL (eds) Aquatic ecosystems interactivity of dissolved organic matter, pp 139–157

Nagata T (2000) Production mechanisms of dissolved organic matter. In: Kirchman DL (ed) Microbial ecology of the oceans. Wiley, pp 121–152

Oliver AA, Tank SE, Giesbrecht I, Korver MC, Floyd WC, Sanborn P, Bulmer C, Lertzman KP (2017) A global hotspot for dissolved organic carbon in hypermaritime watersheds of coastal British Columbia. Biogeosciences 14(15):3743–3762

Pace ML, Prairie YT (2005) Respiration in lakes. In: Del Giorgio PA, Williams PJ (eds) Respiration in aquatic ecosystems. Oxford University Press, New York, pp 103–121

Pollard PC (2013a) In situ rapid measures of total respiration rate capture the super labile DOC bacterial substrates of freshwater. Limnol Oceanogr Methods 11(11):584–593. https://doi.org/10.4319/lom.2013.11.584

Pollard PC (2013b) Freshwater estimates of prokaryotic respiration in tropical Central America. J Aquat Ecosyst Health Manag 16(2):172–176

Pollard PC, Ducklow H (2011) Ultrahigh bacterial production in a eutrophic subtropical Australian river: Does viral lysis short-circuit the microbial loop? Limnol Oceanogr 56:1115–1129. https://doi.org/10.4319/lo.2011.56.3.1115

Pollard PC, Young LM (2010) Lake viruses lyse cyanobacteria, Cylindrospermopsis raciborskii, enhances filamentous-host dispersal in Australia. Acta Oecologica 36(1):114–119. https://doi.org/10.1016/j.actao.2009.10.007

Pradeep Ram AS, Colombet J, Perriere F, Thouvenot A, Sime-Ngando T (2016) Viral regulation of prokaryotic carbon metabolism in a hypereutrophic freshwater reservoir ecosystem (Villerest, France). Front Microbiol 7(81):1–13. https://doi.org/10.3389/fmicb.2016.00081

Roy K, Ghosh D, DeBruyn JM, Dasgupta T, Wommack KE, Liang X, Wagner RE, Radosevich M (2020) Temporal dynamics of soil virus and bacterial populations in agricultural and early plant successional soils. Front Microbiol 11(1494). https://doi.org/10.3389/fmicb.2020.01494

Sebastián M, Gasol JM (2019) Visualization is crucial for understanding microbial processes in the ocean. Philos Trans R Soc B 374(20190083):1–7. https://doi.org/10.1098/rstb.2019.0083

Suttle CA (2005) Viruses in the sea. Nature 437(7057):356–361. https://doi.org/10.1038/nature04160

Swap R, Garstang M, Greco S, Talbot R, Kållberg P (1992) Saharan dust in the Amazon Basin. Tellus B: Chem Phys Meteorol 44(2):133–149. https://doi.org/10.3402/tellusb.v44i2.15434

Thingstad T, Lignell R (1997) Theoretical models for the control of bacterial growth rate, abundance, diversity and carbon demand. Aquat Microb Ecol 13:19–27. https://doi.org/10.3354/ame013019

Thorp JH, Delong MD (2002) Dominance of autochthonous autotrophic carbon in food webs of heterotrophic rivers. Oikos 96(3):543–550. https://doi.org/10.1034/j.1600-0706.2002.960315.x

Ward ND, Keil RG, Medeiros PM, Brito DC, Cunha AC, Dittmar T et al (2013) Degradation of terrestrially derived macromolecules in the Amazon River. Nat Geosci 6:530–533

Ward ND et al (2017) Where carbon goes when water flow: carbon cycling across the Aquatic Continuum. Front Mar Sci 4:1–27

Weinbauer M, Middelboe M, Mulholland PJ (2002) Regeneration of dissolved organic matter by viral lysis in marine microbial communities. Aquat Microb Ecol 27:187–194. https://doi.org/10.3354/ame027187

Weitz JS, Stock CA, Wilhelm SW, Bourouiba L, Coleman ML, Buchan A, Follows MJ, Fuhrman JA, Jover LF, Lennon JT, Middelboe M, Sonderegger DL, Suttle CA, Taylor BP, Frede Thingstad T, Wilson WH, Eric Wommack K (2015) A multitrophic model to quantify the effects of marine viruses on microbial food webs and ecosystem processes. ISME J 9 (6):1352–1364. https://doi.org/10.1038/ismej.2014.220

White PA, Kalff J, Rasmussen JB, Gasol JM (1991) The effect of temperature and algal biomass on bacterial production and specific growth rate in freshwater and marine habitats. Microb Ecol 21:99–118

Whitman WB, Coleman DC, Wiebe WJ (1998) Prokaryotes: the unseen majority. Proc Natl Acad Sci U S A 95(12):6578–6583. https://doi.org/10.1073/pnas.95.12.6578

Wilhelm SW, Suttle CA (1999) Viruses and nutrient cycles in the sea: viruses play critical roles in the structure and function of aquatic food webs. BioScience 49(10):781–788. https://doi.org/10.2307/1313569

Williamson KE, Fuhrmann JJ, Wommack KE, Radosevich M (2017) Viruses in soil ecosystems: an unknown quantity within an unexplored territory. Annu Rev Virol 4(1):201–219. https://doi.org/10.1146/annurev-virology-101416-041639

Winget DM, Helton RR, Williamson KE, Bench SR, Williamson SJ, Wommack KE (2011) Repeating patterns of virioplankton production within an estuarine ecosystem. Proc Natl Acad Sci U S A 108(28):11506–11511. https://doi.org/10.1073/pnas.1101907108

Wommack KE, Colwell RR (2000) Virioplankton: viruses in aquatic ecosystems. Microbiol Mol Biol Rev 64(1):69–114. https://doi.org/10.1128/mmbr.64.1.69-114.2000

Zhang H, Sekar R, Visser PM (2020a) Editorial: microbial ecology in reservoirs and lakes. Front Microbiol 11:1348. https://doi.org/10.3389/fmicb.2020.01348

Zhang R, Weinbauer M, Peduzzi P (2020b) Aquatic viruses and climate change. In: Marxsen J (ed) Climate change and microbial ecology. Current issues in molecular biology, vol 41, 2nd edn. Caister Academic Press, pp 357–380. https://doi.org/10.21775/9781913652579.07

Zimmerman AEH-VC, Needham DM, John SG, Worden AZ, Sullivan MB, Waldbauer JR, Coleman ML (2020) Metabolic and biogeochemical consequences of viral infection in aquatic ecosystems. Nat Rev Microbiol 18(1):21–34. https://doi.org/10.1038/s41579-019-0270-x

Part II
Understanding the Genetic Partnership Between a Host and Its Viruses

Chapter 3
Cataloging the Presence of Endogenous Viruses

Christon J. Hurst

Abstract Forty three viral families are known to have forged endogenous relationships with eukaryotes. There also are 11 groupings of viruses for which endogenous sequences have been found in eukaryotes but without identification of those sequences at the level of viral family. This chapter presents 20 summary tables that list eukaroytic hosts, defined to the taxonomic levels of genus and species, and those tables name the viral families or viral groupings with which each eukaryote is known to have its endogenous relationships. The tables represent endogeneous viruses hosted by algae, amoeba, amphibians, annelids, arachnids, avians, cnidarians, collembolids, crustaceans, echinoderms, fish, fungi, ichthyosporeans, insects, mammals, molluscs, nematodes, oomycetes, plants, platyhelminths, reptiles, tunicates, and unspecified heterokonts. This chapter also lists some basic information about five viral families which are known to form lysogenous relationships with prokaryotes.

3.1 Introduction: Defining Endogeny and Lysogeny

Coevolution has formed many types of functional relationships between viruses and their natural hosts (Hurst 2021a). Some of these relationships have included developing a shared genomic fate by processes which forge together the genomes of virus and host in a way that makes the viral genetic sequences genetically heritable from a host parent to their offspring. Two such categories of relationships are endogeny and lysogeny. Endogenous and lysogenous viral genetic material variously either may or may not be transcriptionally active while stably present within the host cells.

Endogeny, for this presentation, describes the condition that exists when genetic sequences that originally derived from a viral genome have become vertically inherited in a eukaryotic host. This information passes from parent to offspring. Two main possibilities exist for the inheritance of such viral information. Those are,

C. J. Hurst (✉)
Cincinnati, OH, USA

Universidad del Valle, Cali, Colombia
e-mail: christonjhurst@fuse.net

C. J. Hurst (ed.), *The Biological Role of a Virus*, Advances in Environmental Microbiology 9, https://doi.org/10.1007/978-3-030-85395-2_3

that the viral information can be contained either as a DNA sequence that is integrated within the cells chromosomal material, or the viral information can exist in the form of a DNA plasmid. Another option is that the viral information may be incorporated into the genetic material of a plastid. Some of these endogenous viral sequences are proviruses, which means that they would be able to form infectious progeny viruses, and if their inheritance has been within the host chromosomal material then possibly those sequences can excise themselves from the host genome. However, if given a long enough period of time, endogenous viral genetic sequences tend to become grounded which means the viruses cannot excise themselves to leave the host genome. Grounded viruses generally seem unable to form intact progeny virions, although some of the grounded viruses do retain transcriptional capability. It is presumed that most endogenous viral sequences originated from viral infections that occurred in a host which may have been an evolutionarily predecessor of the species in which those sequences now are found. But, another possibility is that some of these sequences were moved by horizontal, often termed lateral, gene transfer into the genome of an evolutionarily predecessor. The time estimates for how many millions of years some of the known endogenous viral sequences have remained forged into the genomes of their eukaryotic hosts often ranges from tens of millions to more than a hundred million.

It is easy to understand how some of the viruses which have become endogenized, such as the Retroviridae, found their homes in a host genome. The Retroviridae have single-stranded, positive-sense, linear RNA genomes. Retroviridae code for a reverse transcriptase which produces a DNA copy of the viral RNA genome. Retroviruses also code for an integrase enzyme activity which installs their reverse transcribed DNA copy into the host cell genomic material. Those steps are a natural, required part of Retroviridae replication. Some of the endogenous Retroviridae still have retained the capacity for excising from the host genome, but most of the endogenous Retroviridae sequences have lost that capability. The common estimate is that for humans, as an example species, 8% of our genomic material represents endogenous Retroviridae sequences (Griffiths 2001). For house mice, the estimates are that 10% of the genome represents endogenous Retroviridae sequences (Broecker and Moelling 2019). The estimate for crocodiles is that endogenous Retroviridae sequences represent less than 2% of their genome (Chong et al. 2014). These endogenous Retroviridae are one class of retrotransposons, meaning that they form RNA copies which then can be reverse transcribed and those new copies reintegrate into the host genome. Less clear is how members of some other viral families became endogenized. For example, viruses with RNA genomes that do not produce DNA intermediates during their replicative cycle seemingly would have required assistance from either the host cell or some other virus in order to generate a DNA copy to represent their viral RNA genome.

Even a seemingly small amount of endogenous viral coding can have a powerful effect upon the genetics of its host. One example of this importance is represented by the fact that regulation of *wing development in the pea aphid Acyrthosiphon pisum is associated with two endogenous Parvoviridae genes (*Parker and Brisson 2019*). Another example is the fact that primates require usage of their endogenous* *retroviral genes in order to generate syncytins, without which their embryonic blastocysts could not initiate formation of a placenta (*Hurst 2021b*).*

Summary information for each of the viral families that have endogenous representation in eukaryotic hosts is presented in Sect. 3.3.

Lysogeny is a term approximately equivalent to endogeny. For this presentation, the term lysogeny describes viral genomic information which either has entered a prokaryotic host cells genome or developed a stable episomal, plasmid-like, presence within the cytosol of that prokaryotic host. The lysogenized viral genetic information may not yet have become vertically inherited by replication of the host cell. This contrasts with the fact that endogenized viral sequences must, by definition, have been vertically inherited. The lysogenized viral sequences that are genetically incorporated into the cells of prokaryotic microbial hosts can be genetically inherited during replication of their host cell. Once inherited, because of traditional terminology, the viral sequences within these prokaryotic hosts still are termed to represent lysogeny. These viral genomes that are hiding within prokaryotic host cells gained the name of lysogen because their production of progeny virions within a single celled host such as a bacteria or archeae will result in lysis of the host cell.

Please be aware that some people additionally use the term lysogeny to describe non-inherited proviruses which are newly created in the cells of eukaryotic organisms. This type of lysogeny also can occur either by insertion of a viral genome into the chromosome of its host cell, or by the viral genome existing within the cell episomally as a circular plasmid-like structure that is called a replicon. It is considered to be characteristic that if these recently acquired lysogenic viruses are incorporated into the host cell chromosome, then these viruses can excise from the host cells genomic material and produce infectious progeny viral particles.

Summary information for each of the viral families that have lysogenous representation in prokaryotic hosts is presented in Sect. 3.4.

3.2 Endogenous Viruses of Eukaryotic Hosts

Tables 3.1–3.20 present lists of eukaryotic host species and the families of endogenous viruses which those hosts presently are known to contain. If your favorite combination of host and virus does not appear in one of these tables, then possibly either that combination has not been researched or I unfortunately failed to find its mention. These tables are:

Table 3.1: Endogenous viruses of algae
Table 3.2: Endogenous viruses of amoeba
Table 3.3: Endogenous viruses of amphibians
Table 3.4: Endogenous viruses of annelids and nematodes
Table 3.5: Endogenous viruses of arachnids collembolids and insects
Table 3.6: Endogenous viruses of avians
Table 3.7: Endogenous viruses of cnidarians

Table 3.8: Endogenous viruses of crustaceans
Table 3.9: Endogenous viruses of echinoderms
Table 3.10: Endogenous viruses of fish
Table 3.11: Endogenous viruses of fungi
Table 3.12: Endogenous viruses of unspecified heterokonts
Table 3.13: Endogenous viruses of ichthyosporeans
Table 3.14: Endogenous viruses of mammals
Table 3.15: Endogenous viruses of molluscs
Table 3.16: Endogenous viruses of oomycetes
Table 3.17: Endogenous viruses of plants
Table 3.18: Endogenous viruses of platyhelminths
Table 3.19: Endogenous viruses of reptiles
Table 3.20: Endogenous viruses of tunicates

3.3 Groups of Viruses for Which Endogenous Sequences Have Been Identified at the Level of Viral Family

I have listed in this section the names of 43 viral families for which endogenous sequences have been found in eukaryotes. I also have listed many of the endogenous sequences found in eukaryotes that have been identified as representing viral taxonomic levels that were less specific, and I have designated these as viral groups. Quite often the normal infectious host range of a virus family does not match with the range of its known endogenous presence. It is anyone's guess as to why, for any particular viral family, the endogenous presence of that viral family does not extend to all of the eukaryotic groups contained within that viral family's known infectious host range. Perhaps the more interesting question is why many viral families have endogenous presence in eukaryotic groups that are beyond the normal infectious host range of those individual virus families.

3.3.1 Amalgaviridae

The members of the Amalgaviridae viral family have double stranded RNA genomes and they utilize cytoplasmic replication. The known normal hosts of Amalgaviridae are plants. The Southern tomato virus presently is known to be endogenous in seven plant species. Suggested references for endogenous Amalgaviridae in plants would be Chu et al. (2014), Liu et al. (2010), and Takahashi et al. (2019).

For host specific information on endogenous Amalgaviridae see Table 3.17 Endogenous viruses of plants.

Table 3.1 Endogenous viruses of algae

Class	Order	Family	Genus	Species	Common name	Known Endogenous viruses
Cercozoans						
Chlorarachniophyceae	Not assigned	Not assigned	*Bigelowiella*	*natans*		Lavidaviridae Phycodnaviridae
Chlorarachniophyceae	Not assigned	Not assigned	*Lotharella*	*globosa*		Phycodnaviridae
Chlorophytes						
Chlorophyceae	Chlamydomonadales	Chlamydomonadaceae	*Carteria*	*crucifera*		Mimiviridae
Chlorophyceae	Chlamydomonadales	Chlamydomonadaceae	*Chlamydomonas*	*applanata*		Mimiviridae
Chlorophyceae	Chlamydomonadales	Chlamydomonadaceae	*Chlamydomonas*	*asymmetrica*		Mimiviridae Phycodnaviridae
Chlorophyceae	Chlamydomonadales	Chlamydomonadaceae	*Chlamydomonas*	*eustigma*		Mimiviridae
Chlorophyceae	Chlamydomonadales	Chlamydomonadaceae	*Chlamydomonas*	*sphaeroides*		Mimiviridae Phycodnaviridae
Chlorophyceae	Chlamydomonadales	Haematococcaceae	*Haematococcus*	*lacustris*		Mimiviridae Phycodnaviridae
Chlorophyceae	Chlamydomonadales	Tetrabaenaceae	*Tetrabaena*	*socialis*		Mimiviridae
Chlorophyceae	Chlamydomonadales	Volvocaceae	*Yamagishiella*	*unicocca*		Mimiviridae
Chlorophyceae	Sphaeropleales	Cylindrocapsaceae	*Cylindrocapsa*	*geminella*		Mimiviridae
Chlorophyceae	Sphaeropleales	Scenedesmaceae	*Coelastrella*	UTEX B 3026		Phycodnaviridae
Chlorophyceae	Sphaeropleales	Scenedesmaceae	*Tetradesmus*	*obliquus*		Phycodnaviridae
Cryptophytes						
Cryptophyceae	Cryptomonadales	Hemiselmidaceae	*Hemiselmis*	not reported		Nucleocytoviricota
Cryptophyceae	Pyrenomonadales	Geminigeraceae	*Guillardia*	not reported		Mimiviridae
Cryptophyceae	Pyrenomonadales	Geminigeraceae	*Hanusia*	not reported		Nucleocytoviricota
Diatoms						
Bacillariophyceae	Bacillariales	Bacillariaceae	*Fragilariopsis*	*cylindrus*		Endornaviridae

(continued)

Table 3.1 (continued)

Class	Order	Family	Genus	Species	Common name	Known Endogenous viruses
Bacillariophyceae	Naviculales	Phaeodactylaceae	*Phaeodactylum*	*tricornutum*		Totiviridae
Coscinodiscophyceae	Thalassiosirales	Thalassiosiraceae	*Thalassiosira*	*pseudonana*		Endornaviridae
Euglenids						
Euglenida	Euglenales	Euglenaceae	*Euglena*	not reported		Nucleocytoviricota
Eustigmatophytes						
Eustigmatophyceae	Eustigmatales	Monodopsidaceae	*Nannochloropsis*	*limnetica*		Phycodnaviridae
Haptophytes						
Haptophyta	Isochrysidales	Isochrysidaceae	*Chrysotila*	*carterae*		Nucleocytoviricota
Haptophyta	Isochrysidales	Isochrysidaceae	*Isochrysis*	not reported		Nucleocytoviricota
Haptophyta	Isochrysidales	Noelaerhabdaceae	*Emiliania*	*huxleyi*		Phycodnaviridae
Haptophyta	Isochrysidales	Noelaerhabdaceae	*Emiliania*	not reported		Mimiviridae
Haptophyta	Prymnesiales	Prymnesiaceae	*Prymnesium*	*polylepis*		Nucleocytoviricota
Mamiellophytes						
Mamiellophyceae	Mamiellales	Mamiellaceae	*Micromonas*	*pusilla*		Geminiviridae
Phaeophytes						
Phaeophyceae	Ectocarpales	Chordariaceae	*Cladosiphon*	*okamuranus*		Phycodnaviridae
Phaeophyceae	Ectocarpales	Ectocarpaceae	*Ectocarpus*	*siliculosus*		Mimiviridae Phycodnaviridae
Phaeophyceae	Laminariales	Laminariaceae	*Saccharina*	*japonica*		Phycodnaviridae
Pyramimonadophytes						
Pyramimonadophyceae	Pyramimonadales	Pyramimonadaceae	*Cymbomonas*	*tetramitiformis*		Phycodnaviridae
Streptophytes						
Klebsormidiophyceae	Klebsormidiales	Klebsormidiaceae	*Interfilum*	*paradoxum*		Phycodnaviridae
Klebsormidiophyceae	Klebsormidiales	Klebsormidiaceae	*Klebsormidium*	*flaccidum*		Phycodnaviridae
Klebsormidiophyceae	Klebsormidiales	Klebsormidiaceae	*Klebsormidium*	*subtile*		Phycodnaviridae

Zygnemophyceae	Zygnematales	Zygnemataceae	*Entransia*	*fimbriata*		Phycodnaviridae
Trebouxiophytes						
Trebouxiophyceae	Chlorellales	Chlorellaceae	*Chlorella*	ArM0029B		Phycodnaviridae
Trebouxiophyceae	Chlorellales	Chlorellaceae	*Chlorella*	not reported		Nucleocytoviricota
Trebouxiophyceae	Trebouxiales	Trebouxiaceae	*Asterochloris*	*glomerata*		Phycodnaviridae
Trebouxiophyceae	Not assigned	Not assigned	*Coccomyxa*	LA000219		Mimiviridae

Table 3.2 Endogenous viruses of amoeba

Class	Order	Family	Genus	Species	Common name	Known endogenous viruses
Ameoba						
Echinamoebida	Not assigned	Not assigned	*Vermamoeba*	*vermiformis*		Endornaviridae
Eumycetozoa	Dictyosteliales	Dictyosteliaceae	*Dictyostelium*	not reported		Mimiviridae
Eumycetozoa	Dictyosteliales	Dictyosteliaceae	*Polysphondylium*	not reported		Mimiviridae
Not assigned	Longamoebia	Acanthamoebidae	*Acanthamoeba*	*mauritaniensis*		Molliviridae[a]
Not assigned	Longamoebia	Acanthamoebidae	*Acanthamoeba*	not reported		Mimiviridae Nucleocytoviricota[b]
Not assigned	Mastigamoebida	Entamoebidae	*Entamoeba*	*histolytica*		Partitiviridae

[a]These sequences represent the viral species Mollivirus kamchatka which has Molliviridae as its commonly recognized tentative viral family assignment. I could not verify official status of the viral family name Molliviridae

[b]Some of the endogenous Nucleocytoviricota sequences found in this host were identified as being Mimiviridae. Others of the endogenous Nucleocytoviricota sequences found in this host were not specified to the level of viral family

Table 3.3 Endogenous viruses of amphibians

Class	Order	Family	Genus	Species	Common name	Known Endogenous viruses
Amphibia	Anura	Dicroglossidae	*Nanorana*	*parkeri*		Retroviridae
Amphibia	Anura	Pipidae	*Xenopus*	*tropicalis*	Tropical clawed frog, also called Western clawed frog	Circoviridae Retroviridae
Amphibia	Anura	Ranidae	*Lithobates*	*catesbeianus*	American bullfrog	Circoviridae
Amphibia	Anura	Ranidae	*Lithobates*	not reported	Leopard frog	Retroviridae
Amphibia	Anura	Ranidae	*Pelophylax*	*lessonae*	Edible frog	Retroviridae
Amphibia	Caudata	Ambystomatidae	*Ambystoma*	*tigrinum*	Tiger salamander	Retroviridae
Amphibia	Caudata	Salamandridae	*Lissotriton*	*helveticus*	Palmate newt	Retroviridae

Table 3.4 Endogenous viruses of annelids and nematodes

Class	Order	Family	Genus	Species	Common name	Known endogenous viruses
Annelids						
Clitellata	Enchytraeida	Enchytraeidae	*Enchytraeus*	*crypticus*		Parvoviridae
Polychaeta	Sabellida	Siboglinidae	*Lamellibrachia*	not reported		Parvoviridae
Nematodes						
Chromadorea	Rhabditida	Aphelenchoididae	*Bursaphelenchus*	*xylophilus*	Pine wood nematode (also called Pine wilt nematode)	Nodaviridae
Chromadorea	Rhabditida	Ascarididae	*Ascaris*	*suum*	Pig roundworm	Parvoviridae
Chromadorea	Rhabditida	Onchocercidae	*Brugia*	*malayi*		Rhabdoviridae
Chromadorea	Rhabditida	Rhabditidae	*Caenorhabditis*	*elegans*		Metaviridae[a]
Chromadorea	Rhabditida	Strongyloididae	*Strongyloides*	*ratti*		Totiviridae

[a]*Caenorhabditis elegans* contains Long terminal repeat (LTR) Gypsy retrotransposons that are presumed to represent Metaviridae

Table 3.5 Endogenous viruses of arachnids collembolids and insects

Class	Order	Family	Genus	Species	Common name	Known endogenous viruses
Arachnida	Araneae	Araneidae	*Araneus*	*ventricosus*		Parvoviridae
Arachnida	Araneae	Theridiidae	*Latrodectus*	*hesperus*	Western black widow	Parvoviridae
Arachnida	Ixodida	Ixodidae	*Amblyomma*	*americanum*	Lone Star tick	Parvoviridae
Arachnida	Ixodida	Ixodidae	*Ixodes*	*ricinus*	Castor bean tick	Flaviviridae Unspecified Mononegavirales-like virus [a]
Arachnida	Ixodida	Ixodidae	*Ixodes*	*scapularis*	Black legged tick	Nairoviridae Orthomyxoviridae Partitiviridae Parvoviridae Phenuiviridae Rhabdoviridae Totiviridae
Arachnida	Ixodida	Ixodidae	*Rhipicephalus*	*appendiculatus*		Parvoviridae
Arachnida	Ixodida	Ixodidae	*Rhipicephalus*	*pulchellus*		Parvoviridae
Arachnida	Mesostigmata	Phytoseiidae	*Galendromus*	*occidentalis*	Western predatory mite	Parvoviridae
Arachnida	Trombidiformes	Tetranychidae	*Tetranychus*	*urticae*	Two-spotted spider mite	Parvoviridae
Arachnida	Trombidiformes	Trombidiidae	*Dinothrombium*	*tinctorium*		Parvoviridae
Collembola	Entomobryomorpha	Orchesellidae	*Orchesella*	*cincta*	Springtail	Parvoviridae
Insecta	Coleoptera	Carabidae	*Pogonus*	*chalceus*		Parvoviridae
Insecta	Coleoptera	Cupedidae	*Priacma*	*serrata*		Parvoviridae
Insecta	Coleoptera	Curculionidae	*Ips*	*typographus*		Parvoviridae
Insecta	Coleoptera	Curculionidae	*Pissodes*	*strobi*		Parvoviridae
Insecta	Coleoptera	Nitidulidae	*Brassicogethes*	*aeneus*		Parvoviridae
Insecta	Coleoptera	Scarabaeidae	*Onthophagus*	*nigriventris*		Parvoviridae
Insecta	Coleoptera	Scarabaeidae	*Oryctes*	*rhinoceros*	Rhinoceros beetle	Nudiviridae

(continued)

Table 3.5 (continued)

Class	Order	Family	Genus	Species	Common name	Known endogenous viruses
Insecta	Diptera	Culicidae	*Aedes*	*aegypti*	Yellow fever mosquito	Chuviridae Flavivridae Potyviridae Reoviridae Rhabdoviridae Totiviridae
Insecta	Diptera	Culicidae	*Aedes*	*albopictus*	Mosquito	Chuviridae Flavivridae Totiviridae
Insecta	Diptera	Culicidae	*Aedes*	not reported	Mosquito	Flavivridae
Insecta	Diptera	Culicidae	*Anopheles*	*stephensi*	Asian malaria mosquito	Virga-like[b]
Insecta	Diptera	Culicidae	*Culex*	*pipiens*	Northern house mosquito	Parvoviridae
Insecta	Diptera	Culicidae	*Culex*	*quinquefasciatus*	Southern house mosquito	Chuviridae Rhabdoviridae
Insecta	Diptera	Diopsidae	*Teleopsis*	*dalmanni*		Parvoviridae
Insecta	Diptera	Drosophilidae	*Drosophila*	*grimshawi*	Fruit fly	Partitiviridae
Insecta	Diptera	Drosophilidae	*Drosophila*	*melanogaster*	Fruit fly	Metaviridae [c] Pseudoviridae [a]
Insecta	Diptera	Drosophilidae	*Drosophila*	*persimilis*	Fruit fly	Parvoviridae
Insecta	Diptera	Drosophilidae	*Drosophila*	*sechellia*	Fruit fly	Parvoviridae Rhabdoviridae
Insecta	Diptera	Drosophilidae	*Drosophila*	*willistoni*	Fruit fly	Unclassified Riboviria
Insecta	Diptera	Drosophilidae	*Drosophila*	*yakuba*	Fruit fly	Rhabdoviridae
Insecta	Diptera	Syrphidae	*Eristalis*	*tenax*	Drone fly	Parvoviridae
Insecta	Diptera	Tephritidae	*Bactrocera*	*dorsalis*	Oriental fruit fly	Parvoviridae
Insecta	Diptera	Tephritidae	*Rhagoletis*	*pomonella*	Apple maggot	Parvoviridae
Insecta	Hemiptera	Aleyrodidae	*Bemisia*	*tabaci*	Whiteflies	Parvoviridae

Insecta	Hemiptera	Aphalaridae	*Pachypsylla*	*venusta*	Hackberry petiole gall psyllid	Parvoviridae
Insecta	Hemiptera	Aphididae	*Acyrthosiphon*	*pisum*	Pea aphid	Partitiviridae Parvoviridae
Insecta	Hemiptera	Aphididae	*Aphis*	*craccivora*	Cowpea aphid	Parvoviridae
Insecta	Hemiptera	Aphididae	*Aphis*	*glycines*	Soybean aphid	Parvoviridae
Insecta	Hemiptera	Aphididae	*Myzus*	*persicae*	Green peach aphid	Parvoviridae
Insecta	Hemiptera	Aphididae	*Sitobion*	*avenae*	English grain aphid	Parvoviridae
Insecta	Hemiptera	Aphididae	*Sitobion*	*miscanthi*		Parvoviridae
Insecta	Hemiptera	Cicadellidae	*Graminella*	*nigrifrons*	Black-faced leafhopper	Parvoviridae
Insecta	Hemiptera	Coreidae	*Clavigralla*	*tomentosicollis*		Parvoviridae
Insecta	Hemiptera	Delphacidae	*Nilaparvata*	*lugens*	Brown planthopper	Nudiviridae Parvoviridae
Insecta	Hemiptera	Kerriidae	*Kerria*	*lacca*	Common lac scale	Parvoviridae
Insecta	Hemiptera	Lachnidae	*Cinara*	*cedri*		Parvoviridae
Insecta	Hemiptera	Liviidae	*Diaphorina*	*citri*	Asian citrus psyllid	Parvoviridae
Insecta	Hemiptera	Miridae	*Lygus*	*hesperus*	Lygus bug	Parvoviridae
Insecta	Hemiptera	Pentatomidae	*Chinavia*	*ubica*	Stink bug	Parvoviridae
Insecta	Hemiptera	Pentatomidae	*Halyomorpha*	*halys*	Brown marmorated stink bug	Parvoviridae
Insecta	Hemiptera	Reduviidae	*Rhodnius*	*prolixus*	Triatome	Partitiviridae Parvoviridae
Insecta	Hemiptera	Reduviidae	*Triatoma*	*infestans*	Triatome	Parvoviridae
Insecta	Hymenoptera	Apidae	*Bombus*	*impatiens*	Common eastern bumble bee	Parvovirdae
Insecta	Hymenoptera	Braconidae	*Chelonus*	*inanitus*	Parasitoid wasp	Polydnaviridae
Insecta	Hymenoptera	Braconidae	*Cotesia*	*congregata*	Parasitoid wasp	Polydnaviridae
Insecta	Hymenoptera	Braconidae	*Cotesia*	*glomerata*	Parasitoid wasp	Polydnaviridae
Insecta	Hymenoptera	Braconidae	*Cotesia*	*vestalis*	Diamondback moth parasitoid	Polydnaviridae

(continued)

Table 3.5 (continued)

Class	Order	Family	Genus	Species	Common name	Known endogenous viruses
Insecta	Hymenoptera	Braconidae	*Glyptapanteles*	*flavicoxis*	Parasitoid wasp	Polydnaviridae
Insecta	Hymenoptera	Braconidae	*Glyptapanteles*	*indiensis*	Parasitoid wasp	Polydnaviridae
Insecta	Hymenoptera	Braconidae	*Microplitis*	*demolitor*	Parasitoid wasp	Nudiviridae Polydnaviridae
Insecta	Hymenoptera	Formicidae	*Acromyrmex*	*echinatior*	Panamanian leafcutter ant	Parvoviridae
Insecta	Hymenoptera	Formicidae	*Atta*	*cephalotes*		Parvoviridae
Insecta	Hymenoptera	Formicidae	*Camponotus*	*floridanus*	Florida carpenter ant	Rhabdoviridae
Insecta	Hymenoptera	Formicidae	*Lasius*	*niger*		Parvoviridae
Insecta	Hymenoptera	Formicidae	*Messor*	*barbarus*		Parvoviridae
Insecta	Hymenoptera	Formicidae	*Messor*	*concolor*		Parvoviridae
Insecta	Hymenoptera	Formicidae	*Monomorium*	*pharaonis*	Pharaoh ant	Parvoviridae
Insecta	Hymenoptera	Formicidae	*Pogonomyrmex*	*barbatus*		Parvoviridae
Insecta	Hymenoptera	Formicidae	*Tetramorium*	*bicarinatum*		Parvoviridae
Insecta	Hymenoptera	Ichneumonidae	*Apophua*	*simplicipes*	Parasitoid wasp	Polydnaviridae
Insecta	Hymenoptera	Ichneumonidae	*Campoletis*	*sonorensis*	Parasitoid wasp	Polydnaviridae
Insecta	Hymenoptera	Ichneumonidae	*Glypta*	*fumiferanae*	Parasitoid wasp	Polydnaviridae
Insecta	Hymenoptera	Ichneumonidae	*Hyposoter*	*didymator*	Parasitoid wasp	Polydnaviridae
Insecta	Hymenoptera	Ichneumonidae	*Hyposoter*	*fugitivus*	Parasitoid wasp	Polydnaviridae
Insecta	Hymenoptera	Ichneumonidae	*Tranosema*	*rostrale*	Parasitoid wasp	Polydnaviridae
Insecta	Hymenoptera	Ichneumonidae	*Venturia*	*canescens*	Parasitoid wasp	Nudiviridae
Insecta	Lepidoptera	Crambidae	*Chilo*	*suppressalis*	Striped riceborer	Parvovirdae
Insecta	Lepidoptera	Micropterigidae	*Micropterix*	*calthella*		Parvovirdae
Insecta	Lepidoptera	Noctuidae	*Agrotis*	*segetum*	Turnip moth	Parvovirdae
Insecta	Lepidoptera	Noctuidae	*Autographa*	*californica*	Alfalfa looper	Baculoviridae
Insecta	Lepidoptera	Noctuidae	*Chrysodeixis*	*includens*	Soybean looper	Parvoviridae
Insecta	Lepidoptera	Noctuidae	*Helicoverpa*	*armigera*	Cotton bollworm	Parvoviridae

Insecta	Lepidoptera	Noctuidae	*Helicoverpa*	*zea*	Corn earworm	Nudiviridae
Insecta	Lepidoptera	Noctuidae	*Heliothis*	*virescens*	Tobacco budworm	Metaviridae
Insecta	Lepidoptera	Noctuidae	*Spodoptera*	*frugiperda*	Fall armyworm	Metaviridae Rhabdoviridae
Insecta	Lepidoptera	Noctuidae	*Trichoplusia*	*ni*	Cabbage looper	Metaviridae
Insecta	Lepidoptera	Notodontidae	*Thaumetopoea*	*pityocampa*	Pine processionary moth	Parvovirdae
Insecta	Lepidoptera	Nymphalidae	*Bicyclus*	*anynana*	Squinting bush brown	Parvovirdae
Insecta	Lepidoptera	Nymphalidae	*Junonia*	*coenia*	Common buckeye	Parvoviridae
Insecta	Lepidoptera	Pyralidae	*Galleria*	*mellonella*	Greater wax moth	Parvoviridae
Insecta	Lepidoptera	Saturniidae	*Antheraea*	*pernyi*	Chinese oak silkmoth	Metaviridae Pseudoviridae
Insecta	Lepidoptera	Saturniidae	*Antheraea*	*yamamai*	Japanese oak silkmoth	Metaviridae Pseudoviridae
Insecta	Mecoptera	Nannochoristidae	*Nannochorista*	*philpotti*		Parvovirdae
Insecta	Megaloptera	Corydalidae	*Corydalinae*	not reported	Dobson flies	Parvovirdae
Insecta	Neuroptera	Chrysopidae	*Chrysopa*	*pallens*		Parvovirdae
Insecta	Orthoptera	Acrididae	*Schistocerca*	*gregaria*		Parvovirdae
Insecta	Orthoptera	Gryllidae	*Gryllus*	*bimaculatus*	Two-spotted cricket	Nudiviridae Parvovirdae
Insecta	Phasmatodea	Heteropterygidae	*Aretaon*	*asperrimus*	Thorny stick insect	Parvovirdae
Insecta	Phasmatodea	Lonchodidae	*Sipyloidea*	*sipylus*	Walking stick	Parvovirdae
Insecta	Phasmatodea	Phasmatidae	*Extatosoma*	*tiaratum*	Giant prickly stick insect	Parvovirdae
Insecta	Phasmatodea	Phasmatidae	*Medauroidea*	*extradentata*	Walking stick	Parvovirdae
Insecta	Raphidioptera	Raphidiidae	*Raphidia*	*ariadne*		Parvovirdae

[a] *Ixodes ricinus* contains both endogenous Flaviviridae sequences and also endogenous sequences of an unspecified Mononegavirales-like virus
[b] *Anopheles stephensi* contains endogenous viral seqiences that were described as being Virga-like
[c] *Drosophila melanogaster* contains endogenous Long Terminal Repeat (LTR) Copia retrotransposons that are presumed to represent Pseudoviridae. *Drosophila melanogaster* also contains endogenous Long Terminal Repeat (LTR) Gypsy retrotransposons that are presumed to represent Metaviridae

Table 3.6 Endogenous viruses of avians

Class	Order	Family	Genus	Species	Common name	Known endogenous viruses
Aves	Falconiformes	Falconidae	*Falco*	*peregrinus*	Peregrine falcon	Retroviridae
Aves	Gruiformes	Gruidae	*Grus*	*japonensis*	Red-crowned crane	Parvoviridae
Aves	Galliformes	Phasianidae	*Gallus*	*gallus*	Chicken	Parvoviridae Retroviridae
Aves	Galliformes	Phasianidae	*Meleagris*	*gallopavo*	Turkey	Retroviridae
Aves	Galliformes	Phasianidae	*Pavo*	*cristatus*	Indian peafowl	Parvoviridae
Aves	Gruiformes	Rallidae	*Gallirallus*	*okinawae*	Okinawa rail	Circoviridae
Aves	Passeriformes	Estrildidae	*Taeniopygia*	*guttata*	Zebra finch	Hepadnaviridae Retroviridae
Aves	Passeriformes	Fringillidae	*Serinus*	*canaria*	Common canary	Circoviridae
Aves	Passeriformes	Parulidae	*Setophaga*	*coronata*	Yellow-rumped warbler	Circoviridae
Aves	Passeriformes	Passerellidae	*Zonotrichia*	*albicollis*	White-throated sparrow	Circoviridae
Aves	Passeriformes	Thraupidae	*Geospiza*	*fortis*	Medium ground-finch	Circoviridae
Aves	Passeriformes	Thraupidae	*Sporophila*	*hypoxantha*	Tawny-bellied seedeater	Circoviridae
Aves	Pelecaniformes	Ardeidae	*Egretta*	*garzetta*	Little egret	Circoviridae
Aves	Piciformes	Picidae	*Picoides*	*pubescens*	Downy woodpecker	Circoviridae
Aves	Psittaciformes	Psittacidae	*Agapornis*	*roseicollis*	Peach-faced lovebird	Circoviridae
Aves	Psittaciformes	Psittacidae	*Amazona*	*aestiva*	Blue-fronted amazon	Circoviridae
Aves	Psittaciformes	Psittacidae	*Nestor*	*notabilis*	Kea	Circoviridae
Aves	Tinamiformes	Tinamidae	*Tinamus*	*guttatus*	White-throated tinamou	Circoviridae

Table 3.7 Endogenous viruses of cnidarians

Class	Order	Family	Genus	Species	Common name	Known endogenous viruses
Anthozoa	Actiniaria	Actiniidae	*Actinia*	*equina*		Parvoviridae
Anthozoa	Actiniaria	Aiptasiidae	*Exaiptasia*	*diaphana*		Parvoviridae
Anthozoa	Actiniaria	Aiptasiidae	*Exaiptasia*	*pallida*	Sea anemone	Asfarviridae
Anthozoa	Alcyonacea	Gorgoniidae	*Eunicella*	*cavolini*	Soft coral	Parvoviridae
Hydrozoa	Anthoathecata	Hydridae	*Hydra*	*vulgaris*		Mimiviridae

3.3.2 *Asfarviridae*

The members of the Asfarviridae viral family have double stranded DNA genomes and they utililze cytoplasmic replication. The known normal hosts of Asfarviridae are insects and mammals. A suggested reference for endogenous Asfarviridae in cnidaria, fungi, unspecified heterokonts and oomycetes would be Gallot-Lavallée and Blanc (2017).

For host specific information on endogenous Asfarviridae see Table 3.7 Endogenous viruses of cnidarians, Table 3.11 Endogenous viruses of fungi, Table 3.12 Endogenous viruses of unspecified heterokonts, and Table 3.16 Endogenous viruses of oomycetes.

3.3.3 *Baculoviridae*

The members of the Baculoviridae viral family have double stranded DNA genomes and they utilize nuclear replication. The known normal hosts of Baculoviridae are insects. Suggested references for endogenous Baculoviridae in insects would be Herniou et al. (2013) and specifically in ants would be Flynn and Moreau (2019).

For host specific information on endogenous Baculoviridae see Table 3.5 Endogenous viruses of arachnids collembolids and insects.

I did not list in a table the information about endogenous Baculoviridae of ants because the publication by Flynn and Moreau (2019) did not seem to match up the viral sequences with specific host genera and species.

3.3.4 *Betaflexiviridae*

The members of the Betaflexiviridae viral family have positive sense single stranded RNA genomes and they utilize cytoplasmic replication. The known normal hosts of

Table 3.8 Endogenous viruses of crustaceans

Class	Order	Family	Genus	Species	Common name	Known endogenous viruses
Branchiopoda	Anostraca	Artemiidae	*Artemia*	*franciscana*	Brine shrimp	Parvoviridae
Branchiopoda	Diplostraca	Daphniidae	*Daphnia*	*pulex*	Water flea	Bunyaviridae Parvoviridae
Branchiopoda	Diplostraca	Daphniidae	*Daphnia*	not reported	Water flea	Nucleocytoviricota
Branchiopoda	Diplostraca	Daphniidae	*Daphnia*	*pulicaria*	Water flea	Bunyaviridae Circoviridae Parvoviridae
Hexanauplia	Calanoida	Temoridae	*Eurytemora*	*affinis*	Copepod	Rhabdoviridae
Hexanauplia	Siphonostomatoida	Caligidae	*Caligus*	*rogercresseyi*	Copepod	Parvoviridae Rhabdoviridae
Hexanauplia	Siphonostomatoida	Caligidae	*Lepeophtheirus*	*salmonis*	Salmon louse	Bunyaviridae Circoviridae Parvoviridae Rhabdoviridae
Ichthyostraca	Arguloida	Argulidae	*Argulus*	*siamensis*		Parvoviridae
Malacostraca	Amphipoda	Ampeliscidae	*Ampelisca*	*abdita*	Amphipod	Parvoviridae
Malacostraca	Amphipoda	Hyalellidae	*Hyalella*	*azteca*	Amphipod	Circoviridae Rhabdoviridae
Malacostraca	Decapoda	Astacidae	*Astacus*	*leptodactylus*	Narrow clawed crayish	Parvoviridae
Malacostraca	Decapoda	Palaemonidae	*Macrobrachium*	*nipponense*		Parvoviridae
Malacostraca	Decapoda	Penaeidae	*Penaeus*	*monodon*	Black tiger shrimp	Parvoviridae
Malacostraca	Decapoda	Portunidae	*Portunus*	*trituberculatus*	**swimming crab**	Parvoviridae
Malacostraca	Decapoda	Portunidae	*Scylla*	*olivacea*	Orange mud crab	Parvoviridae
Malacostraca	Isopoda	Armadillidiidae	*Armadillidium*	*nasatum*	Woodlouse	Bunyaviridae Circoviridae Parvoviridae Totiviridae Mononegavirales (possibly Rhabdoviridae [a])

Malacostraca	Isopoda	Armadillidiidae	*Armadillidium*	*vulgare*	Pllbug	Bunyaviridae Circoviridae Parvoviridae Totiviridae Mononegavirales (presumably Nyamiviridae [b]) Nimaviridae [c]

[a]The endogenous Mononegavirales sequences found in this host crustacean possibly represent Rhabdoviridae
[b]The endogenous Mononegavirales sequences found in this host crustacean were presumed to represent Nyamiviridae
[c]The Nimaviridae sequences found were suggested as representing either a virus naturally endogenized into this animal or sequences that were horizontally transferred into this animals genome from another host that presumably would have been more naturally associated with this virus family

Table 3.9 Endogenous viruses of echinoderms

Class	Order	Family	Genus	Species	Common name	Known endogenous viruses
Asteroidea	Forcipulatida	Asteriidae	*Asterias*	*rubens*	European starfish	Parvoviridae
Asteroidea	Paxillosida	Luidiidae	*Luidia*	*clathrata*	Starfish	Parvoviridae
Asteroidea	Spinulosida	Echinasteridae	*Echinaster*	*spinulosus*		Parvoviridae
Asteroidea	Valvatida	Acanthasteridae	*Acanthaster*	*brevispinus*		Parvoviridae
Asteroidea	Valvatida	Acanthasteridae	*Acanthaster*	*planci*	Crown-of-thorns starfish	Parvoviridae
Asteroidea	Valvatida	Asterinidae	*Patiria*	*pectinifera*		Parvoviridae
Asteroidea	Valvatida	Ophidiasteridae	*Linckia*	*laevigata*		Parvoviridae

Table 3.10 Endogenous viruses of fish

Class	Order	Family	Genus	Species	Common name	Known endogenous viruses
Agnatha						
Myxini	Myxiniformes	Myxinidae	*Eptatretus*	*burgeri*	Inshore hagfish	Circoviridae
Bony Fish—ray finned fish						
Actinopteri	Anguilliformes	Anguillidae	*Anguilla*	*anguilla*	European eel	Circoviridae Retroviridae
Actinopteri	Anguilliformes	Anguillidae	*Anguilla*	*japonica*	Japanese eel	Retroviridae
Actinopteri	Anguilliformes	Anguillidae	*Anguilla*	*rostrata*	American eel	Retroviridae
Actinopteri	Beloniformes	Adrianichthyidae	*Oryzias*	*latipes*	Japanese medaka	Bornaviridae Retroviridae
Actinopteri	Centrarchiformes	Centrarchidae	*Micropterus*	*floridanus*	Florida bass	Circoviridae
Actinopteri	Characiformes	Characidae	*Astyanax*	*mexicanus*	Mexican tetra	Retroviridae
Actinopteri	Characiformes	Serrasalmidae	*Pygocentrus*	*nattereri*	Red-bellied piranha	Circoviridae
Actinopteri	Cichliformes	Cichlidae	*Amphilophus*	*citrinellus*	Midas cichlid	Retroviridae
Actinopteri	Cichliformes	Cichlidae	*Haplochromis*	*burtoni*	Burton's mouthbrooder	Retroviridae
Actinopteri	Cichliformes	Cichlidae	*Maylandia*	*zebra*	Zebra mbuna	Retroviridae
Actinopteri	Cichliformes	Cichlidae	*Neolamprologus*	*brichardi*	Lyretail cichlid	Circoviridae
Actinopteri	Cichliformes	Cichlidae	*Oreochromis*	*niloticus*	Nile tilapia	Retroviridae
Actinopteri	Cichliformes	Cichlidae	*Pundamilia*	*nyererei*		Retroviridae
Actinopteri	Clupeiformes	Clupeidae	*Clupea*	*harengus*	Atlantic herring	Circoviridae Retroviridae
Actinopteri	Cypriniformes	Cyprinidae	*Cyprinus*	*carpio*	Common carp	Circoviridae Retroviridae
Actinopteri	Cypriniformes	Cyprinidae	*Sinocyclocheilus*	*anshuiensis*		Retroviridae
Actinopteri	Cypriniformes	Cyprinidae	*Sinocyclocheilus*	*grahami*	Golden-line barbel	Circoviridae Retroviridae

(continued)

Table 3.10 (continued)

Class	Order	Family	Genus	Species	Common name	Known endogenous viruses
Actinopteri	Cypriniformes	Cyprinidae	*Sinocyclocheilus*	*rhinocerous*		Retroviridae
Actinopteri	Cypriniformes	Danionidae	*Danio*	*rerio*	Zebra fish	Nyamiviridae Retroviridae Rhabdoviridae
Actinopteri	Cypriniformes	Leuciscidae	*Pimephales*	*promelas*	Fathead minnow	Retroviridae
Actinopteri	Cyprinodontiformes	Cyprinodontidae	*Cyprinodon*	*nevadensis*	Amargosa pupfish	Retroviridae
Actinopteri	Cyprinodontiformes	Cyprinodontidae	*Cyprinodon*	*variegatus*	Sheepshead minnow	Retroviridae
Actinopteri	Cyprinodontiformes	Fundulidae	*Fundulus*	*grandis*	Gulf killifish	Parvoviridae
Actinopteri	Cyprinodontiformes	Fundulidae	*Fundulus*	*heteroclitus*	Mummichog	Retroviridae
Actinopteri	Cyprinodontiformes	Nothobranchiidae	*Nothobranchius*	*furzeri*	Turquoise killifish	Retroviridae
Actinopteri	Cyprinodontiformes	Poeciliidae	*Poecilia*	*formosa*	Amazon molly	Retroviridae
Actinopteri	Cyprinodontiformes	Poeciliidae	*Poecilia*	*latipinna*	Sailfin molly	Retroviridae
Actinopteri	Cyprinodontiformes	Poeciliidae	*Poecilia*	*reticulata*	Guppy	Retroviridae
Actinopteri	Cyprinodontiformes	Poeciliidae	*Xiphophorus*	*couchianus*	Monterrey platyfish	Retroviridae
Actinopteri	Cyprinodontiformes	Poeciliidae	*Xiphophorus*	*hellerii*	Green swordtail	Retroviridae
Actinopteri	Cyprinodontiformes	Poeciliidae	*Xiphophorus*	*maculatus*	Southern platyfish	Retroviridae
Actinopteri	Cyprinodontiformes	Rivulidae	*Austrofundulus*	*limnaeus*		Retroviridae
Actinopteri	Cyprinodontiformes	Rivulidae	*Kryptolebias*	*marmoratus*	Mangrove rivulus	Circoviridae Retroviridae
Actinopteri	Esociformes	Esocidae	*Esox*	*lucius*	Northern pike	Retroviridae
Actinopteri	Gadiformes	Gadidae	*Gadus*	*morhua*	Atlantic cod	Retroviridae
Actinopteri	Gobiiformes	Gobiidae	*Periophthalmodon*	*schlosseri*	Giant mudskipper	Retroviridae
Actinopteri	Labriformes	Labridae	*Labrus*	*bergylta*	Ballan wrasse	Retroviridae
Actinopteri	Osteoglossiformes	Osteoglossidae	*Scleropages*	*formosus*	Asian bonytongue	Retroviridae
Actinopteri	Perciformes	Anoplopomatidae	*Anoplopoma*	*fimbria*	Sablefish	Retroviridae

Actinopteri	Perciformes	Gasterosteidae	*Gasterosteus*	*aculeatus*	Three-spined stickleback	Retroviridae
Actinopteri	Perciformes	Sebastidae	*Sebastes*	*rubrivinctus*	Flag rockfish	Retroviridae
Actinopteri	Pleuronectiformes	Cynoglossidae	*Cynoglossus*	*semilaevis*	Tongue sole	Retroviridae
Actinopteri	Salmoniformes	Salmonidae	*Coregonus*	*lavaretus*	Common whitefish (sometimes called Freshwater houting)	Retroviridae
Actinopteri	Salmoniformes	Salmonidae	*Salmo*	*salar*	Atlantic salmon	Circoviridae Retroviridae
Actinopteri	Salmoniformes	Salmonidae	*Salvelinus*	*fontinalis*	Brook trout	Retroviridae
Actinopteri	Scombriformes	Scombridae	*Thunnus*	*orientalis*	Pacific bluefin tuna	Retroviridae
Actinopteri	Semionotiformes	Lepisosteidae	*Lepisosteus*	*oculatus*	Spotted gar	Retroviridae
Actinopteri	Tetraodontiformes	Tetraodontidae	*Takifugu*	*flavidus*	Sansaifugu	Retroviridae
Actinopteri	Tetraodontiformes	Tetraodontidae	*Takifugu*	*rubripes*	Torafugu	Retroviridae
Actinopteri	Tetraodontiformes	Tetraodontidae	*Tetraodon*	*nigroviridis*	Spotted green pufferfish	Retroviridae
Actinopteri	Not assigned	Centropomidae	*Lates*	*calcarifer*	Barramundi perch	Retroviridae
Actinopteri	Not assigned	Moronidae	*Dicentrarchus*	*labrax*	European seabass	Retroviridae
Actinopteri	Not assigned	Pomacentridae	*Acanthochromis*	*polyacanthus*	Spiny damselfish	Circoviridae
Actinopteri	Not assigned	Pomacentridae	*Stegastes*	*partitus*	Bicolor damselfish	Retroviridae
Actinopteri	Not assigned	Sciaenidae	*Miichthys*	*miiuy*	Miiuy croaker	Retroviridae
Cartilaginous fish						
Chondrichthyes	Chimaeriformes	Callorhinchidae	*Callorhinchus*	*milii*	Elephant shark	Retroviridae
Chondrichthyes	Rajiformes	Rajidae	*Leucoraja*	*erinacea*	Little skate	Retroviridae

Table 3.11 Endogenous viruses of fungi

Class	Order	Family	Genus	Species	Common name	Known endogenous viruses
Blastocladiomycetes	Blastocladiales	Blastocladiaceae	*Allomyces*	*macrogynus*		Phycodnaviridae
Eurotiomycetes	Eurotiales	Trichocomaceae	*Talaromyces*	*marneffei*		Totiviridae
Glomeromycetes	Glomerales	Glomeraceae	*Rhizophagus*	*irregularis*		Asfarviridae
Monoblepharidomycetes	Monoblepharidales	Gonapodyaceae	*Gonapodya*	*prolifera*		Phycodnaviridae
Pucciniomycetes	Pucciniales	Pucciniaceae	*Uromyces*	*appendiculatus*		Totiviridae
Saccharomycetes	Saccharomycetales	Debaryomycetaceae	*Candida*	*albicans*		Metaviridae[a] Pseudoviridae[a]
Saccharomycetes	Saccharomycetales	Debaryomycetaceae	*Candida*	*parapsilosis*		Totiviridae
Saccharomycetes	Saccharomycetales	Debaryomycetaceae	*Debaryomyces*	*hansenii*		Totiviridae Metaviridae[a] Pseudoviridae[a]
Saccharomycetes	Saccharomycetales	Debaryomycetaceae	*Scheffersomyces*	*coipomoensis*		Totiviridae
Saccharomycetes	Saccharomycetales	Debaryomycetaceae	*Scheffersomyces*	*segobiensis*		Totiviridae
Saccharomycetes	Saccharomycetales	Debaryomycetaceae	*Scheffersomyces*	*stipitis*	Budding yeast	Totiviridae
Saccharomycetes	Saccharomycetales	Pichiaceae	*Pichia*	*membranifaciens*		Totiviridae
Saccharomycetes	Saccharomycetales	Saccharomycetaceae	*Saccharomyces*	*cerevisiae*	Bakers yeast	Metaviridae[a] Pseudoviridae[a]
Schizosaccharomycetes	Schizosaccharomycetales	Schizosaccharomycetaceae	*Schizosaccharomyces*	*pombe*		Totiviridae
Sordariomycetes	Diaporthales	Cryphonectriaceae	*Cryphonectria*	*parasitica*	Chestnut blight fungus	Hypoviridae

Sordariomycetes	Xylariales	Xylariaceae	*Rosellinia*	*necatrix*		Partitiviridae

[a]*Candida albicans, Saccharomyces cerevisiae* and *Debaryomyces hansenii* contain both endogenous Long Terminal Repeat (LTR) Copia retrotransposons and LTR Gypsy retrotransposons. The Copia retrotransposons are presumed to represent Pseudoviridae. The LTR Gypsy retrotransposons are presumed to represent Metaviridae. *Debaryomyces hansenii* additionally contains endogenous Totiviridae

Table 3.12 Endogenous viruses of unspecified heterokonts

Class	Order	Family	Genus	Species	Common name	Known endogenous viruses
Bigyra	Bicosoecida	Cafeteriaceae	*Cafeteria*	*burkhardae*		Lavidaviridae
Bigyra	Bicosoecida	Cafeteriaceae	*Cafeteria*	*roenbergensis*		Lavidaviridae
Bigyra	Bicosoecida	Not assigned	*Halocafeteria*	*seosinensis*		Nucleocytoviricota
Bigyra	Thraustochytrida	Thraustochytriaceae	*Aurantiochytrium*	*limacinum*		Phycodnaviridae
Bigyra	Thraustochytrida	Thraustochytriaceae	*Schizochytrium*	*aggregatum*		Nucleocytoviricota
Bigyra	Thraustochytrida	Thraustochytriaceae	*Thraustochytrium*	not reported		Nucleocytoviricota
Hyphochytriomycetes	Not assigned	Hyphochytriaceae	*Hyphochytrium*	*catenoides*		Asfarviridae

Table 3.13 Endogenous viruses of ichthyosporeans

Class	Order	Family	Genus	Species	Common name	Known endogenous viruses
Ichthyosporea	Ichthyophonida	Not assigned	*Sphaeroforma*	*arctica*		Iridoviridae/Marseilleviridae[a]
Ichthyosporea	Ichthyophonida	Not assigned	*Sphaeroforma*	*sirkka*		Iridoviridae/Marseilleviridae[a]

[a]The endogenous viral sequences found in these host species were indicated to represent either Iridoviridae or Marseilleviridae without further specification

Table 3.14 Endogenous viruses of mammals

Class	Order	Family	Genus	Species	Common name	Known endogenous viruses
Mammalia	Artiodactyla	Bovidae	*Bos*	*taurus*	Cow	Bornaviridae Parvoviridae Retroviridae
Mammalia	Artiodactyla	Bovidae	*Capra*	*hircus*	Goat	Parvoviridae
Mammalia	Artiodactyla	Bovidae	*Ovis*	*aries*	Sheep	Retroviridae
Mammalia	Artiodactyla	Camelidae	*Vicugna*	*pacos*	Alpaca	Parvoviridae
Mammalia	Artiodactyla	Delphinidae	*Cephalorhynchus*	*commersonii*	Commerson's dolphin	Retroviridae
Mammalia	Artiodactyla	Delphinidae	*Delphinus*	*delphis*	Saddleback dolphin also called common dolphin	Retroviridae
Mammalia	Artiodactyla	Delphinidae	*Globicephala*	*macrorhynchus*	Short-finned pilot whale	Retroviridae
Mammalia	Artiodactyla	Delphinidae	*Grampus*	*griseus*	Risso's dolphin	Retroviridae
Mammalia	Artiodactyla	Delphinidae	*Lagenorhynchus*	*obliquidens*	Pacific white-sided dolphin	Retroviridae
Mammalia	Artiodactyla	Delphinidae	*Orcinus*	*orca*	Killer whale	Retroviridae
Mammalia	Artiodactyla	Delphinidae	*Pseudorca*	*crassidens*	False killer whale	Retroviridae
Mammalia	Artiodactyla	Delphinidae	*Steno*	*bredanensis*	Rough-toothed dolphin	Retroviridae
Mammalia	Artiodactyla	Delphinidae	*Tursiops*	*truncatus*	Bottlenose dolphin	Parvoviridae Retroviridae
Mammalia	Artiodactyla	Phocoenidae	*Phocoena*	Not reported	Harbor porpoises	Retroviridae
Mammalia	Artiodactyla	Physeteridae	*Kogia*	*breviceps*	Pygmy sperm whale	Retroviridae
Mammalia	Artiodactyla	Physeteridae	*Kogia*	*sima*	Dwarf sperm whale	Retroviridae
Mammalia	Artiodactyla	Suidae	*Sus*	*scrofa*	Pig	Circoviridae Filoviridae Parvoviridae Retroviridae
Mammalia	Carnivora	Ailuridae	*Ailurus*	*fulgens*	Lesser panda	Circoviridae

Mammalia	Carnivora	Canidae	*Canis*	*lupus familiaris*	Domestic Dog	Circoviridae Parvoviridae Retroviridae
Mammalia	Carnivora	Canidae	*Lycaon*	*pictus*	African hunting dog	Circoviridae
Mammalia	Carnivora	Felidae	*Acinonyx*	*jubatus*	Cheetah	Circoviridae
Mammalia	Carnivora	Felidae	*Felis*	*catus*	Domestic cat	Circoviridae Retroviridae
Mammalia	Carnivora	Felidae	*Panthera*	*tigris altaica*	Amur tiger	Circoviridae
Mammalia	Carnivora	Mustelidae	*Enhydra*	*lutris*	Sea otter	Circoviridae
Mammalia	Carnivora	Mustelidae	*Mustela*	*putorius furo*	Domestic ferret	Circoviridae
Mammalia	Carnivora	Odobenidae	*Odobenus*	*rosmarus*	Walrus	Circoviridae
Mammalia	Carnivora	Phocidae	*Leptonychotes*	*weddellii*	Weddell seal	Circoviridae
Mammalia	Carnivora	Phocidae	*Neomonachus*	*schauinslandi*	Hawaiian monk seal	Circoviridae
Mammalia	Carnivora	Ursidae	*Ailuropoda*	*melanoleuca*	Giant panda	Circoviridae
Mammalia	Carnivora	Ursidae	*Ursus*	*maritimus*	Polar bear	Circoviridae
Mammalia	Chiroptera	Phyllostomidae	*Desmodus*	*rotundus*	Common vampire bat	Parvoviridae
Mammalia	Chiroptera	Pteropodidae	*Pteropus*	*vampyrus*	Large flying fox	Parvoviridae
Mammalia	Chiroptera	Vespertilionidae	*Eptesicus*	*fuscus*	Big brown bat	Bornaviridae
Mammalia	Chiroptera	Vespertilionidae	*Eptesicus*	*nilssonii*	Northern bat	Bornaviridae
Mammalia	Chiroptera	Vespertilionidae	*Eptesicus*	*serotinus*	Common serotine	Bornaviridae
Mammalia	Chiroptera	Vespertilionidae	*Myotis*	*davidii*	David's myotis	Bornaviridae
Mammalia	Chiroptera	Vespertilionidae	*Myotis*	*lucifugus*	Little brown bat	Bornaviridae Filoviridae Parvoviridae
Mammalia	Cingulata	Dasypodidae	*Dasypus*	*novemcinctus*	Nine–banded armadillo	Parvoviridae
Mammalia	Dasyuromorphia	Dasyuridae	*Sarcophilus*	*harrisii*	Tasmanian devil	Circoviridae Parvoviridae
Mammalia	Dermoptera	Cynocephalidae	*Galeopterus*	*variegatus*	Sunda flying lemur	Circoviridae
Mammalia	Didelphimorphia	Didelphidae	*Monodelphis*	*domestica*	Grey short-tailed opossum	Bornaviridae Circoviridae Filoviridae Parvoviridae

(continued)

Table 3.14 (continued)

Class	Order	Family	Genus	Species	Common name	Known endogenous viruses
Mammalia	Diprotodontia	Macropodidae	*Notamacropus*	*eugenii*	Tammar wallaby	Filoviridae Parvoviridae
Mammalia	Diprotodontia	Phalangeridae	*Trichosurus*	*vulpecula*	Common brushtail	Parvoviridae
Mammalia	Diprotodontia	Phascolarctidae	*Phascolarctos*	*cinereus*	Koala	Circoviridae
Mammalia	Eulipotyphla	Erinaceidae	*Erinaceus*	*europaeus*	European hedgehog	Bornaviridae
Mammalia	Eulipotyphla	Soricidae	*Sorex*	*araneus*	Euopean shrew	Bornaviridae Filoviridae
Mammalia	Hyracoidea	Procaviidae	*Procavia*	*capensis*	Cape hyrax	Bornaviridae Parvoviridae
Mammalia	Lagomorpha	Leporidae	*Lepus*	*europaeus*	European hare	Retroviridae
Mammalia	Lagomorpha	Leporidae	*Oryctolagus*	*cuniculus*	Rabbit	Parvoviridae Retroviridae
Mammalia	Lagomorpha	Ochotonidae	*Ochotona*	*princeps*	American pika	Bornaviridae Parvoviridae
Mammalia	Monotremata	Ornithorhynchidae	*Ornithorhynchus*	*anatinus*	Platypus	Parvoviridae
Mammalia	Perissodactyla	Equidae	*Equus*	*caballus*	Horse	Parvoviridae Retroviridae
Mammalia	Pholidota	Manidae	*Manis*	*pentadactyla*	Chinese pangolin	Circoviridae
Mammalia	Pilosa	Megalonychidae	*Choloepus*	*hoffmanni*	Hoffmann's two-fingered sloth	Circoviridae
Mammalia	Primates	Cebidae	*Callithrix*	*jacchus*	Marmoset	Bornaviridae Retroviridae
Mammalia	Primates	Cebidae	*Cebus*	*imitator*	Panamanian white-faced capuchin	Parvoviridae
Mammalia	Primates	Cebidae	*Saimiri*	not reported	Squirrel monkey	Retroviridae
Mammalia	Primates	Cercopithecidae	*Macaca*	*fascicularis*	Crab-eating macaque	Retroviridae
Mammalia	Primates	Cercopithecidae	*Macaca*	*mulatta*	Rhesus macaque	Bornaviridae Retroviridae
Mammalia	Primates	Cercopithecidae	*Papio*	*hamadryas*	Baboon	

						Bornaviridae Parvoviridae
Mammalia	Primates	Cheirogaleidae	*Microcebus*	*murinus*	Mouse lemur	Bornaviridae
Mammalia	Primates	Daubentoniidae	*Daubentonia*	*madagascariensis*	Aye aye	Retroviridae
Mammalia	Primates	Galagidae	*Otolemur*	*garnettii*	Galago (also called Bush Baby)	Bornaviridae Retroviridae
Mammalia	Primates	Hominidae	*Gorilla*	*gorilla*	Western gorilla	Bornaviridae
Mammalia	Primates	Hominidae	*Gorilla*	not reported	Gorilla	Retroviridae
Mammalia	Primates	Hominidae	*Homo*	*sapiens*	Human	Bornaviridae Metaviridae[a] Parvoviridae Retroviridae
Mammalia	Primates	Hominidae	*Pan*	*paniscus*	Bonobo	Retroviridae
Mammalia	Primates	Hominidae	*Pan*	*troglodytes*	Chimpanzee	Bornaviridae Retroviridae
Mammalia	Primates	Hominidae	*Pongo*	*abelii*	Sumatran orangutan	Bornaviridae
Mammalia	Primates	Hominidae	*Pongo*	not reported	Orangutan	Retroviridae
Mammalia	Primates	Hylobatidae	*Hylobates*	not reported	Gibbon	Retroviridae
Mammalia	Primates	Hylobatidae	*Nomascus*	*leucogenys*	Northern white-cheeked gibbon	Bornaviridae
Mammalia	Primates	Lemuridae	*Lemur*	not reported	Lemur	Retroviridae
Mammalia	Primates	Tarsiidae	*Carlito*	*syrichta*	Tarsier	Bornaviridae Filoviridae Retroviridae
Mammalia	Proboscidea	Elephantidae	*Loxodonta*	*africana*	African savanna elephant	Bornaviridae Parvoviridae
Mammalia	Rodentia	Bathyergidae	*Heterocephalus*	*glaber*	Naked mole rat	Circoviridae
Mammalia	Rodentia	Caviidae	*Cavia*	*porcellus*	Guinea pig	Bornaviridae Filoviridae Parvoviridae

(continued)

Table 3.14 (continued)

Class	Order	Family	Genus	Species	Common name	Known endogenous viruses
Mammalia	Rodentia	Chinchillidae	*Chinchilla*	*lanigera*	Long tailed chinchilla	Parvoviridae
Mammalia	Rodentia	Cricetidae	*Ellobius*	*lutescens*	Transcaucasian mole vole	Parvoviridae
Mammalia	Rodentia	Heteromyidae	*Dipodomys*	*ordii*	Kangaroo rat	Bornaviridae Filoviridae
Mammalia	Rodentia	Muridae	*Mus*	*caroli*	Ryuku mouse	Circoviridae
Mammalia	Rodentia	Muridae	*Mus*	*musculus*	House mouse	Bornaviridae Filoviridae Parvoviridae Retroviridae
Mammalia	Rodentia	Muridae	*Rattus*	*norvegicus*	Norway rat	Bornaviridae Filoviridae Parvoviridae
Mammalia	Rodentia	Octodontidae	*Octodon*	*degus*	Degu	Parvoviridae
Mammalia	Rodentia	Sciuridae	*Ictidomys*	*tridecemlineatus*	Ground squirrel	Bornaviridae
Mammalia	Scandentia	Tupaiidae	*Tupaia*	*belangeri*	Northern tree shrew	Retroviridae
Mammalia	Scandentia	Tupaiidae	*Tupaia*	not reported		Retroviridae
Mammalia	Scandentia	Tupaiidae	*Tupaia*	*chinensis*	Chinese tree shrew	Retroviridae
Mammalia	Sirenia	Trichechidae	*Trichechus*	*manatus latirostris*	florida manatee	Bornaviridae
Mammalia	Tubulidentata	Orycteropodidae	*Orycteropus*	*afer*	Aardvark	Bornaviridae Parvoviridae
Mammalia	Not assigned	Chrysochloridae	*Chrysochloris*	*asiatica*	Cape golden mole	Bornaviridae Circoviridae
Mammalia	Not assigned	Tenrecidae	*Echinops*	*telfairi*	Madagascar hedgehog (also called Tenrec)	Bornaviridae Parvoviridae

[a] I am mentioning here this finding of endogenous Metaviridae sequences in human for the sake of being inclusive. However, the uniqueness of this discovery brings a question to my mind as to the correctness of its detection and reporting, particularly when giving consideration to the fact that Metaviridae sequences thus far have not been reported for any other vertebrates

Table 3.15 Endogenous viruses of molluscs

Class	Order	Family	Genus	Species	Common name	Known endogenous viruses
Bivalvia	Adapedonta	Pharidae	*Ensis*	*directus*	Atlantic razor clam	Metaviridae[a]
Bivalvia	Adapedonta	Pharidae	*Siliqua*	*patula*	Pacific razor clam	Metaviridae[a]
Bivalvia	Cardiida	Tellinidae	*Limecola*	*balthica*	Baltic clam	Metaviridae[a]
Bivalvia	Myida	Hiatellidae	*Panopea*	*generosa*	Geoduck	Metaviridae[a]
Bivalvia	Myida	Myidae	*Mya*	*arenaria*	Soft-shell clam	Metaviridae[a]
Bivalvia	Mytiloida	Mytilidae	*Mytilus*	*galloprovincialis*	Mediterranean mussel	Metaviridae
Bivalvia	Ostreida	Ostreidae	*Crassostrea*	*ariakensis*	Suminoe oyster	Parvoviridae
Bivalvia	Ostreida	Ostreidae	*Crassostrea*	*gigas*	Pacific oyster	Metaviridae Pseudoviridae
Bivalvia	Pterioida	Pteriidae	*Pinctada*	*fucata*	Akoya pearl oyster	Metaviridae
Cephalopoda	Octopoda	Octopodidae	*Octopus*	*bimaculoides*	California two-spot octopus	Metaviridae Parvoviridae Pseudoviridae
Cephalopoda	Octopoda	Octopodidae	*Octopus*	*vulgaris*	Common octopus	Parvoviridae
Cephalopoda	Sepiida	Sepiidae	*Sepia*	*officinalis*	Common cuttlefish	Parvoviridae
Cephalopoda	Sepiida	Sepiolidae	*Euprymna*	*scolopes*	Hawaiian bobtail squid	Parvoviridae
Gastropoda	Aplysiida	Aplysiidae	*Aplysia*	*californica*	California sea hare	Metaviridae Pseudoviridae
Gastropoda	Littorinimorpha	Bithyniidae	*Bithynia*	*siamensis*		Parvoviridae
Gastropoda	Littorinimorpha	Calyptraeidae	*Crepidula*	*fornicata*	American slipper limpet	Parvoviridae
Gastropoda	Neogastropoda	Conidae	*Conus*	*tribblei*	Tribble's cone	Metaviridae Pseudoviridae
Gastropoda	Not assigned	Lottiidae	*Lottia*	*gigantea*	Owl limpet	Metaviridae Pseudoviridae
Gastropoda	Not assigned	Lymnaeidae	*Lymnaea*	*stagnalis*	Great pond snail	Metaviridae Pseudoviridae
Gastropoda	Not assigned	Planorbidae	*Biomphalaria*	*glabrata*	Ram's horn snail	Metaviridae Pseudoviridae

[a]These mollusc species contain Ty3 Steamer Long Terminal Repeat (LTR) retrotransposons that are presumed to represent Metaviridae. Many of the other listed mollusc species contain either endogenous LTR Gypsy retrotransposons that are presumed to represent Metaviridae or Bel/pao LTR retrotransposon sequences which I am including within the family Metaviridae. Some of these listed host species also contain endogenous LTR Copia retrotransposons that are presumed to represent Pseudoviridae

Table 3.16 Endogenous viruses of oomycetes

Class	Order	Family	Genus	Species	Common name	Known endogenous viruses
Oomycota	Peronosporales	Peronosporaceae	*Phytophthora*	*agathidicida*		Asfarviridae
Oomycota	Peronosporales	Peronosporaceae	*Phytophthora*	*alni*		Asfarviridae
Oomycota	Peronosporales	Peronosporaceae	*Phytophthora*	*cambivora*		Asfarviridae
Oomycota	Peronosporales	Peronosporaceae	*Phytophthora*	*cryptogea*		Asfarviridae
Oomycota	Peronosporales	Peronosporaceae	*Phytophthora*	*nicotianae*		Asfarviridae
Oomycota	Peronosporales	Peronosporaceae	*Phytophthora*	*parasitica*		Asfarviridae
Oomycota	Peronosporales	Peronosporaceae	*Phytophthora*	*taxon totara*		Asfarviridae
Oomycota	Pythiales	Pythiaceae	*Globisporangium*	*irregulare*		Asfarviridae
Oomycota	Pythiales	Pythiaceae	*Globisporangium*	*ultimum*		Asfarviridae
Oomycota	Pythiales	Pythiaceae	*Pythium*	*oligandrum*		Asfarviridae

Table 3.17 Endogenous viruses of plants

Class	Order	Family	Genus	Species	Common name	Known endogenous viruses
Non-vascular plants						
Bryopsida	Funariales	Funariaceae	*Physcomitrium*	*patens*		Endornaviridae Pithoviridae
Bryopsida	Funariales	Funariaceae	*Physcomitrella*	not reported		Mimiviridae
Vascular plants						
Magnoliopsida	Arecales	Arecaceae	*Phoenix*	*dactylifera*	Date palm	Partitiviridae
Magnoliopsida	Asparagales	Amaryllidaceae	*Allium*	not reported		Amalgaviridae
Magnoliopsida	Asparagales	Orchidaceae	*Phalaenopsis*	*aphrodite*	Aphrodite's Phalaenopsis	Endornaviridae
Magnoliopsida	Asterales	Asteraceae	*Artemisia*	*annua*	Sweet Annie	Chrysoviridae
Magnoliopsida	Asterales	Asteraceae	*Lactuca*	*sativa*	Cultivated lettuce (also called Garden lettuce)	Endornaviridae
Magnoliopsida	Asterales	Asteraceae	*Lactuca*	*serriola*	Compass plant (also called Prickly lettuce)	Endornaviridae
Magnoliopsida	Asterales	Asteraceae	*Zinnia*	*elegans*	Garden zinnia	Chrysoviridae
Magnoliopsida	Brassicales	Brassicaceae	*Arabidopsis*	*lyrata*	Lyrate rockcress	Endornaviridae Partitiviridae
Magnoliopsida	Brassicales	Brassicaceae	*Arabidopsis*	*thaliana*	Mouse ear cress (also called Thale cress)	Endornaviridae Partitiviridae
Magnoliopsida	Brassicales	Brassicaceae	*Brassica*	*oleracea*	Wild mustard	Endornaviridae Partitiviridae
Magnoliopsida	Brassicales	Brassicaceae	*Brassica*	*rapa*	Field mustard (subspecies include Chinese cabbage)	Partitiviridae Rhabdoviridae
Magnoliopsida	Brassicales	Brassicaceae	*Brassica*	not reported	Cabbages	Amalgaviridae Totiviridae
Magnoliopsida	Brassicales	Brassicaceae	*Capsella*	not reported		Partitiviridae
Magnoliopsida	Brassicales	Brassicaceae	*Olimarabidopsis*	not reported		Partitiviridae

(continued)

Table 3.17 (continued)

Class	Order	Family	Genus	Species	Common name	Known endogenous viruses
Magnoliopsida	Brassicales	Brassicaceae	*Turritis*	not reported		Partitiviridae
Magnoliopsida	Brassicales	Caricaceae	*Carica*	*papaya*	Papaya	Endornaviridae
Magnoliopsida	Caryophyllales;	Tamaricaceae	*Tamarix*	*androssowii*	Tamarisk (also called Salt cedar)	Totiviridae
Magnoliopsida	Cucurbitales	Cucurbitaceae	*Cucumis*	*sativus*	Cucumber	Betaflexiviridae Bromoviridae Endornaviridae Rhabdoviridae
Magnoliopsida	Dioscoreales	Dioscoreaceae	*Dioscorea*	*cayenensis subsp. rotundata*	Guinea yam	Caulimoviridae
Magnoliopsida	Ericales	Theaceae	*Camellia*	*sinensis*	Black tea	Endornaviridae
Magnoliopsida	Fabales	Fabaceae	*Cajanus*	*cajan*	Pigeon pea	Endornaviridae
Magnoliopsida	Fabales	Fabaceae	*Glycine*	*max*	Soybean	Endornaviridae Partitiviridae
Magnoliopsida	Fabales	Fabaceae	*Lotus*	*japonicus*	Birdsfoot trefoil	Endornaviridae Partitiviridae Rhabdoviridae Totiviridae
Magnoliopsida	Fabales	Fabaceae	*Medicago*	*truncatula*	Barrel medic	Bromoviridae Endornaviridae Partitiviridae Totiviridae
Magnoliopsida	Fabales	Fabaceae	*Phaseolus*	*vulgaris*	French bean (also called Kidney bean, String bean)	Endornaviridae
Magnoliopsida	Fabales	Fabaceae	*Vigna*	*unguiculata*	Cowpea	Endornaviridae Partitiviridae
Magnoliopsida	Fagales	Fagaceae	*Quercus*	*petraea*	Sessile oak	Endornaviridae

Magnoliopsida	Fagales	Fagaceae	*Quercus*	*robur*	English oak (also called Pedunculate oak, Truffle oak)	Endornaviridae
Magnoliopsida	Lamiales	Orobanchaceae	*Triphysaria*	*pusilla*	Dwarf owl's clover	Endornaviridae
Magnoliopsida	Lamiales	Orobanchaceae	*Triphysaria*	not reported	Owl's clover	Amalgaviridae
Magnoliopsida	Lamiales	Pedaliaceae	*Sesamum*	*indicum*	Sesame	Endornaviridae
Magnoliopsida	Lamiales	Phrymaceae	*Erythranthe*	*guttata*	Spotted monkey flower	Endornaviridae Partitiviridae Rhabdoviridae
Magnoliopsida	Malpighiales	Euphorbiaceae	*Manihot*	*esculenta*	Cassava	Partitiviridae
Magnoliopsida	Malpighiales	Euphorbiaceae	*Ricinus*	*communis*	Castor bean	Endornaviridae
Magnoliopsida	Malpighiales	Rhizophoraceae	*Rhizophora*	*apiculata*	Mangrove	Metaviridae[a]
Magnoliopsida	Malpighiales	Salicaceae	*Populus*	*trichocarpa*	Black cottonwood	Amalgaviridae Endornaviridae Geminiviridae Rhabdoviridae Totiviridae
Magnoliopsida	Malvales	Malvaceae	*Gossypium*	*hirsutum*	Cotton	Endornaviridae
Magnoliopsida	Malvales	Malvaceae	*Theobroma*	*cacao*	Cacao (also called Chocolate, Cocoa)	Endornaviridae Rhabdoviridae
Magnoliopsida	Myrtales	Myrtaceae	*Eucalyptus*	*grandis*	Rose gum	Endornaviridae
Magnoliopsida	Poales	Poaceae	*Brachypodium*	*distachyon*	Annual false brome	Endornaviridae
Magnoliopsida	Poales	Poaceae	*Festuca*	not reported	Fescues	Amalgaviridae
Magnoliopsida	Poales	Poaceae	*Lolium*	not reported	Ryegrass	Amalgaviridae
Magnoliopsida	Poales	Poaceae	*Oryza*	*sativa*	Rice	Caulimoviridae
Magnoliopsida	Poales	Poaceae	*Sorghum*	*bicolor*	Broomcorn (also called Milo)	Partitiviridae
Magnoliopsida	Poales	Poaceae	*Triticum*	*aestivum*	Bread wheat	Endornaviridae

(continued)

Table 3.17 (continued)

Class	Order	Family	Genus	Species	Common name	Known endogenous viruses
Magnoliopsida	Poales	Poaceae	*Zea*	*mays*	Corn	Endornaviridae Metaviridae[b] Partitiviridae Pseudoviridae[b]
Magnoliopsida	Poales	Poaceae	*Zea*	not reported	Teosinte	Metaviridae[b] Pseudoviridae[b]
Magnoliopsida	Ranunculales	Ranunculaceae	*Aquilegia*	*coerulea*	Colorado blue columbine	Endornaviridae Rhabdoviridae
Magnoliopsida	Rosales	Rosaceae	*Malus*	*domestica*	Apple	Endornaviridae Partitiviridae Rhabdoviridae
Magnoliopsida	Rosales	Rosaceae	*Prunus*	*mume*	Japanese apricot	Endornaviridae
Magnoliopsida	Rosales	Rosaceae	*Prunus*	*persica*	Peach	Endornaviridae
Magnoliopsida	Sapindales	Rutaceae	*Citrus*	*clementina*	Clementine	Endornaviridae
Magnoliopsida	Sapindales	Rutaceae	*Citrus*	*sinensis*	Sweet orange	Endornaviridae
Magnoliopsida	Sapindales	Rutaceae	*Citrus*	*trifoliata*	Hardy orange (also called Trifoliate orange)	Endornaviridae
Magnoliopsida	Solanales	Convolvulaceae	*Ipomoea*	*nil*	Japanese morning glory	Endornaviridae
Magnoliopsida	Solanales	Solanaceae	*Capsicum*	*annuum*	Sweet and chili peppers	Endornaviridae
Magnoliopsida	Solanales	Solanaceae	*Nicotiana*	*tabacum*	Common tobacco	Caulimoviridae Geminiviridae Partitiviridae Rhabdoviridae
Magnoliopsida	Solanales	Solanaceae	*Petunia*	not reported	Petunia	Caulimoviridae
Magnoliopsida	Solanales	Solanaceae	*Physalis*	*peruviana*	Cape-gooseberry (also called Goldenberry, Gooseberry-tomato)	Endornaviridae
Magnoliopsida	Solanales	Solanaceae	*Solanum*	*lycopersicum*	Tomato	Caulimoviridae

Magnoliopsida	Solanales	Solanaceae	*Solanum*	*phureja*	Chaucha (also called Phureja, Wild potato)	Partitiviridae
Magnoliopsida	Solanales	Solanaceae	*Solanum*	*tuberosum*	Potato	Caulimoviridae Partitiviridae
Magnoliopsida	Vitales	Vitaceae	*Vitis*	*vinifera*	Wine grape	Endornaviridae Potyviridae
Magnoliopsida	Vitales	Vitaceae	*Vitis*	not reported	Grapevines	Potyviridae
Magnoliopsida	Zingiberales	Musaceae	*Musa*	*acuminata*	Banana	Caulimoviridae
Magnoliopsida	Zingiberales	Musaceae	*Musa*	not reported	Plantain	Caulimoviridae
Magnoliopsida	Zingiberales	Zingiberaceae	*Zingiber*	not reported	Ginger	Amalgaviridae
Pinopsida	Pinales	Pinaceae	*Picea*	*glauca*	White spruce	Endornaviridae
Pinopsida	Pinales	Pinaceae	*Picea*	not reported	Spruce	Partitiviridae
Pinopsida	Pinales	Pinaceae	*Pseudotsuga*	not reported	Douglas fir	Partitiviridae

[a]*Rhizophora apiculata* contains endogenous Long Terminal Repeat (LTR) Gypsy retrotransposons. LTR Gypsy retrotransposons are presumed to represent Metaviridae

[b]*Zea mays* and its genetic ancestors in the genus *Zea* that are called Teosinte contain endogenous LTR Copia retrotransposons and LTR Gypsy retrotransposons. LTR Copia retrotransposons are presumed to represent Pseudoviridae. LTR Gypsy retrotransposons are presumed to represent Metaviridae. *Zea mays* also contains endogenous Endornaviridae and endogenous Partitiviridae

Table 3.18 Endogenous viruses of platyhelminths

Class	Order	Family	Genus	Species	Common name	Known endogenous viruses
Cestoda	Cyclophyllidea	Anoplocephalidae	*Moniezia*	*expansa*	Sheep tapeworm	Parvoviridae
Cestoda	Cyclophyllidea	Hymenolepididae	*Hymenolepis*	*microstoma*		Parvoviridae
Cestoda	Cyclophyllidea	Hymenolepididae	*Rodentolepis*	*nana*		Parvoviridae
Cestoda	Cyclophyllidea	Taeniidae	*Echinococcus*	*granulosus*		Parvoviridae
Cestoda	Cyclophyllidea	Taeniidae	*Echinococcus*	*multilocularis*		Parvoviridae
Cestoda	Cyclophyllidea	Taeniidae	*Hydatigera*	*taeniaeformis*		Parvoviridae
Cestoda	Cyclophyllidea	Taeniidae	*Taenia*	*asiatica*		Parvoviridae
Cestoda	Cyclophyllidea	Taeniidae	*Taenia*	*multiceps*		Parvoviridae
Monogenea	Capsalidea	Capsalidae	*Neobenedenia*	*melleni*		Parvoviridae
Rhabditophora	Tricladida	Dugesiidae	*Schmidtea*	*mediterranea*		Parvoviridae
Trematoda	Opisthorchiida	Opisthorchiidae	*Clonorchis*	*sinensis*		Parvoviridae
Trematoda	Opisthorchiida	Opisthorchiidae	*Opisthorchis*	*felineus*		Parvoviridae
Trematoda	Opisthorchiida	Opisthorchiidae	*Opisthorchis*	*viverrini*		Parvoviridae
Trematoda	Plagiorchiida	Dicrocoeliidae	*Dicrocoelium*	*dendriticum*		Parvoviridae
Trematoda	Plagiorchiida	Echinostomatidae	*Echinostoma*	*caproni*		Parvoviridae
Trematoda	Strigeidida	Schistosomatidae	*Schistosoma*	*mansoni*		Parvoviridae
Trematoda	Strigeidida	Schistosomatidae	*Schistosoma*	*margrebowiei*		Parvoviridae

Table 3.19 Endogenous viruses of reptiles

Class	Order	Family	Genus	Species	Common name	Known endogenous viruses
Lepidosauria	Sphenodontia	Sphenodontidae	*Sphenodon*	*punctatus*	Tuatara	Retroviridae
Lepidosauria	Squamata	Colubridae	*Pantherophis*	*guttatus*	Corn snake	Circoviridae
Lepidosauria	Squamata	Dactyloidae	*Anolis*	*carolinensis*	Green anole	Retroviridae
Lepidosauria	Squamata	Elapidae	*Ophiophagus*	*hannah*	King cobra	Circoviridae
Lepidosauria	Squamata	Gekkonidae	*Gekko*	*japonicus*	Schlegel's Japanese gecko	Retroviridae
Lepidosauria	Squamata	Gekkonidae	*Paroedura*	*picta*	Panther gecko	Retroviridae
Lepidosauria	Squamata	Lacertidae	*Lacerta*	*viridis*	European green lizard	Retroviridae
Lepidosauria	Squamata	Pythonidae	*Python*	*bivittatus*	Burmese python	Retroviridae
Lepidosauria	Squamata	Pythonidae	*Python*	*molurus*	Indian rock python	Circoviridae Retroviridae
Lepidosauria	Squamata	Scincidae	*Mabuya*	not reported		Retroviridae
Lepidosauria	Squamata	Viperidae	*Crotalus*	*horridus*	Timber rattlesnake	Circoviridae
Lepidosauria	Squamata	Viperidae	*Crotalus*	*mitchellii*	Speckled rattlesnake	Bornaviridae Circoviridae Hepadnaviridae
Lepidosauria	Squamata	Viperidae	*Crotalus*	*pyrrhus*	Mitchell's rattlesnake	Circoviridae
Lepidosauria	Squamata	Viperidae	*Protobothrops*	*mucrosquamatus*	Chinese habu(also called Brown spotted pitviper)	Circoviridae Parvoviridae Retroviridae
Not assigned	Crocodylia	Crocodylidae	*Crocodylus*	*porosus*	Australian saltwater crocodile	Retroviridae
Not assigned	Crocodylia	Gavialidae	*Gavialis*	*gangeticus*	Gharial	Retroviridae
Not assigned	Testudines	Cheloniidae	*Chelonia*	*mydas*	Green sea turtle	Retroviridae
Not assigned	Testudines	Emydidae	*Chrysemys*	*picta*	Painted turtle	Retroviridae
Not assigned	Testudines	Emydidae	*Malaclemys*	*terrapin*	Diamondback terrapin	Retroviridae
Not assigned	Testudines	Emydidae	*Trachemys*	*scripta*	Red eared slider turtle	Retroviridae
Not assigned	Testudines	Trionychidae	*Apalone*	*spinifera*	Spiny softshell turtle	Retroviridae
Not assigned	Testudines	Trionychidae	*Pelodiscus*	*sinensis*	Chinese soft-shelled turtle	Retroviridae

Table 3.20 Endogenous viruses of tunicates

Class	Order	Family	Genus	Species	Common name	Known endogenous viruses
Ascidiacea	Phlebobranchia	Cionidae	*Ciona*	*intestinalis*	vase tunicate	Parvoviridae

Betaflexiviridae are plants. A suggested reference for endogenous Betaflexiviridae in plants would be Chu et al. (2014).

For host specific information on endogenous Betaflexiviridae see Table 3.17 Endogenous viruses of plants.

3.3.5 Bornaviridae

The members of the Bornaviridae viral family have negative sense single stranded RNA genomes and they utilize nuclear replication. The known normal hosts of Bornaviridae are birds, mammals and reptiles. Suggested references for endogenous Bornaviridae in fish would be Belyi et al. (2010), and for endogenous Bornaviridae in mammals would be Belyi et al. (2010), Horie et al. (2016), Katzourakis and Gifford (2010) plus Kobayashi et al. (2016). Specific information for endogenous Bornaviridae in human can be found in the publication by Honda (2017). A suggested reference for endogenous Bornaviridae in reptiles would be Gilbert et al. (2014).

For host specific information on endogenous Bornaviridae see Table 3.10 Endogenous viruses of fish, Table 3.14 Endogenous viruses of mammals, and Table 3.19 Endogenous viruses of reptiles.

3.3.6 Bromoviridae

The members of the Bromoviridae viral family have positive sense single stranded RNA genomes and they utilize cytoplasmic replication. The known normal hosts of Bromoviridae are plants. Suggested references for endogenous Bromoviridae in plants would be Chu et al. (2014) and Takahashi et al. (2019).

For host specific information on endogenous Bromoviridae see Table 3.17 Endogenous viruses of plants.

3.3.7 *Bunyaviridae*

This no longer is used as a viral family name. Members of the Bunyaviridae viral family have negative sense single stranded RNA genomes and they utilize cytoplasmic replication. The known normal hosts for Bunyaviridae are crustaceans, insects, and mammals. There is information on endogenous Bunyaviridae sequences that unfortunately seems to offer no suggestion as to how those sequences should fit into the more recently assigned viral family names of Nairoviridae, Peribunyaviridae, and Phenuiviridae. Suggested references for endogenous Bunyaviridae in crustaceans would be Metegnier et al. (2015) and Thézé et al. (2014).

For host specific information on endogenous Bunyaviridae see Table 3.8 Endogenous viruses of crustaceans.

3.3.8 *Caulimoviridae*

The members of the Caulimoviridae viral family have double stranded DNA genomes. They use a combination of cytoplasmic and nuclear replication that includes reverse transcription. Caulimoviridae often are called plant pararetroviruses. The known normal hosts of Caulimoviridae are plants. Suggested references for endogenous Caulimoviridae in plants would be Chen et al. (2018), Chu et al. (2014), Kuriyama et al. (2020), Staginnus et al. (2007), Takahashi et al. (2019), Tripathi et al. (2019) and Umber et al. (2014).

For host specific information on endogenous Caulimoviridae see Table 3.17 Endogenous viruses of plants.

3.3.9 *Chrysoviridae*

The members of the Chrysoviridae viral family have double stranded RNA genomes and they utilize cytoplasmic replication. The known normal hosts of Chrysoviridae are plants. Suggested references for endogenous Chrysoviridae in plants would be Chu et al. (2014) and Takahashi et al. (2019).

For host specific information on endogenous Chrysoviridae see Table 3.17 Endogenous viruses of plants.

3.3.10 *Chuviridae (and Unspecified Mono-Chu)*

The members of the viral family Chuviridae have negative-sense single-stranded genomes and presumably they use cytoplasmic replication. The known normal hosts for Chuviridae are insects. Suggested references for endogenous Chuviridae in mosquitos would be Dezordi et al. (2020) and for Mono-Chu sequences in ants would be Flynn and Moreau (2019). I did not list the information from Flynn and Moreau (2019) in a table because the publication by Flynn and Moreau (2019) did not seem to match up the viral sequences with specific host genera and species.

For host specific information on endogenous Chuviridae see Table 3.5 Endogenous viruses of arachnids collembolids and insects.

3.3.11 Circoviridae

The members of the Circoviridae viral family have ambisense (partially positive sense and partially negative sense) single stranded RNA genomes and they utilize nuclear replication. The known normal hosts of Circoviridae are amphibians, avians, crustaceans, fish, insects, mammals and reptiles. Suggested references for endogenous Circoviridae in amphibians would be Dennis et al. (2019), for endogenous Circoviridae in avians would be Dennis et al. (2019), for endogenous Circoviridae in crustaceans would be Metegnier et al. (2015) and Thézé et al. (2014), for endogenous Circoviridae in fish would be Dennis et al. (2019), for endogenous Circoviridae in insects (specifically ants) would be Flynn and Moreau (2019), for endogenous Circoviridae in mammals would be Dennis et al. (2019) and Pénzes et al. (2018), and for endogenous Circoviridae in reptiles would be Dennis et al. (2019), Gilbert et al. (2014) plus Katzourakis and Gifford (2010).

For host specific information on endogenous Circoviridae see Table 3.3 Endogenous viruses of amphibians, Table 3.6 Endogenous viruses of avians, Table 3.8 Endogenous viruses of crustaceans, Table 3.10 Endogenous viruses of fish, Table 3.14 Endogenous viruses of mammals, and Table 3.19 Endogenous viruses of reptiles.

I did not list in a table the information about endogenous Circoviridae of ants because the publication by Flynn and Moreau (2019) did not seem to match up the viral sequences with specific host genera and species.

3.3.12 Endornaviridae

The members of the Endornaviridae viral family have positive sense single stranded RNA genomes and they utilize cytoplasmic replication. The known normal hosts of Endornaviridae are algae, fungi, oomycetes, and plants. It is important to note that the Endornaviridae do not have true capsids. Suggested references for endogenous Endornaviridae in algae would be Song et al. (2013), for endogenous Endornaviridae in amoeba would be Song et al. (2013), and for endogenous Endornaviridae in plants would be Chu et al. (2014), Song et al. (2013) plus Takahashi et al. (2019).

For host specific information on endogenous Endornaviridae see Table 3.1 Endogenous viruses of algae, Table 3.2 Endogenous viruses of amoeba, and Table 3.17 Endogenous viruses of plants.

3.3.13 *Filoviridae*

The members of the Filoviridae viral family have negative sense single stranded RNA genomes and they utilize cytoplasmic replication. The known normal hosts of Filoviridae are mammals. Suggested references for endogenous Filoviridae in mammals would be Edwards et al. (2018), Katzourakis and Gifford (2010) and Pénzes et al. (2018).

For host specific information on endogenous Filoviridae see Table 3.14 Endogenous viruses of mammals.

3.3.14 *Flavivridae*

The members of the Flavivridae viral family have positive sense single stranded RNA genomes and they use cytoplasmic replication. The known normal hosts of Flavivridae are insects and mammals. Suggested references for endogenous Flaviviridae in arachnids would be Supplemental Table 2 from ter Horst et al. (2019), and for endogenous Flaviviridae in insects would be Katzourakis and Gifford (2010), Suzuki et al. (2017) plus Supplemental Table 2 from ter Horst et al. (2019).

For host specific information on endogenous Flavivridae see Table 3.5 Endogenous viruses of arachnids collembolids and insects.

3.3.15 *Geminiviridae*

The members of the Geminiviridae viral family have single stranded DNA genomes and they use nuclear replication. The known normal hosts of Geminiviridae are algae and plants. Suggested references for endogenous Geminiviridae in algae would be Chu et al. (2014), and for endogenous Geminiviridae in plants would be Chu et al. (2014) plus Takahashi et al. (2019).

For host specific information on endogenous Geminiviridae see Table 3.1 Endogenous viruses of algae, and Table 3.17 Endogenous viruses of plants.

3.3.16 *Hepadnaviridae*

The members of the Hepadnaviridae viral family have partially double stranded DNA genomes, they use a combination of cytoplasmic and nuclear replication. The known normal hosts of Hepadnaviridae are avians, mammals and reptiles. Suggested references for endogenous Hepadnaviridae in avians would be Supplemental

Table S7 of Katzourakis and Gifford (2010), and for endogenous Hepadnaviridae in reptiles would be Gilbert et al. (2014).

For host specific information on endogenous Hepadnaviridae see Table 3.6 Endogenous viruses of avians, and Table 3.19 Endogenous viruses of reptiles.

3.3.17 *Hypoviridae*

The members of the Hypoviridae viral family have double stranded RNA genomes and they use cytoplasmic replication. They do not form capsid structures. The known normal hosts of Hypoviridae are fungi. A suggested reference for endogenous Hypoviridae in fungi would be Hillman and Milgroom (2021).

For host specific information on endogenous Hypoviridae see Table 3.11 Endogenous viruses of fungi.

3.3.18 *Lavidaviridae*

The members of the Lavidaviridae viral family have double stranded DNA genomes and their replication strategy is at least partly cytoplasmic (Duponchel and Fischer 2019). The known normal hosts of Lavidaviridae are algae and heterokonts. Several of the Lavidaviridae can have endogenous presence as proviruses that become reactivated if their host is infected by a member of the viral family Mimiviridae. These Lavidaviridae are considered to be satellite viruses of Mimiviridae. Suggested references for endogenous Lavidaviridae in algae would be Blanc et al. (2015), and for endogenous Lavidaviridae in unspecified heterokonts would be Fischer and Hackl (2016) plus Hackl et al. (2020).

For host specific information on endogenous Lavidaviridae see Table 3.1 Endogenous viruses of algae, and Table 3.12 Endogenous viruses of unspecified heterokonts.

3.3.19 *Metaviridae*

The members of the Metaviridae viral family have positive sense single stranded RNA genomes and they use nuclear replication. Their genomes encode both reverse transcriptase and integrase capabilities. The known normal hosts of Metaviridae are amoeba, fish, fungi, insects, nematodes, and plants. Metaviridae form Long terminal repeat (LTR) retrotransposons and the LTR 'Gypsy' elements found in genomes are presumed to represent Metaviridae. Gypsy is a member of the Ty3 retrotransposons. The LTR retrotransposon named Steamer also belongs to the Ty3 group (Metzger et al. 2018) and I am considering that to be a member of the Metaviridae family. The

Steamer retrotransposons have moved horizontally to other bivalve species and also to animals of completely different phyla (Metzger et al. 2018). I also have included Bel/pao LTR retransposons (Thomas-Bulle et al. 2018) under Metaviridae which traditionally has been their viral family designation. It is possible that Bel/pao LTR retransposons eventually will be considered as a separate viral family to be named Belpaoviridae. Suggested references for endogenous Metaviridae in fungi would be Liu et al. (2010) Wickner (1989) and Zhang et al. (2014), for endogenous Metaviridae in insects would be Breitenbach et al. (2011), Feng et al. (2019), Rohrmann (2019), Touret et al. (2014) plus Roossinck and Bazán (2017) and specifically in ants would be Flynn and Moreau (2019), for endogenous Metaviridae in mammals would be Vargiu et al. (2016), for endogenous Metaviridae in molluscs would be Metzger et al. (2018) and Thomas-Bulle et al. (2018), for endogenous Metaviridae in nematodes would be Britten (1995) and Kapulkin (2016), and for endogenous Metaviridae in plants would be SanMiguel and Vitte (2009), Wang et al. (2018) plus Zhang and Qi (2019).

For host specific information on endogenous Metaviridae see Table 3.4 Endogenous viruses of annelids and nematodes, Table 3.5 Endogenous viruses of arachnids collembolids and insects, Table 3.11 Endogenous viruses of fungi, Table 3.14 Endogenous viruses of mammals, Table 3.15 Endogenous viruses of molluscs, and Table 3.17 Endogenous viruses of plants.

I did not list in a table the information about endogenous Metaviridae of ants because the publication by Flynn and Moreau (2019) did not seem to match up the viral sequences with specific host genera and species. Vargiu et al. (2016) identified in human some endogenous virus sequences that most closely represented the genus Errantivirus (family Metaviridae). That discovery by Vargiu et al. (2016) is a point of curiosity for me because, although Metaviridae are known to infect fish, endogenous Metaviridae sequences thus far seem not to have been found in any other vertebrates aside from this one reporting in human. I am mentioning here this finding of endogenous Metaviridae in human for the sake of being inclusive, but the uniqueness of this discovery brings a possible question as to the correctness of its detection and reporting.

3.3.20 Mimiviridae

The members of the Mimiviridae viral family have double stranded DNA genomes and they use cytoplasmic replication. The known normal hosts of Mimiviridae are amoeba. Suggested references for endogenous Mimiviridae in algae would be Filée (2014) Gallot-Lavallée and Blanc (2017) and Moniruzzaman et al. (2020), for endogenous Mimiviridae in amoeba would be Filée (2014), for endogenous Mimiviridae in cnidarians would be Filée (2014) plus Gallot-Lavallée and Blanc (2017), and for endogenous Mimiviridae in plants would be Filée (2014).

For host specific information on endogenous Mimiviridae see Table 3.1 Endogenous viruses of algae, Table 3.2 Endogenous viruses of amoeba, Table 3.7 Endogenous viruses of cnidarians, and Table 3.17 Endogenous viruses of plants.

3.3.21 *Molliviridae*

The members of the tentatively named Molliviridae viral family have double stranded DNA genomes, they use a combination of nuclear and cytoplasmic replication. The known normal hosts of Molliviridae are amoeba. A suggested reference for endogenous sequences that are presumed to represent the viral species Mollivirus kamchatka as specifically found in amoeba would be Gallot-Lavallée and Blanc (2017).

For host specific information on endogenous Molliviridae see Table 3.2 Endogenous viruses of amoeba.

3.3.22 *Nairoviridae*

The members of the Nairoviridae viral family have negative sense single stranded RNA genomes and they use cytoplasmic replication. The known normal hosts of Nairoviridae are insects and mammals. A suggested reference for endogenous Nairoviridae in arachnids would be Katzourakis and Gifford (2010).

For host specific information on endogenous Nairoviridae see Table 3.5 Endogenous viruses of arachnids collembolids and insects.

3.3.23 *Nanoviridae*

The members of the Nanoviridae viral family have single stranded DNA genomes and they use nuclear replication. The known normal hosts of Nanoviridae are plants. Nanoviridae have been mentioned as being endogenous in plants (Chu et al. 2014) but without accompanying information which connected the endogenous sequences with specific host genera or species, and thusly Nanoviridae are not listed in Tables 3.1–3.20.

3.3.24 *Nodaviridae*

The members of the Nodaviridae viral family have positive sense single stranded RNA genomes and they use cytoplasmic replication. The known normal hosts of Nodaviridae are fish and insects. A suggested reference for endogenous Nodaviridae in nematodes would be Cotton et al. (2016).

For host specific information on endogenous Nodaviridae see Table 3.4 Endogenous viruses of annelids and nematodes..

3.3.25 *Nudiviridae*

The members of the Nudiviridae viral family have double stranded DNA genomes and they use nuclear replication. The known normal hosts of Nudiviridae are crustaceans and insects. Suggested references for endogenous Nudiviridae in insects would be Burke and Strand (2012), Herniou et al. (2013) and Leobold et al. (2018).

For host specific information on endogenous Nudiviridae see Table 3.5 Endogenous viruses of arachnids collembolids and insects.

3.3.26 *Nimaviridae*

The members of the Nimaviridae viral family have double stranded DNA genomes and they use nuclear replication. The known normal hosts of Nimaviridae are crustaceans. A suggested reference for endogenous Nimaviridae in crustaceans would be Thézé et al. (2014).

For host specific information on endogenous Nyamiviridae see Table 3.8 Endogenous viruses of crustaceans.

3.3.27 *Nyamiviridae*

The members of the Nyamiviridae viral family have negative sense single stranded RNA genomes and they use nuclear replication. The known normal hosts of Nyamiviridae are arachnids, avians, cestodes, crustaceans, echinoderms, insects, nematodes, and sipunculids. A suggested reference for Nyamiviridae in fish would be Belyi et al. (2010). A suggested reference for endogenous Mononegavirales in crustaceans that presumably represent Nyamiviridae would be Thézé et al. (2014).

For host specific information on endogenous Nyamiviridae see Table 3.8 Endogenous viruses of crustaceans, and Table 3.10 Endogenous viruses of fish.

3.3.28 *Orthomyxoviridae*

The members of the Orthomyxoviridae viral family have negative sense single stranded RNA genomes and they use nuclear replication. The known normal hosts of Orthomyxoviridae are birds and mammals. Suggested references for endogenous Orthomyxoviridae in arachnids would be Liu et al. (2010) and for endogenous Orthomyxoviridae in insects would be Katzourakis and Gifford (2010).

For host specific information on endogenous Orthomyxoviridae see Table 3.5 Endogenous viruses of arachnids collembolids and insects.

3.3.29 *Partitiviridae*

The members of the Partitiviridae viral family have double stranded RNA genomes and they use cytoplasmic replication. The known normal hosts of Partitiviridae are fungi and plants. Suggested references for endogenous Partitiviridae in amoeba would be Liu et al. (2010), for endogenous Partitiviridae in arachnids and insects would be Liu et al. (2010), for endogenous Partitiviridae in fungi would be Chu et al. (2014), and for endogenous Partitiviridae in plants would be Chu et al. (2014), Liu et al. (2010) and Takahashi et al. (2019).

For host specific information on endogenous Partitiviridae see Table 3.2 Endogenous viruses of amoeba, Table 3.5 Endogenous viruses of arachnids collembolids and insects, Table 3.11 Endogenous viruses of fungi, and Table 3.17 Endogenous viruses of plants.

3.3.30 *Parvoviridae*

The members of the Parvoviridae viral family have single stranded DNA genomes and they use nuclear replication. The known normal hosts of Parvoviridae are crustaceans, echinoderms, insects, and mammals. Suggested references for endogenous Parvoviridae in annelids would be François et al. (2016), for endogenous Parvoviridae in arachnids would be François et al. (2016) and Jackson et al. (2021), for endogenous Parvoviridae in avians would be François et al. (2016) and Jackson et al. (2021), for endogenous Parvoviridae in cnidarians would be François et al. (2016) and Jackson et al. (2021), for endogenous Parvoviridae in collembolids would be François et al. (2016), for endogenous Parvoviridae in crustaceans would be François et al. (2016), Jackson et al. (2021), Metegnier et al. (2015), Supplemental Table 2 from ter Horst et al. (2019) and Thézé et al. (2014), for endogenous Parvoviridae in echninoderms would be François et al. (2016) and Jackson et al. (2021), for endogenous Parvoviridae in fish would be François et al. (2016), for endogenous Parvoviridae in insects would be Clavijo et al. (2016), François et al. (2016), Jackson et al. (2021), Parker and Brisson (2019), Supplemental Table 2 from ter Horst et al. (2019) and specifically in ants would be Flynn and Moreau (2019), for endogenous Parvoviridae in mammals would be François et al. (2016), Jackson et al. (2021), Katzourakis and Gifford (2010) and Pénzes et al. (2018), for endogenous Parvoviridae in molluscs would be François et al. (2016) and Jackson et al. (2021), for endogenous Parvoviridae in nematodes would be François et al. (2016), for endogenous Parvoviridae in platyhelminths would be François et al. (2016) and Jackson et al. (2021), for endogenous Parvoviridae in reptiles would be Pénzes et al. (2018), and for endogenous Parvoviridae in tunicates would be François et al. (2016).

It is interesting to note that Pénzes et al. (2018) reported finding Parvoviridae sequences both in a vole (*Ellobius lutescens*) and a pit-viper snake (*Protobothrops mucrosquamatus*). Viruses will be acquired by ingestion of virally infected animals and a pit viper eagerly would eat a vole. But, it is uncertain as to whether the presence of endogenous Parvoviridae sequences in snakes aroze by predation on infected animals.

For host specific information on endogenous Parvoviridae see: Table 3.4 Endogenous viruses of annelids and nematodes, Table 3.5 Endogenous viruses of arachnids collembolids and insects, Table 3.6 Endogenous viruses of avians, Table 3.7 Endogenous viruses of cnidarians, Table 3.8 Endogenous viruses of crustaceans, Table 3.9 Endogenous viruses of echinoderms, Table 3.10 Endogenous viruses of fish, Table 3.14 Endogenous viruses of mammals, Table 3.15 Endogenous viruses of molluscs, Table 3.18 Endogenous viruses of platyhelminths, Table 3.19 Endogenous viruses of reptiles, and Table 3.20 Endogenous viruses of tunicates.

I did not list in a table the information which Flynn and Moreau (2019) published about endogenous Parvoviridae of ants because the publication by Flynn and Moreau (2019) did not seem to match up the viral sequences with specific host genera and species.

3.3.31 Phenuiviridae

The members of the Phenuiviridae viral family have negative sense single stranded RNA genomes and they use cytoplasmic replication. The known normal hosts of Phenuiviridae are insects and mammals. Suggested references for endogenous Phenuiviridae in arachnids would be Katzourakis and Gifford (2010).

For host specific information on endogenous Phenuiviridae see: Table 3.5 Endogenous viruses of arachnids collembolids and insects.

3.3.32 Phycodnaviridae

The members of the Phycodnaviridae viral family have double stranded DNA genomes and they use nuclear cytoplasmic replication. The known normal hosts of Phycodnaviridae are algae. Suggested references for endogenous Phycodnaviridae in algae would be Gallot-Lavallée and Blanc (2017) and Moniruzzaman et al. (2020), for endogenous Phycodnaviridae in fungi would be Gallot-Lavallée and Blanc (2017), and for endogenous Phycodnaviridae in unspecified heterokonts would be Gallot-Lavallée and Blanc (2017).

For host specific information on endogenous Phycodnaviridae see: Table 3.1 Endogenous viruses of algae, Table 3.11 Endogenous viruses of fungi, and Table 3.12 Endogenous viruses of unspecified heterokonts.

3.3.33 *Pithoviridae*

The members of the Pithoviridae viral family have double stranded DNA genomes and they use cytoplasmic replication. The known normal hosts of Pithoviridae are amoeba. A suggested reference for endogenous Pithoviridae in moss would be Gallot-Lavallée and Blanc (2017).

For host specific information on endogenous Pithoviridae see: Table 3.17 Endogenous viruses of plants.

3.3.34 *Polydnaviridae*

The members of the Polydnaviridae viral family have double stranded DNA genomes an they use nuclear replication. The known normal hosts of Polydnaviridae are insects. Suggested references for endogenous Polydnaviridae in insects would be Bredlau et al. (2019), Desjardins et al. (2008), Heringer et al. (2017), Herniou et al. (2013), Legeai et al. (2020), Louis et al. (2013), Tan et al. (2018), Volkoff and Cusson (2020) plus Zhu et al. (2018), and specifically for endogenous Polydnaviridae in ants would be Flynn and Moreau (2019).

For host specific information on endogenous Polydnaviridae see: Table 3.5 Endogenous viruses of arachnids collembolids and insects.

I did not list in a table the information about endogenous Polydnaviridae of ants that was presented by Flynn and Moreau (2019) because their information did not seem to match up the viral sequences with specific host genera and species.

3.3.35 *Potyviridae*

The members of the Potyviridae viral family have positive sense single stranded RNA genomes and they use cytoplasmic replication. The known normal hosts of Potyviridae are plants. Suggested references for endogenous Potyviridae in insects would be Liu et al. (2010), and for endogenous Potyviridae in plants would be Chu et al. (2014) plus Takahashi et al. (2019).

For host specific information on endogenous Potyviridae see: Table 3.5 Endogenous viruses of arachnids collembolids and insects, and Table 3.17 Endogenous viruses of plants.

3.3.36 *Poxviridae*

The members of the Poxviridae viral family have double stranded DNA genomes and they use cytoplasmic replication. The known normal hosts of Poxviridae are avians, insects, and mammals. A suggested reference for endogenous Poxviridae in ants would be Flynn and Moreau (2019). I did not list the information about endogenous Poxviridae of ants in a table because the publication by Flynn and Moreau (2019) did not seem to match up the viral sequences with specific host genera and species.

3.3.37 Pseudoviridae

The members of the Pseudoviridae viral family have positive sense single stranded RNA genomes and their replication variously seems to be either cytoplasmic or nuclear. Pseudoviridae utilize reverse transcription in their replication strategy. The known normal hosts of Pseudoviridae are fungi. Pseudoviridae form Long terminal repeat (LTR) retrotransposons, and the LTR 'Copia' elements found in genomes are presumed to represent Pseudoviridae. Suggested references for endogenous Pseudoviridae specifically LTR Copia in fungi would be Liu et al. (2010), Wickner (1989) and Zhang et al. (2014), for endogenous Pseudoviridae specifically LTR Copia in insects would be Bryant et al. (1991) and Feng et al. (2019), for endogenous Pseudoviridae specifically LTR Copia in molluscs would be Thomas-Bulle et al. (2018), and for endogenous Pseudoviridae specifically LTR Copia in plants would be Bousios et al. (2012), SanMiguel and Vitte (2009) plus Zhang and Qi (2019).

For host specific information on endogenous Pseudoviridae see: Table 3.5 Endogenous viruses of arachnids collembolids and insects, Table 3.11 Endogenous viruses of fungi, Table 3.15 Endogenous viruses of molluscs, and Table 3.17 Endogenous viruses of plants.

3.3.38 Qinviridae

The members of the Qinviridae viral family have negative sense single stranded RNA genomes and they presumably use cytoplasmic replication. The known normal hosts of Qinviridae are crustaceans and insects. A suggested reference for endogenous Qinviridae in ants would be Flynn and Moreau (2019). I did not list the information about endogenous Qinviridae of ants in a table because the publication by Flynn and Moreau (2019) did not seem to match up the viral sequences with specific host genera and species.

3.3.39 Reoviridae

The members of the Reoviridae viral family have double stranded RNA genomes and they use cytoplasmic replication. The known normal hosts of Reoviridae are flagellated chlorophyte algae (specifically noted is the reovirus of *Micromonas pusilla*), amphibians, avians, crustaceans, fish, fungi, insects, mammals, molluscs, plants, and reptiles. Suggested references for endogenous Reoviridae in insects would be Katzourakis and Gifford (2010) and Supplemental Table 2 from ter Horst et al. (2019).

For host specific information on endogenous Reoviridae see: Table 3.5 Endogenous viruses of arachnids collembolids and insects.

3.3.40 Retroviridae

The members of the Retroviridae viral family have positive sense single stranded RNA genomes and they use nuclear replication. Retroviridae encode reverse transcriptase and integrase activities and they form Long terminal repeat (LTR) retrotransposons. The known normal hosts of Retroviridae are amphibians, avians, fish, mammals, and reptiles. Endogenous Retroviridae seem to be ubiquitous in jawed vertebrates. Suggested references for endogenous Retroviridae in amphibians would be Brown et al. (2014) and Xu et al. (2018), for endogenous Retroviridae in avians would be Bolisetty et al. (2012), Garcia-Etxebarria et al. (2014) and Xu et al. (2018), for endogenous Retroviridae in fish would be Brown et al. (2014) and Xu et al. (2018), for endogenous Retroviridae in mammals would be Brown et al. (2014), Flügel et al. (1978), Garcia-Etxebarria et al. (2014), LaMere et al. (2009) and Xu et al. (2018), for endogenous Retroviridae in reptiles would be Aiewsakun et al. (2019), Brown et al. (2014), Denner (2017) and Xu et al. (2018).

For host specific information on endogenous Retroviridae see: Table 3.3 Endogenous viruses of amphibians, Table 3.6 Endogenous viruses of avians, Table 3.10 Endogenous viruses of fish, Table 3.14 Endogenous viruses of mammals, and Table 3.19 Endogenous viruses of reptiles.

3.3.41 Rhabdoviridae

The members of the Rhabdoviridae viral family have negative sense single stranded RNA genomes and they use cytoplasmic replication. The known normal hosts of Rhabdoviridae are birds, fish, insects, mammals, and plants. Suggested references for endogenous Rhabdoviridae in arachnids would be Fort et al. (2012) plus Katzourakis and Gifford (2010), for endogenous Rhabdoviridae in crustceans would be Fort et al. (2012) and endogenous Mononegavirales in crustaceans that presumably represent Rhabdoviridae would be Metegnier et al. (2015), for endogenous Rhabdoviridae in fish would be Fort et al. (2012), for endogenous Rhabdoviridae in insects would be Fort et al. (2012) Katzourakis and Gifford (2010) and Supplemental Table 2 from ter Horst et al. (2019), for endogenous

Rhabdoviridae in nematodes would be Fort et al. (2012), and for endogenous Rhabdoviridae in plants would be Chu et al. (2014) plus Takahashi et al. (2019).

For host specific information on endogenous Rhabdoviridae see: Table 3.4 Endogenous viruses of annelids and nematodes, Table 3.5 Endogenous viruses of arachnids collembolids and insects, Table 3.8 Endogenous viruses of crustaceans, Table 3.10 Endogenous viruses of fish, and Table 3.17 Endogenous viruses of plants.

3.3.42 *Totiviridae*

The members of the Totiviridae viral family have double stranded RNA genomes and they use cytoplasmic replication. The known normal hosts of Totiviridae are algae, crustaceans, flagellated protozoa, and fungi. Suggested references for endogenous Totiviridae in algae (diatoms) would be Chu et al. (2014), for endogenous Totiviridae in arachnids would be Liu et al. (2010), for endogenous Totiviridae in crustaceans would be Metegnier et al. (2015) and Thézé et al. (2014), for endogenous Totiviridae in fungi would be Liu et al. (2010) plus Taylor and Bruenn (2009) and Taylor et al. (2013), for endogenous Totiviridae in insects would be Liu et al. (2010) and Supplemental Table 2 from ter Horst et al. (2019), for endogenous Totiviridae in nematodes would be Liu et al. (2010), and for endogenous Totiviridae in plants would be Chu et al. (2014) plus Takahashi et al. (2019).

For host specific information on endogenous Totiviridae see: Table 3.1 Endogenous viruses of algae, Table 3.4 Endogenous viruses of annelids and nematodes, Table 3.5 Endogenous viruses of arachnids collembolids and insects, Table 3.8 Endogenous viruses of crustaceans, Table 3.11 Endogenous viruses of fungi, and Table 3.17 Endogenous viruses of plants.

3.3.43 *Virgaviridae (Includes Former Tobamoviridae)*

The members of the Virgaviridae viral family have positive sense single stranded RNA genomes and they use cytoplasmic replication. The known normal hosts of Virgaviridae are plants. A suggested reference for endogenous Virgaviridae in ants would be Flynn and Moreau (2019). I did not list the information about endogenous Virgaviridae of ants in a table because the publication by Flynn and Moreau (2019) did not seem to match up the viral sequences with specific host genera and species.

3.4 Groups of Viruses for Which Endogenous Sequences Were Not Identified at the Level of Viral Family

3.4.1 Unspecified Bunya-Arena

The unspecified Bunya-Arena sequences are presumed to represent either the viral family Peribunyaviridae or the viral family Arenaviridae. Members of the viral family Peribunyaviridae have negative sense single-stranded RNA genomes. Members of the viral family Arenaviridae have ambisense single-stranded RNA genomes. Both Peribunyaviridae and Arenaviridae use cytoplasmic replication. The known normal hosts of Peribunyaviridae are avians, insects and mammals. The normal hosts of Arenaviridae are mammals. A suggested reference for endogenous Bunya-Arena sequences in ants would be Flynn and Moreau (2019). I did not list the information about endogenous Bunya-Arena of ants in a table because the publication by Flynn and Moreau (2019) did not seem to match up the viral sequences with specific host genera and species.

3.4.2 Unspecified Hepe-Virga

The unspecified Hepe-Virga sequences are presumed to represent either the viral family Hepeviridae or the viral family Virgaviridae. Members of the viral families Hepeviridae and Virgaviridae have positive sense single stranded RNA genomes. Hepeviridae replicate in association with the host cell endoplasmic reticulum. Virgaviridae use cytoplasmic replication. The known normal hosts of Hepeviridae are avians, fish, and mammals. The known normal hosts of Virgaviridae are plants. A suggested reference for endogenous Hepe-Virga sequences in ants would be Flynn and Moreau (2019). I did not list the information about endogenous Hepe-Virga of ants in a table because the publication by Flynn and Moreau (2019) did not seem to match up the viral sequences with specific host genera and species.

3.4.3 Unspecified Iridoviridae/Marseilleviridae Group

The unspecified Iridoviridae/Marseilleviridae sequences are presumed to represent either the viral family Iridoviridae or the viral family Marseilleviridae. Members of the viral families Iridoviridae and Marseilleviridae have double stranded DNA genomes. Iridoviridae use nucleo-cytoplasmic replication. Marseilleviridae replication has yet to be completely understood. The known normal hosts of Iridoviridae are amphibians, crustaceans, insects, and fish. The natural hosts of Marseilleviridae are amoeba. A suggested references for endogenous Iridoviridae/Marseilleviridae in Ichthyosporea would be Gallot-Lavallée and Blanc (2017).

For host specific information on endogenous Iridoviridae/Marseilleviridae see: Table 3.13 Endogenous viruses of ichthyosporeans.

3.4.4 *Unspecified Mononegavirales*

The Mononegavirales group includes numerous viral families that have negative sense single stranded RNA genomes and use cytoplasmic replication. Collectively, the Mononegavirales have an enormously wide normal host range. Some of the sequences that I found attributed as unspecified Mononegavirales possibly represent Rhabdoviridae and others of those sequences presumably represent Nyamiviridae. I have described the viral families Nyamiviridae and Rhabdoviridae earlier in this chapter. A suggested reference for endogenous Mononegavirales sequences in crustceans that possibly represent Rhabdoviridae would be Metegnier et al. (2015), and for endogenous Mononegavirales sequences in crustaceans that presumably represent Nyamiviridae would be Thézé et al. (2014).

For host specific information on endogenous Unspecified Mononegavirales see: Table 3.8 Endogenous viruses of crustaceans.

3.4.5 *Unspecified Mononegavirales-Like Virus*

The unspecified Mononegavirales-like virus could represent any one of several different viral families all of which have negative sense single stranded RNA genomes. A suggested reference for endogenous Mononegavirales-like sequences in an arachnid would be Supplemental Table 2 from ter Horst et al. (2019).

For host specific information on endogenous unspecified Mononegavirales-like virus see: Table 3.5 Endogenous viruses of arachnids collembolids and insects.

3.4.6 *Unspecified Narna-Levi*

The unspecified Narna-Levi sequences are presumed to represent either the viral family Narnaviridae or the viral family Leviviridae. Members of the viral family Narnaviridae have positive sense single stranded RNA genomes and use cytoplasmic replication. The known normal hosts of Narnaviridae are fungi. A suggested reference for endogenous unspecified Narna-Levi squences in ants would be Flynn and Moreau (2019). I did not list the information about endogenous Narna-Levi of ants in a table because the publication by Flynn and Moreau (2019) did not seem to match up the viral sequences with specific host genera and species. The Leviviridae also have positive sense single stranded RNA genomes. Leviviridae likely would not infect insects because their known normal host range entirely is prokaryotes.

However, horizontal gene transfer mechanisms certainly might be capable of transferring information from prokaryotes into eukaryotes.

3.4.7 *Unspecified Nucleocytoviricota*

The unspecified members of the viral phylum Nucleocytoviricota have double stranded DNA genomes and they use nucleocytoplasmic replication. The known normal host ranges mentioned for these unspecified Nucleocytoviricota are algae, amoeba, crustaceans, and fungi, although the host range of Nucleocytoviricota certainly can extend well beyond that. A fair number of eukaryotes contain core protein homologs that suggest presence of endogenous sequences of the phylum Nucleocytoviricota, and often those sequences are found in the absence of evidence that their hosts naturally are infected by this group of viruses. Many of these sequences seem to represent either the viral families Asfarviridae, Mimiviridae, or Phycodnaviridae. I have described earlier in this chapter those three viral families under their separate names Asfarviridae, Mimiviridae, and Phycodnaviridae. Iridoviridae and Mimiviridae, which also are viral families belonging to the Nucleocytoviricota, have been described earlier in this chapter as an Unspecified Iridoviridae/Marseilleviridae group. A suggested reference for Unspecified Nucleocytoviricota in several host groups is Gallot-Lavallée and Blanc (2017) and there was a phylogenetic correlation with closely related eukaryotes having a tendency to possess closely related endogenous viral sequences, suggesting coevolution of the endogenous viruses and their hosts.

For host specific information on endogenous unspecified Nucleocytoviricota see: Table 3.1 Endogenous viruses of algae, Table 3.2 Endogenous viruses of amoeba, Table 3.8 Endogenous viruses of crustaceans, and Table 3.12 Endogenous viruses of unspecified heterokonts.

3.4.8 *Unclassified Riboviria*

The unclassified members of the viral clade Riboviria have RNA genomes that may be either positive or negative sense single stranded RNA, or double stranded RNA. A suggested reference for unclassified Riboviria in insects would be Supplemental Table 2 from ter Horst et al. (2019).

For host specific information on endogenous Unclassified Riboviria see: Table 3.5 Endogenous viruses of arachnids collembolids and insects.

3.4.9 *Unspecified Partiti-Picobirna*

The unspecified Partiti-Picobirna sequences are presumed to represent either the viral family Partitiviridae or the viral family Picobirnaviridae. The viral family Partitiviridae has been described earlier in this chapter. Members of the viral family Picobirnaviridae have double stranded RNA genomes and they utilize cytoplasmic replication. The known normal hosts of Picobirnaviridae are avians and mammals. A suggested reference for endogenous unspecified Partiti-Picobirna sequences in ants would be Flynn and Moreau (2019). I did not list the information about endogenous unspecified Partiti-Picobirna of ants in a table because the publication by Flynn and Moreau (2019) did not seem to match up the viral sequences with specific host genera and species.

3.4.10 *Unspecified Toti-Chryso*

The unspecified Toti-Chryso sequences are presumed to represent either the viral family Chrysoviridae or the viral family Totiviridae. The viral families Chrysoviridae and Totiviridae have been described earlier in this chapter. A suggested reference for endogenous unspecified Toti-Chryso sequences in ants would be Flynn and Moreau (2019). I did not list the information about endogenous unspecified Toti-Chryso of ants in a table because the publication by Flynn and Moreau (2019) did not seem to match up the viral sequences with specific host genera and species.

3.4.11 *Unspecified Virga-Like*

The unspecified Virga-like sequences are presumed to represent the viral family Virgaviridae as has been described earlier in this chapter. A suggested reference for endogenous unspecified Virga-like sequences in an insect would be Supplemental Table 2 from ter Horst et al. (2019).

For host specific information on endogenous Unspecified Virga-like virus see: Table 3.5 Endogenous viruses of arachnids collembolids and insects.

3.5 Groups of Viruses for Which Lysogenous Sequences Have Been Identified at the Level of Viral Family

Lysogeny is conceptually similar to endogeny. However, the term lysogeny describes not only those viruses whose genomes are present as inherited genetic sequences, but also those viruses which have recently developed that genetic presence without the host cell yet having divided. Endogeny, by definition, signifies that the viral genetic sequences were vertically inherited when a host replicated after it had incorporated the viral genetic material. The Siphoviridae include Escherichia virus Lambda (phage λ), which must remain transcriptionaly active to maintain its lysogenic state. That requirement for Escherichia virus Lambda and related phages, which are termed to be lambdoid, to continue transcriptional activity in order to sustain their lysogenic status marks a huge difference in molecular biology because endogenous viruses of eukaryotes are presumed capable of maintaining their endogenous status without requirement that the viral genetic material be transcriptionally active.

3.5.1 Inoviridae

Members of the Inoviridae viral family have single stranded DNA genomes. The known normal hosts of Inoviridae are archaea and bacteria. Some selected references regarding endogenous Inoviridae would be Davis et al. (1999), Pant et al. (2020), Faruque and Mekalanos (2012), Hurst (2019), and Roux et al. (2019).

3.5.2 Microviridae

Members of the Microviridae viral family have single stranded DNA genomes. The known normal hosts of Microviridae are bacteria. Some selected references regarding endogenous Microviridae would be Kirchberger and Ochman (2020) plus Krupovic and Forterre (2011).

3.5.3 Myoviridae

Members of the Myoviridae viral family have double stranded DNA genomes. The known normal hosts of Myoviridae are archaea and bacteria. Some selected references regarding endogenous Myoviridae would be Bordenstein and Bordenstein (2016), Harshey (2014), and Wang et al. (2016).

3.5.4 *Podoviridae*

Members of the Podoviridae viral family have double stranded DNA genomes. The known normal hosts of Podoviridae are archaea and bacteria. A suggested reference regarding endogenous Podoviridae would be Campbell (1994).

3.5.5 *Siphoviridae*

Members of the Siphoviridae viral family have double stranded DNA genomes. The known normal hosts of Siphoviridae are archaea and bacteria. Some selected references regarding endogenous Siphoviridae would be Benzer (1955), Burmeister et al. (2016), Campbell (1994), Hammerl et al. (2007), Ravin (2015), Schubert et al. 2007), Scott et al. (2008), Stokar-Avihail et al. (2019), Zajdowicz and Holmes (2016), and Ziegelin et al. (2005).

3.6 Summary Thoughts

Endogenous viruses are a fascinating subject of curiosity and they have been found in all eukaryotes. How that endogeny has arisen is obvious for the Metaviridae, Pseudoviridae and Retroviridae, because the infective cycle of those viral families involves generating DNA copies of their genomes and then integrating those DNA copies into the host cell chromosomal material. The mechanisms by which other viruses have evolved an endogenous presence may be related to natural molecular mechanisms of their hosts.

Maintaining the endogenous presence of these viral genomes does not seem to require their active transcription. Some of these endogenous viruses have, however, remained transcriptionally active and do generate viral proteins. Many others of the endogenous viruses contain transcription initiation sites that can be activated to result in the transcription of adjacent non-viral DNA sequences. A few of these endogenous viruses are proviruses, meaning that they can generate infectious viral particles. The lysogenous viruses of prokaryotes often seem to actively seek a non-infectious presence within a host organism and some of them must remain transcriptionally active in order to sustain their lysogenic existance. There is a companion chapter in this same book: Chap. 4, Pages 113–154, "Do the biological roles of endogenous and lysogenous viruses represent Faustian bargains?" by Christon J. Hurst.

References

Aiewsakun P, Simmonds P, Katzourakis A (2019) The first co-opted endogenous foamy viruses and the evolutionary history of reptilian foamy viruses. Viruses 11(7):641. https://doi.org/10.3390/v11070641

Belyi VA, Levine AJ, Skalka AM (2010) Unexpected inheritance: multiple integrations of ancient bornavirus and ebolavirus/marburgvirus sequences in vertebrate genomes. PLoS Pathog 6(7): e1001030. https://doi.org/10.1371/journal.ppat.1001030

Benzer S (1955) Fine structure of a genetic region in bacteriophage. Proc Natl Acad Sci U S A 41:344–354. https://doi.org/10.1073/pnas.41.6.344

Blanc G, Gallot-Lavallée L, Maumus F (2015) Provirophages in the *Bigelowiella* genome bear testimony to past encounters with giant viruses. Proc Natl Acad Sci U S A 112(38):E5318–E5326. https://doi.org/10.1073/pnas.1506469112

Bolisetty M, Blomberg J, Benachenhou F et al (2012) Unexpected diversity and expression of avian endogenous retroviruses. mBio 3(5):e00344–e00312. https://doi.org/10.1128/mBio.00344-12

Bordenstein S, Bordenstein S (2016) Eukaryotic association module in phage WO genomes from *Wolbachia*. Nat Commun 7:13155. https://doi.org/10.1038/ncomms13155

Bousios A, Kourmpetis YA, Pavlidis P et al (2012) The turbulent life of Sirevirus retrotransposons and the evolution of the maize genome: more than ten thousand elements tell the story. Plant J 69(3):475–488. https://doi.org/10.1111/j.1365-313X.2011.04806.x

Bredlau JP, Kuhar D, Gundersen-Rindal DE et al (2019) The parasitic wasp, *Cotesia congregata* (Say), consists of two incipient species isolated by asymmetric reproductive incompatibility and hybrid inability to overcome host defenses. Front Ecol Evol 7:187. https://doi.org/10.3389/fevo.2019.00187

Breitenbach JE, Shelby KS, Popham HJ (2011) Baculovirus induced transcripts in hemocytes from the larvae of *Heliothis virescens*. Viruses 3(11):2047–2064. https://doi.org/10.3390/v3112047

Britten RJ (1995) Active *gypsy*/Ty*3* retrotransposons or retroviruses in *Caenorhabditis elegans*. Proc Natl Acad Sci U S A 92:599–601. https://doi.org/10.1073/pnas.92.2.599

Broecker F, Moelling K (2019) Evolution of immune systems from viruses and transposable elements. Front Microbiol 10:51. https://doi.org/10.3389/fmicb.2019.00051

Brown K, Emes RD, Tarlinton RE (2014) Multiple groups of endogenous epsilon-like retroviruses conserved across primates. J Virol 88:12464–12471. https://doi.org/10.1128/JVI.00966-14

Bryant LA, Brierley C, Flavell AJ et al (1991) The retrotransposon *copia* regulates *Drosophila* gene expression both positively and negatively. Nucleic Acids Res 19(20):5533–5536. https://doi.org/10.1093/nar/19.20.5533

Burke GR, Strand MR (2012) Deep sequencing identifies viral and wasp genes with potential roles in replication of Microplitis demolitor Bracovirus. J Virol 86:3293–3306. https://doi.org/10.1128/JVI.06434-11

Burmeister AR, Lenski RE, Meyer JR (2016) Host coevolution alters the adaptive landscape of a virus. Proc R Soc B 283:20161528. https://doi.org/10.1098/rspb.2016.1528

Campbell A (1994) Comparative molecular biology of lambdoid phages. Annu Rev Microbiol 48:193–222. https://doi.org/10.1146/annurev.mi.48.100194.001205

Chen S, Saito N, Encabo JR et al (2018) Ancient endogenous pararetroviruses in *Oryza* genomes provide insights into the heterogeneity of viral gene macroevolution. Genome Biol Evol 10:2686–2696. https://doi.org/10.1093/gbe/evy207

Chong AY, Kojima KK, Jurka J et al (2014) Evolution and gene capture in ancient endogenous retroviruses - insights from the crocodilian genomes. Retrovirology 11:71. https://doi.org/10.1186/s12977-014-0071-2

Chu H, Jo Y, Cho WK (2014) Evolution of endogenous non-retroviral genes integrated into plant genomes. Curr Plant Biol 1:55–59

Clavijo G, van Munster M, Monsion B et al (2016) Transcription of densovirus endogenous sequences in the *Myzus persicae* genome. J Gen Virol 97:1000–1009. https://doi.org/10.1099/jgv.0.000396

Cotton JA, Steinbiss S, Yokoi T et al (2016) An expressed, endogenous Nodavirus-like element captured by a retrotransposon in the genome of the plant parasitic nematode *Bursaphelenchus xylophilus*. Sci Rep 6:39749. https://doi.org/10.1038/srep39749

Davis BM, Kimsey HH, Chang W et al (1999) The *Vibrio cholerae* O139 Calcutta bacteriophage CTXphi is infectious and encodes a novel repressor. J Bacteriol 181:6779–6787. https://doi.org/10.1128/JB.181.21.6779-6787.1999

Denner J (2017) Function of a retroviral envelope protein in the placenta of a viviparous lizard. Proc Natl Acad Sci U S A 114(51):13315–13317. https://doi.org/10.1073/pnas.1719189114

Dennis TPW, de Souza WM, Marsile-Medun S et al (2019) The evolution, distribution and diversity of endogenous circoviral elements in vertebrate genomes. *Virus Res* 262:15–23. https://doi.org/10.1016/j.virusres.2018.03.014

Desjardins CA, Gundersen-Rindal DE, Hostetler JB et al (2008) Comparative genomics of mutualistic viruses of *Glyptapanteles* parasitic wasps. Genome Biol 9:R183. https://doi.org/10.1186/gb-2008-9-12-r183

Dezordi FZ, Vasconcelos CRDS, Rezende AM et al (2020) In and outs of chuviridae endogenous viral elements: origin of a potentially new retrovirus and signature of ancient and ongoing arms race in mosquito genomes. Front Genet 11:542437. https://doi.org/10.3389/fgene.2020.542437

Duponchel S, Fischer MG (2019) Viva lavidaviruses! Five features of virophages that parasitize giant DNA viruses. PLoS Pathog 15(3):e1007592. https://doi.org/10.1371/journal.ppat.1007592

Edwards MR, Liu H, Shabman RS et al (2018) Conservation of structure and immune antagonist functions of filoviral VP35 homologs present in microbat genomes. Cell Rep 24(4):861–872.e6. https://doi.org/10.1016/j.celrep.2018.06.045

Faruque SM, Mekalanos JJ (2012) Phage-bacterial interactions in the evolution of toxigenic *Vibrio cholerae*. Virulence 3(7):556–565. https://doi.org/10.4161/viru.22351

Feng M, Ren F, Zhou Y, Zhang N, Lu Q, Swevers L, Sun J (2019) Correlation in expression between LTR retrotransposons and potential host *Cis*-targets during infection of antherea pernyi with ApNPV baculovirus. Viruses 11(5):421. https://doi.org/10.3390/v11050421

Filée J (2014) Multiple occurrences of giant virus core genes acquired by eukaryotic genomes: the visible part of the iceberg? Virology 466–467:53–59. https://doi.org/10.1016/j.virol.2014.06.004

Fischer MG, Hackl T (2016) Host genome integration and giant virus-induced reactivation of the virophage mavirus. Nature 540(7632):288–291. https://doi.org/10.1038/nature20593

Flügel RM, Zentgraf H, Munk K, Darai G (1978) Activation of an endogenous retrovirus from *Tupaia* (tree shrew). Nature 271(5645):543–545. https://doi.org/10.1038/271543a0

Flynn PJ, Moreau CS (2019) Assessing the diversity of endogenous viruses throughout ant genomes. Front Microbiol 10:1139. https://doi.org/10.3389/fmicb.2019.01139

Fort P, Albertini A, Van-Hua A et al (2012) Fossil rhabdoviral sequences integrated into arthropod genomes: ontogeny, evolution, and potential functionality. Mol Biol Evol 29(1):381–390. https://doi.org/10.1093/molbev/msr226

François S, Filloux D, Roumagnac P et al (2016) Discovery of parvovirus-related sequences in an unexpected broad range of animals. Sci Rep 6:30880. https://doi.org/10.1038/srep30880

Gallot-Lavallée L, Blanc G (2017) A glimpse of nucleo-cytoplasmic large DNA virus biodiversity through the eukaryotic genomics window. Viruses 9(1):17. https://doi.org/10.3390/v9010017

Garcia-Etxebarria K, Sistiaga-Poveda M, Jugo BM (2014) Endogenous retroviruses in domestic animals. Curr Genom 15(4):256–265. https://doi.org/10.2174/1389202915666140520003503

Gilbert C, Meik JM, Dashevsky D et al (2014) Endogenous hepadnaviruses, bornaviruses and circoviruses in snakes. Proc R Soc B 281:20141122. https://doi.org/10.1098/rspb.2014.1122

Griffiths DJ (2001) Endogenous retroviruses in the human genome sequence. Genome Biol 2(6): Reviews 1017. https://doi.org/10.1186/gb-2001-2-6-reviews1017

Hackl T, Duponchel S, Barenhoff K et al (2020) Endogenous virophages populate the genomes of a marine heterotrophic flagellate. bioRxiv. https://doi.org/10.1101/2020.11

Hammerl JA, Klein I, Appel B (2007) Interplay between the temperate phages PY54 and N15, linear plasmid prophages with covalently closed ends. J Bacteriol 189:8366–8370. https://doi.org/10.1128/JB.01066-07

Harshey RM (2014) Transposable phage Mu. Microbiol Spectr 2(5). https://doi.org/10.1128/microbiolspec.MDNA3-0007-2014

Heringer P, Dias GB, Kuhn GCS (2017) A horizontally transferred autonomous helitron became a full polydnavirus segment in *Cotesia vestalis*. G3 (Bethesda) 7:3925–3935. https://doi.org/10.1534/g3.117.300280

Herniou EA, Huguet E, Thézé J et al (2013) When parasitic wasps hijacked viruses: genomic and functional evolution of polydnaviruses. Philos Trans R Soc Lond Ser B Biol Sci 368 (1626):20130051. https://doi.org/10.1098/rstb.2013.0051

Hillman BI, Milgroom MG (2021) The ecology and evolution of fungal viruses. In: Hurst CJ (ed) Studies in viral ecology (2nd edn). Wiley, Chichester, pp 139–182

Honda T (2017) Potential links between hepadnavirus and bornavirus sequences in the host genome and cancer. Front Microbiol 8:2537. https://doi.org/10.3389/fmicb.2017.02537

Horie M, Kobayashi Y, Honda T et al (2016) An RNA-dependent RNA polymerase gene in bat genomes derived from an ancient negative-strand RNA virus. Sci Rep 6:25873. https://doi.org/10.1038/srep25873

Hurst CJ (2019) Briefly summarizing our understanding of *Vibrio cholerae* and the disease cholera. In: Hurst CJ (ed) The structure and function of aquatic microbial communities. Advances in environmental microbiology, vol 7. Springer, Cham. https://doi.org/10.1007/978-3-030-16775-2_7

Hurst CJ (2021a) Defining the ecology of viruses. In: Hurst CJ (ed) Studies in viral ecology, 2nd edn. Wiley, Chichester, pp 3–48

Hurst CJ (2021b) Do the biological roles of endogenous and lysogenous viruses represent Faustian bargains? In: Hurst CJ (ed) The biological role of a virus. Advances in environmental microbiology, vol 9. Springer, Cham

Jackson EW, Wilhelm RC, Johnson MR et al (2021) Diversity of sea star-associated densoviruses and transcribed endogenous viral elements of densovirus origin. J Virol 95:e01594–e01520. https://doi.org/10.1128/JVI.01

Kapulkin WJ (2016) Retroviral origins of the *Caenorhabditis elegans* orphan gene F58H7.5. bioRxiv. https://doi.org/10.1101/073510

Katzourakis A, Gifford RJ (2010) Endogenous viral elements in animal genomes. PLoS Genet 6 (11):e1001191

Kirchberger PC, Ochman H (2020) Resurrection of a global, metagenomically defined gokushovirus. elife 9:e51599. https://doi.org/10.7554/eLife.51599

Kobayashi Y, Horie M, Nakano A et al (2016) Exaptation of bornavirus-like nucleoprotein elements in afrotherians. PLoS Pathog 12(8):e1005785. https://doi.org/10.1371/journal.ppat.1005785

Krupovic M, Forterre P (2011) Microviridae goes temperate: microvirus-related proviruses reside in the genomes of Bacteroidetes. PLoS One 6(5):e19893. https://doi.org/10.1371/journal.pone.0019893

Kuriyama K, Tabara M, Moriyama H et al (2020) Disturbance of floral colour pattern by activation of an endogenous pararetrovirus, petunia vein clearing virus, in aged petunia plants. Plant J 103:497–511. https://doi.org/10.1111/tpj.14728

LaMere SA, St Leger JA, Schrenzel MD et al (2009) Molecular characterization of a novel gammaretrovirus in killer whales (*Orcinus orca*). J Virol 83:12956–12967. https://doi.org/10.1128/JVI.01354-09

Legeai F, Santos BF, Robin S et al (2020) Genomic architecture of endogenous ichnoviruses reveals distinct evolutionary pathways leading to virus domestication in parasitic wasps. BMC Biol 18:89. https://doi.org/10.1186/s12915-020-00822-3

Leobold M, Bézier A, Pichon A et al (2018) The domestication of a large DNA Virus by the Wasp *Venturia canescens* involves targeted genome reduction through pseudogenization. Genome Biol Evol 10:1745–1764. https://doi.org/10.1093/gbe/evy127

Liu H, Fu Y, Jiang D et al (2010) Widespread horizontal gene transfer from double-stranded RNA viruses to eukaryotic nuclear genomes. J Virol 84:11879–11887. https://doi.org/10.1128/JVI.00955-10

Louis F, Bézier A, Periquet G et al (2013) The bracovirus genome of the parasitoid wasp *Cotesia congregata* is amplified within 13 replication units, including sequences not packaged in the particles. J Virol 87:9649–9660. https://doi.org/10.1128/JVI.00886-13
Metegnier G, Becking T, Chebbi MA et al (2015) Comparative paleovirological analysis of crustaceans identifies multiple widespread viral groups. Mob DNA 6:16. https://doi.org/10.1186/s13100-015-0047-3
Metzger MJ, Paynter AN, Siddall ME et al (2018) Horizontal transfer of retrotransposons between bivalves and other aquatic species of multiple phyla. Proc Natl Acad Sci U S A 115(18):E4227–E4235. https://doi.org/10.1073/pnas.1717227115
Moniruzzaman M, Weinheimer AR, Martinez-Gutierrez CA et al (2020) Widespread endogenization of giant viruses shapes genomes of green algae. Nature. https://doi.org/10.1038/s41586-020-2924-2
Pant A, Das B, Bhadra RK (2020) CTX phage of *Vibrio cholerae*: genomics and applications. Vaccine 38(Suppl 1):A7–A12. https://doi.org/10.1016/j.vaccine.2019.06.034
Parker BJ, Brisson JA (2019) A laterally transferred viral gene modifies aphid wing plasticity. Curr Biol 29:2098–2103. https://doi.org/10.1016/j.cub.2019.05.041
Pénzes JJ, Marsile-Medun S, Agbandje-McKenna M et al (2018) Endogenous amdoparvovirus-related elements reveal insights into the biology and evolution of vertebrate parvoviruses. *Virus Evol* 4(2):vey026. https://doi.org/10.1093/ve/vey026
Ravin NV (2015) Replication and maintenance of linear phage-plasmid N15. Microbiol Spectrum 3(1):PLAS-0032-2014. https://doi.org/10.1128/microbiolspec.PLAS-0032-2014
Rohrmann GF (2019) Baculovirus molecular biology, 4th edn. National Center for Biotechnology Information, Bethesda
Roossinck MJ, Bazán ER (2017) Symbiosis: viruses as intimate partners. Annu Rev Virol 4:123–139. https://doi.org/10.1146/annurev-virology-110615-042323
Roux S, Krupovic M, Daly RA et al (2019) Cryptic inoviruses revealed as pervasive in bacteria and archaea across Earth's biomes. Nat Microbiol 4:1895–1906. https://doi.org/10.1038/s41564-019-0510-x
SanMiguel P, Vitte C (2009) The LTR-retrotransposons of maize. In: Bennetzen JL, Hake S (eds) Handbook of maize. Springer, New York, NY. https://doi.org/10.1007/978-0-387-77863-1_15
Schubert RA, Dodd IB, Egan JB et al (2007) Cro's role in the CI Cro bistable switch is critical for λ's transition from lysogeny to lytic development. Genes Dev 21(19):2461–2472. https://doi.org/10.1101/gad.1584907
Scott J, Thompson-Mayberry P, Lahmamsi S et al (2008) Phage-associated mutator phenotype in group A streptococcus. J Bacteriol 190:6290–6301. https://doi.org/10.1128/JB.01569-07
Song D, Cho WK, Park S-H et al (2013) Evolution of and horizontal gene transfer in the endornavirus genus. PLoS One 8(5):e64270. https://doi.org/10.1371/journal.pone.0064270
Staginnus C, Gregor W, Mette MF et al (2007) Endogenous pararetroviral sequences in tomato (*Solanum lycopersicum*) and related species. BMC Plant Biol 7:24. https://doi.org/10.1186/1471-2229-7-24
Stokar-Avihail A, Tal N, Erez Z et al (2019) Widespread utilization of peptide communication in phages infecting soil and pathogenic bacteria. Cell Host Microbe 25(5):746–755.e5. https://doi.org/10.1016/j.chom.2019.03.017
Suzuki Y, Frangeul L, Dickson LB et al (2017) Uncovering the repertoire of endogenous flaviviral elements in *Aedes* mosquito genomes. J Virol 91(15):e00571–e00517. https://doi.org/10.1128/JVI.00571-17
Takahashi H, Fukuhara T, Kitazawa H et al (2019) Virus latency and the impact on plants. Front Microbiol 10:2764. https://doi.org/10.3389/fmicb.2019.02764
Tan C-W, Peiffer M, Hoover K et al (2018) Symbiotic polydnavirus of a parasite manipulates caterpillar and plant immunity. Proc Natl Acad Sci U S A 115:5199–5204. https://doi.org/10.1073/pnas.1717934115
Taylor DJ, Bruenn J (2009) The evolution of novel fungal genes from non-retroviral RNA viruses. BMC Biol 7:88. https://doi.org/10.1186/1741-7007-7-88
Taylor DJ, Ballinger MJ, Bowman SM et al (2013) Virus-host co-evolution under a modified nuclear genetic code. PeerJ 1:e50. https://doi.org/10.7717/peerj.50

ter Horst AM, Nigg JC, Dekker FM et al (2019) Endogenous viral elements are widespread in arthropod genomes and commonly give rise to PIWI-interacting RNAs. J Virol 93(6):e02124–e02118. https://doi.org/10.1128/JVI.02124-18

Thézé J, Leclercq S, Moumen B et al (2014) Remarkable diversity of endogenous viruses in a crustacean genome. Genome Biol Evol 6:2129–2140. https://doi.org/10.1093/gbe/evu163

Thomas-Bulle C, Piednoël M, Donnart T et al (2018) Mollusc genomes reveal variability in patterns of LTR-retrotransposons dynamics. BMC Genomics 19:821. https://doi.org/10.1186/s12864-018-5200-1

Touret F, Guiguen F, Terzian C (2014). *Wolbachia* influences the maternal transmission of the *gypsy* endogenous retrovirus in *Drosophila melanogaster*. mBio 5(5):e01529–14. doi:https://doi.org/10.1128/mBio.01529-14

Tripathi JN, Ntui VO, Ron M et al (2019) CRISPR/Cas9 editing of endogenous *banana streak virus* in the B genome of *Musa* spp. overcomes a major challenge in banana breeding. Commun Biol 2:46. https://doi.org/10.1038/s42003-019-0288-7

Umber M, Filloux D, Muller E et al (2014) The genome of African yam (*Dioscorea cayenensis-rotundata* complex) hosts endogenous sequences from four distinct Badnavirus species. Mol Plant Pathol 15(8):790–801. https://doi.org/10.1111/mpp.12137

Vargiu L, Rodriguez-Tomé P, Sperber GO et al (2016) Classification and characterization of human endogenous retroviruses; mosaic forms are common. *Retrovirology* 13:7. https://doi.org/10.1186/s12977-015-0232-y

Volkoff AN, Cusson M (2020) The unconventional viruses of ichneumonid parasitoid wasps. Viruses 12(10):1170. https://doi.org/10.3390/v12101170

Wang GH, Sun BF, Xiong TL et al (2016) Bacteriophage WO can mediate horizontal gene transfer in endosymbiotic *Wolbachia* genomes. Front Microbiol 7:1867. https://doi.org/10.3389/fmicb.2016.01867

Wang Y, Liang W, Tang T (2018) Constant conflict between *Gypsy* LTR retrotransposons and CHH methylation within a stress-adapted mangrove genome. New Phytol 220(3):922–935. https://doi.org/10.1111/nph.15209

Wickner RB (1989) Yeast virology. FASEB J 3:2257–2265. https://doi.org/10.1096/fasebj.3.11.2550303

Xu X, Zhao H, Gong Z et al (2018) Endogenous retroviruses of non-avian/mammalian vertebrates illuminate diversity and deep history of retroviruses. PLoS Pathog 14(6):e1007072. https://doi.org/10.1371/journal.ppat.1007072

Zajdowicz SLW, Holmes RK (2016) Phage conversion and the role of bacteriophage and host functions in regulation of diphtheria toxin production by *Corynebacterium diphtheriae*. In: Hurst CJ (ed) The mechanistic benefits of microbial symbionts. Advances in environmental microbiology, vol 2. Springer, Cham. https://doi.org/10.1007/978-3-319-28068-4_2

Zhang X, Qi Y (2019) The landscape of copia and gypsy retrotransposon during maize domestication and improvement. Front Plant Sci 10:1533. https://doi.org/10.3389/fpls.2019.01533

Zhang L, Yan L, Jiang J et al (2014) The structure and retrotransposition mechanism of LTR-retrotransposons in the asexual yeast *Candida albicans*. Virulence 5(6):655–664. https://doi.org/10.4161/viru.32180

Zhu F, Cusumano A, Bloem J et al (2018) Symbiotic polydnavirus and venom reveal parasitoid to its hyperparasitoids. Proc Natl Acad Sci U S A 115:5205–5210. https://doi.org/10.1073/pnas.1717904115

Ziegelin G, Tegtmeyer N, Lurz R et al (2005) The repA gene of the linear *Yersinia enterocolitica* prophage PY54 functions as a circular minimal replicon in *Escherichia coli*. J Bacteriol 187:3445–3454. https://doi.org/10.1128/JB.187.10.3445-3454.2005

Chapter 4
Do the Biological Roles of Endogenous and Lysogenous Viruses Represent Faustian Bargains?

Christon J. Hurst

Abstract Genetic transference between a host species and its viruses results in development of a collective genome which has fluidity and influences evolution. Endogeny and lysogeny represent a forging together of that shared genomic fate which obligates partnership of the virus and its host. Faustian bargains are agreements made with the devil, by which a human receives benefits in exchange for granting away their soul. This chapter explores the benefits and risks of genomic partnerships which exist between hosts and their viruses. The host tries to gain benefits from those genetic relationhips by harnessing its viral genomic partners, while simultaneously trying to avoid the perilous detriments including death which could result if one of its viruses undergoes complete reactivation. The topics covered in this chapter include the importance of endogenous viral sequences for health and disease of unicellular eukaryotes, importance for morphological development as well as health and disease of animals, a comparison of how non-endogenous and endogenous viral sequences affect health and disease of plants and fungi, plus some understanding of how partnership with an endogenous virus can reduce the phytopathogenicity of its fungal host. I also have included information comparing viral versus non-viral retrotransposons. My discussion on lysogenous viruses of prokaryotic hosts includes some of the benefits and detriments associated with lysogenic archaelphage and bacteriophage, plus a summary of how tailocins, type VI secretion systems, and gene transfer agents represent the concept of retaining and subsequently using the genomic coding for only part of a lysogenic phage.

4.1 Introduction

Long ago, someone asked me "What is it that viruses do? They must do something because otherwise evolution surely would have found a way to eliminate them". My answer would be that viruses and cellular entities likely coevolved. The ecologies of

C. J. Hurst (✉)
Cincinnati, OH, USA

Universidad del Valle, Cali, Colombia

C. J. Hurst (ed.), *The Biological Role of a Virus*, Advances in Environmental Microbiology 9, https://doi.org/10.1007/978-3-030-85395-2_4

the viruses and their hosts are so intertwined that viruses could not exist without their hosts, and hosts owe an evolutionary debt to the viruses. The interactions that occur between viruses and cells are basic life processes (Villarreal and Witzany 2019). Some of those associations are antagonistic with the viruses clearly causing disease and death. And yet, the symbiotic relationships which viruses and their hosts have established also include a range of commensal and mutualistic associations (Hurst 2021; Roossinck 2015; Roossinck and Bazán 2017). Some of these interactions have been symbiogenic, and there has even been offered a suggestion that the eukaryotic nucleus is of viral origin (Bell 2020).

Genetic information flows from viruses into their host genomes. Interestingly, many of the endogenous viral elements found in animals and plants represent viral families that do not code for an integrase enzyme, which might suggest that either a host cell integrase, or an integrase from another virus genome, has been responsible for installing bits of unrelated viral coding into the host cells genetic material. Genetic information also flows in the reverse direction with gene fragments from the host being inserted into viral genomes (Swanstrom et al. 1983). Examples of host genes being incorporated into viruses include: Dicistroviridae that contain genes from bees (Gilbert and Cordaux 2017); Flaviviridae that contain genes from mammals including cattle, giraffe and sheep (Gilbert and Cordaux 2017); Hepeviridae that contain genes from human (Gilbert and Cordaux 2017); Orthomyxoviridae that contain genes from chicken (Gilbert and Cordaux 2017); Picornaviridae that contain genes from the HeLa human cell line (McClure and Perrault 1985); Polydnaviridae that contain genes from wasps (Desjardins et al. 2008) and also contain transposable elements from hosts that are used by parasitoid wasps (Heringer et al. 2017); Potyviridae that contain genes from tobacco (Gilbert and Cordaux 2017); Retroviridae that contain genes from human (Gilbert and Cordaux 2017); and Togaviridae that contain genes from chicken (Gilbert and Cordaux 2017). Transduction can then transfer those host gene fragments from the virus into another cell, and viruses that infect more than a single host species can transport genetic material between the genomes of different species (Gilbert and Cordaux 2017; Heringer et al. 2017).

I address in this chapter the contributions that endogenous and lysogenous viruses have made to their hosts by the transference of genetic information into their hosts genomes. I also mention the genetic contributions that the hosts have made to the viruses when genetic information moves oppositely into the virus genome. The host species and its viruses thus develop a collective genome.

4.1.1 Viruses as Genomic Partners of Their Hosts

Endogeny and lysogeny are similar, but not identical concepts. Endogeny signifies that the viral genome is incorporated within the hosts genetic material and was vertically inherited from predecessors of that host. The presumption is that an endogenous virus can sustain its genomic presence without requirement for the endogenous sequences to be transcriptionally active. However, endogenized sequences can be transcriptionally active, sometimes behaving beneficially for the host and at other times seemingly detrimental to the host. Some lysogenic viruses must remain partially active by achieving transcription of their genes that code for repressor molecules, because those repressor molecules maintain the lysogenized status of that virus. Most endogenized or lysogenized viral sequences are not perceived as threats to the life of their host. The age estimates for when individual endogenous viral sequences took up residence within their host genomes often range to tens of millions of years in the past, and even well beyond one hundred million years. I have not seen published age estimates for lysogenous viral sequences. Taxonomically related host species often will carry either identical or similar endogenous or lysogenous viral sequences. I will provide additional discussion regarding endogeny and lysogeny later in this chapter.

4.1.2 Viruses as Transfer Agents for Genomic Information

One of the functions that viruses fulfill is their valuable service as agents of horizontal gene transfer (Durzyńska and Goździcka-Józefiak 2015) as will be discussed further in Sect. 4.4 of this chapter. Any non-viral DNA that is introduced to cells via transfection or viral transduction is termed to be an exogenous factor. Incorporation of exogenous DNA into the genetic information of a cell and subsequent inheritance of that DNA by progeny cells means that the newly added DNA has become endogenous. Tassetto et al. (2019) have suggested that endogenous retrotransposons might facilitate the movement of non-retrovirus RNA sequences into the hosts genetic material, with those sequences then becoming incorporated as a DNA copy in the hosts genome. Retrotransposons are discussed further in Sect. 4.5 of this chapter. Homologous recombination may provide yet a different mechanism by which genetic movement could occur without the necessity of endogenization (Gilbert and Cordaux 2017). The skillfullness with which gene transfer occurs from virus to host has been harnessed to create recombinant viruses which variously serve as vectors for gene therapy and in the development of vaccines (Lundstrom 2019).

It is indeed very interesting to notice that many eukaryotes contain endogenous sequences from Nucleocytoviricota, which includes the Mimiviridae family, and those endogenous sequences often are found even in the absence of evidence that the hosting eukaryotes natually are infected by members of this viral group (Filée 2014; Gallot-Lavallée and Blanc 2017) suggesting the possibility that these Nucleocytoviricota sequences were transferred by a horizontally vectoring mechanism such as transduction.

If the question becomes which came first, the virus or the cell, then a starting point for answering that question might be provided by the suggestion of Jalasvuori et al. (2015) that capsid proteins may have originated to facilitate gene transfer prior to the evolution of viruses. Horizontal gene transfer certainly can occur by several different processes. For bacteria, transfer of insertion sequences is more effectively done by plasmid rather than by bacteriophage because plasmids have a greater 'cargo capacity' than do bacteriophage (Leclercq et al. 2012). Also, plasmids seem more tolerant as transfer agents in comparison with bacteriophage (Leclercq et al. 2012). Transferring information by bacteriophage does provide a long term environmental storage benefit for that genetic information relative to plasmid-mediated gene transfer because, while plasmid transfer is presumed to require the recipient cell to have contact with a living donor cell, the genomic sequences contained in bacteriophage can survive and be transferred to a new host even after death of their parental host microbe (see Chap. 5, pages 155–172, "Einstein's Capsid: Bacteriophage Solve the Problems of Space and Time for Bacteria" by author Leigh Owens). In some ways, having a host cell die in order to thereby release and send away its genetic material inside of transducing bacteriophage would seem like being the executor of ones own genetic will, scattering your cells genetic information at death in hopes of passing the benefit of your knowledge to other cells. Unicellular beings face a total loss if their one infected host cell dies. The change to multicellularity valuably allowed for the loss of some cells due to viral attack without complete loss of the individual.

4.2 What Is a Faustian Bargain

A Faustian bargain is an agreement in which someone receives either wealth or knowledge from the devil in exchange for granting their soul to the devil. The concept of a Faustian bargain presumably is named for Johann Georg Faust (c. 1480–1540) from Knittlingen, in southwest Germany. Faust was considered damned for preferring human knowledge of medicine rather than the divine knowledge present in holy scriptures. Figure 4.1 is a imaginative painting of how Faust might have appeared.

I first learned about the concept of Faustian Bargains when I read "The Devil and Daniel Webster" by Stephen Vincent Benét (Benét 1936). Benét's fictional story tells about an American farmer named Jabez Stone who barters his soul to the devil in exchange for some years of prosperity. The biggest difference between most Faustian bargains and the story by Stephen Vincent Benét is that, in most stories the devil eventually claims the soul. Benét has the devil losing, with the human keeping both their soul and the advantageous benefits which they received from the bargain.

4.2.1 *The Devil and Daniel Webster*

During 1936, The Saturday Evening Post magazine published a story titled "The Devil and Daniel Webster" by author Stephen Vincent Benét. I will summarize that story for you with the following paragraph.

A perpetually unfortunate farmer named Jabez Stone, who lives in the state of New Hampshire, one day declares "I vow it's enough to make a man want to sell his

Fig. 4.1 Dr. Fausto by Jean Paul Laurens. The title of this image is "Jean Paul Laurens—Dr. Fausto.jpg". It is an oil on canvas painting by Jean Paul Laurens and is part of the Rio Grande do Sul Museum of Art collection in Porto Alegre, Brazil. The painting is considered to be in the public domain because of its age and its photograph likewise is considered to be in the public domain. https://commons.wikimedia.org/wiki/File:Jean_Paul_Laurens_-_Dr._Fausto.jpg

soul to the devil. And I would, too, for two cents!". A stranger shows up late on the next day and Jabez Stone signs a seven-year contract with the stranger. The farm of Jabez Stone then prospers and Jabez Stone gains tremendous community respect. Jabez Stone experiences a good life, seemingly without worry, until the stranger shows up at the end of six years. The stranger declares the mortgage will be due in a year. At that time, the stranger expresses a wish that he could add to his collection the soul of the noted lawyer, lawmaker and statesman Daniel Webster. Jabez Stone receives an extension of three years. Nearly four years later, Jabez Stone travels to find Daniel Webster and Daniel agrees to defend Jabez Stone for his mortgage legal case. At midnight, at the end of the agreed three year extension, the stranger arrives while Jabez Stone and Daniel Webster are seated in Stone's kitchen. Daniel Webster announces himself as being the "Attorney of record for Jabez Stone". The stranger has come to take possession of his property. When Daniel asks to know the stranger's name, the stranger says that he has been cited by many names but "Perhaps Scratch will do for the evening". Daniel Webster insists upon a trial with a jury. The stranger picks the jurors who instantly appear and enter through the door. The jurors are dead villians who "...came into the room with the fires of hell still upon them, ...". An equally villified justice appears. The trial proceeds for the rest of the night, after which the jury determines that because of Daniel Webster's eloquence, Jabez Stone has won and the jury members then disappear. The stranger signs an agreement stating his payment as settlement for the costs of the case will be that he never again bother either Jabez Stone or the descendants or the inheritors of Jabez.

4.2.2 Do Endogeny and Lysogeny Represent Faustian Bargains that Have Been Forged with Viruses?

I think that perhaps they do! Most of the endogenous viruses are permanently tied into their hosts' genomes, and the lysogenic viruses have nearly that same level of absolute committment to their host. These viruses have done much in terms of creating components of their hosts' genome and their hosts' molecular machinery. We may, as vertebrates, owe parts of our immune systems to the viruses (Broecker and Moelling 2019). Similarly, as placental mammals, we owe a debt to the endogenous retroviral envelope genes which we use at the beginning of placental development. Without that successful placental development, which allows us to be carried to term in the womb, we would be hatched from an eggshell.

Cellular beings must recognize that they cannot get rid of the viruses, and at the same time cellular beings must admit their biological debt to viruses. The cellular beings try to balance an ability to gain benefit from the viruses while keeping the harmful aspects of endogenous viruses and lysogenous viruses repressed. Perhaps one of our evolutionary goals as hosts, is that with respect to the viruses we can have the same accomplishment as did Jabez Stone, that we gain the benefits of the bargain while not losing the ultimate battle.

4.3 General Benefits Versus Detriments of Endogenous and Lysogenous Interactions

Endogeny can be perceived as a viral strategy for achieving either a non-productive, or virtually non-productive, pattern of infection. Achieving an endogenous state implies that the genome of the virus is passed through the host's reproductive process to offspring of the infected host (van der Kuyl et al. 1995; Villarreal 2016). Both endogeny and lysogeny represent a high degree of coevolution between the virus and host species. Endogenous and lysogenous infections may never be life threatening to the host. Were an endogenous or lysogenous virus to enter a replicative cycle that produces progeny virions, then it might be that neither the individual virus nor its host would survive. Mutations occurring in the endogenous or lysogenous viral genomic information can result in that viral information becoming incapable of leaving the genome of its host, and incapable of even producing an infectious virus particle, in which case that viral genomic information is said to have become 'grounded'.

Those viruses that become a part of their hosts genomic material through either endogeny or lysogeny can serve both to the benefit and detriment of their hosts. It also is important to note that coevolution with its host can alter the adaptive landscape of a virus (Burmeister et al. 2016).

When viewed from the perspective of a virus, endogeny and lysogeny certainly help the viruses because these relationships provide a means of facilitating viral survival, and the main goal of a virus always must be survival without destroying the last accessible host individual. By using either endogeny or lysogeny as a means of sharing their common genomic existance with a host, the virus cannot accidently become separated from its host. The host provides an intracellular shelter for the viral genomic information. The host also provides transportation for the virus, because the host always must take with it the viral sequences which either endogenously or lysogenously have become part of the hosts genetic material.

The main goal from the perspective of a host must be controlling the endogenous or lysogenous viral sequences. The host would like to gain whatever benefits it can from the endogenous or lysogenous viral sequences, while suppressing any lurking dangers which those sequences represent.

Perceiving benefits that the host gains from endogeny and lysogeny can begin with a most basic perception that endogeny and lysogeny serve to maintain viability of a host and allow host propagation during conditions that otherwise might limit the possibility of an infecting virus achieving transmission to a new host individual.

Benefits to be found from the perspective of the host also include the fact that achieving endogeny or lysogeny will reduce the impact which a virus infection imposes upon the energy budget of its host. Inheritance of endogenous and lysogenous infections, and successfully maintaining the status of endogeny and lysogeny in the subsequent offspring of those hosts, increases not only the short term survival of each infected host individual but also the long term survival of the host species. Some of these viruses have evolved to provide clearly defined vital functions for the host as will be discussed later in this chapter, and those functions serve as inducements for the host to retain the virus.

It has been suggested that endogenous viral sequences might be used to generate PIWI-interacting RNAs (piRNAs), which are part of the innate immune response and function as one of the mechanisms by which hosts achieve post-transcriptional RNA silencing (Tassetto et al. 2019; ter Horst et al. 2019). The piRNAs combine with argonaut proteins, after which the argonaut proteins lead the piRNAs to locations where the piRNA sequence matches with the seqences of targeted messenger RNA (mRNA) molecules. The targeted mRNAs subsequently are either cleaved or their translation silenced. This silencing of mRNA molecules might contribute to anti-viral defenses if sequences from endogenous viruses prime host cells to attack mRNA that is generated by new viral infections.

Lysogeny provides the host bacteria with resistance against attack by viruses which are related to those prophage that the host already contains (Ramisetty and Sudhakari 2019). This resistance can be effected by CRISPR-Cas, which is a form of acquired immunity system by which archaea and bacteria use endogenized viral sequences as a means of identifying and targeting for destruction related viruses that subsequently infect the cell (Broecker and Moelling 2019).

The ability of a virus to enter latency similarly results from coevolution between a virus species and its host species, but the benefit which a host derives by surviving latent infections must be established anew consequent to each generation of the host encountering and being infected by the virus. Endogeny does not require that each new generation of the host establish anew the endogenous relationships, only that the new host generation must be able to sustain and survive the genomic integrations that it has inherited. Some lysogenic relationships are inherited, but others must be established anew by successive generations of the host.

4.4 Viruses as a Means of Horizontal Gene Transfer

Gallot-Lavallée and Blanc (2017) discerned that endogenous viruses have transferred both expansin genes and cyclin genes into the streptophyte algae *Klebsormidium flaccidum*. The expansin genes mediate cell wall expansion by disrupting non-covalent bonding of cell wall polysaccharides. The cyclin genes drive cell cycle transitions of the host cell. The aquatic fungus *Gonapodya prolifera* is another example of a species that contains several genes which seemingly have been contributed to its genome by an endogenous virus, and in that case the endogenous viral source was identified as being a member of the double stranded DNA viral family Phycodnaviridae (Gallot-Lavallée and Blanc 2017). Filée (2014) has suggested that genes of the double stranded DNA viral family Mimiviridae can be laterally transferred into eukaryotic host genomes. Wang et al. (2016) have found that Wolbachia phage WO, which belongs to the double stranded DNA viral family Myoviridae, can mediate horizontal gene transfer in endosymbiotic genomes of the bacterial genus Wolbachia. Chu et al. (2014) have suggested that endogenous members of the positive sense single stranded *Endornaviridae* viral family may have facilitated the movement of genetic information into plants. Liu et al. (2010) found evidence for horizontal gene transfer having occurred from double stranded RNA viruses of the viral families Partitiviridae and Totiviridae into the nuclear genomes of several plant hosts.

These valuable examples of knowledge have prompted the use of engineered viruses in gene therapy as described both by Finer and Glorioso (2017) and Kenneth Lundstrom (2019). Also, please see Chap. 9, pages 285–341 "Application of Viruses as Delivery Vehicles for Gene Therapy and Vaccine Development" by author Kenneth Lundstrom. I will again mention the suggestion by Chu et al. (2014) and the discovery by Wang et al. (2016) later in this chapter. I also will present information regarding Gene transfer agents later in this chapter.

4.5 What Is the Functional Role of Retrotransposons?

Genomes contain a broad range of transposable elements (Platt et al. 2018) among which are retrotransposons. The retrotransposons are double stranded DNA sequences that have been created by reverse transcription from RNA, and those DNA sequences then inserted into the host cell chromosome. There are two main types of retrotransposons. The distinction between these types is whether a transposon either lacks or possesses long repeat sequences at its termini. Retrotransposons which lack these long terminal repeating stuctures are indicated as being Non-LTR retrotransposons and their origin remains uncertain. Retrotransposons which possess long terminal repeating stuctures are indicated as being LTR retrotransposons and presumably originated from viruses.

4.5.1 Non-LTR Retrotransposons

Non-LTR retrotransposons also are termed Long interspersed nuclear elements (LINE elements) and they are independent of the LTR retrotansposons. Non-LTR retrotransposons account for approximately 17 percent of the human genome. Those non-LTR retrotransposons found in the human genome are actively transposing and have been found to cause many clinical disorders (Katoh and Kurata 2013). It has been suggested that Non-LTR retrotransposons might have a role in diseases such as cancer (Kano et al. 2009). The process by which these Non-LTR retrotransposons copy themselves presently seems not to be understood, although clearly there must be a mechanism by which they have been copied and those copies inserted within the host genome.

4.5.2 The LTR Retrotransposons

The LTR retrotransposons are presumed to have originated from single stranded RNA viruses whose replicative process, following their infection of a host cell, included the use of reverse transcription to produce a DNA copy of their viral genome. The virus then would have integrated that DNA copy into the chromosomes of its host cell as an open reading frame bracketed at its ends by long terminal repeat sequences. During a normal process of infection by this type of virus, the genomically integrated DNA subsequently is transcribed to produce RNA copies that serve as progeny viral genomes. These new genomes are packaged into progeny virus particles and released to infect other cells.

The replicative process of these reverse transcribing viruses offers benefits to both the virus and its host. Endogenous integration into the chromosomes of their host cells assures that the viral genome gets copied into all of the descendents produced by that cell, and this accomplishment can be perceived as an initial advantage for the virus (Pistello and Antonelli 2016) because a virus that would instead only attack and then soon afterwards leave its host cell subsequently must either find a new host or lose viability. That integration of the virus also can, upon reflection, be perceived as representing either a temporary reprieve for the host cell, because the host cell temporarily survives, or a permanent sense of impending peril for the host cell because the virus could undergo a replicative cycle that lyses the cell. Mutations of the integrated viral genomic sequence which result in the viral genome becoming unable to leave the host cell may shift the advantage of genomic incorporation by providing safety to the harboring host cell. The grounded viral genomic sequences either may become useful to the host cell or subtly present a different and deferred type of danger to the host. What might some of the benefits be? It is possible that grounded retroviral genomes may provide the host cell with resistance against subsequent reinfection by either the same or a related virus as certainly occurs for bacteria that carry lysogenic viruses. Tassetto et al. (2019) have suggested that endogenous retrotransposons might offer a genetic benefit by facilitating the movement of non-retrovirus RNA sequences into the host genome.

If the process of transcribing an LTR retrotransposon to create single stranded RNA copies of its sequence is then cyclically followed within that same cell by reverse transcription of the single stranded RNA copy, thus producing a new DNA copy, and a subsequent chromosomal integration of the new DNA copy occurs, then the result will be host cells that contain multiple copies of the same virus genome. When this process occurs repeatedly over a time period of millions of years, there gradually can arise mutations in the genomically incorporated viral sequences which result in a genome containing variants of the same LTR retrotransposon. The result will be what is described as a 'family' of related LTR retrotransposons. The same host cell can contain many different families of LTR retrotransposons, with each family of LTR retrotransposons possibly representing the eventual result of a different initiating viral infection.

The known viral families that generate LTR retrotransposons as an obligatory part of their replication process are Metaviridae, Pseudoviridae, and Retroviridae. Metaviridae are represented by Ty3 'Gypsy' LTR retrotransposons. There can be a biological competition involving Gypsy LTR retrotransposons and host DNA methylation processes, as has been found within a mangrove host (Wang et al. 2018). The LTR retrotransposon named Steamer belongs to this Ty3 group (Metzger et al. 2018) and I therefore am considering Steamer to be a member of the Metaviridae family. The Steamer retrotransposon is highly expressed in cells of a transmissible cancer which affects the bivalve *Mya arenaria* clam (Metzger et al. 2015) and that clam species also has the common name 'Steamer'. The Steamer retrotransposons have moved horizontally to other bivalve species and also to animals of completely different phyla (Metzger et al. 2018). I additionally have included Bel/pao LTR retransposons (Thomas-Bulle et al. 2018) under the viral family name Metaviridae, which traditionally has been their viral family designation. It is possible that Bel/pao LTR retrotansposons eventually will be considered to represent a new viral family named Belpaoviridae. Pseudoviridae are represented by Ty1 'Copia'

Retrotransposons. The 'Ty' is an abbreviation for 'Transposons of yeast', and the designation numbers originate from the fact that five types of retrotransposons were identified in the yeast *Saccharomyces cerevisiae*. The LTR retrotransposons generated by Retroviridae do not seem to have been given a group name such as were the Gypsy and Copia elements.

It is important to note that, although the term retrovirus once was applied to only members of the viral family Retroviridae, now the Metaviridae and Pseudoviridae viral familes also are referred to as being retroviruses. The LTR retrotransposons that represent Retroviridae are ubiquitous in jawed vertebrates but seem not to have been identified in other groups of eukaryotes. The LTR retrotransposons that represent Metaviridae thus far have been found in fungi, nematodes, insects, and plants. There also has been one suggestion made that endogenous sequences most closely resembling the genus Errantivirus (family Metaviridae) exist in a mammal (Vargiu et al. 2016) and that would represent a unique departure from the known host range of endogenous Metaviridae. Those LTR retrotransposons that represent Pseudoviridae thus far have been found in fungi, insects, and plants. The genomes of many host species contain both 'Gypsy' and 'Copia' endogenous sequences, indicating that these two groups of LTR retrotransposons do not mutually exclude one another. Thus far, except for the mention by Vargiu et al. (2016), hosts which contain LTR retrotransposons representing Retroviridae have not been found to also contain either Ty3 'Gypsy' LTR retrotransposons or Ty1 'Copia' LTR retrotransposons.

I will be providing some additional information about Metaviridae in Sects. 4.7.3 and 4.8.2.3 of this chapter. The subject of Pseudoviridae is further examined in Sect. 4.8.2.3 of this chapter. Endogenous Retroviridae again will be the subject in Sect. 4.7.7. For information that catalogs the presence of endogenous Metaviridae, Pseudoviridae, and Retroviridae in their eukaryotic hosts, please see Chap. 3 Cataloging the Presence of Endogenous Viruses, which was written by me and appears on pages 47–112 of this volume.

4.5.3 The Role of Mammalian Apparent LTR Retrotransposons (MaLRs) in Humans

Mammals contain in their chromosomal DNA a significant category of LTR retrotransposons that have been designated Mammalian apparent (MaLRs) and these have been associated with retroviruses, presumably Retroviridae. It has been suggested that the MaLRs may have influenced a number of host genes in various modes during human evolution (Katoh and Kurata 2013). Solitary LTRs derived from HERVs (the endogenous Retroviridae sequences contained in humans often are identified by this abbreviation which represents "Human endogenous retroviruses") and MaLRs numerically dominate the Retroviridae provirus forms. Lamprecht et al. (2010) reported that derepression of the transposon-like human element 1 (THE1) subfamily of MaLR LTRs is widespread in the genome of Hodgkin's lymphoma cells and is associated with impaired epigenetic control due to loss of expression of the corepressor protein CBFA2T3. Protein CBFA2T3 (core-binding factor, runt domain, alpha subunit 2; translocated to, 3) is a master transcriptional coregulator in hematopoiesis. Corepressor CBFA2T3 belongs to a family of ubiquitously expressed transcriptional repressors that interact with transcription factors bound to promoters of target genes. It has been suggested that the functions of CBFA2T3 include suppression of breast tumors, and thus losing normal function of CBFA2T3 may be a key event in the early stage of breast cancer (Kochetkova et al. 2002). Lamprecht et al. (2010) concluded that derepression of these LTR genetic elements is involved in the pathogenesis of human lymphomas. Transcription of the Colony Stimulating Factor 1 Receptor (CSF1R) in lymphoma cells initiates at transposon-like human element 1B (THE1B), which represents an aberrant activation of this endogenous LTR of the MaLR family (Lamprecht et al. 2010). The CSF1R functions as a receptor for the cytokine CSF1, which controls the differentiation, function and production of macrophages. Lamprecht et al. (2010) also indicated that aberrantly expressed LTR-driven CSF1R transcripts have been detected in anaplastic large cell lymphoma. Not all of the health aspects associated with this group of LTR retrotransposons are negative. For example, humans and anthropoid apes produce the peptide Corticotropin-releasing hormone (CRH) in the placenta, where its concentration rises just prior to birth. Production of CRH in the placenta is contolled by the endogenous retroviral element THE1B (Dunn-Fletcher et al. 2018).

Future research efforts in the fields of oncology and immunogenetics are certain to unveil more details about the involvement of endogenous LTR retrotransposons in human pathogenesis (Katoh and Kurata 2013).

4.6 An Example Regarding the Importance of Endogenous Viral Sequences in Health and Disease of Unicellular Eukaryotes

4.6.1 Lavidaviridae

Lavidaviridae sequences can exist endogenously in some flagellates including unicellular algae. These endogenous Lavidaviridae genomes (Hackl et al. 2020) are proviruses that, by themselves, seem to be replicatively incompetent. The Lavidavididae (Fischer 2020) are satellite viruses of Mimiviridae (Fischer et al. 2010) and the normal host range of Lavidaviridae presumably would be equal to that of their Mimiviridae helpers. If a cell containing the Lavidaviridae genome becomes infected by a suitable helper virus, then the endogenous Lavidaviridae genome becomes activated resulting in the infected cell producing and releasing progeny lavidaviruses in addition to progeny mimiviruses (Blanc et al. 2015; Duponchel and Fischer 2019). Replication of the endogenous Lavidaviridae seems to reduce the number of helper virus progeny that are produced (Blanc et al. 2015), and simultanous infection of another host cell by progeny of both the lavidavirus and progeny of its corresponding mimivirus may also suppresses replication of the mimivirus (Fischer and Hackl 2016). Thus, the presence of an endogenous lavidavirus can be protective of the host population by reducing the risk that neighboring cells will succumb to infection by progeny Mimiviridae.

4.7 Examples Regarding the Importance of Endogenous Viral Sequences in Morphological Development, Health and Disease of Animals

Some of the endogenous viruses have evolved to offer a survival-related benefit to their natural host, and this can give an added measure of stability to their mutual relationship. Examples of such benefits have been found by studying Bornaviridae, Filoviridae, Metaviridae, Nudiviridae, Parvoviridae, Polydnaviridae, and Retroviridae.

4.7.1 Bornaviridae

Bornaviridae sequences that are endogenized in human genomes detrimentally may be associated with a predisposition for lung adenocarcinoma among nonsmokers (Honda 2017). There also may be benefits associated with endogenous Bornaviridae. One example of a benefit would be the possibility that endogenous Borna-like sequences in ground squirrel inhibit the replication of infecting Bornaviridae (Fujino et al. 2014). Endogenous bornavirus-like nucleoprotein elements (EBLNs) of afrotherian mammals (superorder afrotheria) encode functional genes which are used by the host cells, representing an exaptation process (Kobayashi et al. 2016). Exaptation is the use of a viral protein to serve a species which did not design that protein. Hosts frequently exapt proteins whose coding has been imported into the host by endogenous and lysogenous viruses.

4.7.2 Filoviridae

Filoviridae sequences that have been endogenously conserved in bats may act as regulators that suppress the innate immune system (Edwards et al. 2018). This finding suggests that genomically integrated sequences from non-retro RNA viruses had been coopted.

4.7.3 *Metaviridae*

Metaviridae sequences, specifically those known as gypsy elements, are endogenous in many groups of eukaryotes including insects. Gypsy elements may serve as transporters of genes (Kapulkin 2016) although little else seems to be known about the possible role of gypsy sequences in the ecology of their hosts. Most insects also contain intracellular endosymbiont bacteria of the genus *Wolbachia* that have been maternally inherited. Touret et al. (2014) discovered that the *Wolbachia* which *Drosophila melanogaster* offspring inherit seem to reduce maternal transmission of gypsy sequences.

4.7.4 *Nudiviridae*

Nudiviridae sequences that have been endogenized are used by some parasitoid wasps, with an example being *Venturia canescens*, to aid the ability of their eggs to hatch, and their offspring develop, within the bodies of their parasitized insect hosts. Many wasps inject infectious Polydnaviridae into their insect host during oviposition (Drezen et al. 2017) as will be discussed in Sect. 4.7.6 of this chapter, and those injected Polydnaviridae infect the parasitized host. The Nudiviridae association has offered a different, hypothetically non-infectious option. The Nudiviridae genetic information which has become endogenized into *Venturia canescens* includes pseudo-genes of the Nudiviridae capsid protein. Those pseudogenes are viral genomic remnants that no longer code for a capsid protein, although the pseudogenes do indicate that the captured virus initially did encode a capsid protein. The wasp incorporates its own immunosuppressive virulence proteins inside of liposomes that bear the nudiviral envelope protein (Drezen et al. 2017). These liposomes are termed virus-like particles and injected along with the wasps eggs into its host. Viral DNA is not incorporated into these viral-like particles (Drezen et al. 2017; Leobold et al. 2018; Volkoff and Cusson 2020) and thus, if the criterion of infection is considered by a strict definition, then injection of these virus-like particles into the host insect does not represent an infection of the host insect. If the definition of infectiveness is allowed to change such that it does not require the ability of an infecting organism to direct self replication, then perhaps the use of these viral-like particles eventually will be judged a type of infection (Drezen et al. 2017).

4.7.5 *Parvoviridae*

Simultaneous infection of the rosy apple aphid, *Dysaphis plantaginea*, by two viral species Dysaphis plantaginea densovirus (genus Ambidensovirus, family Parvoviridae) and Acyrthosiphon pisum virus (unclassified above the level of species) is necessary in order to achieve induction of wing development in asexual clones of that aphid species. Winged morphs enable aphid species to successfully disperse and then colonize new plants. The Acyrthosiphon pisum virus alone does not fulfill that function for *Dysaphis plantaginea*, and the parvovirus infection results in a replicative cost which the aphid pays in association with the gained benefit of wing production (Ryabov et al. 2009). The pea aphid, *Acyrthosiphon pisum*, relies upon upregulation of endogenous Parvoviridae sequences to assist with its generation of winged morphs and that upregulation similarly can involve a reproductive cost to the aphid. The endogenous Parvoviridae genes which provide this function for *Acyrthosiphon pisum* have been presumed to represent the same virus that infectiously is important for induction of wing formation by *Dysaphis plantaginea* (Parker and Brisson 2019).

4.7.6 Polydnaviridae

Polydnaviridae sequences that are endogenous and fully capable of generating progeny virions play a mutualistic life cycle role for the Braconidae and Ichneumonidae families of parasitoid wasps (Legeai et al. 2020; Volkoff and Cusson 2020) including the braconid species *Cotesia congregata*. That wasp species functions as a biological control agent for the tobacco hornworm, *Manduca sexta*. The wasp and its host are in an evolutionary arms race where failure equals death. From the perspective of this wasp species, which is an obligate parasite, usage of a symbiotic virus has provided the best strategy for overcoming the defenses of its host. The endogenous polydnavirus genome exists as a provirus both in male and female *Cotesia congregata* and may first have incorporated into this evolutionary line of wasps around 100 million years ago (Murphy et al. 2008). The provirus generates assembled virions only in the nuclei of calyx cells contained within the female wasp ovaries. Eventually, the membranes of those cells breakdown and the progeny virions enter the oviduct lumen. Both venom and progeny virions are injected along with wasp eggs into the host caterpillar during oviposition. The venom produced by *Cotesia* wasps reduces the possibility of superparasitism, which could result if the same host animal were later attacked by another wasp seeking to inject its own eggs (Chen et al. 2020). The injected virions contain circular viral DNA sequences that encode virulence genes. Subsequent expression of these virulence genes within the caterpillar's body suppresses the caterpillar immune system, which allows successful hatching and development of the wasp inside of its caterpillar host (Beckage and Gelman 2004; Bredlau et al. 2019; Chevignon et al. 2014; Louis et al. 2013). The polydnavirus also interferes with development of the caterpillar host in several ways, with one example being that the virus prevents the caterpillar from entering its underground pupation. The virus additionally suppresses secretion of glucose oxidase into the caterpillars saliva. Glucose oxidase elicits plant defenses, and therefore suppressing the level of glucose oxidase in the caterpillars saliva may result in the caterpillars growing at a faster rate

Fig. 4.2 These images show a female parasitic wasp *Cotesia congregata* injecting eggs into the hemocoel of a tobacco hornworm *Manduca sexta*. The images appear courtesy of their author Justin Bredlau

and thus being more suitable hosts to the wasp. There likely are many additional endocrinological and behavioral effects associated with presence of the virus in the caterpillar. *Cotesia congregata* also can replicate in the catalpa sphinx, *Ceratomia catalpae*. The dramatic interaction between *Cotesia congregata* and its caterpillar prey are shown in Figs. 4.2, 4.3, and 4.4. This parasitoid wasp replication cycle seemingly can result in horizontal gene transfers between animals and viruses, with an example having been a Helitron transposable element possibly moving first from a *Drosophila* host ancestor into the genome of a polydnavirus that became endogenous within the parasitoid wasp *Cotesia vestalis*, and that transposable element then moving from *Cotesia vestalis* into a host *Bombyx mori* (Heringer et al. 2017).

Fig. 4.3 The upper image shows larvae of *Cotesia congregata* after they have emerged through the cuticle of their host tobacco hornworm *Manduca sexta*. Some of the larvae shown in this image have begun to spin their individual cocoons. The lower image shows larvae of the *Cotesia congregata* wasp within their cocoons, still attached to the host from which the larvae had emerged. These images appear courtesy of their author Justin Bredlau

Helitron transposons seem to capture and mobilize gene fragments in eukaryotes. *Cotesia vestalis* will parasitically attack *Bombyx mori*, but with only a ten percent rate of replicative success and no production of parasitoid cocoons (Hiroyoshi et al. 2017).

The larvae and pupae of *Cotesia congregata*, and of many other parasitoid wasps, can in turn be parasitized by several different groups of wasps that are considered to be hyperparasitoids. Those hyperparasitoids use the parasitoid progeny as hosts (Harvey 2008; Zhu et al. 2018). A chain of interactions initiated by the symbiotic Polydnaviridae reveals the presence of parasitoid larvae to their hyperparasitoid enemies. Those hyperparasitoids include members of the wasp genus *Hypopteromalus* (family Pteromalidae) and other members of the wasp families

Fig. 4.4 This image shows a *Cotesia congregata* wasp and *Cotesia congregata* cocoons on their tobacco hornworm *Manduca sexta* host after some of the offspring wasps had emerged. The lower image is an enlargement of the upper image. These images appear courtesy of their author Beatriz Moisset

Chalcididae, Ichneumonidae, and Pteromalidae. Some of those hyperparasitoids parasitize the pre-pupal stage of *Cotesia congregata*, which occurs while the developing *Cotesia* are in their cocoons, and one species of Ichnuemonid has been found to parasitize *Cotesia* larvae while they are still inside the caterpillar.

The caterpillars of *Manduca sexta* are not limited to feeding on tobacco and in fact these caterpillars are very common on garden tomatoes. As suggested above, there are additional *Cotesia* species which carry and use endogenous Polydnaviridae. The endogenous Polydnaviridae also play a similar role for numerous other genera of parasitic wasps, and it is possible that the effects of endogenous Polydnaviridae may differ among the many other genera of parasitoids that mutualistically use this virus family. Tan et al. (2018), as an example, studied a wasp of the *Microplitis* genus in their experiments. It is important to note that although the caterpillars used as hosts by *Cotesia congregata* pupate underground, many of the other caterpillar species which serve as hosts for parasitic wasps do instead make cocoons during the caterpillar pre-pupal stage before the wasp larvae emerge.

4.7.7 Retroviridae

Retroviridae sequences respectively belonging to the Alpharetrovirus, Betaretrovirus, Gammaretrovirus and Spumavirus genera have been found in avians. Chickens express genes from approximately twenty percent of their endogenous Retroviridae, meaning that within the host cells there are viral proteins being produced by transcription of the endogenous Retroviridae sequences and subsequent translation of those transcripts. This process of producing viral proteins from the endogenous Retroviridae sequences follows tissue-specific patterns in chickens. Also, it is important to note that the number of endogenous Retroviridae viruses expressed in chicken embryo fibroblasts is greater than is the expression of endogenous Retroviridae which occurs in older chickens (Bolisetty et al. 2012). The transmembrane proteins of endogenous retroviruses can have a detrimental immunosuppressive role which they fulfill by modulating cytokine release, and that unfortunately can facilitate the formation of tumors (Denner 2017).

Denner (2017) has published a summary of information about the different endogenous Retroviridae envelope gene sequences which various mammalian species use beneficially as syncytins during the initiation of a placenta. When used for this purpose of placentation, the envelope proteins encoded by endogenous Retroviridae are activated and employed by eutherian (placental) and also by metatherian (marsupial) embryos during that early stage of placental generation (Denner 2017). These endogenous viral proteins are called syncytins because they function by causing embryo-derived cells to fuse their outer membranes resulting in formation of a multinuclear syncytium. There also are numerous groups of viviparous lizards, including the *Mabuya* which are a genus of long-tailed skink, that similarly use the envelope protein of an endogenous Retroviridae to create a syncytium during initial development of their transient organ which serves equivalently to a placenta.

It also is possible that endogenized genes from Retroviridae provide their host with some protection against infection by other members of that same virus family (Broecker and Moelling 2019).

4.7.7.1 The Importance of Endogenous Retroviridae in Health and Disease of Their Human Hosts

Perhaps eight percent of the human genome represents sequences derived from Retroviridae (Griffiths 2001). The endogenous Retoviridae represent enveloped viruses, those which have an outer membrane surrounding their nucleoprotein. Infectious Retroviridae natually enter their new host cells by fusing together their viral outer membrane and the host cells membrane (Denner 2017) after which the viral RNA genomes are released into the new host cell and the viral genomes will be reverse transcribed. The produced DNA copies of the viral genomes are integrated into the host cell genome.

The endogenous Retroviridae sequences contained in humans have been grouped into 31 families (Denner 2016). These families of human endogenous retroviruses (often abbreviated HERV) have been clustered into three broad classes based upon analysis of their DNA genomes (Griffiths 2001; Nelson et al. 2003; Vargiu et al. 2016). Class 1 are most closely related to the Retroviridae genera Epsilonretrovirus and Gammaretrovirus, and Class 1 notably includes HERV-E, HERV-F including HEP, the HERV-Fc groups, HERV-FRD, HERV-H, HERV-I, HERV-R, HERV-T, HERV-W, plus the HUERSP groups, MER including MER41, PABL, and PRIMA. The family HERV-W does, by itself, account for about 1 percent of the human genome (Denner 2016). Class 2 are most closely related to the Retroviridae genus Betaretrovirus and notably include both HERV-K and HML. Class 3 are more closely related to the Retroviridae genus Spumavirus and include both HERV-L and HERV-S. It seems that HERV-L and HUERSP3 represent the oldest integrations (Vargiu et al. 2016), of which the HERV-L sequences are believed to have integrated into the mammalian genome approximately 100 to 150 million years ago. All placental mammals have been found to contain HERV-L sequences. The primate specific endogenous Retroviridae virus MER41 entered the ancestral primate genome approximately 45 million to 60 million years ago and there now are an estimated 7190 long terminal repeat elements of this virus in the human genome (Chuong et al. 2016). The endogenous Retroviridae groups HERV-K (HML2) and HERV-Fc represent the most recent integrations (Vargiu et al. 2016). Some of the Retroviridae sequences that represent HERV-K (HML-10) may have entered the genome of ancestral Old World monkeys about 35 million years ago (Broecker et al. 2016).

Most of the endogenous LTR retrotransposon sequences produced from Retroviridae seem to be ancient relics that may either partially transcribe or not transcribe at all. Perhaps 90 percent of endogenous Retroviridae elements consist only of LTR (long terminal repeat) segments that are not associated with open reading frames, and thus these elements do not code for proteins. Some of the LTR retrotransposons are capable of moving throughout their host genome (Chuong et al. 2016), which is why they are called transposons. However, it has been suggested that most of the HERVs as a representative group pose no immediate risk as transposable elements and most of the LTR retrotransposons that represent human endogenous Retroviridae lost their infectivity and transposing ability prior to the human-chimpanzee speciation (Katoh and Kurata 2013). The HERV-L sequences are an example of endogenous Retroviridae sequences that are not known to produce any proteins and so far have not been linked to disease.

It potentially is possible for HERVs and also solitary LTRs derived from HERVs that normally are suppressed to be reactivated and then act as extra transcriptional initiation points for nearby cellular genes (Katoh and Kurata 2013). Some of the LTR retrotransposons generated from Retroviridae have found usage by their host for different functions, including times when these retrotransposon sequences serve as promoters (Thompson et al. 2016), and this repurposing is termed exaptation. Indeed, there may be hundreds of thousands of LTR sequences in a genome that can provide binding sites for transcription factors (Katoh and Kurata 2013). There also are numerous endogenous Retroviridae sequences that are relatively younger and have remained competent as proviruses. The term provirus is a signification of their ability to produce progeny virions. Serving as examples, the human genome has 3167 Retroviridae proviruses, zebra finch has 1221, chicken has 492, and turkey has 150 (Bolisetty et al. 2012). Hosts have developed restriction pathways to suppress the activity of Retroviridae proviruses at the points of transcription and post-transcription (Thompson et al. 2016).

4.7.7.1.1 The Beneficial Role of Endogenous Retroviridae in Placentation and Gestation

Humans and other primates are among the animals which express endogenous Retroviridae genes in their placenta. The envelope gene of viruses which belong to the family group HERV-W intentionally is activated by a human embryo and that envelope protein serves as syncytin-1, which is critically important for achieving cell-cell fusion in order to initiate development of a placenta. All primates use this same endogenous viral protein for that purpose early in the development of a placenta. In those other primates, this virus family is designated as ERVW-1 (Endogenous retrovirus W-1) rather than HERV-W. Humans and other primates additionally express in their placentas a reactivation of endogenous viruses belonging to the HERV-FRD group, whose envelope genes also serve as syncytins and likewise help to mediate the cell fusion which occurs early in placental development. Syncytin-2 represents expression of HERV-FRD. Blastocyst implantation critically requires both of those two syncytins (Soygur and Sati 2016). Expression of HERV-K additionally occurs during the human embryos early developmental stages of pre-implantation but that expression is then silenced (Küry et al. 2018). Luis Villareal presented in 1997 a hypothesis that endogenous retroviral elements suppress maternal immunity during pregnancy and that this viral action facilitated the evolution of placental mammals (Villareal 1997). Expression of the transmembrane (TM) protein of HERV-K indeed may be immunosuppressive and its activation by the embryo thereby helpful to prevent maternal rejection of the fetus (Denner 2016). Genes of HERV-R (also identified as ERV-3) and HERV-E additionally are expressed in the placenta. The reactivation of at least some of those endogenous Retroviridae occurs in conjunction with changing hormonal levels.

4.7.7.1.2 Other Involvements of Endogenous Retroviridae in Health Versus Disease

Unfortunately, the same endogenous retrovirus envelope proteins which are crucially important as syncytins 1 and 2 during the intiation of a primate placenta also induce the establishment of human immunodeficiency virus (HIV) reservoirs in the placenta and facilitate the placental transmission of HIV by fusing together infected and uninfected cells (Tang et al. 2020).

Some of the endogenous Retroviridae sequences in humans do remain capable of transcriptional activity at other times during the life cycle of their host. One example of these would be the MER41.AIM2 endogenous virus which regulates transcription of AIM2 (Absent in Melanoma 2) that encodes a sensor of foreign cytosolic DNA necessary for an inflammatory response to infection. The endogenous Retroviridae of humans also have been associated with and in some ways implicated as being causal for several autoimmune related diseases including amyotrophic lateral sclerosis (ALS), multiple sclerosis (MS), rheumatoid arthritis, and systemic lupus erythematosus (Brütting et al. 2017; Katoh and Kurata 2013; Tokuyama et al. 2018). If members of the endogenous virus family HERV-W, which so valuably serve us at the embryo stage of our life, should accidently become active at some other point in a human life cycle then the same HERV-W envelope gene product which functioned as a syncytin may instead cause multiple sclerosis (Katoh and Kurata 2013; Ruprecht and Mayer 2019). The HERV-W env protein also induces a release of inflammatory cytokines by monocytes (Küry et al. 2018) which can upset the healthy balance between inflammatory cytokines and anti-inflammatory cytokines. The family of endogenous Retroviridae sequences designated HERV-K (HML2), which has been estimated to represent less than one percent of the human endogenous Retroviridae sequences, is active and associated with the initiation of amyotrophic lateral sclerosis (Denner 2016; Li et al. 2015). Endogenized Retroviridae sequences have been associated with human tumors (Honda 2017; Soygur and Sati 2016) and specifically the HERV-K family of sequences has been implicated in germ cell tumors including cancer of the testis (Denner 2016). The virus group HERV-K10 has been linked to juvenile rheumatoid arthritis (Brütting et al. 2017). Human endogenous retroviruses additionally may be involved in schizophrenia and type 1 diabetes mellitus, and some of these illnesses for which human endogenous Retroviridae have been implicated seem to cooccur (Brütting et al. 2017). Activation of endogenous Retroviridae sequences may be triggered by human Herpesviridae infections including cytomegalovirus, epstein-barr virus, and varicella-zoster virus. Infections by Retroviridae such as HTLV-1 (human T-cell lymphotropic virus type 1), and HIV-1 (human immunodeficiency virus 1) also may trigger endogenous Retroviridae into reactivation (Küry et al. 2018).

Endogenous Retroviridae sequences have become regulators that enhance interferon regulation, they are induced by interferon gamma (IFNG), and specifically are involved in activation of the AIM2 inflammasome which activates inflammatory responses (Chuong et al. 2016). When the DNA methylation-mediated suppression system becomes compromised, HERVs and other LTR retrotransposons can cause detrimental and self-protecting effects. Two examples of clinically significant HERV or LTR activation are CSF1R oncogene activation by a MaLR LTR in Hodgkin's lymphoma, and renal cell carcinoma specific novel expression of an HERV-E antigen facilitating immunotherapy.

4.7.7.2 The Parallel Roles of Endogenous Retroviridae in Health and Disease of Animals

Endogenus Retroviridae seem to exist in all mammals. We know that metatherians use the envelope genes of their endogenous Retroviridae during initiation of their placentas, just as do eutherians. Presumably there are many additional commonalities that will be found between the ways in which endogenous Retroviridae function beneficially and detrimentally for the health of humans and our fellow mammals. One possible concern is that transplanting animal tissues into humans will result in detrimental activation of those animals endogenous Retroviridae. Swine often are used as a source of tissue for transplantation into human, and thusly there is an accompanying concern about those porcine endogenous Retrovirudae that exist as proviruses. Joachim Denner has studied this subject area. His review of the literature about transmission of endogenous Retroviridae during transplantation of porcine tissues into mammals did fortunately suggest the risk to be very limited (Denner 2018).

4.8 Examples Regarding the Importance of Non-endogenous and Endogenous Viral Sequences in Health and Disease of Plants and fungi

Many of the viruses that infect plants and fungi cause disease in their host, and yet some seem to benefit their host.

4.8.1 The Role of Non-endogenous Viruses in Plant Health

Many non-endogenous viruses of plants certainly seem to offer survival benefits to the health their hosts. Examples of these benefits as summarized from Takahashi et al. (2019) include:

Alteration of predator preference

Bromoviridae (genus Cucumovirus)

Cold tolerance

Bromoviridae (genus Cucumovirus)

Drought tolerance:

Bromoviridae (genera Bromovirus and Cucumovirus)
Partitiviridae (genera Alphapartitivirus and Deltapartitivirus) Tobamoviridae (genus Tobamovirus)
Virgaviridae (genus Tobravirus)

Heat tolerance

Bromoviridae (genus Cucumovirus)

Increased plant productivity

Amalgaviridae (genus Amalgavirus)
Endornaviridae (genus Alphaendornavirus)

Reduce plants attractiveness to insects (some of the insects are fungal vectors, others may vector acutely infective viruses)

Alphaflexiviridae (genus Potexvirus)
Partitiviridae (genus Deltapartitivirus)
Potyviridae (genus Potyvirus)

Reduced susceptibility to fungi

Alphaflexiviridae (genus Potexvirus)
Potyviridae (genus Potyvirus)

Suppression of root nodule activity when sufficient nitrogen is present

Partitiviridae (genus Alphapartitivirus)

Some of these benefits to plants involve viruses which do not directly infect plants, but instead the viruses infect fungi. A three-way mutualistic beneficial effect is evidenced when those virally-infected fungi simultaneously are infecting plants. Two examples of non-endogenous viruses which infect fungi and thereby exert a beneficial effect that is evidenced when the fungus subsequently infects a plant are Sclerotinia sclerotiorum hypovirulence associated DNA virus 1-A and Curvularia thermal tolerance virus. Of these two viruses, Sclerotinia sclerotiorum hypovirulence associated DNA virus 1-A (viral family Genomoviridae; genus Gemycircularvirus) reduces the virulence of its fungal host *Sclerotinia sclerotiorum* (Hillman and Milgroom 2021) which infects a wide range of host plants.

Curvularia thermal tolerance virus, which is an unclassified double stranded RNA virus, exerts its three-way mutualistic association by infecting the fungus *Curvularia protuberata*. *Curvularia protuberata* is an endophyte of the tropical panic grass *Dichanthelium lanuginosum* and that grass is found in geothermal soils of Yellowstone park. The plant and fungus cannot grow individually at temperatures above 38 °C. However, when the fungus *Curvularia protuberata* lives as an endophyte in the grass plants and this fungus is infected by Curvularia thermal tolerance virus, the virus confers heat tolerance to the grass. The three-way mutualism allows both the plant and its fungal endophyte to survive at root zone temperatures up to 65 °C. Isolates of that fungus which lack the virus do not provide the plant with heat tolerance (Márquez et al. 2007).

4.8.2 The Role of Endogenous Viruses in Plant Health

The viral family groups whose endogenous representation in vascular plants includes having their viral sequences being integrated into the plants chromosomal material are: Amalgaviridae, Betaflexiviridae, Bromoviridae, Caulimoviridae, Chrysoviridae, Endornaviridae, Geminiviridae, Metaviridae, Partitiviridae, Potyviridae, Pseudoviridae, Rhabdoviridae, and Totiviridae. Of these viral families, the Caulimoviridae can also have an episomal existance within the cells of their host plant. The endogenous Mimiviridae and Pithoviridae sequences found in plants presumably are integrated into the plants chromosomal material.

Most of the endogenous viral elements in plants originate from viruses that lack integrase genes, and so their presence as endogenous elements within the chromosomal material of a plant presumably results from either host cell integrase functions or integrase activity by other viruses. The endogenous viral genomes present in plants generally are considered to be grounded, meaning that they cannot excise from the host cell genome. When considered as a group, the endogenous viral elements of plants often consist only of partial viral genomes as the result of fragmentation, although entire viral genomes also can be present. Whether viral integration into host genomes is ultimately of net benefit versus harm to the host remains to be determined (Takahashi et al. 2019).

4.8.2.1 Caulimoviridae

Caulimoviridae are the endogenous pararetroviruses most represented in plants. In general, the endogenous pararetroviruses of plants have become grounded, meaning that they no longer can generate progeny virions. However, some of the Caulimoviridae sequences have retained the ability to produce virions and thus are considered to be proviruses. These proviral Caulimoviridae can exist either chromosomally or episomally and many of the episomal viral forms seemingly are able to transition between the status of latency and the active production of progeny virions.

The viral species Banana streak virus (genus Badnavirus, family Caulimoviridae) is an endogenous pararetrovirus in plantain (genus *Musa*) and presumably this virus normally is integrated into its hosts chromosomal material. When the host plant is under stress, this endogenous virus activates by recombination to create an episomal viral genome and produces infectious progeny virions. Editing of the endogenous Banana streak virus genome by a CRISPR/Cas9 technique can suppress the ability of that viral genome to produce viral proteins, thus allowing the plant to successfully be cultivated (Tripathi et al. 2019).

The viral species Petunia vein clearing virus (PVCV, genus Petuvirus, family Caulimoviridae) is found as a provirus of the plant genus *Petunia*. This virus affects the coloration pattern of the host plant and can produce leaf malformation. Activation of this virus often happens as the plant ages and characteristically blotched flowers resultingly occur in aging *Petunia* plants. It has been suggested that activation of the Petunia vein clearing virus results from the plants having a natural age-related decrease in maintenance of DNA methylation, and consequently decreased methylation of the viral promoter. Both transcripts and episomal elements of Petunia vein clearing virus accumulate in blotched areas of the plants flowers. Blotching is caused by post-transcriptional gene silencing of an enzyme which is important to anthocyanin biosynthesis. Because interference of anthocyanin biosynthesis corresponds to activation of PVCV, it has been suggested that the virus has a capability for post-transcriptional gene silencing (Kuriyama et al. 2020). There also are elevated levels of viral suppressors of RNA silencing in those blotched areas.

4.8.2.2 Endornaviridae

Endornaviridae seem to be represented in plants including Bell pepper, *Capsicum annuum*, by sequences that are homologous to the glycosyltransferase 28 domain. Related sequences also are present in bacteria and fungi. It has been suggested that these endogenous Endornaviridae sequences may have originated from bacteria (Chu et al. 2014) and valuably serve by bringing this gene into the genome of their host plant.

4.8.2.3 Metaviridae and Pseudoviridae

Metaviridae and Pseudoviridae are abundantly represented by Long terminal repeat retrotransposons in many plant species. As one example, the genome of *Zea mays* contains approximately 150,000 to 250,000 Long terminal repeat (LTR) retrotransposons which collectively represent 50 to 80 percent of the maize nuclear DNA (San Miguel and Vitte 2009). Most of the LTR retrotransposons in corn are either Copia, which represent the viral family Pseudoviridae (Bousios et al. 2012; Zhang and Qi 2019), or Gypsy which represent the viral family Metaviridae (Zhang and Qi 2019). The importance of these groups of retrotransposons for plant health still seems to remain a mystery, although they may serve to facilitate genetic rearrangements and could be valuable as transcription initiation sites.

4.9 Understanding How Partnership with an Endogenous Fungal Virus Can Reduce the Phytopathogenicity of its Fungal Host

Some of the endogenous viruses of fungi have evolved to provide a survival-related hypovirulence benefit to the association between their natural fungal host and plant species which are infected by those fungi. This can give an added measure of stability to their mutual relationship. An example of this type of relationship has been found with a species of the viral family Hypoviridae which exists as endogenous hypovirulence elements associated with some strains of the Chestnut blight fungus *Cryphonectria parasitica*.

The fungal species *Cryphonectria parasitica*, commonly named chestnut blight fungus, causes cankers of chestnut trees. Those cankers have proven fatal for, and virtually eliminated, the American chestnut tree, *Castanea dentata,* which once was a predominant woodland forest species in parts of the United States. When the virus Cryphonectria hypovirus 1 (genus Hypovirus, viral family Hypoviridae) endogenously infects *Cryphonectria parasitica* the virus reduces virulence of the fungus with a result that all three participants, the virus, its host fungus, and in turn the fungal infected tree, manage to survive (Hillman and Milgroom 2021).

4.10 Lysogenous Viruses of Prokaryotic Hosts

All prokaryotes seem to be lysogenized by viruses and the goal for the host must be surviving that situation.

4.10.1 Some of the Benefits and Detriments Associated with Lysogenic Archaelphage and Bacteriophage

Archael and bacterial viruses play an integral role in the ecology of their hosts, including the fact that these viruses affect prokaryotic population diversity and effect prokaryote population control. These viruses also are important as means of transferring genomic information between prokaryotes (Weinbauer 2004). The presence of eukaryotic association modules (Bordenstein and Bordenstein 2016) within bacteriophage genomes suggests that lateral transfers occur between phage genomes and the eukaryotes with which their host bacteria are associated.

Lysogenic bacteriophage, also called temperate phage, are viruses that naturally infect bacteria. There also are lysogenic viruses associated with archaea, and those have been termed archaealphage. When these viruses infect a host cell, the virus has a choice between either initiating a lytic replication cycle or entering a lysogenic state. The result of a lytic cycle would be generation of progeny viruses which subsequently are liberated into the surrounding environment by rupturing and resulting death of the host cell. That rupturing is termed lysis. Alternatively, entering lysogeny involves viruses either inserting their genetic material into the genome of the host cell or forming a circular copy of their genome that will exist independently within the host cell's cytoplasm. That decision as to where the lysogenic phage genome will reside represents an evolved characteristic of the viral family to which the virus belongs. The independent cytoplasmic form of a phage genome is called either a plasmid, or a replicon, or an episome. There are lysogenic viruses that can switch their choice of lodging, sometimes being integrated into the host genome and at other times existing as a cytoplasmic replicon (Scott et al. 2008). These housed viral genomes, whether in the host chromosome or cytoplasm, often are termed 'prophage' which indicates that the viral genome has retained capability for generating progeny virus particles. Lysogeny offers the virus an opportunity to survive without killing its host cell, and that can be very important under conditions where newly released progeny viruses might not easily encounter a susceptible host.

Lysogeny may involve a fitness cost to the host (Broecker and Moelling 2019). Most bacteria, and presumably also most archaea, possess prophage and genetic recombination can occur between coinfecting prophage (Fortier and Sekulovic 2013).

The prophage can contain genes that remain transcriptionally active, producing products that are beneficial to the host and changing the phenotype of the host. Some of these viral genes code for production of toxins and enzymes which may increase pathogenicity of the host (Fortier and Sekulovic 2013; Pant et al. 2020). Prophage can possess genes which provide the host with resistance against antimicrobial compounds and resistance against attack by other microbes (Fortier and Sekulovic 2013). Other phenotypic traits of host prokaryotes which can be effected by the prophage are ability of the host to develop bioflims, and ability of the host to disperse, both of which are traits that can affect the environmental survival and also pathogenicity of the microbial host (Fortier and Sekulovic 2013). These changes in phenotypic expression by the host archaea and bacteria which result from transcriptional activity of their prophage are termed to be either 'phenotypic conversion' or 'lysogenic conversion' (Fortier and Sekulovic 2013). Some prophage seem to promote sporulation of their host bacteria which provides the prophage with additional environmental protection and long term persistence if the prophage successfully becomes sheltered within the sporulated host cell, and yet other prophage seem to suppress sporulation (Fortier and Sekulovic 2013).

Some antibiotics can trigger the induction of a prophage, which means that the prophage has ended its installation as a genetic element within the microbial host and instead the prophage produces a lytic cycle. If those prophage encode a toxin, then release of progeny viruses during lysis of their microbial host may be accompanied by a release of toxins into the surrounding material. Production and release of such toxins could cause clinical harm to animals if the microbial hosts being lysed are either naturally colonizing microflora or pathogenic invaders of those animals (Zhang et al. 2000).

Mutations in a prophage can result in the prophage being unable to enter a lytic cycle, and that inability is described as grounding of the prophage. Ramisetty and Sudhakari (2019) have suggested that grounding of a prophage offers two benefits to the host. Those benefits would be: firstly, allowing the host to not risk being destroyed by the prophage either intentionally or accidentally becoming lytic, and secondly allowing the host to maintain resistance against lytic attack by other viruses that are related to the prophage. Those benefits respectively are examples of viruses effecting genotypic and phenotypic changes of their host.

4.10.2 *Examples Regarding the Ecology of Lysogenous Viral Sequences in Prokaryotes*

4.10.2.1 Inoviridae

Vibrio virus CTXphi is a notable example of lysogenic Inoviridae. Vibrio virus CTXphi provides the coding for a toxin associated with the pathogenicity of *Vibrio cholerae* (Pant et al. 2020). My suggestions for information about the ecology of this virus would be Davis et al. (1999) plus Faruque and Mekalanos (2012). Information about the repressor and operator genes of Vibrio virus CTXphi can be found in the publication by Davis et al. (1999). I previously have described the role of this lysogenic virus in the pathogenicity of *Vibrio cholerae* (Hurst 2019).

4.10.2.2 Microviridae

The known normal hosts of Microviridae are bacteria and the most widely recognized lytic member of this viral family is Escherichia virus phiX174. The role of Microviridae as lysogens and the actions which determine their change from lytic to lysogenic lifestyle have not yet been resolved. As with the Inoviridae, the fact that members of this viral family do not code for an integrase activity had led to the presumption that Microviridae were limited to having only a lytic lifestyle. Krupovic and Forterre's (2011) discovery of lysogenic Microviridae was therefore of interest. The host cell Chromosomally Encoded tyrosine Recombinases XerC and XerD seem responsible for incorporation of Inoviridae and also Microviridae prophages into the chromosome of their host cells. The replicative viability of Microviridae prophage has been confirmed by Kirchberger and Ochman (2020).

4.10.2.3 Myoviridae

Escherichia phage Mu is a notable lysogenic member of the viral family Myoviridae. Escherichia phage Mu encodes a transposase and can transpose within the genome of its host bacteria. That transposition can result in host sequences either becoming inactivated because those host sequences are adjacent to the inserted prophage, or host sequences returning to activity when the prophage excises to move elsewhere in the hosts genome (Harshey 2014). The name designation Mu represents the properties of this prophage as a cause of phenotypic mutations.

Wolbachia phage WO which affects the genus *Wolbachia* carries genes related to the black widow spider toxin (Bordenstein and Bordenstein 2016) and may mediate horizontal gene transfer in its bacterial host (Wang et al. 2016).

4.10.2.4 Podoviridae

Salmonella virus P22 is a well studied lysogenic member of the viral family Podoviridae and differs from many of the lambdoid (Lambda—like, and presumed relatives of Escherichia virus Lambda) viruses in that P22 has a second operator-repressor system which controls its primary repressor, both of which serve to maintain the lysogenic state of this virus within its host organism (Campbell 1994). Salmonella virus P22 is used as a transduction tool for inducing mutations of cultured bacteria and fulfills that same role in nature.

4.10.2.5 Siphoviridae

Bacillus virus SPbeta is a notable lysogenic member of the viral family Siphoviridae and encodes an integrase. Coordination of the decision as to whether this virus will be lytic versus lysogenic is done by a peptide-based communication system termed "arbitrium". Stokar-Avihail et al. (2019) have very nicely described this communication system by stating that "During lytic infection, each phage produces a measured amount of a communication peptide, which is released into the growth medium and internalized by nearby bacteria. In subsequent infections, progeny phages sense the concentration of this peptide and preferably enter the lysogenic cycle when its concentration is high. Such communication allows the presently infecting phage to evaluate the extent of recent infections and lysis events by predecessor phage infections, and according the infecting phage can switch to the lysogenic cycle when its host bacterial population dwindles. During its lysogenic mode, Bacillus virus SPbeta genetically provides its host with protection against superinfection by related phage (McLaughlin et al. 1986).

Corynephage beta (phage β) also is a notable lysogenic member of the viral family Siphoviridae, and this virus provides the toxic activity associated with *Corynebacterium diptheriae*. The role of bacteriophage in the ecology and toxicology of *Corynebacterium diptheriae* has been described by Zajdowicz and Holmes (2016).

Escherichia virus Lambda (phage λ) is another notable lysogenic member of the viral family Siphoviridae. Escherichia virus Lambda encodes an integrase plus an excisionase. The maintenance of lysogeny by Escherichia virus Lambda is controlled by two competing repressor genes, one of which maintains lysogeny and the other regulates the change from lysogeny to a lytic cycle (Campbell 1994). Lysogeny by Escherichia virus Lambda can offer a benefit to the host population by inhibiting the release of progeny generated by coinfecting viruses (Benzer 1955). The change from a lysogenic existance to a lytic existance for Escherichia virus Lambda is triggered by the host cell experiencing environmental stress (Schubert et al. 2007). Escherichia virus Lambda has been studied as an example of how coevolution with its host alters the adaptive landscape of a virus (Burmeister et al. 2016).

Escherichia virus N15 produces a linear plasmid prophage (Ravin 2015). During its lysogenic existance approximately half of the Escherichia virus N15 viral genes are transcribed, which is far greater than the amount of transcriptional activity that is required by many other prophage such as Escherichia virus Lambda. The high extent of transcriptional activity which is required in order to control lysogeny by the Escherichia virus N15 prophage suggests that maintenance of its lysogeny, and perhaps similarly the lysogeny achieved by other groups of viruses which produce plasmid prophage, involves a far greater complexity than does maintenance of those lysogenous viruses such as Escherichia virus Lambda whose prophage instead reside chromosomally within their host. Part of this increased complexity may relate to the fact that prophage which exist as plasmids must have greater involvement in assuring that they will replicate in coordination with their host, and will not become left behind when their host cell divides.

Streptococcus virus Sfi21 (Prophage SF370.4 of *Streptococcus pyogenes* strain SF370) also is a notable lysogenic member of the viral family Siphovridae. The prophage of Streptococcus virus Sfi21 integrates into its host cells genome at a site which results in this virus exercising partial control over the host cells DNA mismatch repair. That integration site is between two host cell genes, and this insertion blocks expression of the downstream gene mutL that codes for DNA mismatch repair protein MutL. When the population density of its host cells is low during, for example, early logarithmic bacterial growth the prophage excises from the host chromosome to form a replicating episome. That excision allows expression of mutL. When the host population density is high, the prophage reintegrates into the chromosome and mutL gene expression ends (Scott et al. 2008).

Yersinia phage PY54 produces a linear plasmid prophage. The normal host for this virus is *Yersinia enterocolitica*, although the virus also can replicate in *Escherichia coli* (Ziegelin et al. 2005). The prophages produced by Yersinia phage PY54 and Escherichia virus N15 are compatible plasmids meaning that they can coexist within a host cell, and they simultaneously can infect either *Escherichia coli* or *Yersinia enterocolitica*. Replication of this virus within a host cell involves generation of a circular form with covalently closed ends and its replication has been very well characterized. Doubly infected lysogenic *E. coli* can release three types of virions and those are normal N15, normal PY54, and PY54 particles that contain the N15 genome (Hammerl et al. 2007).

4.10.3 Tailocins, Type VI Secretion Systems, and Gene Transfer Agents Represent the Concept of Retaining and Subsequently Using Only Part of a Lysogenic Phage

Two types of tailocins, grouped as R-type and F-type, are produced by bacteria and these represent the tail and tailplate structures of bacterial viruses. The R-tailocins represent the rigid contractile structure of Myoviridae tails. The F-tailocins represent the flexible but non-contractile structure of Siphoviridae tails. These types of tailocins are created from bacterial genomic sequences that presumably had viral origin. Tailocins are released by lysis of the bacteral cell which produced them (Ghequire and De Mot 2015; Patz et al. 2019). Tailocins then attach to receptor sites on susceptible bacterial target cells and subsequently kill those target cells. The generation of these tailocins thus eases the competition that exists within a mixed bacterial population. Type VI secretion systems (T6SS) also employ structures that resemble phage tail plates and phage tails. The type VI systems are used for injecting toxins into other cells and function without requiring lysis of their host cell (Patz et al. 2019). Thus, the host cell survives use of its T6SS structures. Tailocins contrastingly are much shorter than the physical structures used by T6SS, and tailocins only become effective when they are released by lysis of the cell in which they were produced. Gene transfer agents physically resemble tailed bacteriophage and are created using pieces of grounded lysogenic bacteriophage genomes. These transfer agents are used as a technique for dispersing fragments of the host cells genome in a form that can be aquired by neighboring cells. Release of the gene transfer agents requires lysis of the cell which generated those agents. Usage of tailocins, type VI secretion systems, and gene transfer agents, all represent instances of the host exercising control of its lysogenic viral sequences.

4.10.3.1 R-Type Tailocins

The R-tailocins will bind to targeted liposaccharides which serve as the tailocin receptor sites on susceptible bacteria, after which contraction of the R-tailocin depolarizes the cell membrane of its target bacteria in association with pore formation. The cell which produced the R-tailocins dies due to its lysing which released those tailocins. However, if a cell produces tailocins but does not lyse to liberate the tailocins, then the produced tailocins remain within the still viable cell which had produced those tailocins. R-tailocins are identified sequentially as types R1 through R5, and the genomes of many bacterial strains can code for more than a single type of R-tailocin. The R-tailocins of *Pseudomonas aeruginosa* seem to have been derived from a common ancestor of Escherichia virus P2 (phage P2, Family Myoviridae) (Nakayama et al. 2000). *Pseudomonas chlororaphis* produces as many as 2 types of R-tailocin particles which differ in their ancestral origin (Dorosky et al. 2017). The ability to produce both of those two R-tailocins broadens the killing spectrum of this bacterial species and contributes to the persistence of *Pseudomonas chlororaphis* in mixed rhizosphere communities that contain susceptible bacterial strains (Dorosky et al. 2017). *Pseudomonas fluorescens* typically produces one R-tailocin although some strains of this bacterial species can produce two R-Tailocins (Dorosky et al. 2017). *Burkholderia cenocepacia* also produces an R-tailocin (Yao et al. 2017). I did not find information that genetically linked the R-tailocins of either *Pseudomonas chlororaphis*, or *Pseudomonas fluorescens*, or *Burkholderia cenocepacia* to any specific viruses as having been their possible origins.

4.10.3.2 F-Type Tailocins

The F-tailocins similarly attack target cells as would R-tailocins. At least some of the F-tailocins seem to be derived from a common ancestor of Escherichia virus Lambda (phage λ, family Siphoviridae) as discussed by Nakayama et al. (2000). *Listeria monocytogenes* produces F-tailocins that seem to have been derived from Lactococcus phage TP901-1 (Family Siphoviridae) as discussed by Lee et al. (2016). A suggestion has been offered by Lee et al. (2016) that parallel coevolution occurred between tailocins and tailed bacteriophage.

4.10.3.3 Type VI Secretion Systems

Type VI secretion systems (T6SS) are used by bacteria (Coulthurst 2019) to inject toxins into other prokaryotic cells and even into eukaryotic cells. They function by means of the bacteria producing a structure which resembles a bacteriophage tail (Navarro-Garcia et al. 2019), and that structure can protrude up to several micrometers from the host cell (Patz et al. 2019). The genetic coding which is used to produce Type VI secretion systems presumably has been derived from genomes of lysogenic phage. Examples of their ecological roles include the delivery of toxins by *Acidovorax avenae* (Masum et al. 2017), *Acinetobacter baylyi* (Smith et al. 2020), *Escherichia coli* (Navarro-Garcia et al. 2019), and *Vibrio cholerae* (Logan et al. 2018).

4.10.3.4 Gene Transfer Agents

Bininda-Emonds et al. (2016) have published some discussion of the fact that, while sexual reproduction serves as a mechanism for genetic exchange and genetic recombination, species that do not have sexual reproduction may use other means of horizontal gene transfer to achieve a similar genetic goal. Gene transfer agents, which are generated by host cell control of lysogenized bacteriophage capsid genes, represent the concept of retaining the transducing benefit provided by a bacteriophage without keeping the risk associated with the remainder of the viral genome (Redfield and Soucy 2018; Shakya et al. 2017). An example of this would be the discovery by Bárdy et al. (2020) that phage-like gene transfer agents (GTA) are produced by many alphaproteobacteria, such as *Rhodobacter capsulatus*, in response to the depletion of nutrients that occurs during times of high host population density. The GTA particles physically resemble intact tailed bacteriophage (Sherlock et al. 2019) and they are produced by the cell using genes that have been derived from incomplete phage genomes. Those genes used for generating GTA particles are integrated into the host chromosome and regulated by the cell. Figure 4.5 shows a member of the viral family Myoviridae, and that viral family may have provided the genes which are used to generate GTA particles. Figures 4.6 and 4.7 show the appearance of GTA particles.

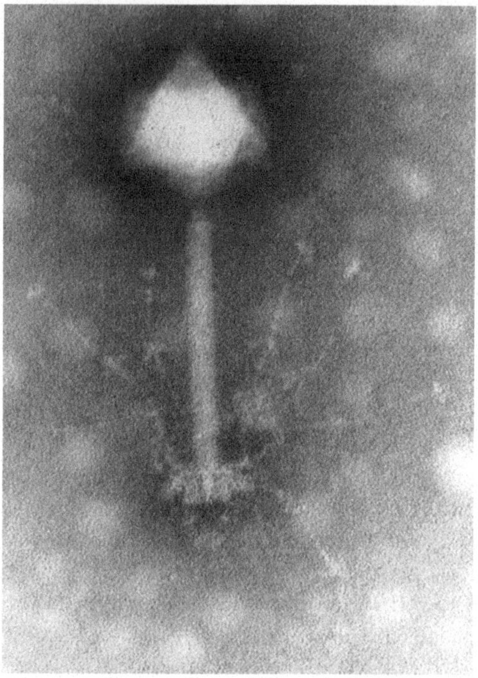

Fig. 4.5 This is a transmission electron microscope image of the Synechococcus Phage S-PM2, which belongs to the viral family Myoviridae and infects members of the cyanobacterial genus *Synechococcus* (Mann et al. 2005). The title of this image is Phage S-PM2 by Hans-Wolfgang Ackermann

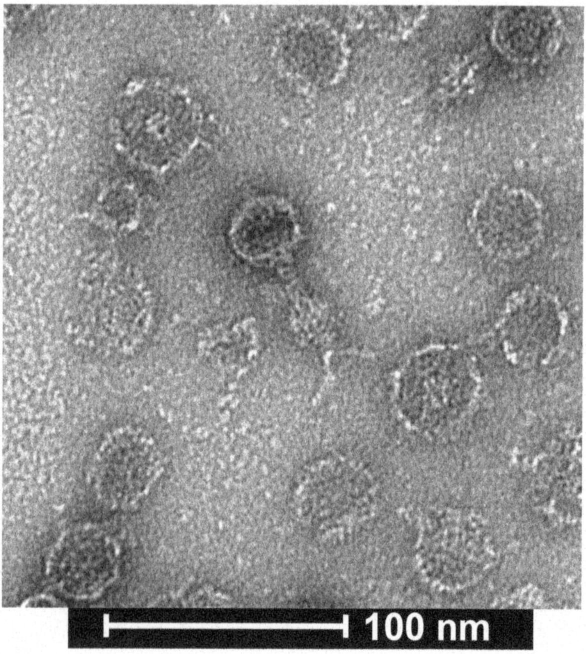

Fig. 4.6 This is a transmission electron microscope image of gene transfer agents (GTAs) visible in various stages of assembly. These GTAs were produced by *Rhodobacter capsulatus* and the image appears courtesy of Paul C. M. Fogg. Purification of these GTA capsids was done by heterologous expression of His6-capsid and affinity purification

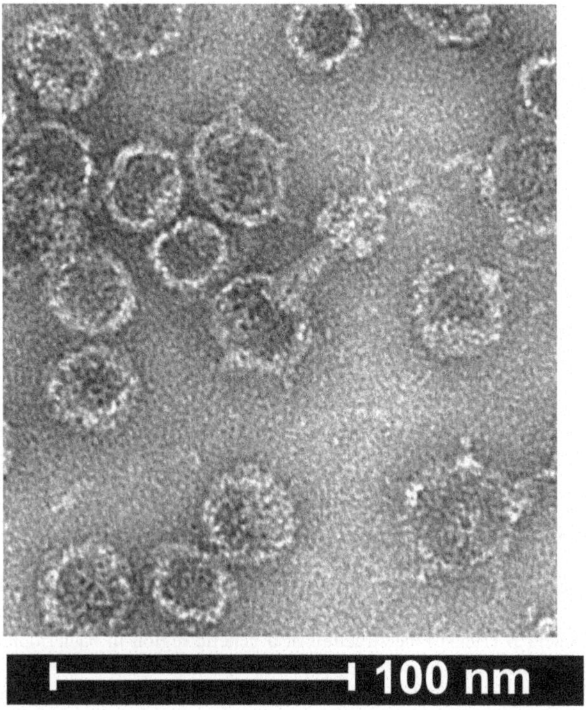

Fig. 4.7 This is a transmission electron microscope image of gene transfer agents (GTAs) visible in various stages of assembly. These GTAs were produced by *Rhodobacter capsulatus* and the image appears courtesy of Paul C. M. Fogg. Purification of these GTA capsids was done by heterologous expression of His6-capsid and affinity purification

Release of the gene transfer agents only occurs by lysis of their producing host cell. The GTA serve their bacterial host population by encapsidating short segments of host DNA that can become injected into a new host cell in the same way that a virus would inject its genome into a new host cell. This transfer of genetic material via GTA represents a type of transduction process. However, GTA are not considered to be parasitic because the DNA that these agents contain does not include the coding which is required for creating them. An infectious virus does include coding which is required for its creation.

4.11 Summary

Do endogenous and lysogenous viruses represent Faustian bargains? The goal of the host is to survive and optimistically to gain what benefits might be derived from presence of the viral genomic material without succumbing to a fatal outcome which the virus might impose. The goal of the virus is to survive and that means not killing its last potential host individual. Most stories of Faustian Bargains result in the devil claiming the soul of the person who made that agreement. The story written by Stephen Vincent Benét (Benét 1936) has the person who made the bargain gaining the agreed benefits without consequently losing their soul. All living beings seem to host viral genomic sequences, whether we term those sequences to be endogenous versus lysogenous. Hopefully, as with Jabez Stone, the hosts will be able to both gain the benefits offered by housing those viral sequences and also win the final judgement.

Acknowledgement I wish to thank Justin Bredlau, Paul Fogg, and Beatriz Moisset for the kindness of allowing me to use their images.

References

Bárdy P, Füzik T, Hrebík D et al (2020) Structure and mechanism of DNA delivery of a gene transfer agent. Nat Commun 11(1):3034. https://doi.org/10.1038/s41467-020-16669-9

Beckage NE, Gelman DB (2004) Wasp parasitoid disruption of host development: implications for new biologically based strategies for insect control. Annu Rev Entomol 49:299–330. https://doi.org/10.1146/annurev.ento.49.061802.123324

Bell PJL (2020) Evidence supporting a viral origin of the eukaryotic nucleus. Virus Res 289:198168. https://doi.org/10.1016/j.virusres.2020.198168

Benét SV (1936) The Devil and Daniel Webster. The Saturday Evening Post, 24 October 1936, Curtis, Philadelphia. http://gutenberg.net.au/ebooks06/0602901h.html

Benzer S (1955) Fine structure of a genetic region in bacteriophage. Proc Natl Acad Sci U S A 41:344–354. https://doi.org/10.1073/pnas.41.6.344

Bininda-Emonds ORP, Hinz C, Ahlrichs WH (2016) Evidence supporting the uptake and genomic incorporation of environmental DNA in the "ancient asexual" Bdelloid rotifer *Philodina roseola*. Life 6:38. https://doi.org/10.3390/life6030038

Blanc G, Gallot-Lavallée L, Maumus F (2015) Provirophages in the *Bigelowiella* genome bear testimony to past encounters with giant viruses. Proc Natl Acad Sci U S A 112(38):E5318–E5326. https://doi.org/10.1073/pnas.1506469112

Bolisetty M, Blomberg J, Benachenhou F et al (2012) Unexpected diversity and expression of avian endogenous retroviruses. MBio 3(5):e00344–e00312. https://doi.org/10.1128/mBio.00344-12

Bordenstein S, Bordenstein S (2016) Eukaryotic association module in phage WO genomes from *Wolbachia*. Nat Commun 7:13155. https://doi.org/10.1038/ncomms13155

Bousios A, Kourmpetis YA, Pavlidis P et al (2012) The turbulent life of Sirevirus retrotransposons and the evolution of the maize genome: more than ten thousand elements tell the story. Plant J 69(3):475–488. https://doi.org/10.1111/j.1365-313X.2011.04806.x

Bredlau JP, Kuhar D, Gundersen-Rindal DE et al (2019) The parasitic wasp, *Cotesia congregata* (say), consists of two incipient species isolated by asymmetric reproductive incompatibility and hybrid inability to overcome host defenses. Front Ecol Evol 7:187. https://doi.org/10.3389/fevo.2019.00187

Broecker F, Moelling K (2019) Evolution of immune systems from viruses and transposable elements. Front Microbiol 10:51. https://doi.org/10.3389/fmicb.2019.00051

Broecker F, Horton R, Heinrich J et al (2016) The intron-enriched HERV-K (HML-10) family suppresses apoptosis, an indicator of malignant transformation. Mob DNA 7:25. https://doi.org/10.1186/s13100-016-0081-9

Brütting C, Emmer A, Kornhuber ME et al (2017) Cooccurrences of putative endogenous retrovirus-associated diseases. Biomed Res Int 2017:7973165. https://doi.org/10.1155/2017/7973165

Burmeister AR, Lenski RE, Meyer JR (2016) Host coevolution alters the adaptive landscape of a virus. Proc R Soc B 283:20161528. https://doi.org/10.1098/rspb.2016.1528

Campbell A (1994) Comparative molecular biology of lambdoid phages. Annu Rev Microbiol 48:193–222. https://doi.org/10.1146/annurev.mi.48.100194.001205. https://www.semanticscholar.org/paper/Comparative-molecular-biology-of-lambdoid-phages.-Campbell/8776674248324c052016017fb72a10962be0221f

Chen W-B, Vasseur L, Zhang S-Q et al (2020) Mechanism and consequences for avoidance of superparasitism in the solitary parasitoid *Cotesia vestalis*. Sci Rep 10:11463. https://doi.org/10.1038/s41598-020-67050-1

Chevignon G, Thézé J, Cambier S et al (2014) Functional annotation of *Cotesia congregata* bracovirus: identification of viral genes expressed in parasitized host immune tissues. J Virol 88:8795–8812. https://doi.org/10.1128/JVI.00209-14

Chu H, Jo Y, Cho WK (2014) Evolution of endogenous non-retroviral genes integrated into plant genomes. Curr Plant Biol 1:55–59. https://doi.org/10.1016/j.cpb.2014.07.002

Chuong EB, Elde NC, Feschotte C (2016) Regulatory evolution of innate immunity through co-option of endogenous retroviruses. Science 351(6277):1083–1087. https://doi.org/10.1126/science.aad5497

Coulthurst S (2019) The type VI secretion system: a versatile bacterial weapon. Microbiology 165:503–515. https://doi.org/10.1099/mic.0.000789

Davis BM, Kimsey HH, Chang W et al (1999) The *Vibrio cholerae* O139 Calcutta bacteriophage CTXphi is infectious and encodes a novel repressor. J Bacteriol 181:6779–6787. https://doi.org/10.1128/JB.181.21.6779-6787.1999

Denner J (2016) Expression and function of endogenous retroviruses in the placenta. APMIS 124:31–43. https://doi.org/10.1111/apm.12474

Denner J (2017) Function of a retroviral envelope protein in the placenta of a viviparous lizard. Proc Natl Acad Sci U S A 114(51):13315–13317. https://doi.org/10.1073/pnas.1719189114

Denner J (2018) Why was PERV not transmitted during preclinical and clinical xenotransplantation trials and after inoculation of animals? Retrovirology 15:28. https://doi.org/10.1186/s12977-018-0411-8

Desjardins CA, Gundersen-Rindal DE, Hostetler JB et al (2008) Comparative genomics of mutualistic viruses of *Glyptapanteles* parasitic wasps. Genome Biol 9:R183. https://doi.org/10.1186/gb-2008-9-12-r183

Dorosky RJ, Yu JM, Pierson LS et al (2017) Pseudomonas chlororaphis produces two distinct R-Tailocins that contribute to bacterial competition in biofilms and on roots. Appl Environ Microbiol 83(15):e00706–e00717

Drezen JM, Leobold M, Bézier A et al (2017) Endogenous viruses of parasitic wasps: variations on a common theme. Curr Opin Virol 25:41–48. https://doi.org/10.1016/j.coviro.2017.07.002

Dunn-Fletcher CE, Muglia LM, Pavlicev M et al (2018) Anthropoid primate-specific retroviral element THE1B controls expression of CRH in placenta and alters gestation length. PLoS Biol 16(9):e2006337. https://doi.org/10.1371/jou

Duponchel S, Fischer MG (2019) Viva lavidaviruses! Five features of virophages that parasitize giant DNA viruses. PLoS Pathog 15(3):e1007592. https://doi.org/10.1371/journal.ppat.1007592

Durzyńska J, Goździcka-Józefiak A (2015) Viruses and cells intertwined since the dawn of evolution. Virol J 12:169. https://doi.org/10.1186/s12985-015-0400-7

Edwards MR, Liu H, Shabman RS et al (2018) Conservation of structure and immune antagonist functions of Filoviral VP35 homologs present in microbat genomes. Cell Rep 24(4):861–872.e6. https://doi.org/10.1016/j.celrep.2018.06.045

Faruque SM, Mekalanos JJ (2012) Phage-bacterial interactions in the evolution of toxigenic *Vibrio cholerae*. Virulence 3(7):556–565. https://doi.org/10.4161/viru.22351

Filée J (2014) Multiple occurrences of giant virus core genes acquired by eukaryotic genomes: the visible part of the iceberg? Virology 466–467:53–59. https://doi.org/10.1016/j.virol.2014.06.004

Finer M, Glorioso J (2017) A brief account of viral vectors and their promise for gene therapy. Gene Ther 24(1):1–2. https://doi.org/10.1038/gt.2016.71

Fischer MG (2020) The Virophage family Lavidaviridae. Curr Issues Mol Biol 40:1–24. https://doi.org/10.21775/cimb.040.001

Fischer MG, Hackl T (2016) Host genome integration and giant virus-induced reactivation of the virophage mavirus. Nature 540(7632):288–291. https://doi.org/10.1038/nature20593

Fischer MG, Allen MJ, Wilson WH et al (2010) Giant virus with a remarkable complement of genes infects marine zooplankton. Proc Natl Acad Sci U S A 107:19508–19513. https://doi.org/10.1073/pnas.1007615107

Fortier LC, Sekulovic O (2013) Importance of prophages to evolution and virulence of bacterial pathogens. Virulence 4(5):354–365. https://doi.org/10.4161/viru.24498

Fujino K, Horie M, Honda T et al (2014) Inhibition of Borna disease virus replication by an endogenous bornavirus-like element in the ground squirrel genome. Proc Natl Acad Sci U S A 111(36):13175–13180. https://doi.org/10.1073/pnas.1407046111

Gallot-Lavallée L, Blanc G (2017) A glimpse of Nucleo-cytoplasmic large DNA virus biodiversity through the eukaryotic genomics window. Viruses 9(1):17. https://doi.org/10.3390/v9010017

Ghequire MGK, De Mot R (2015) The Tailocin tale: peeling off phage tails. Trends Microbiol 23 (10):587–590. https://doi.org/10.1016/j.tim.2015.07.011

Gilbert C, Cordaux R (2017) Viruses as vectors of horizontal transfer of genetic material in eukaryotes. Curr Opin Virol 25:16–22. https://doi.org/10.1016/j.coviro.2017.06.005

Griffiths DJ (2001) Endogenous retroviruses in the human genome sequence. Genome Biol 2 (6) Reviews 1017. https://doi.org/10.1186/gb-2001-2-6-reviews1017

Hackl T, Duponchel S, Barenhoff K et al (2020) Endogenous virophages populate the genomes of a marine heterotrophic flagellate. bio Rxiv. https://www.biorxiv.org/content/10.1101/2020.11.30.404863v1.full

Hammerl JA, Klein I, Appel B (2007) Interplay between the temperate phages PY54 and N15, linear plasmid prophages with covalently closed ends. J Bacteriol 189:8366–8370. https://doi.org/10.1128/JB.01066-07

Harshey RM (2014) Transposable Phage Mu. Microbiol Spectr 2(5). https://doi.org/10.1128/microbiolspec.MDNA3-0007-2014

Harvey JA (2008) Comparing and contrasting development and reproductive strategies in the pupal hyperparasitoids *Lysibia nana* and *Gelis agilis* (Hymenoptera: Ichneumonidae). Evol Ecol 22:153–166. https://doi.org/10.1007/s10682-007-9164-x

Heringer P, Dias GB, Kuhn GCS (2017) A horizontally transferred autonomous Helitron became a full Polydnavirus segment in *Cotesia vestalis*. G3 (Bethesda) 7:3925–3935. https://doi.org/10.1534/g3.117.300280

Hillman BI, Milgroom MG (2021) The ecology and evolution of fungal viruses. In: Hurst CJ (ed) Studies in viral ecology, 2nd edn. Wiley, Chichester, pp 139–182

Hiroyoshi S, Harvey JA, Nakamatsu Y et al (2017) Potential host range of the larval Endoparasitoid *Cotesia vestalis* (= *plutellae*) (Hymenoptera: Braconidae). Int J Insect Sci 9:1–12. https://doi.org/10.1177/1179543317715623

Honda T (2017) Potential links between Hepadnavirus and Bornavirus sequences in the host genome and cancer. Front Microbiol 8:2537. https://doi.org/10.3389/fmicb.2017.02537

Hurst CJ (2019) Briefly summarizing our understanding of *Vibrio cholerae* and the disease cholera. In: Hurst CJ (ed) The structure and function of aquatic microbial communities. Advances in environmental microbiology, vol 7. Springer, Cham. https://doi.org/10.1007/978-3-030-16775-2_7

Hurst CJ (2021) Defining the ecology of viruses. In: Hurst CJ (ed) Studies in viral ecology, 2nd edn. Wiley, Chichester, pp 3–48

Jalasvuori M, Mattila S, Hoikkala V (2015) Chasing the origin of viruses: capsid-forming genes as a life-saving preadaptation within a Community of Early Replicators. PLoS One 10(5):e0126094. https://doi.org/10.1371/journal.pone.0126094

Kano H, Godoy I, Courtney C et al (2009) L1 retrotransposition occurs mainly in embryogenesis and creates somatic mosaicism. Genes Dev 23(11):1303–1312. https://doi.org/10.1101/gad.1803909

Kapulkin WJ (2016) Retroviral origins of the *Caenorhabditis elegans* orphan gene F58H7.5. bioRxiv. https://doi.org/10.1101/073510

Katoh I, Kurata S (2013) Association of endogenous retroviruses and long terminal repeats with human disorders. Front Oncol 3:234. https://doi.org/10.3389/fonc.2013.00234

Kirchberger PC, Ochman H (2020) Resurrection of a global, metagenomically defined gokushovirus. elife 9:e51599. https://doi.org/10.7554/eLife.51599

Kobayashi Y, Horie M, Nakano A et al (2016) Exaptation of Bornavirus-like nucleoprotein elements in Afrotherians. PLoS Pathog 12(8):e1005785. https://doi.org/10.1371/journal.ppat.1005785

Kochetkova M, McKenzie OL, Bais AJ et al (2002) CBFA2T3 (MTG16) is a putative breast tumor suppressor gene from the breast cancer loss of heterozygosity region at 16q24.3. Cancer Res 62:4599–4604

Krupovic M, Forterre P (2011) Microviridae goes temperate: microvirus-related proviruses reside in the genomes of Bacteroidetes. PLoS One 6(5):e19893. https://doi.org/10.1371/journal.pone.0019893

Kuriyama K, Tabara M, Moriyama H et al (2020) Disturbance of floral colour pattern by activation of an endogenous pararetrovirus, petunia vein clearing virus, in aged petunia plants. Plant J 103:497–511. https://doi.org/10.1111/tpj.14728

Küry P, Nath A, Créange A et al (2018) Human endogenous retroviruses in neurological diseases. Trends Mol Med 24(4):379–394. https://doi.org/10.1016/j.molmed.2018.02.007

Lamprecht B, Walter K, Kreher S et al (2010) Derepression of an endogenous long terminal repeat activates the CSF1R proto-oncogene in human lymphoma. Nat Med 16(5):571–579, 1p following 579. https://doi.org/10.1038/nm.2129

Leclercq S, Gilbert C, Cordaux R (2012) Cargo capacity of phages and plasmids and other factors influencing horizontal transfers of prokaryote transposable elements. Mob Genet Elem 2 (2):115–118. https://doi.org/10.4161/mge.20352

Lee G, Chakraborty U, Gebhart D et al (2016) F-type Bacteriocins of *Listeria monocytogenes*: a new class of phage tail-like structures reveals broad parallel coevolution between tailed bacteriophages and high-molecular-weight Bacteriocins. J Bacteriol 198:2784–2793. https://doi.org/10.1128/JB.00489-16

Legeai F, Santos BF, Robin S et al (2020) Genomic architecture of endogenous ichnoviruses reveals distinct evolutionary pathways leading to virus domestication in parasitic wasps. BMC Biol 18:89. https://doi.org/10.1186/s12915-020-00822-3

Leobold M, Bézier A, Pichon A et al (2018) The domestication of a large DNA virus by the wasp *Venturia canescens* involves targeted genome reduction through Pseudogenization. Genome Biol Evol 10:1745–1764. https://doi.org/10.1093/gbe/evy127

Li W, Lee M-H, Henderson L et al (2015) Human endogenous retrovirus-K contributes to motor neuron disease. Sci Transl Med 7(307):307ra153. https://doi.org/10.1126/scitranslmed.aac8201

Liu H, Fu Y, Jiang D et al (2010) Widespread horizontal gene transfer from double-stranded RNA viruses to eukaryotic nuclear genomes. J Virol 84:11879–11887. https://doi.org/10.1128/JVI.00955-10

Logan SL, Thomas J, Yan J et al (2018) The *Vibrio cholerae* type VI secretion system can modulate host intestinal mechanics to displace gut bacterial symbionts. Proc Natl Acad Sci U S A 115: E3779–E3787. https://doi.org/10.1073/pnas.1720133115

Louis F, Bézier A, Periquet G et al (2013) The bracovirus genome of the parasitoid wasp *Cotesia congregata* is amplified within 13 replication units, including sequences not packaged in the particles. J Virol 87:9649–9660. https://doi.org/10.1128/JVI.00886-13

Lundstrom K (2019) RNA viruses as tools in gene therapy and vaccine development. Genes 10 (3):189. https://doi.org/10.3390/genes10030189

Mann NH, Clokie MR, Millard A et al (2005) The genome of S-PM2, a "photosynthetic" T4-type bacteriophage that infects marine *Synechococcus* strains. J Bacteriol 187:3188–3200. https://doi.org/10.1128/JB.187.9.3188-3200.2005

Márquez LM, Redman RS, Rodriguez RJ, Roossinck MJ (2007) A virus in a fungus in a plant: three-way symbiosis required for thermal tolerance. Science 315(5811):513–515. https://doi.org/10.1126/science.1136237

Masum MMI, Yang Y, Li B et al (2017) Role of the genes of type VI secretion system in virulence of Rice bacterial Brown stripe pathogen *Acidovorax avenae* subsp. *avenae* strain RS-2. Int J Mol Sci 18(10):2024. https://doi.org/10.3390/ijms18102024

McClure MA, Perrault J (1985) Poliovirus genome RNA hybridizes specifically to higher eukaryotic rRNAs. Nucleic Acids Res 13:6797–6816. https://doi.org/10.1093/nar/13.19.6797

McLaughlin JR, Wong HC, Ting YE et al (1986) Control of lysogeny and immunity of *Bacillus subtilis* temperate bacteriophage SP beta by its d gene. J Bacteriol 167:952–959. https://doi.org/10.1128/jb.167.3.952-959.1986

Metzger MJ, Reinisch C, Sherry J et al (2015) Horizontal transmission of clonal cancer cells causes leukemia in soft-shell clams. Cell 161(2):255–263. https://doi.org/10.1016/j.cell.2015.02.042

Metzger MJ, Paynter AN, Siddall ME et al (2018) Horizontal transfer of retrotransposons between bivalves and other aquatic species of multiple phyla. Proc Natl Acad Sci U S A 115(18):E4227–E4235. https://doi.org/10.1073/pnas.1717227115

Murphy N, Banks JC, Whitfield JB et al (2008) Phylogeny of the parasitic microgastroid subfamilies (Hymenoptera: Braconidae) based on sequence data from seven genes, with an improved time estimate of the origin of the lineage. Mol Phylogenet Evol 47:378–395. https://doi.org/10.1016/j.ympev.2008.01.022

Nakayama K, Takashima K, Ishihara H et al (2000) The R-type pyocin of *Pseudomonas aeruginosa* is related to P2 phage, and the F-type is related to lambda phage. Mol Microbiol 38(2):213–231. https://doi.org/10.1046/j.1365-2958.2000.02135.x

Navarro-Garcia F, Ruiz-Perez F, Cataldi Á et al (2019) Type VI secretion system in pathogenic *Escherichia coli*: structure, role in virulence, and acquisition. Front Microbiol 10:1965. https://doi.org/10.3389/fmicb.2019.01965

Nelson PN, Carnegie PR, Martin J et al (2003) Demystified … Human endogenous retroviruses. Mol Pathol 56(1):11–18. https://doi.org/10.1136/mp.56.1.11

Pant A, Das B, Bhadra RK (2020) CTX phage of *Vibrio cholerae*: genomics and applications. Vaccine 38(Suppl 1):A7–A12. https://doi.org/10.1016/j.vaccine.2019.06.034

Parker BJ, Brisson JA (2019) A laterally transferred viral gene modifies aphid wing plasticity. Curr Biol 29:2098–2103. https://doi.org/10.1016/j.cub.2019.05.041

Patz S, Becker Y, Richert-Pöggeler KR et al (2019) Phage tail-like particles are versatile bacterial nanomachines – a mini-review. J Adv Res 19:75–84. https://doi.org/10.1016/j.jare.2019.04.003

Pistello M, Antonelli G (2016) Integration of the viral genome into the host cell genome: a double-edged sword. Clin Microbiol Infect 22:296–298. https://doi.org/10.1016/j.cmi.2016.01.022

Platt RN 2nd, Vandewege MW, Ray DA (2018) Mammalian transposable elements and their impacts on genome evolution. Chromosom Res 26(1-2):25–43. https://doi.org/10.1007/s10577-017-9570-z

Ramisetty BCM, Sudhakari PA (2019) Bacterial 'grounded' prophages: hotspots for genetic renovation and innovation. Front Genet 10:65. https://doi.org/10.3389/fgene.2019.00065

Ravin NV (2015) Replication and maintenance of linear phage-plasmid N15. Microbiol Spectrum 3(1):PLAS-0032-2014. https://doi.org/10.1128/microbiolspec.PLAS-0032-2014

Redfield RJ, Soucy SM (2018) Evolution of bacterial gene transfer agents. Front Microbiol 9:2527. https://doi.org/10.3389/fmicb.2018.02527

Roossinck MJ (2015) Move over, bacteria! Viruses make their mark as mutualistic microbial symbionts. J Virol 89:6532–6535. https://doi.org/10.1128/JVI.02974-14

Roossinck MJ, Bazán ER (2017) Symbiosis: viruses as intimate partners. Annu Rev Virol 4:123–139. https://doi.org/10.1146/annurev-virology-110615-042323

Ruprecht K, Mayer J (2019) On the origin of a pathogenic HERV-W envelope protein present in multiple sclerosis lesions. Proc Natl Acad Sci U S A 116:19791–19792. https://doi.org/10.1073/pnas.1911703116

Ryabov EV, Keane G, Naish N et al (2009) Densovirus induces winged morphs in asexual clones of the rosy apple aphid, Dysaphis plantaginea. Proc Natl Acad Sci U S A 106:8465–8470. https://doi.org/10.1073/pnas.0901389106

San Miguel P, Vitte C (2009) The LTR-retrotransposons of maize. In: Bennetzen JL, Hake S (eds) Handbook of maize. Springer, New York. https://doi.org/10.1007/978-0-387-77863-1_15

Schubert RA, Dodd IB, Egan JB et al (2007) Cro's role in the CI Cro bistable switch is critical for λ's transition from lysogeny to lytic development. Genes Dev 21(19):2461–2472. https://doi.org/10.1101/gad.1584907

Scott J, Thompson-Mayberry P, Lahmamsi S et al (2008) Phage-associated mutator phenotype in group a streptococcus. J Bacteriol 190:6290–6301. https://doi.org/10.1128/JB.01569-07

Shakya M, Soucy SM, Zhaxybayeva O (2017) Insights into origin and evolution of α-proteobacterial gene transfer agents. Virus Evol 3(2):vex036. https://doi.org/10.1093/ve/vex036

Sherlock D, Leong JX, Fogg PCM (2019) Identification of the first gene transfer agent (GTA) small terminase in *Rhodobacter capsulatus* and its role in GTA production and packaging of DNA. J Virol 93:e01328–e01319. https://doi.org/10.1128/JVI.01328-19

Smith WPJ, Vettiger A, Winter J et al (2020) The evolution of the type VI secretion system as a disintegration weapon. PLoS Biol 18(5):e3000720. https://doi.org/10.1371/journal.pbio.3000720

Soygur B, Sati L (2016) The role of syncytins in human reproduction and reproductive organ cancers. Reproduction 152:R167–R178. https://doi.org/10.1530/REP-16-0031

Stokar-Avihail A, Tal N, Erez Z et al (2019) Widespread utilization of peptide communication in phages infecting soil and pathogenic Bacteria. Cell Host Microbe 25(5):746–755.e5. https://doi.org/10.1016/j.chom.2019.03.017

Swanstrom R, Parker RC, Varmus HE et al (1983) Transduction of a cellular oncogene: the genesis of Rous sarcoma virus. Proc Natl Acad Sci U S A 80:2519–2523. https://doi.org/10.1073/pnas.80.9.2519

Takahashi H, Fukuhara T, Kitazawa H et al (2019) Virus latency and the impact on plants. Front Microbiol 10:2764. https://doi.org/10.3389/fmicb.2019.02764

Tan C-W, Peiffer M, Hoover K et al (2018) Symbiotic polydnavirus of a parasite manipulates caterpillar and plant immunity. Proc Natl Acad Sci U S A 115:5199–5204. https://doi.org/10.1073/pnas.1717934115

Tang Y, Woodward BO, Pastor L et al (2020) Endogenous retroviral envelope Syncytin induces HIV-1 spreading and establishes HIV reservoirs in placenta. Cell Rep 30:4528–4539.e4. https://doi.org/10.1016/j.celrep.2020.03.016

Tassetto M, Kunitomi M, Whitfield ZJ et al (2019) Control of RNA viruses in mosquito cells through the acquisition of vDNA and endogenous viral elements. elife 17(8):e41244. https://doi.org/10.7554/eLife.41244

ter Horst AM, Nigg JC, Dekker FM et al (2019) Endogenous viral elements are widespread in arthropod genomes and commonly give rise to PIWI-interacting RNAs. J Virol 93(6):e02124–e02118. https://doi.org/10.1128/JVI.02124-18. Link with supplemental tables. https://www.biorxiv.org/content/biorxiv/early/2018/08/20/396382.full.pdf

Thomas-Bulle C, Piednoël M, Donnart T et al (2018) Mollusc genomes reveal variability in patterns of LTR-retrotransposons dynamics. BMC Genomics 19:821. https://doi.org/10.1186/s12864-018-5200-1

Thompson PJ, Macfarlan TS, Lorincz MC (2016) Long terminal repeats: from parasitic elements to building blocks of the transcriptional regulatory repertoire. Mol Cell 62(5):766–776. https://doi.org/10.1016/j.molcel.2016.03.029

Tokuyama M, Kong Y, Song E et al (2018) ERVmap analysis reveals genome-wide transcription of human endogenous retroviruses. Proc Natl Acad Sci U S A 115:12565–12572. https://doi.org/10.1073/pnas.1814589115

Touret F, Guiguen F, Terzian C (2014) Wolbachia influences the maternal transmission of the Gypsy endogenous retrovirus in Drosophila melanogaster. MBio 5(5):e01529–e01514. https://doi.org/10.1128/mBio.01529-14

Tripathi JN, Ntui VO, Ron M et al (2019) CRISPR/Cas9 editing of endogenous *banana streak virus* in the B genome of *Musa* spp. overcomes a major challenge in banana breeding. Commun Biol 2:46. https://doi.org/10.1038/s42003-019-0288-7

van der Kuyl AC, Dekker JT, Goudsmit J (1995) Distribution of baboon endogenous virus among species of African monkeys suggests multiple ancient cross-species transmissions in shared habitats. J Virol 69:7877–7887. https://doi.org/10.1128/JVI.69.12.7877-7887.1995

Vargiu L, Rodriguez-Tomé P, Sperber GO et al (2016) Classification and characterization of human endogenous retroviruses; mosaic forms are common. Retrovirology 13:7. https://doi.org/10.1186/s12977-015-0232-y

Villareal LP (1997) On viruses, sex, and motherhood. J Virol 71:859–865. https://doi.org/10.1128/JVI.71.2.859-865.1997

Villarreal LP (2016) Viruses and the placenta: the essential virus first view. APMIS 124 (1–2):20–30. https://doi.org/10.1111/apm.12485

Villarreal LP, Witzany G (2019) That is life: communicating RNA networks from viruses and cells in continuous interaction. Ann N Y Acad Sci 1447(1):5–20

Volkoff AN, Cusson M (2020) The unconventional viruses of Ichneumonid parasitoid wasps. Viruses 12(10):1170. https://doi.org/10.3390/v12101170

Wang GH, Sun BF, Xiong TL et al (2016) Bacteriophage WO can mediate horizontal gene transfer in endosymbiotic *Wolbachia* genomes. Front Microbiol 7:1867. https://doi.org/10.3389/fmicb.2016.01867

Wang Y, Liang W, Tang T (2018) Constant conflict between *Gypsy* LTR retrotransposons and CHH methylation within a stress-adapted mangrove genome. New Phytol 220(3):922–935. https://doi.org/10.1111/nph.15209

Weinbauer MG (2004) Ecology of prokaryotic viruses. FEMS Microbiol Rev 28(2):127–181. https://doi.org/10.1016/j.femsre.2003.08.001

Yao GW, Duarte I, Le TT et al (2017) A broad-host-range Tailocin from *Burkholderia cenocepacia*. Appl Environ Microbiol 83(10):e03414–e03416. https://doi.org/10.1128/AEM.03414-16

Zajdowicz SLW, Holmes RK (2016) Phage conversion and the role of bacteriophage and host functions in regulation of diphtheria toxin production by *Corynebacterium diphtheriae*. In: Hurst CJ (ed) The mechanistic benefits of microbial symbionts advances in environmental microbiology, vol 2. Springer, Cham. https://doi.org/10.1007/978-3-319-28068-4_2

Zhang X, Qi Y (2019) The landscape of Copia and Gypsy retrotransposon during maize domestication and improvement. Front Plant Sci 10:1533. https://doi.org/10.3389/fpls.2019.01533

Zhang X, McDaniel AD, Wolf LE et al (2000) Quinolone antibiotics induce Shiga toxin-encoding bacteriophages, toxin production, and death in mice. J Infect Dis 181:664–670. https://doi.org/10.1086/315239

Zhu F, Cusumano A, Bloem J et al (2018) Symbiotic polydnavirus and venom reveal parasitoid to its hyperparasitoids. Proc Natl Acad Sci U S A 115:5205–5210. https://doi.org/10.1073/pnas.1717904115

Ziegelin G, Tegtmeyer N, Lurz R et al (2005) The repA gene of the linear *Yersinia enterocolitica* prophage PY54 functions as a circular minimal replicon in *Escherichia coli*. J Bacteriol 187:3445–3454. https://doi.org/10.1128/JB.187.10.3445-3454.2005

Chapter 5
Einstein's Capsid: Bacteriophages Solve the Problems of Space and Time for Bacteria with Emergency Dead to Alive Horizontal Gene Transfer (EDA-HGT)

Leigh Owens

Abstract Bacteriophages evolved to solve three major problems faced by bacterial populations. Phages have allowed the emergency horizontal transfer of genes from dead cells to alive cells (EDA-HGT) so that unique genetic diversity is not lost to the bacterial species' genosphere. Secondly, they help maintain genetic diversity in clonal bacterial cells that have moved away from effective conjugational distance allowing less hazardous mobility across space. Thirdly, they allow a delayed temporal component to gene transfer for future generations. Bacteriophages have been fundamental in allowing bacteria to exploit hostile, rapidly changing environments including patchiness in resource availability. Prime examples are the Enterobacteriaceae and the Vibrionaceae. Derivatives such as gene transfer agents (GTA) are viable descendants of phages that deliver more varied genetic material but at the cost of a slower rate as they cannot self-liberate and therefore are reliant on other causes of lysis. Why the alphaproteobacteria have such an affinity for dead to alive GTA is unknown but it is possible that their environment is so hazardous that high genetic flexibility is an advantage. In addition, bacteria actively participate in their own infection by bacteriophage. Once bacteriophages evolved, selection pressure has pushed the host-phage interactions in many directions, but the underlying advantage of emergency, diverse horizontal gene transfer has meant that viruses are an integral part of life and they have not been removed by natural selection.

5.1 Introduction

In terms of genetic diversity, the importance of horizontal gene transfer for particularly a clonal organism like bacteria is a well-established paradigm of biology (Lawrence 1999, 2002). This is especially true for opportunist r-selective organisms

L. Owens (✉)
College of Public Health, Medical and Veterinary Sciences, James Cook University, Townsville, Australia
e-mail: Leigh.Owens@jcu.edu.au

C. J. Hurst (ed.), *The Biological Role of a Virus*, Advances in Environmental Microbiology 9, https://doi.org/10.1007/978-3-030-85395-2_5

(environmental exploitative, rapid reproduction and short generation time) or those in rapidly changing environments. Whilst sexual gene transfer through sex pili, conjugative plasmids and homologous recombinations of high frequency recombinants (hfr) are possible for relatively closely related species in close proximity, horizontal gene transfer agents (HGTA) allows genetic material exchange across broader horizons (McDaniel et al. 2010). Bacteriophages can transcend the tyrannies of distance, time and mortality. All other methods of gene transfer (e.g. mitosis, conjugation, sexual pili, and transformation from free DNA) mean that individual cells must be alive (or at least, very recently alive for free DNA), in very close proximity, preferably in physical contact, perhaps in a biofilm. Bacteriophages allow gene transfer across relatively large tracts of both space, as close proximity is not necessary, and time, as receipt of genetic information will always post-cede its' development, perhaps by years. This evolvement allows planktonic-phase bacterial cells to find new zones of nutrient and minimise the disadvantage of genetic isolation from leaving the "clonal mother organism" in the biofilm.

I examine the hypothesis that bacteria, when about to die, evolved viruses as an emergency, remote genetic inheritance mechanism to save unique genes for their close relatives. It is really a delayed, horizontal gene transfer between a dead and a live cell (Emergency, Dead to Alive Horizontal Gene Transfer, EDA-HGT); a true phoenix. Once EDA-HGT evolved, bacteria became free of constraints of the need for proximity for gene exchange allowing them to exploit the massive advantages of mobility "to boldly go where no bacterium had gone before". Once bacteriophages formed, evolutionary pressures exerted their own influences selecting other various traits as necessitated by co-evolution with host cells.

The review uses the analogy of humans wishing to pass on their own legacy when their demise is imminent as the parallels are instructive and perhaps a little light hearted to help digest a radical re-think of our current paradigms. Bacteria provide "a last will and testament" to their closest, living relatives leaving their most precious heirlooms, their unique genes, to their relatives. When a bacterial cell senses that it is about to die, it is time to "write the (genetic) will". What could trigger the preparing of genes for inheritance? Mortal level stress including DNA damage caused by environmental triggers like UV light, pollutants, and high temperature denaturing proteins inducing the SOS/*RecA* DNA repair system which often triggers the bacteriophage lytic cycle over the lysogenic cycle. UV light might be strong enough to fuse adjacent pyrimidines beyond repair or damage beyond the reading by mechanisms designed to ignore DNA knots, thus silencing critical genes and condemning bacterial cells to a slow death. To pass on important genes, bacteria had to produce "a time capsule", the bacteriophage capsid, so the genes would have some chance of surviving the hostile external environment until the relatives could "read the will". Following the precepts of the "The Selfish Gene" (Dawkins 2016), bacteria "want" their precious inheritance to have a higher probability of being picked up by relatives which are genetically closest to the dying donor, hence the specificity of the tail fibre receptors to immediate sibling cells or closely related strains and species sharing strong genetic homology giving rise to phage specificity.

Occasionally inheritance goes awry, and non-intended recipients gain the benefit and perhaps also the cost of new genetic capabilities.

Like in the human analogy, not all the inheritance you receive is unique and therefore wanted. How many copies of the book, "The Selfish Gene" do you need? Immunity to superinfection by bacteriophages could have evolved to prevent unnecessary clogging of the internal, intracellular genetic library. To dispose of all the excess copies of genes that you already have in your library and just keep the unique ones, then a filter system like clustered regularly interspaced short palindromic repeats (CRISPR), which by definition, can only knock out gene sequences it has already met, could develop. Therefore, *all novel* genes pass the CRISPR filter and potentially become available for use by the new bacterial host cell. Indeed, Kalatzis et al. (2017) found phage sequences widely distributed among CRISPR-Cas arrays of publicly available sequences of vibrios. Also, like a true library, many books (aka genes) can be in storage on the shelves (extracellular in phages) for later reading. With an estimate of 10^{31} phages on the planet (Dion et al. 2020), the library for exploitation is vast.

All biological systems are leaky and not 100% efficient. Once a system of inheritance like this has evolved, it is very hard to stop parasites and mistakes from happening. Evolution of soft membraned organisms like eukaryotes and mycoplasma rather than rigid cell-walled organisms meant the need for the delivery apparatus (tail constructs etc.) was redundant allowing evolution of just the "time capsule", naked icosahedral virions. This freed up a huge amount of intracapsid space (up to 50% for the structural genes of phages) in the capsid for more genes to be inherited. Thus the viruses were free of their internal space constraints, their original specific target and perhaps, purpose. Indeed, "the gen(e)ie was out of the bottle". There is a vast amount of evidence that most, if not all, eukaryotic viruses evolved from phages, perhaps polyphyleticly, but that argument is beyond the scope of this paper. For excellent dealings with this topic, see the many papers by Krupovic, Bamford and colleagues including Krupovic and Bamford (2008).

5.2 Approach

Language can constrain the way people think about issues. The term "bacteriophage" containing the stem "-phage" which is the common abbreviation for bacteriophages immediately makes scientists prejudiced in the way they think about the relationship between bacteriophage and bacteria—"a virus that eats bacteria". Immediately, thoughts are biased to lysis being the common state over lysogeny. Even the term "virus"—"a poisonous substance" brings its own constraints. Language suggests bacteriophages are "bad for you" and this has had an impact on how we view lysogeny and lysis. However, the terms are so entrenched in the literature and microbiologist's psyche, it would be near impossible to change. The present author believes that lysogeny is a normal state for the relationship between a bacterial cell

Table 5.1 Bacteriophages mentioned in the text and the families (if known) they are assigned to

Family	Bacteriophage
Autographiviridae	*Eschericia coli* T7
Inoviridae	*Vibrio* virus CTXphi
Myoviridae	Mu, P1, P2, VHML, Vp58.5
Siphoviridae	Czyszczon1, *Eschericia coli* lambda
Tectiviridae	PRD1
Pelagiphage	vSAG 37-F6

and its' bacteriophage. Indeed, both states are critical to the survival of the population of unique genes belonging to and indeed defining, a species of bacteria.

By analysing modern abundance, Ackermann (2003) suggests that the *Siphoviridae* are the ancestors of the tailed phages and evolved around 3.5 billion years ago with the *Myoviridae* evolving not long after before the split between the eubacteria and the archaea. Recently, it was estimated that >85% of records in public genome databases are members of the *Order Caudovirales*, the tailed phages with *Siphoviridae* being over 50% (Dion et al. 2020). We should perhaps find caution that common, successful bacteria would by sheer weight of numbers, swamp other relationships and this can be seen by the number of pelagiphages that infect the SAR11/*Pelagibacteria*/SAR116 (Mizuno et al. 2013) which host the single most abundant phage species, vSAG 37-F6 (Table 5.1) in marine waters (Dion et al. 2020). Nevertheless, I have taken the siphovirus, *Escherichia coli* lambda bacteriophage, as our default model to reflect the eminently sensible suggestions of Ackermann and because much is known about the control of lysogeny and lysis in Lambda virus.

5.3 Defining the Hypothesis

5.3.1 Sensing Impending Death

Most human adults know when their physiology and anatomy is starting to show wear. We call it aging and this signals a time to write our will. However, bacterial cells by their continued binary division and clonality are in effect, continually renewed, never ageing. What signal could a forever-renewing cell use to recognise impending death? The induction of the SOS/*RecA* DNA repair system could be a premonition of impending demise. Mitogens like ultraviolet (UV) light and Mitomycin C damage the DNA and turn on the DNA repair machinery. In the case of UV light, fusion of near opposing, adjacent pyrimidines means repair is often impossible as the DNA strands cannot be unzipped to be repaired by replication-blocked replicases and fusing of adjacent pyrimidines means the gene is effectively silenced. In many cases, cell death will follow or mutated cells will be compromised from the loss of gene function.

In the SOS/*RecA* system, the protein RecA becomes activated as a specific co-protease that causes lexA and the homodimer cI (clear I) to undergo self-cleavage thus dropping the concentration of cl, inducing the lytic cascade. cl is a repressor protein encoded by the *cl* gene of lambda, and the cl protein functions as a transcription inhibitor which maintains Lambda phage in a lysogenic state. This repression is achieved when two dimers of cl bind cooperatively to the integrated lambda phage DNA at adjacent operator sites. That binding both activates the phage *cl* gene and represses the phage *cro* gene, which codes for the protein cro. *cro* is a gene that in some ways exists as the counterbalance to *cl*, with the cro protein being a repressor active during the Lambda lytic cycle and cro prevents transcription of the *cl* gene. The hairpin loop needed for RNAase III degradation of the Lambda prophage mRNA is cleaved off and therefore the now full length mRNA is patent. Consequently, the *excis*ion/*int*egration proteins xis and int are produced in equal amounts, so excision of the template prophage occurs allowing further DNA replication. Thus any genes between the chromosomal attachment sites (att sites, where the phage DNA binds) of the prophage will be magnified and prepared for "inheritance" by genetically near relatives. If there have been any genetic changes to the prophage by say, homologous recombination, also these changes will be "prepared for inheritance" as "the will is written".

5.3.2 The Probability of Lysis over Lysogeny

Logically, when cell biomass is high, then it is safest to undergo lysis as there are plenty cells to keep the strain of bacteria alive to continue the species/strain existence. Paul (2008) provides a review of the evidence that lysis is high in nutrient rich waters (high cell numbers) whilst lysogeny is high in oligotrophic waters. Additional support comes from two indicators of high cell density that exist, absolute cell numbers itself and long cells (large cells) which occur as cells grow in the polar directions before division when there is sufficient nutrient. Papers by St Pierre and Endy (2008) and Zeng et al. (2010) showed that as cell size increased, a prelude to cellular division, the probability of lysis increased. Similarly, when there are plenty of nutrients around, there will be many bacterial cells in the environment, so lysis is a safe option from population-evolutionary viewpoint. In Lambda phage, when there is an abundance of proteins (nutrients) available, then protease production of the bacterial cell is increased and this breaks down Clear 2 (cll) which reduces the binding (silencing) of three promoters (P_{RE}, P_I, P_{antiq}) that stabilise lysogeny, thus leading to the lytic cycle.

5.3.3 Superinfection Immunity and CRISPR

Superinfection immunity is where a lysogen is producing lots of cl to retain the prophage. When other viral nucleic acid is injected by more phages using the same Lambda cl control system, cl immediately binds to the integrated Lambda preventing transcription and effectively silences the incoming phage.

Separately, a second system, CRISPR could be activated. The incoming phage genome can be processed by CRISPR to silence already acquired gene sequences. CRISPR-associated (*cas*) genes process viral DNA to produce ~24–48 bp, mostly ~30 bp lengths of DNA called spacers that are inserted into the genome (5′ to 3′ direction) following the *cas* genes after a leader sequence that becomes the template for the production of interfering-like RNAs. The mechanism for production of spacers and whether it is from transcripts or the original viral template and how they are integrated is currently unknown. The RNAs transcribed from the spacers plus imperfect terminal repeats combine with other cas proteins to produce a cas-crRNA complex that acts as a RNA-guided restriction enzyme, and that complex functions by cutting and thereby inactivating viral DNA or RNA. Any already known phage genes can be immediately silenced, whilst novel sequences have to be processed first, thus giving a chance for alternate processing to allow time for genome or prophage integration to occur.

5.3.4 Other Supporting Evidence

EDA-HGT is not the only dead-to-alive messaging system found in bacteria. Bhattacharyya et al. (2020) have found that when swarming morphs of bacteria (the equivalent of scouting bacteria going at warp-speed "to boldly go where no bacterium had gone before") are killed by antibiotics, a "necrosignal" protein AcrA, non-genetically turns conspecific bacteria in the biofilms into a more antibiotic-resistant population. This was found in five species of both Gram-negative and Gram-positive bacteria tested. It will be fascinating to see how widespread these dead-to-alive communications (EDTA-HGT, necrosignaling) turn out to be.

5.4 What Is the Advantage of EDA-HGT (Viruses)?

The positive selection pressure for the persistence of EDA-HGT consists of three main advantages. Firstly, that it is the most robust method of gene transfer that allows genes from dying and dead cells to be retrieved by live cells. Genes and partial genes are not lost from the species' gene population or even the greater virosphere. Secondarily, as the capsid protects the genes from environmental and biotic degradation; there is also the possibility of delayed acquisition of genes;

therefore a later time component to inheritance. The death of a bacterial cell thousands of years ago can still have viable phages and hence viable genes (e.g. Czyszczon1 "mycobacterium" *Siphoviridae* phage; from the last ice age ~ 10,000 years ago which were still viable).

The third major advantage is bacteriophage can circumvent the tyranny of distance. All other methods of gene transfer (e.g. mitosis, conjugation, sexual pili, transformation from free DNA) need individual cells in very close proximity, preferably in physical contact. An intriguing supportive example for emergency gene recovery comes from the Ups (*U*V-induced type IV *p*ili of the *Sulfolobus*) system in the crenarchaeotan *Sulfolobus* (Ajon et al. 2011, also see below). When UV damage occurs in *Sulfolobus*, a strong, type IV pili system is induced which produces massive amounts of pili that cause species-specific aggregations, conjugative transfer of chromosomal markers and repair. The system is incredibly similar to SOS/*RecA* system except it needs both partners to be alive, mobile and close enough for the induced pili to form aggregates (biofilm) for DNA transfer and repair.

5.5 Potential Ontogeny of the Bacteriophage Related Systems

5.5.1 Progressive Capture of Supplementary DNA by EDA-HGT

Lambda *Siphoviridae* integrate into a single site, the *attB* site between gal and bio operons. Upon excision, only small bits of extraneous DNA from the integration site are sometimes accidentally incorporated into virions. However, any extra sequences of DNA will always only represent the same two genes at either end of the prophage. Clearly, this is not an efficient way of increasing genetic diversity in HGT. However, Mu-like myoviruses have a 62% preference for certain insertional sequences, with a randomness to the remaining insertions. That randomness allows a transposon-like quality to the resulting insertions (Haapa-Paananen et al. 2002) which, upon induction, may remove and then transfer a wider selection of DNA sequence information variability. Due to the head full or tape measure mechanism which seems to govern filling the capsid with DNA, on average, Mu phage packages into each capsid 50–150 bp of host DNA on the left hand end (5′) of the phage genome and approximately 2 kb of host DNA on the right hand end (3′). In addition, when heat-sensitive repressor mutants (42 °C) of Mu phage are triggered (under heat stress leading to protein denaturing; a near death experience), then 50–100 new host chromosomal integration sites are utilised, greatly increasing genetic diversity and HGT.

5.5.2 The Effects of Plasmids

Many linear phages form a plasmid as the first act upon entering a cell. This allows them to easily utilize the rolling cycle replication machinery of the host cell which is optimised to operate on a bacterial circular genome. In Lambda phage and many other phages, it is the sticky *cos* ends that allows this to happen. Falling into this category are the telomere phages which consist of what scientists consider a mosaic of characteristics. The so-called telomere phages exist as a linear plasmid in the cell but are circularised upon initial entry and can be circularised during replication and include the gene, *RepA* (Mardanov and Ravin 2007), the hallmark of rolling cycle replication from circular templates. *RepA* is a replicative helicase with a nucleoside triphosphate (NTP) binding site for duplicating DNA. Telomere phages have the *Ori* site (the site of the initiation of replication), plasmid partitioning genes and many other genes associated with plasmids but are controlled and triggered into lysis by a lambda-like system of UV-induced *RecA-LexA* system. Morphologically and genetically, telomere phages have both family *Siphoviridae* lambda structural genes (N15, Ravin 2011) and family *Myoviridae* structure for the tail-associated genes but lambda-like head genes (VHML, Oakey et al. 2002; Vp58.5, Zabala et al. 2009). Their main hosts recognised at present are the Vibrionaceae, Enterobacteriaceae and alphaproteobacteria which are major utilisers of the EDA-HGT. Is it possible that telomere phages are not a HGT virus to virus chimera of all these genes but progenitors of the two families?

The myovirus P1 exists as a single or very low copy number plasmid with terminal redundancy and circularly permuted DNA. Within the capsid, the genome has identical extra sequences at each end being equivalent of up to ~130% of the genome. The Cre-lox (*c*auses *re*combination—*lo*cus *o*f *X*-over P1) system is used to circularise the linear genome upon cellular entry and to separate daughter plasmids in rolling cycle propagation. The phage genome encodes a "plasmid addiction system" that kills daughter cells that fail to get a copy of the plasmid upon cellular division. The plasmid addiction system consists of a stable protein toxin and often two molecules of the antitoxin that dynamically bind together which neutralizes the toxin (Gazit and Sauer 1999). Bacterial cells that lose or fail to get the plasmid get killed as the antitoxin degrades faster than the toxin. Despite many modifications to the basic scheme above, there are strong arguments for a common ancestor (see review Rawlings 1999). Interestingly, *relBE* toxin/antitoxin and similar genes are present on the chromosome of *E. coli, Haemophilus influenzae, Vibrio cholerae* and an enterotoxin-encoding plasmid p307, widespread on the chromosomes of Gram-positive bacteria and Archaea (Rawlings 1999).

By using plasmids or chromosomes to encode any necessary bacteriophage functions, these genes do not have to be placed in the capsid, thus usable space is created in the capsid. Under a "head full -tape measure" system of filling capsids with nucleic acid, terminal redundancy can occur. This has two major advantages. Extraneous DNA can be packaged as HGT for recombination thus increasing diversity and the possibility of novel recombinations particularly as one copy of the duplicated DNA can undergo massive mutational change with no functional phenotypic change to the virus. Paul (2008) suggests that one of the advantages of lysogeny is the introduction of new fitness factors by conversion or transduction.

Secondarily, every virion is slightly different as the starting and end points of the DNA in the capsid is different which may mean that some virions have dual, near terminal ORFs of some genes which may be beneficial on occasions. For example, myoviruses are known to up-regulate haemolysin activity when they infect members of the *Vibrio harveyi* clade (Munro et al. 2003; Vidgen et al. 2006; Buscido-Salcedo and Owens 2013). Most isolates of *Vibrio harveyi* have a single copy of a haemolysin gene that hybridises with *V. parahaemolyticus* thermolabile haemolysin which partially confers virulence. However, two of 13 isolates examined had two very near to identical copies of the haemolysin gene (Zhang et al. 2001) which could have come about from the terminal redundancy scenario above. It would be interesting to examine the haemolysin gene cassettes of VIB645, an isolate with two near identical haemolysin genes, to see if they are embedded at the ends of an ancient prophage pathogenicity island or plasmid or whether they are flanked by excision sites that might support the above supposition. Indeed, VIB645 does contain pVH1 a 16,427 bp conjugative virulence plasmid (GenBank: HM752272, accessed 18/12/2013) and numerous entries of haemolytic *V. parahaemolyticus* entered in Genbank have plasmids with plasmid addiction-type genes (e.g. *Par*A-*B*; plasmid stabilisation proteins) lending support to the above argument.

5.6 Bacteria Encoding for Their Own Infection with HGTA

Bacteria actively participate in their own infection by bacteriophages suggesting a mutualistic relationship rather than a pathogenic relationship. As bacteria have kept this attribute, then evolutionary logic says it in their own best interest to be infected by phages, otherwise it would be selected against and reduced or removed. Kalatzis et al. (2017) state specific genes like N6-adenine methyltransferase, lambda-like repressor and $tRNA^{Arg}$ demonstrate both a mutualistic and parasitic relationship between phages and host bacteria.

5.6.1 Integration Host Factor

A well-known example of bacteria aiding infection by bacteriophage is bacterial encoded integration host factor (IHF) also known as host-encoded DNA bending protein, HU (Friedman 1988). IHF attaches to the minor DNA grove, and bends the DNA to aid attachment of other proteins including Sigma factors. Amongst other actions, this facilitates integration of the phage genome (Friedman 1988). "It is noteworthy that many genetic free agents (e.g. phages, transposons, plasmids) use IHF in their independent DNA transactions (e.g. recombination, replication, partitioning and transfer)" (Friedman 1988). Some of the phage-related genes that IHF influences include ones for the production of *fimA* fimbriae, flagellin, phenylalanyl-tRNA synthetase and IHFa, *tra* transfer of plasmid DNA, cII in lysogeny, F-plasmid exclusion of phage T7, Mu phage transposition and replication repressor. "It is striking that both *int* and *Nu7* can be altered to produce IHF-independent proteins by single base pair changes. The fact that these simple changes have not occurred in wild-type Lambda suggests that maintenance of IHF's involvement in these functions is under strong positive selection to be maintained. What is confounding is that one reaction is part of the lysogenic pathway and the other is far along in the lytic pathway." (Friedman 1988). Friedman's statement demonstrates bacteria are not positively selected to remove their involvement in bacteriophage infection and therefore we must conclude both lysogeny and lysis are advantageous states for bacteria.

5.6.2 Type IV Secretion Systems and Type IV Pili

Many P-type plasmids encode both type IV secretion systems and type IV pili/fimbrae on the same plasmid (Bruno Gomez-Gil, CIAD, Mazatlan, Sinaloa, Mexico 82,000, personal communication). Also, the sex (F) pilus assembly machinery shares extensive similarities with the type IV secretion system (Filloux 2010). Members of the family *Tectiviridae*, PDR1 can only infect *E. coli* that have the conjugative P, N, or W type plasmid (all P-type plasmids) that encodes for the genes *Tra*1, 2–3 necessary for transfer (Kotilainen et al. 1993). Originally, PRD1 was believed to enter via a pilus tip encoded on plasmids including Resistance Plasmid 1 (RP1). However, Kotilainen et al. (1993) concluded that the functional P-type pilus did not need to elongate like an F sex pilus. Due to widespread visualisation of attached PRD1 phages to the planar surface of bacterial hosts, it was deduced that the highly expressed surface complex of the P-type pilus was responsible for both PRD1 adsorption and for conjugation ability.

This is similar to *Vibrio cholerae* phage CTXφ which in part is highly homologous to myovirus P2 (Nakayama et al. 2000). CTXφ needs the chromosomal vibrio pathogenicity island (VPI)φ containing genes for the toxin-coregulated pilus (TCP) to enter the cell for toxigenic conversion. Here, though, the relationship has gone one step more intimate in that the VPIφ has integrated by homologous recombination into classic lambda-like bacteriophage, site specific *att* sites (Karaolis et al. 1999). The TCP gene cluster encodes a type IV pilus that functions both as an essential colonization factor and as a CTXφ receptor. The TCP-A subunit is in fact, a coat protein of VPIφ.

Likewise, Carter et al. (2010) demonstrated that type IV pili are the components of *Pseudomonas aeruginosa* pathogenicity island (PAPI-1) conjugational machinery. It matters not if the pilus is exchanging DNA through a hollow tube or via pili retraction into the cell (a current controversy) as the pilus brings the cells close enough for conjugational exchange to occur anyway (Filloux 2010). Similarly, in the hyperthermic crenarchaeotan, order *Sulfolobales*, a chromosomal, conjugational *U*V-induced type IV *p*ili of *Sulfolobus* (Ups) exists that is used for DNA exchange after UV damage (Ajon et al. 2011). The genes encoding the Ups system are not recognised as showing a former integration event (AT rich zones) or components currently recognizable as the SOS/*RecA* system despite a similar, damaging environmental trigger, UV light.

The type IV secretion system is intriguing as it can excrete DNA, protein or nucleoprotein complexes and take up free dsDNA for transformation e.g. the *Campylobacter jejuni* Cjp/VirB system and the *Helicobacter pylori* ComB system (Cascales and Christie 2003) destroying one DNA strand to produce ssDNA which protein RecA can homologously recombine into the bacterial chromosome. It is not hard to see how the type IV secretion system could become specialised for just one function in certain bacteria. For example, it secretes nucleoprotein from *Agrobacterium tumefaciens* (VirB system) in which bacterial DNA is bound to at least one protein, the VirD2 transesterase, which transfers into the genomes of plants. This could be the ultimate reductive evolution of a viral nucleocapsid, one protein and nucleic acid. Alternatively, for secretion of chromosomal DNA extracellularly, *Neisseria gonorrhoeae* uses the type IV secretory system which is encoded on the gonococcal genetic island (GGI). Many of the genes are similar to *Tra* genes (see above) found on the *Escherichia coli* F-plasmid. The secreted DNA is available for uptake from the environment for transformation. Thus reduction here is to a *Tra* F- plasmid-like entity. The last type IV secretion method is to secrete just protein alone (no nucleotides) as a toxin like in *Bartonella* and *Brucella* VirB system (Cascales and Christie 2003).

Also interestingly, Pell et al. (2009) provide evidence that the tube protein Hcp1 from type VI bacterial secretion system has a common evolutionary origin to long tailed (Lambda) phage major tail protein gpV. This provides further evidence of the link between phage and secretion systems might even be a general rule rather than an exception.

5.7 Derivatives of EDA-HGT

5.7.1 Gene Transfer Agents

Gene Transfer Agents (GTA) can be defined as indisputably phage-like entities containing a random piece of DNA that are insufficient to transfer the structural blueprint to reproduce itself into a recipient cell (Lang et al. 2012). They are DNA-containing particles generated by some archaea and bacteria (Bárdy et al. 2020). "...the notion that GTAs predate phages is not substantiated by the current evidence" (Lang et al. 2012). The genes for the GTA are usually encoded on the major chromosome in an incomplete prophage-like manner. The bacterial cell has to be lysed by other means to release the GTA as they do not have the complete ability to "self-release" but may still have some of these genes. Therefore, efficiency of

GTA transfer is correlated to activity of other incumbent phages that are needed to release them. Multiplicity of Infection, increased release with higher nutrient and salinity are likely factors for release (McDaniel et al. 2010). One would predict that GTA will also be prevalent with other indicators of bacteriophage abundance like high cell density leading to nutrient depletion (Bárdy et al. 2020), high UV, high temperature, high pollutants and reactive oxygen stress.

Are there selective advantages of GTA over phages, particularly as the GTA are libero-parasitic (coined word, meaning they have a parasitic need to be released, incapable of their own release) on phages or other means of cell rupture? There seems to be three advantages. Firstly, by GTA not having all the structural genes or lysis machinery, the GTA have more capsid space to carry DNA (more HGT), although, on average, they appear to carry less genetic material than do similar capsid sized ancestral phages (Lang et al. 2012). Is much of this lessening of capacity for transporting genetic material due to the GTA lacking structural genes? Secondly, an advantage may come from the fact that GTA's never lyse a host that might have recovered from DNA damage, even though that same type of damage event in other bacteria might trigger a lysogenic phage into lysis. Thus, some percentage of "unnecessary accidental lysis" might be prevented by the host relying upon GTA instead of phage. Thirdly, the DNA encapsidated by GTA's is often randomly selected, allowing for horizontal transfer of a great diversity of genetic material that can be located far away from the insertion site of either the GTA or a prophage. Most of the material transferred by GTA from one bacterial cell to another cell will not simply be identical DNA copies of a common genomic sequence. GTA are only released on cell death, so they are a phage-derived, dead to alive HGT mechanism.

GTA are particularly common in the alphaproteobacteria, with ~65% of orders showing homology of GTA-associated genes, but they are found in the deltaproteobacterium *Desulfovibrio desulfuricans*, the spirochaete *Brachyspira hyodysenteriae* and interestingly the euryarchaeota *Methanococcus voltae* (Lang et al. 2012). Within the alphaproteobacteria, GTA have been shown to be effective (10–30%) in the natural environment at transduction of antibiotic resistance genes (McDaniel et al. 2010, 2012). Why do the alphaproteobacteria have such a proclivity for GTA? Perhaps the probability of cell lysis was so high in their ancestral environment, that there was no need to code for those liberation functions which freed up space for more varied gene transfer. Indeed, the alphaproteobacteria sequences dominate the prokaryotic sequences that have been identified from the photic zone of the Sargasso Sea (Venter et al. 2004). The photic zone is known to have high levels of UV, reactive oxygen and shear stress all of which damage cells leading to death and lysis.

5.7.2 *Bacteriocins: Type F and R Pyocins of* Pseudomonas aeruginosa

The F and R pyocins found in 90% of *Pseudomonas aeruginosa* have been extremely well characterised as the concatenated complete tail components and lysis gene cassettes of phages that are still under strong lambda-like, SOS/*RecA* control (Nakayama et al. 2000). Type F pyocin is derived from a lambda-like phage whilst type R pyocin with the lytic genes is derived from myovirus-like P2 phage and very similar to CTXφ. The pyocin cassette is integrated into the gene cassette for tryptophan metabolism between *trpE* and *trpGCD*. Both pyocins can be expressed by the application of mutagens like UV light, mitomycin C and acridine orange where the activated protein RecA releases the repressor prtP (~cI) allowing the transcription activator PrtN (~Cro) to transcribe the pyocin genes. Furthermore, Chang et al. (2005) showed reactive oxygen species damaging the DNA could also trigger the up-regulation of mRNA of the pyocin cluster. Nakayama et al. (2000) state that "The gene organization of the R2 and F2 pyocin gene cluster, however, suggested that both pyocins are not simple defective phages, but are phage tails that have been evolutionarily specialized as bacteriocins." As these pyocins are produced as proteins only, they have no function in HGT except in direct transformation. The very low frequency of the phage-evolved F and R pyocins only being found in a single species so far, *P. aeruginosa* shows this only evolved rarely. Perhaps it was for a special function (bacteriocin) in a specialised niche and was not advantageous enough to become widespread.

5.8 Biological Implications if EDA-HGT Theory Is Correct

5.8.1 *Benthic Biofilm as the Ancestral State for Bacteria?*

I am hypothesising that EDA-HGT allowed bacteria to transcend effective conjugational distance and therefore some testable predictions on the role of biofilm should be able to be made. The need for phages in biofilm should be reduced as all but the escaping planktonic cells will be able to undergo HGT easily via DNA transformation or conjugation. It should be possible to review the literature on the rate of phage induction in biofilm. Unfortunately, it becomes a circular argument as the biofilm protects cells from most DNA insults, thus the induction rate is decreased independent of the need of HGT. Furthermore, the amount of literature on biofilms and phage is enormous as many investigations on using phages to attack cells in the biofilm have been conducted. Pursuit of this idea is unfortunately outside the scope of this chapter.

5.8.2 Lysogeny vs Lysis Revisited

If this EDA-HGT theory is correct then one of the corollaries is that lysogeny is a partner to lysis between opportunistic bacteria and bacteriophages. There may be a perception that lysis is a more frequent occurrence than lysogeny but "your science is only as good as your assays". Recent publications have suggested a greater level of lysogeny than previously suggested, as lysis/lysogeny assays have become more sophisticated and independent methods have been used to cross-verify older techniques. Initially, mitogen-induced lysis on bacterial lawns showed low levels of lysogeny. This suffered from the problem that not all phages can be induced by mitomycin C which is the most common method used and also the mitomycin dose necessary for induction varies dramatically between bacteria/prophage couplets. The comprehensive section in Paul (2008) reviews the commonness of lysogeny, so we will paraphrase his review here with a few additions and the reader is directed to that review. Inducible prophage-like entities have been detected in marine bacteria at rates of 43% (Jiang and Paul 1994; Jiang et al. 1998); 51% (Oakey and Owens 2000); 71% (Stopar et al. 2004) and 28% (Leitet et al. 2006). With sequencing, it is estimated that 60–70% of all bacterial genomes (Casjens 2003) and ~ 67% (Brussow et al. 2004) contain prophage. It is possible that the choice of techniques used for cultivating medically significant and other non-marine bacteria may change the correctness with which we assess their possession of prophage. Surveying a subset of cultivation data suggested 47% contained prophages (Ackermann and DuBow 1987). A recent study mining bacterial genomes in the NCBI database for inovirus-like prophage sequences found 10,295 sequences (Dion et al. 2020). Following the methodology of Martinez-Hernandez et al. (2017), Dion et al. (2020) state that if we represent phage relationships as a network, then lysogenic (temperate) phages are at its' centre connecting to lytic (virulent) phages at the periphery and therefore lysogenic phages act as banks of genes for HGT. There is a trend for more modern or sophisticated studies to find higher prophage occurrence rates. Overall, the modern data suggests that prophage (lysogeny) are in the majority of domain *Bacteria* and therefore this is the dominate state.

Due to the technical difficulties of working with archaea, prophage data for these organisms is scarce. But, at present, it would seem that archaea do not follow this trend of high prophage possession rates. Perhaps, the tightly defined archaeal niche, which seems to be a hostile environment for other life forms, favours k-selection for stability (c.f. extremophiles) as being paramount over the need for variability that would facilitate rapidly exploiting a transient resource. Perhaps, we do not recognise all the methods of EDA-HGT in archaea. Indeed, in *Sulfolobus* the UV triggered, induced pili linked to conjugational chromosomal repair (Ups) acts like SOS/*RecA*, has the same HGT results but lacks the dead to alive and transiting space components. At the same time, *Sulfolubus* turreted icosahedral virus is only a lytic virus, never lysogenic (Snyder et al. 2011) which may suggest these two activities, lysogeny and lysis may have become decoupled in *Sulfolobus*, thus providing a pathway to eukaryotic virus lifestyles. On that note is that eukaryotes are more genetically closely related to archae (e.g. *Sulfolobus*) than true bacteria. It will be fascinating to follow the future trends in prophages detection studies as more archaea are fully sequenced.

5.9 Testing the Hypothesis

Many of the logical ways to test this hypothesis end up in circular or competing arguments and so they are unworkable. e.g. lysis should be less in deeper, non-photic zones away from the dangerous mitogen of UV light, dangerous levels of oxygen and the nutrient rich photic zone. However, bacterial cell biomass is less in deeper water, so phages would also be less common as their hosts are less common. It is not possible to separate the effect of there being fewer bacterial cells due to increasing aquatic depth from there possibly being less cell death due to induced phages in a more hostile environment. See also the section on biofilm (above).

It may be possible to use mathematical modelling to estimate the increased fitness attributable to HGT over other factors that contribute fitness in a clonal organism. Indeed, Raz and Tannenbaum (2010) using modelling showed that conjugational HGT does not confer an advantage in static environments and was slightly detrimental due to possible damage to necessary genes, which seemed to represent the "if it is not broken, do not fix it" principle. However HGT promotes faster adaptation in dynamic environments consistent with environmental stresses and patchiness of resources on a population. This latter conclusion, which was based on a totally different mathematical approach, was remarkably congruent with the arguments developed herein this chapter.

5.10 Conclusions

Bacteriophages evolved to solve three major problems for bacteria. They have allowed the emergency transfer of genes from dead cells to live cells so that unique genetic diversity is not lost to the bacterial genosphere. Secondly, they help maintain genetic diversity in clonal bacterial cells that have moved beyond effective conjugational distance. Thirdly, they allow a delayed temporal component to gene transfer for future generations. Bacteriophages have been fundamental in allowing survival advantages for those bacteria that wish to exploit hostile, rapidly changing environments including patchiness in resource availability like carbon sources for heterotrophic bacteria. Prime examples are the Enterobacteriaceae and the Vibrionaceae. Derivatives such as gene transfer agents are viable descendants of phages that deliver more varied genetic material but at a slower rate as they cannot self-liberate. Why the alphaproteobacteria have such a penchant for GTA is unknown unless it is a reflection of an event that occurred in their last universal common ancestor. Possibly it's due to the ancestral environment of alphaproteobacteria having a high probability of cell lysis, therefore there was no need to encode for cellular escape, which in turn allowed more capsid space for larger random gene transfers. Once bacteriophage evolved, selection pressure seemingly pushed the host-phage interactions in many directions, but the underlying advantage of diverse horizontal gene transfer has meant that viruses have not been removed by natural selection.

Could HGT be why viruses have not been selectively removed from the biosphere by evolutionary pressure? Viruses are often disadvantageous at the individual level causing illness or even death but at the population level (group selection), survivors have the viral associated ability to have their genes added to, rearranged or the speed of translation altered. For most organisms, this maybe the first such chance since fertilisation for shuffling genetic material; excluding rogue transposons (also a derivative of viruses). Examples of acquiring useful viral genes in the germ line include placental endogenous retroviruses that have allowed the age of eutherian mammals to arise 250 million years ago and polydnaviruses in the wasps which allow circumvention of the immune system of the host insect. It is somewhat ironic in that we are using viral capsids to deliver genes into humans for gene therapy and immunization trials against COVID-19 (SARS-CoV-2), toted as wonderful scientific breakthroughs but in fact, it is HGT using a technology, a capsid, perhaps 3.5 billion years old which is solving familiar problems. We are conversant with other examples of individual sacrifice for the survival of the population (group selection) e.g. social insects, colonial medusa, even human soldiers at war. Also, in species that are dioecious, single individuals are not important and quite useless by themselves but the population of even two opposite sexes is a unit on which selection and evolution acts. To clonal organisms like bacteria, the occasional loss of individual cells to phage lysis is of no more importance than shedding of a skin cell from ourselves (dandruff) and even less important if unique genes are salvaged by EDA-HGT.

Acknowledgements Professor Colin Munn, University of Plymouth, U.K. is thanked for many hours of discussion on the ideas presented herein but for personal reasons, did not want to be a co-author. However, all the errors of interpretation are my own. There are no conflicts of interest identified. The majority of this manuscript was written whilst the author was funded by James Cook University for Special Study Leave with Professor Munn.

References

Ackermann HW (2003) Bacteriophage observation and evolution. Res Microbiol 154:245–151

Ackermann HW, DuBow MS (1987) Viruses of prokaryotes. In: General properties of bacteriophages, vol I. CRC Press, Boca Raton, FL. (cited in Paul 2008)

Ajon M, Fröls S, van Wolferen M, Stoecker K, Teichmann D, Driessen AJ, Grogan DW, Albers S-V (2011) UV-inducible DNA exchange in hyperthermophilic archaea mediated by type IV pili. Mol Microbiol 82:807–817

Bárdy P, Füzik T, Hrebík D, Pantůček R, Beatty JT, Plevka P (2020) Structure and mechanism of DNA delivery of a gene transfer agent. Nat Commun 11:3034. https://doi.org/10.1038/s41467-020-16669-9

Bhattacharyya S, Walker DM, Harshe RM (2020) Dead cells release a 'necrosignal' that activates antibiotic survival pathways in bacterial swarm. Nat Commun 11:4157. https://doi.org/10.1038/s41467-020-17709-0

Brussow H, Canchaya C, Hardt W-D (2004) Phages and the evolution of bacterial pathogens: from genomic rearrangements to lysogenic conversion. Microbiol Mol Biol Rev 68:560–602

Buscido-Salcedo N, Owens L (2013) Virulence changes to *Vibrio harveyi* clade bacteria experimentally infected with bacteriophage from *Vibrio owensii*. Indian J Virol 24:180–187

Carter MQ, Chen J, Lory S (2010) The *Pseudomonas aeruginosa* Pathogenicity Island PAPI-1 is transferred via a novel type IV pilus. J Bacteriol 192:3249–3258

Cascales E, Christie PJ (2003) The versatile bacterial type IV secretory systems. Nat Rev Microbiol 1:137–149

Casjens S (2003) Prophages and bacterial genomics: what have we learned so far? Mol Microbiol 49:277–300

Chang W, Small DA, Toghrol F, Bentley WE (2005) Microarray analysis of *Pseudomonas aeruginosa* reveals induction of pyocin genes in response to hydrogen peroxide. BMC Genomics 6:115. https://doi.org/10.1186/1471-2164-6-115

Dawkins R (2016) The selfish gene. Oxford University Press, Oxford

Dion MB, Oechslin F, Moineau S (2020) Phage diversity, genomics and phylogeny. Nat Rev Microbiol 18:125–138

Filloux A (2010) A variety of bacterial pili involved in horizontal gene transfer. J Bacteriol 192:3243–3245

Friedman DI (1988) Integration host factor: a protein for all reasons. Cell 55:545–554

Gazit E, Sauer RT (1999) The doc toxin and Phd antidote proteins of the bacteriophage P1 plasmid addiction system form a heterotrimeric complex. J Biol Chem 274:16813–16818

Haapa-Paananen S, Hannu R, Savilahti (2002) DNA transposition of bacteriophage Mu. A quantitative analysis of target site selection *in vitro*. J Biol Chem 277:2843–2851

Jiang SC, Paul JH (1994) Seasonal and diel abundance of viruses and occurrence of lysogeny/bacteriocinogeny in the marine environment. Mar Ecol Prog Ser 104:163–172

Jiang SC, Kellogg CA, Paul JH (1998) Characterization of marine temperate phage-host systems isolated from Mamala Bay, Oahu, Hawaii. Appl Environ Microbiol 64:535–542

Kalatzis PG, Rørbo N, Castillo D, Mauritzen JJ, Jørgensen J, Kokkari C, Zhang F, Katharios P, Middelboe M (2017) Stumbling across the same phage: comparative genomics of widespread temperate phages infecting the fish pathogen *Vibrio anguillarum*. Viruses 9:122. https://doi.org/10.3390/v9050122

Karaolis DKR, Somara S, Maneval DA Jr, Johnson JA, Kaper JB (1999) A bacteriophage encoding a pathogenicity island, a type-IV pilus and a phage receptor in cholera bacteria. Letters to Nature 399:375–379

Kotilainen MM, Grahn AM, Bamford JKH, Bamford DH (1993) Binding of an *Escherichia coli* double-stranded DNA virus PRD1 to a receptor coded by an IncP-type plasmid. J Bacteriol 175:3089–3095

Krupovic M, Bamford DH (2008) Virus evolution: how far does the double β-barrel viral lineage extend? Nat Rev Microbiol 6:941–948

Lang AS, Zhaxybayeva O, Beatyy JT (2012) Gene transfer agents: phage-like elements of genetic exchange. Nat Rev Microbiol 10:472–482

Lawrence JG (1999) Gene transfer, speciation, and the evolution of bacterial genomes. Curr Opin Microbiol 2:519–523

Lawrence JG (2002) Gene transfer in bacteria: speciation without species? Theor Popul Biol 61:449–460

Leitet C, Riemann B, Hagstrom A (2006) Plasmids and prophages in Baltic Sea bacterioplankton isolates. J Mar Biolog Assoc UK 86:567–575

Mardanov AV, Ravin NV (2007) The antirepressor needed for induction of linear plasmid-prophage N15 belongs to the SOS regulon. J Bacteriol 189:6333–6338

Martinez-Hernandez F, Fornas O, Gomez ML, Bolduc B, de la Cruz Pena MJ, Martinez JM, Anton J, Gasol JM, Rosselli R, Rodriguez-Valera F, Sullivan MB, Acinas SG, Martinez-Garcia M (2017) Single-virus genomics reveals hidden cosmopolitan and abundant viruses. Nat Commun 8(15892):1–13

McDaniel LD, Young E, Delany J, Ruhnau F, Ritchie KB, Paul JH (2010) High frequency of horizontal gene transfer in the oceans. Science 330:50–51

McDaniel LD, Young EC, Ritchie KB, Paul JH (2012) Environmental factors influencing gene transfer agent (GTA) mediated transduction in the subtropical ocean. PLoS One 7(e43506):1–9

Mizuno CM, Rodriguez-Valera F, Kimes NE, Ghai R (2013) Expanding the marine virosphere using metagenomics. PLoS Genet 9(12):e1003987. https://doi.org/10.1371/journal.pgen.1003987

Munro J, Oakey HJ, Bromage E, Owens L (2003) Experimental bacteriophage-mediated virulence in strains of *Vibrio harveyi*. Dis Aquat Org 54:187–194

Nakayama K, Takashima K, Ishihara H, Shinomiya T, Kageyama M, Kanaya S, Ohnishi M, Murata T, Mori H, Hayashi T (2000) The R-type pyocin of *Pseudomonas aeruginosa* is related to P2 phage, and the F-type is related to lambda phage. Mol Microbiol 38:213–231

Oakey J, Owens L (2000) A new bacteriophage, VHML, isolated from a toxin-producing strain of *Vibrio harveyi* in tropical Australia. J Appl Microbiol 89:702–709

Oakey HJ, Cullen BR, Owens L (2002) The complete nucleotide sequence of the *Vibrio harveyi* bacteriophage VHML. J Appl Microbiol 93:1089–1098

Paul JH (2008) Prophages in marine bacteria: dangerous molecular time bombs or the key to survival in the seas? ISME J 2:579–589

Pell LG, Kanelis V, Donaldsond LW, Howella PL, Davidson AR (2009) The phage major tail protein structure reveals a common evolution for long-tailed phages and the type VI bacterial secretion system. Proc Natl Acad Sci 106:4160–4165

Ravin NV (2011) N15: the linear phage-plasmid. Plasmid 65:102–109

Rawlings DE (1999) Proteic toxin-antitoxin, bacterial plasmid addiction systems and their evolution with special reference to the *pas* system of pTF-FC2. FEMS Microbiol Lett 176:269–277

Raz Y, Tannenbaum E (2010) The influence of horizontal gene transfer on the mean fitness of unicellular populations in static environments. Genetics 185:327–337

Snyder J, Brumfield SK, Peng N, She Q, Young MJ (2011) Sulfolobus turreted icosahedral virus c92 protein responsible for the formation of pyramid-like cellular lysis structures. J Virol 85:6287–6292

St Pierre F, Endy D (2008) Determination of cell fate selection during phage lambda infection. PNAS 105:20705–20710

Stopar D, Cerne A, Zigman M, Poljisal-Prijatelj M, Turk V (2004) Viral abundance and high proportion of lysogens suggest that viruses are important members of the microbial community in the Gulf of Trieste. Microb Ecol 47:1–8

Venter JC, Remington K, Heidelberg JF, Halpern AL, Rusch D, Eisen JA, Wu D, Paulsen I, Nelson KE, Nelson W, Fouts D, Levy S, Knap AH, Lomas MW, Nealson K, White O, Petersen J, Hoffman J, Parsons R, Daden-Tillson H, Pfannkoch C, Rogers YH, Smith HO (2004) Environmental genome shotgun sequencing of the Sargasso Sea. Science 304:66–74

Vidgen M, Carson J, Higgins M, Owens L (2006) Changes to the phenotypic profile of *Vibrio harveyi* when infected with the bacteriophage Vibrio harveyi myovirus-like (VHML). J Appl Microbiol 100:481–487

Zabala B, Hammer JA, Espejo RT, Hertwig S (2009) The linear plasmid prophage Vp58.5 of *Vibrio parahaemolyticus* is closely related to the integrating phage VHML and constitutes a new incompatibility group of telomere phages. J Virol 83:9313–9320

Zeng L, Skinner SO, Zong C, Sippy J, Feiss M, Golding I (2010) Decision making at a subcellular level determines the outcome of bacteriophage infection. Cell 141:682–691

Zhang X-H, Meaden PG, Austin B (2001) Duplication of haemolysin genes in a virulent isolate of *Vibrio harveyi*. Appl Environ Microbiol 67:3161–3167

Chapter 6
Diverse Phage-Encoded Toxins and Their Role in Bacterial Ecology

Sheryl L. W. Zajdowicz

Abstract Bacteriophages are the most abundant and diverse microbial entity on the planet. Found in every environment, the influence of phages extends beyond their bacterial host and has a variety of impacts within an ecosystem, including controlling bacterial abundance, affecting community composition, and even influencing biogeochemical cycling; they are also key players in microbial evolution. Lysogenic phages commonly introduce beneficial genes that drive evolution and promote adaptation. The effects of lysogeny are as broad reaching as phages themselves, imparting such benefits as enhanced bacterial fitness, phenotypic plasticity, biofilm formation, antimicrobial compounds or resistance, as well as virulence factors. While pathogens benefit from a multitude of virulence factors through lysogeny, phage-encoded toxins and toxin-conversion are most well-characterized insofar as their contribution to bacterial pathogenesis and their overall impact on the human or animal host. The toxins' effects on hosts are typically detrimental or even lethal; sometimes the expression of the toxin results in lethality to the bacterial host itself. However, not all phage-encoded toxins have a negative effect on the environment in which the toxin is produced. The protection against parasitoid wasps that exists in some forms of aphids is a result of a complex interplay between the aphid, their facultative endosymbiont *Hamiltonella defensa*, and the APSE phage-encoded toxins it produces. Understanding the contributions of phage-encoded toxins to their bacterial hosts and the impacts they have on the various ecosystems where they are expressed is paramount to deciphering the complexity and driving force for the diversity and exchange of phage-encoded toxins within bacterial communities. This chapter gives an overview of the various impacts that prophages and subsequent lysogenic conversion confer to their bacterial hosts, with a focus on toxin production by several medically important pathogens and the subsequent impacts on the human hosts they infect. The beneficial relationship between phage-encoded toxins and their impact on insects is also explored.

S. L. W. Zajdowicz (✉)
Department of Biology, Metropolitan State University of Denver, Denver, CO, USA
e-mail: swaltonz@msudenver.edu

C. J. Hurst (ed.), *The Biological Role of a Virus*, Advances in Environmental Microbiology 9, https://doi.org/10.1007/978-3-030-85395-2_6

6.1 Bacteriophage and Bacteria: A Fine Line Between Friend and Foe

Bacteriophages, also known as phages, represent the most diverse and abundant microbial entity on the planet, having an estimated magnitude of 10^{31} viral particles, which outnumbers bacteria by a factor of 10 to 1 (Wommack and Colwell 2000). They can be found in a multitude of environments, including in the soil, in sea and freshwater, (Suttle 2005, 2007; Srinivasiah et al. 2008), deserts (Fancello et al. 2013), polar regions (Koskella et al. 2011), and within multicellular organisms (Breitbart et al. 2008; Breitbart et al. 2003). Not surprisingly, because of their prevalence, bacteriophages have extraordinary ecological impacts within an ecosystem, including controlling bacterial abundance, affecting community composition, and even influencing biogeochemical cycling; they are also key players in microbial evolution (Krisch 2003; Chaturongakul and Ounjai 2014; Wilhelm and Suttle 1999; Azam 1998; Shelford et al. 2012; Caron 1994). Phages rely on bacterial hosts to complete their life cycle and there are four life cycles used: lytic, lysogenic, pseudolysogenic, and chronic or carrier state (Ackermann and Dubow 1987). While some of these life cycles are detrimental to the bacterial host and result in destruction, other phage life cycle strategies are neutral or provide broad-spanning benefits to their bacterial hosts. The majority of bacteriophages are classified as either lytic or temperate (Guttman et al. 2005). Lytic phage (virulent phage) replication results in lysis of the bacterial host at the end of the phage replication cycle, thereby releasing newly formed phage. In contrast, temperate phages may integrate their genetic material into the bacterial host's chromosome becoming a prophage whose DNA is replicated along with the host chromosome and is vertically transferred to progeny; a bacterial host that contains a prophage is referred to as a lysogen (Guttman et al. 2005). Temperate phages possess the same ability to enter the lytic cycle under certain environmental or stress conditions, thus killing the cell. However, most prophages inhibit the genes required for the lytic cycle and the prophage becomes quiescent. Further, evolutionarily, prophages frequently lose the ability to excise from the bacterial genome, resulting in an inability to form new virions; these prophages are typically considered "cryptic prophages". Prophage and "cryptic prophages" contribute to overall fitness, adaptation to a new environment, and contribute to the pathogenicity of the recipient bacterium. Integration of phage into the bacterial genome also protects the bacterial host from infection by other similar phages (Lynch et al. 2010; Matos et al. 2013).

Through horizontal transfer, phages are a major driving force in bacterial diversification. Evaluation of genomes from gamma-proteobacteria and G+C-rich Gram-positive bacteria revealed that nearly two-thirds of the genomes harbor prophages (Canchaya et al. 2003; Casjens 2003). Additional research showed that the composition of bacterial genomes can be up to 30% phage inserts and remnants of phage genomes (Casjens 2003; Petrov et al. 2010; Comeau et al. 2007). Further pangenomic studies showed that prophage genes comprise approximately 13.5% of *Escherichia coli* and 5% of *Salmonella* genomes (Bobay et al. 2013; Touchon et al. 2009). More astonishing, the global rate by which phages affect genetic composition in bacteria is estimated to be approximately 20×10^{15} gene transfer events per second (Bushman 2002). The integration of prophages typically occurs at transfer RNA (tRNA) genes where the phage *attP* site reconstitutes the tRNA coding sequence upon integration (Campbell 1992); a number of prophages encode their own tRNA genes that complement the disrupted tRNA gene (Ventura et al. 2003). However, phage integration may also occur at non-tRNA sites on the bacterial chromosome and can lead to genetic inactivation, which may or may not result in functional consequences; these mutations are typically not investigated thoroughly (Coleman et al. 1991; Goh et al. 2007; Lee and Iandolo 1986).

This chapter provides an overview of the various impacts that prophages and subsequent lysogenic conversion confer to their bacterial hosts, with a focus on toxin production and their effect on the environments in which the bacterial hosts are found.

6.2 Impact of Lysogenic Phages on Bacterial Hosts

Bacteria must adapt to a multitude of environments and a variety of growth conditions in order to survive. Through lysogenic conversion, bacterial hosts often gain determinants that promote survival in their respective environmental niches and allow transit between niches; phage-encoded benefits include, but are not limited to, enhanced metabolism, stress tolerance, biofilm formation, modulation of sporulation, bacterial warfare, and most notably, virulence factors; a few specific examples of these benefits are listed in Table 6.1.

Through the incorporation of auxiliary metabolic genes, an integrated prophage can confer profound benefits by enhancing bacterial metabolism or by expanding the bacterial host's metabolic capabilities to allow for its survival under various growth conditions (Sekulovic and Fortier 2015; Edlin et al. 1975, 1977). For example, *Escherichia coli* strain BW25113 that contains CPS-53 and CP4-57 prophages exhibited stable metabolism when grown under extreme oxidative, osmotic, and acid-stress conditions; deletion of all the cryptic prophages in this strain resulted in an increased susceptibility to exogenous stresses and reduced the overall growth rate of the bacterium (Wang et al. 2010). *Salmonella enterica* serotype Typhimurium strains lysogenized with the temperate phage Salmonella virus SopEphi (SopEΦ) were found to have increased metabolic rates under anoxic conditions and outcompete non-lysogens in vivo (Lopez et al. 2012; Barrett and Riggs 1982). The SopEΦ phage increases the production of inducible nitric oxide synthase (iNOS), thereby facilitating the production of precursors to the electron acceptor nitrate and promoting growth under anaerobic conditions that may occur in the inflamed intestinal tract during infection (Lopez et al. 2012; Barrett and Riggs 1982).

Prophages have also been shown to influence biofilm formation and dispersal by a number of bacterial pathogens including *Pseudomonas aeruginosa*, *Bacillus anthracis*, *Vibrio cholerae*, *Streptococcus pneumoniae*, and *E. coli*. The role of

Table 6.1 Impacts of bacteriophage on bacterial hosts

	Bacterial host	Phage	Impact	Reference
Metabolism/stress tolerance	*Escherichia coli*	Escherichia virus Lambda (λ) Escherichia virus P1 Escherichia virus P2 Escherichia virus Mu CPS-53	Enhanced metabolism/stress tolerance	Edlin et al. (1975, 1977); Wang et al. (2010)
	Salmonella enterica serovar Typhimurium	Salmonella virus SopEphi (SopEϕ) Salmonella virus Gifsy-2	Nitrate metabolism/stress tolerance	Lopez et al. (2012); Figueroa-Bossi and Bossi (1999); Pilar et al. (2012)
	Clostridium difficile	Clostridium phage CD38	Enhanced metabolism/stress tolerance	Sekulovic and Fortier (2015)
Bacterial signaling/quorum sensing	*Bacillus subtilis*	Bacillus phage Phi3T	Promotes lytic cycle	Erez et al. (2017)
	Clostridium difficile	Clostridium phage phiCDHMI	Agr precursors modulate fitness and pathogenicity	Hargreaves et al. (2014)
	Iodobacteria	Iodobacter phage PhiPLPE (ϕPLPE)	Inhibits the LuxS system	Leblanc et al. (2009)
Biofilm formation and dispersal	*Bacillus anthracis*	Bacillus virus Bcp1 Bacillus virus Wip1 Bacillus virus Wip4 Bacillus virus Frp2	Promotes complex biofilm formation	Schuch and Fischetti (2009)
	Pseudomonas aeruginosa	Pf4	Enhanced biofilm formation and dispersal	Whiteley et al. (2001); Rice et al. (2009)
	Streptococcus pneumoniae	Streptococcus phage SV1	Promotes biofilm formation and dispersal	Carrolo et al. (2010)
Sporulation	*Bacillus subtilis*	PMB12 Bacillus virus SP-10	Enhanced sporulation	Kinney and Bramucci (1981)
	Bacillus anthracis	Bacillus virus Wip1 Bacillus virus Wip4	Inhibits sporulation	Schuch and Fischetti (2009)

(continued)

Table 6.1 (continued)

	Bacterial host	Phage	Impact	Reference
		Bacillus virus Frp2		
Antibiotic resistance	*Staphylococcus xylosis*	Staphylococcus phage PhiJW4341 (ϕJW4341)	Erythromycin resistance	Wipf et al. (2014)
	Escherichia coli	Escherichia phage 933W	Tetracycline resistance	Marinus and Poteete (2013)
		Escherichia virus RCS47	Cephalosporin resistance	Billard-Pomares et al. (2014)
	Salmonella spp.	P1-like	Cephalosporin resistance	Yang et al. (2017)
Resistance to phage superinfection	*Pseudomonas aeruginosa*	Pseudomonas virus D3	Prevents phage absorption	Newton et al. (2001)
		Pseudomonas phage JBD88a	Inhibits CRISPR/CAS system	Pawluk et al. (2017)
	Escherichia coli	Escherichia virus HK97	Prevents DNA entry	Cumby et al. (2012)
		Escherichia virus phiV10	Modification to LPS prevents phage binding	Perry et al. (2009)
	Shigella flexneri	Shigella phage SfII	Modification of LPS prevents phage absorption	Lehane et al. (2005)
Virulence factors	[a]			

[a]An overview of phage-encoded virulence factors can be found in Table 6.2

phages on biofilm formation was assessed in *B. anthracis* utilizing the ΔSterne strain, which is a prophage-free variant that is unable to form biofilms (Schuch and Fischetti 2009). Lysogenization of the ΔSterne strain with phages Burkholderia virus Bcp1, Bacillus phage Wip1, Bacillus phage Wip4, and Bacillus phage Frp2 resulted in formation of complex biofilms; further analysis also showed that phage-encoded RNA polymerase sigma factors can act in *trans* to induce the expression of genes required for *B. anthracis* biofilm formation (Schuch and Fischetti 2009). *P. aeruginosa* forms prolific biofilms in the environment and during infections. The most highly expressed genes in biofilms formed by the *P. aeruginosa* PAO1 strain are those associated with the filamentous phage Pf4 (Whiteley et al. 2001); not surprisingly, free phage have been found from biofilm effluents as well as from clinical isolates. Complete deletion of Pf4 from *P. aeruginosa* PAO1 resulted in a loss of bacterial lysis and death seen in late biofilm formation, a loss of small colony variant (SCV) formation, and greater survival of mice infected with *P. aeruginosa* PAO1 lacking Pf4 in comparison to the wild-type strain (Rice et al. 2009). These findings showcased the importance of filamentous prophages on biofilm dispersal, the formation of drug-resistant SCVs, and the virulence of *P. aeruginosa* in general (Rice et al. 2009). One final example of prophage impact on biofilms can be observed in *S. pneumoniae*. Extracellular DNA (eDNA) is an integral part of the biofilm matrix of biofilms in many bacterial species, including *S. pneumoniae* (Carrolo et al. 2010). The presence of eDNA in biofilms typically arises as a result of bacterial lysis; this is also true in *S. pneumoniae* biofilms. However, the presence of eDNA in *S. pneumoniae* was found to be the result of synergistic effects between host-encoded autolysin LytA and a phage-encoded lysin Sv1 associated with the SV1 phage (Carrolo et al. 2010). Inactivation of either the bacterial autolysin LytA or the phage lysin SV1 resulted in reduced biofilm formation and deletion of both resulted in loss of biofilm formation in *S. pneumoniae* entirely, showing the importance of phage lysin's role in *S. pneumoniae* biofilm formation. Additionally, because free SV1 phage is detected during biofilm formation, with the greatest titers being found at peak biofilm formation, it is proposed that spontaneous prophage induction of SV1 is key to the production of eDNA that is paramount for biofilm formation in *S. pneumoniae* (Carrolo et al. 2010).

Another way in which some prophages contribute to bacterial host adaptation to various environments is through the modulation of sporulation. Spore formation is employed by numerous bacterial species for survival under nutrient limitation or unfavorable environmental conditions; endospores are metabolically dormant and can withstand harsh environmental conditions, including desiccation, ionizing radiation, extreme temperatures, and antimicrobial compounds (Wilcox and Fawley 2000; Nicholson et al. 2000). The process of sporulation, which has been most studied and characterized in *Bacillus subtilis*, is a complex multi-stage process involving gene activation and repression (Higgins and Dworkin 2012; Paredes et al. 2005). Bacteria infected with spore-converting phages have a higher frequency in their sporulation rates than uninfected cells; spore-converting phages also confer a higher frequency of sporulation oligosporogenic or sporulation-negative Spo^{c} mutants (Bramucci et al. 1977a, b; Keggins et al. 1978; Kinney and Bramucci 1981; Perlak et al. 1979). Spore-converting phages have been isolated from various bacteria including *B. subtilis*, *Bacillus cereus*, *Bacillus pumilus*, *Bacillus thuringiensis*, and *Clostridium perfringens* (Bramucci et al. 1977a, b; Hemphill and Whiteley 1975; Keggins et al. 1978; Perlak et al. 1979; Sandman et al. 1987; Stewart and Johnson 1977). Investigation of two spore-converting phages PMB12 and Bacillus phage SP-10 found that lysogenization of *B. subtilis* with either of these phages confer the ability to produce spores in sporulation-negative strains (Bramucci et al. 1977a, b; Kinney and Bramucci 1981; Silver-Mysliwiec and Bramucci 1990). Studies that showed that PMB12 suppress stage 0 sporulation mutations suggest that PMB12 influenced early stages in the sporulation process and at least 3 of the *scn* genes found within PMB12 are required to induce sporulation (Kinney and Bramucci 1981). Further analysis of *B. subtilis* strain 3–13 lysogens harboring either PMB12 or Bacillus phage SP-10 demonstrated an ability to grow in the presence of concentrations glucose that are inhibitory to the sporulation process and the expression of α-amylase in wild-type strains, thereby overcoming the catabolite repression-associated inhibition of sporulation in *B. subtilis* (Takahashi 1979; Henkin et al. 1991; Nicholson et al. 1987). Interestingly, prophages have also been shown to inhibit sporulation in some bacterial species, including in *B. anthracis*. Investigation of sporulation by the unlysogenized ΔSterne strain (described above) had prolific spore production of 5×10^8 spores/mL, whereas lysogens carrying Bacillus phage Wip1, Bacillus phage Wip4, and Bacillus phage Frp2 resulted in the inhibition of sporulation (Schuch and Fischetti 2009), but promoted biofilm formation. In this case, the resulting lysogens survived in the soil for longer periods of time than did the non-lysogenic ΔSterne strain, suggesting an advantageous adaptation for long-term survival in soil environments (Schuch and Fischetti 2009).

As already described, prophages can have profound impacts on the lysogen by contributing to the lifestyle and overall fitness, but they also bolster the virulence of many bacterial pathogens (Table 6.2), including *Salmonella enterica* (Cooke et al. 2007; Figueroa-Bossi et al. 2001; Hermans et al. 2005, 2006; Thomson et al. 2004), *Vibrio cholerae* (Boyd et al. 2000a, b; Boyd and Waldor 1999; Davis et al. 2000; Mekalanos et al. 1997; Waldor and Mekalanos 1994, 1996), *Escherichia coli* (Mead and Griffin 1998; Ohnishi et al. 2001; Hayashi et al. 2001; Ogura et al. 2006; Ohnishi et al. 1999, 2002; Yokoyama et al. 2000), *Streptococcus pyogenes* (Aziz et al. 2005; Banks et al. 2002; Cleary et al. 1998), *Staphylococcus aureus* (Baba et al. 2008; Bae et al. 2006; Goerke et al. 2009; Rahimi et al. 2012), *Clostridium difficile*, and *Corynebacterium diphtheriae* (Freeman and Morse 1952; Trost et al. 2012). Lysogenic conversion provides key mechanisms to invade host tissues, evade immune defenses, and can cause tissue damage, thereby promoting greater survival and selective advantage to the bacterial host; phage-encoded products include, but are not limited to, hydrolytic enzymes, antibiotic resistance factors, superantigens, adhesins, serum resistance, detoxifying enzymes, LPS-modifying enzymes, toxins, and type III effector proteins (Boyd 2012; Brussow et al. 2004; Fortier and Sekulovic 2013). Acquisition of phage-encoded virulence factors promote overall fitness, replication, and survival within their respective environments. While phage-encoded virulence factors provide key functionality and promote the ability of lysogens to navigate the various stages of pathogenesis within the internal environment of the human host, phage-encoded toxigenicity associated with a few medically important pathogens and their overall toxic effect on the human host will be highlighted and will be reviewed in the next section.

6.3 Lysogeny and Toxigenicity in Bacterial Pathogens

During infection, bacterial pathogens secrete a variety of virulence factors, including exotoxins. Many exotoxin genes are carried by phages and are readily transferred between bacterial hosts; phage-associated exotoxin genes can be detected in bacterial and free phage DNA in a multitude of environments, including in the human gut (Casas and Maloy 2011; Casas et al. 2006). Through transduction of these phage-encoded genes, even commensal bacteria may be converted to pathogenic strains due

Table 6.2 Representative phage-encoded virulence factors

	Bacterial host	Phage	Gene (s)	Reference
Adhesion proteins	*Vibrio cholerae*	CTXφ	*tcp*	Karaolis et al. (1999)
	Escherichia coli	Escherichia virus Lambda (λ)	*lom*	Barondess and Beckwith (1990)
	Enterococcus faecalis	Pp1, pp4, pp6	*pblA*/ *pblB*	Matos et al. (2013)
	Streptococcus mitis	Streptococcus phage SM1	*pblA*/ *pblB*	Bensing et al. (2001)
Evasion of immune response and intracellular survival	*Escherichia coli*	Escherichia virus Lambda (λ)	*bor*	Barondess and Beckwith (1990)
	Salmonella enterica	Salmonella virus *Gifsy*-1,2 Salmonella virus *Gifsy*-1 Salmonella virus *Gifsy*-2 Salmonella virus SopEphi (ϕSopE)	*ailT*, *ailF* *gipA*, *gogB* *sodC1*, *SseI* *sopE*	McClelland et al. (2001) Stanley et al. (2000) Figueroa-Bossi et al. (2001) Mirold et al. (1999)
Toxins	*Vibrio cholerae*	CTXφ	*ctxA*/ *ctxB* *ace* *zot*	Waldor and Mekalanos (1996) Trucksis et al. (1993) Fasano et al. (1991)
	Corynebacterium diphtheriae	Corynephage beta (β)	*tox*	Freeman (1951)
	Clostridium botulinum	CEβ DEβ	*bont*	Barksdale and Arden (1974) Eklund et al. (1972)
	Clostridium novyi	NA1	*tcnA*	Eklund et al. (1974); Popoff and Bouvet (2009)
	Escherichia coli	Stx	*stx1*, *stx2*	Newland et al. (1985); O'Brien et al. (1984)
	Staphylococcus aureus	Sa3int	*eta* *pvl*	Yamaguchi et al. (2000) Wirtz et al. (2009)

to their newly acquired means of toxin production. In fact, phage-encoded toxins serve as the primary virulence factor or they contribute to the overall pathogenicity for numerous pathogens, including *Vibrio cholerae*, Shiga-toxin producing

Escherichia coli, *Clostridium botulinum*, *Staphylococcus aureus*, and *Corynebacterium diphtheriae*. While there are many advantages that lysogenic conversion provides, it is a fine balance between benefit and detriment since most phage-encoded toxins are expressed when the prophage is induced to enter the lytic cycle (Wagner et al. 2002; Koudelka et al. 2018). In the following sections, the phage-bacterial host relationship will be reviewed for each of these medically important pathogens and the impact that their phage-conferred toxigenicity has on the human host will be described.

6.3.1 Phage-Conversion and Toxigenicity in Vibrio cholerae

V. cholerae is a facultative bacterium that has both environmental and human life cycles (Faruque et al. 1998). There are over 200 serogroups of *V. cholerae*; however, only two *V. cholerae* serogroups O1 and O139 act as causative agents for cholera, a gastrointestinal disease that is characterized by profuse and explosive watery diarrhea and subsequent dehydration (Faruque 2013; Faruque et al. 1998; Kaper et al. 1995). Strains within these two serogroups of *V. cholerae* have acquired two key virulence factors, the cholera toxin (CT) and toxin-coregulated pilus (TCP) through acquisition of phages or phage-like elements (Karaolis et al. 1999; Waldor and Mekalanos 1996). While TCP is of importance, this section will focus on CT and other phage-associated toxins. Toxigenic *V. cholerae* strains carry the *ctxAB* genes, which encode for the cholera toxin; these genes are acquired through the lysogenic filamentous phage CTXϕ (Waldor and Mekalanos 1996).

CTXϕ is similar in size, structure, and gene order to the M13 and f1 filamentous phages found in *E. coli* (Boyd 2008; Davis and Waldor 2003). CTXϕ has a 6.9-kb genome that is organized into two distinct modules, a 2.4-kb long RS2 (repeat sequence 2) and a 4.5-kb long core (Waldor et al. 1997), which contain necessary genes for phage attachment, chromosomal integration, phage morphogenesis and assembly, as well as the *ctxAB* genes that are required for disease development (Waldor and Mekalanos 1996). Its genome also contains genes encoding Ace (Accessory cholera enterotoxin) and Zot (Zonula occludens toxin) proteins, which are involved in phage production and assembly, respectively, and are also accessory toxins in *V. cholerae* pathogenesis (Trucksis et al. 1993; Fasano et al. 1991). CTXϕ integration occurs at a highly conserved 28-bp dimer resolution site (*dif*) found on chromosome 1 of *V. cholerae*; the process is dependent on two host-encoded tyrosine recombinases, XerC and XerD and cell division protein FtsK (Huber and Waldor 2002; McLeod and Waldor 2004) and is distinct from other lysogenic phages in that it uses its folded ssDNA for integration (Val et al. 2005). While CTXϕ is pervasive in epidemic and pandemic-strains of *V. cholerae*, environmental isolates may also carry variants of CTXϕ; however, there are major genetic differences found between CTXϕ present in *V. cholerae* O1, O139, non-O1, and non-0139 strains, which may result in limited transmission within non-O1 and non-O139 strains (Maiti et al. 2006). The non-O1 and non-O139 strains have only been associated with sporadic diarrhea and not severe disease (Maiti et al. 2006).

Expression of the phage-encoded *ctxAB* results in the production of the potent enterotoxin CT. In human volunteers, CT was shown to induce the diarrhea characteristic of cholera (Levine et al. 1983). Strains of *V. cholerae* that are deficient in cholera toxin production are attenuated in animals and humans (Guinee et al. 1985, 1987, 1988). CT is required to cause severe disease and promotes transmission of the organism through the prolific diarrhea that typically contains upwards of 10^{11} cells per ml (Faruque et al. 1998); however, studies have shown that CT also enhances its bacterial growth by creating an iron-depleted environment in the intestinal environment by way of modulating host cell metabolism that ultimately provides the pathogen with host-derived nutrients such as heme and intestinal long-chain fatty acids (LCFAs) (Rivera-Chavez and Mekalanos 2019). CT is a characteristic hetero-oligomeric AB toxin, consisting of a singular A subunit associated with five identical B subunits and causes diarrhea through an increase in cyclic adenosine-3′, 5-monophosphate (cAMP) production by disrupting the stimulatory G-protein Gsα (Field et al. 1972). The B subunit specifically binds to the GM_1 ganglioside receptor found on enterocytes (Pierce 1973; King and Van Heyningen 1973). Upon binding, the A subunit is translocated and its A_1 subunit activated to catalyze the ADP-ribosylation of the host cell G protein Gsα, thereby resulting in its inactivation; this inactivation of Gsα leads to the permanent activation of adenylate cyclase in the host cell and an overabundance of intracellular cAMP (Field et al. 1972). The increased cAMP disrupts sodium absorption and increases secretion of sodium bicarbonate and chloride, ultimately resulting in excessive water loss by the cells and causing the characteristic symptoms of cholera (as reviewed in Fishmann 1990; Kaper et al. 1994). Regulation of CT is under the control of host-encoded master regulators ToxR, AphA, and AphB as well as by phage encoded RstC, which is found in the RS1 module (Lee et al. 1999, 2001).

While the impact of prophage on *V. cholerae* is well-characterized, little is known regarding the role of prophage in other marine *Vibrio* species. Notably prophage-encoded toxins have been identified in the human pathogen *Vibrio parahaemolyticus* (Nasu et al. 2000), fish pathogens *Vibrio anguillarum* (Kalatzis et al. 2017) and *Vibrio harveyi* (Munro et al. 2003), and coral pathogen *Vibrio coralliilyticus* (Weynberg et al. 2015); *zot*-encoding prophages have widespread distribution in marine *Vibrio* species (Castillo et al. 2018). Further studies of marine *Vibrio* species have highlighted pervasive temperate phage prevalence in pathogenic and non-pathogenic *Vibrio* species within marine communities; and the ways in which phage-encoded traits contribute to genetic diversification, niche adaptation, and virulence in marine *Vibrio* communities and could potentially play a role in the emergence of new pathogenic *Vibrio* species (Castillo et al. 2018; Breitbart et al. 2018; Boyd et al. 2000b).

6.3.2 *The Rise of a Lethal Pathogen: Lysogenic Conversion and Toxigenicity in* Escherichia coli

While *Escherichia coli* is a common commensal bacterium found in the intestinal tract of humans and animals, Shiga toxin-producing *E. coli* (STEC) are a heterogenous group of enteric pathogens that cause diseases ranging from diarrhea to devastating hemorrhagic colitis (HC) and hemolytic uremic syndrome (HUS) (Beutin and Martin 2012; Hunt 2010; Karch et al. 2012; Mellmann et al. 2011). *E. coli* O157:H7 is the causative pathotype for the majority of STEC-associated HC and HUS (Banatvala et al. 2001; Verweyen et al. 1999); however, over 400 non-O157 serotypes are recognized for their involvement in human disease (Bettelheim 2007; Mora et al. 2011). The primary virulence factors produced by STEC are the phage-encoded Shiga toxins (Stx), whose genes are found on Stx phages (Newland et al. 1985); the Shiga toxins are released when phage-mediated lysis occurs. Comparison of the genomes of pathogenic *E. coli* strain O157:H7 to laboratory *E. coli* strain K12, revealed that the majority of the differences observed in the pathogenic strain are due to prophages (Blattner et al. 1997; Hayashi et al. 2001; Ohnishi et al. 2001). Stx phages are a diverse group that present with different morphologies, genome size, and host infectivity range (Rietra et al. 1989; Muniesa et al. 2000; Allison et al. 2003; Karama and Gyles 2008; Karama et al. 2008; Gamage et al. 2004; Muniesa et al. 2004; Park et al. 2013; Osawa et al. 2000). Not surprisingly, integration sites for Stx phages in the bacterial chromosome also show great diversity. In *E. coli* O157:H7 strains, at least five integration sites have been described for Stx phages and these include *wrbA*, *yehV*, *argW*, *sbcB*, and *yecE* (De Greve et al. 2002; Shaikh and Tarr 2003; Besser et al. 2007; Mellor et al. 2012; Shringi et al. 2012); and several integration sites have been described for non-O157 STEC strains (Recktenwald and Schmidt 2002; Koch et al. 2003; Ahmed et al. 2012; Steyert et al. 2012; Cooper et al. 2014). The process by which the phage integrates is not well-defined; however, double lysogens have been detected (Allison et al. 2003). There is also diversity in the sequence of *stx* genes and Shiga toxins produced by STEC are classified into two primary types: Shiga toxin 1 (Stx1) and Shiga toxin 2 (Stx2); each of these groups is comprised of several subtypes, with Stx2 having the most heterogeneity (Scheutz et al. 2012). STEC strains may carry one or more *stx* genes in their genome and strains carrying three or more have been described (Bertin et al. 2001; Eklund et al. 2002; Kruger et al. 2011). The production of Stx1 and Stx2 in O157:H7 and the production of Shiga toxin 2 in the case of O104:H4 is the result of these strains being lysogenzied by more than one of the Stx-phage group of bacteriophages (Allison 2007; Laing et al. 2012).

Despite the genetic diversity of *stx* genes, all Shiga toxins share common structural and enzymatic elements. Shiga toxins are characteristic AB toxins, having a single A-subunit bound to pentameric B subunits (Law 2000). The B subunit binds to glycosphingolipid globotriaosylceramide (Gb3) on the target cell (Schuller 2011). Following binding the Shiga toxin and its receptor are endocytosed, the A subunit is activated, and the A1 subunit translocates into the host cell cytoplasm where it cleaves one adenine residue of the 28s rRNA subunit; this disruption inhibits the binding of aminoacyl-tRNA to the 60s ribosomal subunit and inhibits protein synthesis, thereby triggering endoplasmic reticulum and ribotoxic stress responses and ultimately leading to apoptosis (Endo et al. 1988; Furutani et al. 1992; Saxena et al. 1989; Schuller 2011; Tesh 2012). When Shiga toxin enters the bloodstream, it can result in more systemic symptoms because Gb3 is expressed on microvascular endothelial cells found throughout the body, including in the kidney and brain (Brigotti et al. 2010; te Loo et al. 2000). Damage to the vasculature by Shiga toxin is a hallmark of STEC pathophysiology and its effects can lead to a prothrombogenic environment within the host, which is also characteristic of HUS caused by STEC strains (Chandler et al. 2002).

6.3.3 Lysogenic Conversion and Toxin-Production in **Clostridium botulinum**

Various pathogenic *Clostridium* species benefit through their relationship with temperate phages and through toxin conversion, including *Clostridium botulinum*. *C. botulinum* is the causative agent for botulism, a life-threatening disease that is characterized by the symptom of flaccid paralysis and can affect humans, animals, and birds (Sobel 2005). The virulence of *C. botulinum* is through the production of botulinum neurotoxin (BoNT). BoNT toxins are categorized into seven groups based on antigenicity (A, B, C1, D, E, F, or G) and these groups can be further divided into more than 40 subtypes (Rummel 2015; Smith et al. 2015). BoNT A, B, and F are chromosomally encoded and G is plasmid encoded (Barksdale and Arden 1974; Hutson et al. 1996; Zhou et al. 1995), whereas BoNT/C1 and BoNT/D, and possibly BoNT/E neurotoxins are encoded on CEβ and DEβ phages; curing of these phages results in a loss of virulence in these strains (Barksdale and Arden 1974; Eklund et al. 1971, 1972; Zhou et al. 1993; Inoue and Iida 1970, 1971; Oguma et al. 1973). Analysis of the genetic organization of the C1 and D loci of CEβ and DEβ phages shows genes for both toxin secretion and regulation (Hauser et al. 1992; Tsuzuki et al. 1990). Further analysis of the genetic organization shows that the BoNT *bont* genes are in close proximity to non-haemagglutinin gene *ntnha* and hemagglutinin operon, which encode proteins that form a heterodimer with BoNT and are proposed to protect the toxin from pH denaturation and proteases activity in the gastrointestinal tract and facilitate absorption through the gastric mucosa (Bonventre 1979; Iwasaki et al. 1980; Ohishi et al. 1980; Hambleton 1992).

All BoNT are potent toxins that result in a prevention of acetylcholine release by cholinergic presynaptic receptors (Montecucco et al. 2004). BoNT are expressed as a large 150-kDa polypeptide inactive precursor that are cleaved by bacterial or host proteases to yield the mature BoNT, which consists of a light (L) chain (50-kDa) and a heavy (H) chain (100-kDa) linked by a disulfide bond (Hambleton 1992). The structure of the active form of BoNT includes three domains: the HC domain

(C-terminus of the H chain) is responsible for binding to presynaptic terminals; the HN domain (N-terminus of the H chain) is responsible for translocation of the L chain across the membrane of endocytic vesicles into the neuronal cytosol; the L chain is a Zn^{2+}-dependent metalloprotease that cleaves the SNARE proteins required for neurotransmitter exocytosis (Lacy and Stevens 1999). The HC-C domain binds to a polysialoganglioside (PSG) receptor present on presynaptic neurons (Simpson and Rapport 1971), followed by binding to either synaptotagmin (Syt) or SV2 protein receptor (Nishiki et al. 1994; Dong et al. 2003; Rummel et al. 2007; Peng et al. 2012). Following binding, BoNT is endocytosed and the L chain is ultimately translocated across the synaptic vesicle membrane where the L chain is released from the HN chain and becomes enzymatically activated (Fischer and Montal 2007; Montal 2010; Fischer 2013). BoNT cleave at least one of three proteins involved in acetylycholine exocytosis: synaptic vesicle associated membrane protein (VAMP), 25 kDa synaptosomal-associated protein (SNAP-25), or syntaxin and the target protein varies depending on type of BoNT; BoNT/B, /D, /F, and /G cleave VAMP (Schiavo et al. 1993a, b, 1994); BoNT/A and /E cleave only SNAP-25 (Schiavo et al. 1993a; Simpson 1979, 2004); BoNT/C cleaves both SNAP-25 and syntaxin (Schiavo et al. 1995; Simpson 1979, 2004). The end result of this cleavage is a loss of neurotransmitter release and the characteristic flaccid paralysis associated with botulism occurs.

C. botulinum isn't the only *Clostridium* species to gain toxin production through phage conversion. *Clostridium novyi* is a toxigenic pathogen that causes disease in both humans and animals (Leal et al. 2008; Popoff and Bouvet 2009) and is divided into four types (A-D) based on the production of toxins. *C. novyi* type A produce α-toxin (TcnA), which is a large glucosylating toxin that is encoded on prophage NA1 (Popoff and Bouvet 2009; Eklund et al. 1974). Eklund et al. showed that *C. botulinum* type C1 strain that produces BoNT C could be cured of its phage 3C and reinfected with phage NA1 from *C. novyi* and the resulting *C. botulinum* produced α-toxin, showing interspecies conversion (Eklund et al. 1974). Type B strains of *C. novyi* also harbor various phages (NB1, NB2, NB3, NB5, NB7, NB9) that lead to α-toxin production as well and interspecies toxin conversion conferred by NB phages between *C. novyi* and *C. botulinum* has been reported as well (Eklund et al. 1976).

While toxin conversion in *C. botulinum* types C1 and D and *C. novyi* types A and B are the only pathogenic *Clostridium* spp. in which true phage conversion of toxin has been shown to date, some strains of *Clostridium difficile* (reclassified as *Clostridiodes difficile*) produce binary toxin (CDT), which has actin-specific ADP-ribosyltransferase activity and is composed of two subunits encoded by the *cdtA* and *cdtB* genes (Bauer et al. 2011; Popoff et al. 1988; Snydman et al. 2015; Carman et al. 2011; Gerding et al. 2014). The toxin genes are found together with the *cdtR* gene that encodes a regulator on the 6.2-kb CDT locus (CdtLoc) (Carman et al. 2011; Gerding et al. 2014). This locus was identified on the sequence of an episomic prophage phiSemix9P1 found within the *C. difficile* genome and represents the sole example of a phage-encoded toxin genes in *C. difficile* (Dannheim et al. 2017; Riedel et al. 2017). Interestingly, regulator CdtR was shown to be involved in the regulation of the expression of the main toxins TcdA and TcdB in *C. difficile* (Lyon et al. 2016); however, further investigation is needed.

6.3.4 *Phage-Conversion and Toxigenicity in* Corynebacterium diphtheriae

Perhaps the most well-characterized prototypical example of toxin-conversion is found within diphtheria toxin producing *C. diphtheriae*. *C. diphtheriae* is the etiological agent for the highly communicable respiratory disease, diphtheria (von Graevenitz and Bernard 2006), a disease that is characterized by the formation of a pseudomembrane on the tonsils, uvula, oropharynx, and nasopharynx; *C. diphtheriae* can also cause systemic intoxication whereby the diphtheria toxin (DT) produced by this pathogen may destroy the parenchymal tissues of the heart, liver, and kidneys (Hadfield et al. 2000). The diphtheria toxin is the primary virulence factor for *C. diphtheriae* and is instrumental in the pathogenesis of diphtheria; the gene for DT expression is carried by corynephages. Freeman first proposed a relationship between lysogeny and toxigenicity after studies showed that exposure of non-toxigenic *C. diphtheriae* to a lysogenic corynephage resulted in the recovery of toxigenic strains (Freeman and Morse 1952). Additional studies by Groman (1953) and Barksdale (Barksdale and Pappenheimer 1954) confirmed these findings; Groman showed that toxigenicity is induced and was dependent on bacteriophage involvement (Groman 1953) and Barksdale and Pappenheimer coined the term "conversion" to indicate that production of the diphtheria toxin was conferred via lysogeny (Barksdale and Pappenheimer 1954).

The ability of corynephage β to convert susceptible nontoxigenic strains of *C. diphtheriae* to toxigenic strains is the result of *tox*$^+$ phages (Barksdale 1955; Groman and Eaton 1955). While corynephage β is the prototypical toxigenicity-converting phage, a family of β-related phages, referred to as β family, contains not only toxin-converting phages, but also non-converting phages like γ phage and other *tox* mutants (Buck and Groman 1981a; Groman et al. 1983; Michel et al. 1982; Buck et al. 1985; Groman 1984).

Corynephage β is a temperate phage that has a polyhedral head, a 270 nm long, slender tail (Freeman 1951; Mathews et al. 1966), and contains approximately 34.7 kb of linear, double-double stranded DNA (Buck et al. 1978); as indicated above, its genome contains the gene that encodes the diphtheria toxin. Corynephage β integrates into its host's chromosome in a similar fashion as does Escherichia virus λ (λ phage) in *E. coli* (Laird and Groman 1976; Buck and Groman 1981b; Michel et al. 1982); this integration occurs via site-directed recombination between a phage attachment site (*att*P) and one of the two functionally equivalent bacterial attachment sites (*attB1* and *attB2*) found within Arg-tRNA$_2$ genes on the chromosome of *C. diphtheriae* (Rappuoli and Ratti 1984). The *att*B sites share approximately 93 bp of core sequence that has high homology to that of corynephage β phage *att*P (Ratti et al. 1997; Buck et al. 1985). Integration with a *tox*$^+$ phage results in a toxigenic bacterium that expresses *tox* from a phage-encoded promoter; however, its expression is controlled through the host's regulatory mechanisms and most notably repressed by Diphtheria Toxin Repressor (DtxR) (Welkos and Holmes 1981a, b; Murphy et al. 1976, 1978).

As described above, diphtheria toxin is the predominant virulence factor associated with *C. diphtheriae* and is a potent bacterial toxin for which the minimal lethal dose of DT is less than 0.1 μg/kg of body weight for humans and susceptible animals (Pappenheimer 1984). Secreted DT contains three functional domains in total; DT-A is the catalytically active domain (C-domain) that has ADP-ribosylatase activity and results in intracellular toxicity, whereas DT-B contains the translocation and receptor-binding domains (T-domain and R-domain, respectively) that are responsible for binding and internalization of the toxin (Collier and Kandel 1971; Gill and Dinius 1971; Gill and Pappenheimer 1971). Introduction of a single molecule of DT-A into the cytosol of a eukaryotic cell is sufficient for killing that cell (Yamaizumi et al. 1978). DT-B facilitates binding to HB-EGF, and following binding, DT is endocytosed via clathrin-coated vesicles and enters the endosomal pathway (Morris et al. 1985). In the endosome, acidification occurs that induces a conformational change in the translocation domain of DT (Draper and Simon 1980; Sandvig and Olsnes 1980) which ultimately facilitates the translocation of the DT-A fragment into the cytosol (Hu and Holmes 1984; Moskaug et al. 1988; Olsnes et al. 1988; Kagan et al. 1981). Following entry into the cytoplasm, the DT-A fragment binds to NAD and catalyzes the transfer of an ADP-ribose group from NAD to a diphthamide residue of EF-2 (Van Ness et al. 1980; Chung and Collier 1977a, b). EF-2 mediates the translocation step of protein synthesis by promoting the transfer of peptidyl tRNA from the A site to the P site of the ribosome (Moldave 1985); therefore, inactivation of EF-2 results in termination of protein synthesis and lethality to the cells. The susceptibility of animal cells to the action of DT varies dramatically; studies showed that highly susceptible cells have a greater level of DT-receptors on their surface than do those cells having a lower susceptibility (Dorland et al. 1979; Middlebrook et al. 1978; Middlebrook and Dorland 1977).

Interestingly, phage conversion leading to the ability to produce diphtheria toxin is not solely observed in *C. diphtheriae*, but has also been observed in additional *Corynebacterium* spp. Diphtheria toxin-producing strains of *Corynebacterium ulcerans* and *Corynebacterium pseudotuberculosis* have been isolated from nature (Maximescu et al. 1974a); these species can be lysogenized with phages isolated from *C. diphtheriae*, ultimately conferring toxigenicity (Maximescu 1968; Maximescu et al. 1974a, b). The corynephage β *attB* site is present in numerous other *Corynebacterium* spp. (Cianciotto et al. 1986); therefore, it is not surprising that other *Corynebacterium* spp. such as *C. ulcerans* and *C. pseudotuberculosis* can be lysogenized by a *tox*$^+$ β-like phage (Cianciotto et al. 1986). *C. ulcerans* can be transmitted to humans zoonotically (Lartigue et al. 2005) and has become of increasing clinical importance worldwide as both a human and animal pathogen (Dewinter et al. 2005; Sing et al. 2005; de Carpentier et al. 1992; Wagner et al. 2001; Kaufmann et al. 2002; Hatanaka et al. 2003; von Hunolstein et al. 2003; Komiya et al. 2010; Wagner et al. 2010; Bonnet and Begg 1999). Sequence analysis of *C. ulcerans* clinical isolate 0102 possesses a unique *tox*$^{+}$ prophage ϕCULC0102 in its chromosome (Sekizuka et al. 2012) and may help to explain great genetic diversity observed in *tox* genes found within *C. ulcerans* strains that are in contrast to the conserved *tox* sequence observed in *C. diphtheriae* (Sing et al. 2003, 2005). Further investigation of the diversity of *tox* genes within *C. pseudotuberculosis* is needed.

6.4 Insects, Symbionts, and Phage-Encoded Toxins. . .Oh My!

While this chapter has focused on the benefits of the phage to its bacterial host, as well as on the impacts that toxins have primarily on humans, it would be remiss to not mention the relationship that exists between various insects, their symbionts, and the bacteriophages they harbor. Most notable is the relationship between the facultative endosymbiont *Candidatus Hamiltonella defensa* (*Hamiltonella defensa*), a γ-proteobacterium found in a wide-range of sap-sucking insects including aphids, mealybugs, psyllids, and whiteflies (Clark et al. 1992; Sandstrom et al. 2001; Russell et al. 2003; Moran et al. 2005a, b; Moran 2007) and its phage and the protection they confer against endoparasitoid wasps. APSE (for *Acyrthosiphon pisum* secondary endosymbiont) is a lambda-like phage with an isometric head and a short tail that was originally isolated from an *Acyrthosiphon pisum* (pea aphid) infected by a secondary endosymbiont (van der Wilk et al. 1999) and APSE is associated with *Hamiltonella defensa*. To date, seven types of APSE have been described (APSE-1, APSE-2, APSE-3, APSE-4, APSE-5, APSE-6, APSE-7) and are associated with *Hamiltonella* strains from six aphid species and the whitefly *Bemisia tabaci* (van der Wilk et al. 1999; Moran et al. 2005a; Degnan and Moran 2008a; Oliver et al. 2008, 2009; 2010; Martinez et al. 2014; Oliver and Higashi 2019; Dennis et al. 2017). These varying types of APSE encode homologs of toxins from one of three protein families, including the Shiga-like toxin, cytolethal distending toxin (CdtB), as well as a tyrosine-aspartic acid (YD-repeat)-containing toxin (van der Wilk et al. 1999; Moran et al. 2005a; Degnan and Moran 2008a; Oliver et al. 2008, 2009, 2010; Martinez et al. 2014; Oliver and Higashi 2019; Dennis et al. 2017). In general, pea aphids infected with *Hamiltonella defensa* and APSE are significantly more protected against parasitoid wasps; in these aphids, the wasp larva dies prematurely and the aphid is able to develop to its adult stage and subsequently reproduce (Oliver et al. 2003). However, varying levels of protection have been observed and differ with APSE type (Oliver et al. 2003, 2005; Bensadia et al. 2006). APSE-2 carries a *cdtB* homolog and investigation of aphids harboring APSE-2-*Hamiltonella defensa* showed ~40% mortality of parasitoid wasps, indicating moderate protection of the aphids (Oliver et al. 2003, 2005). In contrast, *Hamiltonella defensa* carrying APSE-3, which encodes a YD-repeat containing toxin, confers a high level of protection to aphids, showing greater than 85% mortality of parasitoid wasps (Oliver et al. 2005); spontaneous loss of APSE-3 eliminates this protection and results in an increase in the intracellular level of *Hamiltonella defensa* that negatively impacts aphid fitness (Oliver et al. 2009; Weldon et al. 2013). Furthermore, introduction of APSE-3 into an APSE-free and non-protective A2C strain of *Hamiltonella defensa* conferred protection against the effects of parasitoid wasps (Brandt et al. 2017). Under these protective measures, the level of expression for the toxin homologs was constitutively expressed at high levels, suggesting an important role by these toxins in the protectivity provided to the insect (Moran et al. 2005a; Oliver et al. 2009). In all, this relationship between the endosymbiont, the phage, and the insect is a fascinating relationship where further investigation into the role of phage-encoded toxins in APSE-mediated *Hamiltonella defensa* parasitoid defense is warranted.

6.5 Summary

Through genetic transfer, bacteriophages are key drivers of bacterial evolution. The prolific genotypic and phenotypic changes that can arise in a bacterial host through the acquisition of new phages can truly have significant impacts on not only the bacterial host itself, but also on its environment. Lysogenic conversion can benefit the bacterial host by increasing bacterial fitness, enhancing virulence, restricting secondary infection by lytic phages, providing antibiotic resistance, or conferring the ability to produce a toxin by an otherwise non-toxigenic strain (Obeng et al. 2016; Bondy-Denomy and Davidson 2014; Argov et al. 2017). Indeed, toxin-conversion in bacterial hosts contributes to their virulence and often results in detrimental impacts to the environment in which these pathogens may be found. While the role of phage-encoded toxins on bacterial virulence has been well-studied in a number of medically important pathogens, further analysis into the genetic diversity of phages within an environment, their rampant recombination and genetic rearrangement events, the possible inter- and intra-species acquisition of phage-associated toxin genes, and the overall potential evolution of new pathogens is needed. Through such investigations, the potential of phage on the evolution of already existing pathogens or new pathogens may be fully appreciated.

Additionally, toxin-conversion in endosymbionts is a realm of possibility insofar as phage research. Obligate intracellular mutualists of insects typically lack phages (Degnan et al. 2005; Akman et al. 2002; Nakabachi et al. 2006); however, phages are commonly found in facultative (secondary) endosymbionts (Hypsa and Dale 1997; Masui et al. 2000; Toh et al. 2006; van der Wilk et al. 1999) . In the example of the relationship between aphids, *Hamiltonella defensa*, and APSE phage there is a profound phage-provided protective measures conferred to the insect (van der Wilk et al. 1999; Oliver and Higashi 2019). Further genomic studies of APSE phages across aphid taxonomic groups have revealed diversity associated with the phage as well as the toxin arsenal (Rouil et al. 2020). Studies have also shown that significant horizontal and vertical genetic transfer drive diversity in the phages and impact not only the endosymbiont, but also their insect hosts (Degnan and Moran 2008b). Many unanswered questions exist with respect to the origin for the diversity observed in these phages and what additional roles phages, by way of facultative endosymbionts, have in horizontal gene transfer in the evolution of animals outside of insects.

References

Ackermann HW, Dubow MS (1987) General properties of bacteriophages. In: Viruses of prokaryotes. CRC, p 202

Ahmed SA, Awosika J, Baldwin C, Bishop-Lilly KA, Biswas B, Broomall S, Chain PS, Chertkov O, Chokoshvili O, Coyne S, Davenport K, Detter JC, Dorman W, Erkkila TH, Folster JP, Frey KG, George M, Gleasner C, Henry M, Hill KK, Hubbard K, Insalaco J, Johnson S, Kitzmiller A, Krepps M, Lo CC, Luu T, McNew LA, Minogue T, Munk CA, Osborne B, Patel M, Reitenga KG, Rosenzweig CN, Shea A, Shen X, Strockbine N, Tarr C, Teshima H, van Gieson E, Verratti K, Wolcott M, Xie G, Sozhamannan S, Gibbons HS, Threat Characterization Consortium (2012) Genomic comparison of *Escherichia coli* O104:H4 isolates from 2009 and 2011 reveals plasmid, and prophage heterogeneity, including shiga toxin encoding phage stx2. PLoS One 7(11):e48228. https://doi.org/10.1371/journal.pone.0048228

Akman L, Yamashita A, Watanabe H, Oshima K, Shiba T, Hattori M, Aksoy S (2002) Genome sequence of the endocellular obligate symbiont of tsetse flies, *Wigglesworthia glossinidia*. Nat Genet 32(3):402–407. https://doi.org/10.1038/ng986

Allison HE (2007) Stx-phages: drivers and mediators of the evolution of STEC and STEC-like pathogens. Future Microbiol 2(2):165–174. https://doi.org/10.2217/17460913.2.2.165

Allison HE, Sergeant MJ, James CE, Saunders JR, Smith DL, Sharp RJ, Marks TS, McCarthy AJ (2003) Immunity profiles of wild-type and recombinant shiga-like toxin-encoding bacteriophages and characterization of novel double lysogens. Infect Immun 71(6):3409–3418. https://doi.org/10.1128/iai.71.6.3409-3418.2003

Argov T, Azulay G, Pasechnek A, Stadnyuk O, Ran-Sapir S, Borovok I, Sigal N, Herskovits AA (2017) Temperate bacteriophages as regulators of host behavior. Curr Opin Microbiol 38:81–87. https://doi.org/10.1016/j.mib.2017.05.002

Azam F (1998) Oceanography: microbial control of oceanic carbon flux. Science 280 (5364):694–696

Aziz RK, Edwards RA, Taylor WW, Low DE, McGeer A, Kotb M (2005) Mosaic prophages with horizontally acquired genes account for the emergence and diversification of the globally disseminated M1T1 clone of *Streptococcus pyogenes*. J Bacteriol 187(10):3311–3318. https://doi.org/10.1128/JB.187.10.3311-3318.2005

Baba T, Bae T, Schneewind O, Takeuchi F, Hiramatsu K (2008) Genome sequence of *Staphylococcus aureus* strain Newman and comparative analysis of staphylococcal genomes: polymorphism and evolution of two major pathogenicity islands. J Bacteriol 190(1):300–310. https://doi.org/10.1128/JB.01000-07

Bae T, Baba T, Hiramatsu K, Schneewind O (2006) Prophages of *Staphylococcus aureus* Newman and their contribution to virulence. Mol Microbiol 62(4):1035–1047. https://doi.org/10.1111/j.1365-2958.2006.05441.x

Banatvala N, Griffin PM, Greene KD, Barrett TJ, Bibb WF, Green JH, Wells JG, Hemolytic Uremic Syndrome Study Collaborators (2001) The United States National Prospective Hemolytic Uremic Syndrome Study: microbiologic, serologic, clinical, and epidemiologic findings. J Infect Dis 183(7):1063–1070. https://doi.org/10.1086/319269

Banks DJ, Beres SB, Musser JM (2002) The fundamental contribution of phages to GAS evolution, genome diversification and strain emergence. Trends Microbiol 10(11):515–521

Barksdale L (1955) [Bacteriophages of *Corynebacterium diphtheriae* and their hosts]. C R Hebd Seances Acad Sci 240(18):1831–1833

Barksdale L, Arden SB (1974) Persisting bacteriophage infections, lysogeny, and phage conversions. Annu Rev Microbiol 28:265–299. https://doi.org/10.1146/annurev.mi.28.100174.001405

Barksdale LW, Pappenheimer AMJ (1954) Phage-host relationships in nontoxigenic and toxigenic diphtheria bacilli. J Bacteriol 67:220–232

Barondess JJ, Beckwith J (1990) A bacterial virulence determinant encoded by lysogenic coliphage lambda. Nature 346(6287):871–874. https://doi.org/10.1038/346871a0

Barrett EL, Riggs DL (1982) Evidence of a second nitrate reductase activity that is distinct from the respiratory enzyme in *Salmonella typhimurium.* J Bacteriol 150(2):563–571. https://doi.org/10.1128/JB.150.2.563-571.1982

Bauer MP, Notermans DW, van Benthem BH, Brazier JS, Wilcox MH, Rupnik M, Monnet DL, van Dissel JT, Kuijper EJ ECDIS Study Group (2011) *Clostridium difficile* infection in Europe: a hospital-based survey. Lancet 377(9759):63–73. https://doi.org/10.1016/S0140-6736(10)61266-4

Bensadia F, Boudreault S, Guay JF, Michaud D, Cloutier C (2006) Aphid clonal resistance to a parasitoid fails under heat stress. J Insect Physiol 52(2):146–157. https://doi.org/10.1016/j.jinsphys.2005.09.011

Bensing BA, Siboo IR, Sullam PM (2001) Proteins PblA and PblB of *Streptococcus mitis*, which promote binding to human platelets, are encoded within a lysogenic bacteriophage. Infect Immun 69(10):6186–6192. https://doi.org/10.1128/IAI.69.10.6186-6192.2001

Bertin Y, Boukhors K, Pradel N, Livrelli V, Martin C (2001) Stx2 subtyping of Shiga toxin-producing *Escherichia coli* isolated from cattle in France: detection of a new Stx2 subtype and correlation with additional virulence factors. J Clin Microbiol 39(9):3060–3065. https://doi.org/10.1128/jcm.39.9.3060-3065.2001

Besser TE, Shaikh N, Holt NJ, Tarr PI, Konkel ME, Malik-Kale P, Walsh CW, Whittam TS, Bono JL (2007) Greater diversity of Shiga toxin-encoding bacteriophage insertion sites among *Escherichia coli* O157:H7 isolates from cattle than in those from humans. Appl Environ Microbiol 73(3):671–679. https://doi.org/10.1128/AEM.01035-06

Bettelheim KA (2007) The non-O157 shiga-toxigenic (verocytotoxigenic) *Escherichia coli;* under-rated pathogens. Crit Rev Microbiol 33(1):67–87. https://doi.org/10.1080/10408410601172172

Beutin L, Martin A (2012) Outbreak of Shiga toxin-producing *Escherichia coli* (STEC) O104:H4 infection in Germany causes a paradigm shift with regard to human pathogenicity of STEC strains. J Food Prot 75(2):408–418. https://doi.org/10.4315/0362-028X.JFP-11-452

Billard-Pomares T, Fouteau S, Jacquet ME, Roche D, Barbe V, Castellanos M, Bouet JY, Cruveiller S, Medigue C, Blanco J, Clermont O, Denamur E, Branger C (2014) Characterization of a P1-like bacteriophage carrying an SHV-2 extended-spectrum beta-lactamase from an *Escherichia coli* strain. Antimicrob Agents Chemother 58(11):6550–6557. https://doi.org/10.1128/AAC.03183-14

Blattner FR, Plunkett G 3rd, Bloch CA, Perna NT, Burland V, Riley M, Collado-Vides J, Glasner JD, Rode CK, Mayhew GF, Gregor J, Davis NW, Kirkpatrick HA, Goeden MA, Rose DJ, Mau B, Shao Y (1997) The complete genome sequence of *Escherichia coli* K-12. Science 277(5331):1453–1462

Bobay LM, Rocha EP, Touchon M (2013) The adaptation of temperate bacteriophages to their host genomes. Mol Biol Evol 30(4):737–751. https://doi.org/10.1093/molbev/mss279

Bondy-Denomy J, Davidson AR (2014) When a virus is not a parasite: the beneficial effects of prophages on bacterial fitness. J Microbiol 52(3):235–242. https://doi.org/10.1007/s12275-014-4083-3

Bonnet JM, Begg NT (1999) Control of diphtheria: guidance for consultants in communicable disease control. World Health Organization. Commun Dis Public Health 2(4):242–249

Bonventre PF (1979) Absorption of botulinal toxin from the gastrointestinal tract. Rev Infect Dis 1(4):663–667. https://doi.org/10.1093/clinids/1.4.663

Boyd F (2008) Filamentous phages of *Vibrio cholerae*. In: Faruque SM, Nair GB (eds) *Vibrio cholerae*: genomics and molecular biology. Horizon Scientific, Norwich, pp 49–66
Boyd EF (2012) Bacteriophage-encoded bacterial virulence factors and phage-pathogenicity island interactions. Adv Virus Res 82:91–118. https://doi.org/10.1016/B978-0-12-394621-8.00014-5
Boyd EF, Waldor MK (1999) Alternative mechanism of cholera toxin acquisition by *Vibrio cholerae*: generalized transduction of CTXPhi by bacteriophage CP-T1. Infect Immun 67 (11):5898–5905
Boyd EF, Heilpern AJ, Waldor MK (2000a) Molecular analyses of a putative CTXphi precursor and evidence for independent acquisition of distinct CTX(phi)s by toxigenic V*ibrio cholerae*. J Bacteriol 182(19):5530–5538
Boyd EF, Moyer KE, Shi L, Waldor MK (2000b) Infectious CTXPhi and the vibrio pathogenicity island prophage in *Vibrio mimicus*: evidence for recent horizontal transfer between *V. mimicus* and V. *cholerae*. Infect Immun 68(3):1507–1513. https://doi.org/10.1128/iai.68.3.1507-1513.2000
Bramucci MG, Keggins KM, Lovett PS (1977a) Bacteriophage conversion of spore-negative mutants to spore-positive in *Bacillus pumilus*. J Virol 22(1):194–202. https://doi.org/10.1128/JVI.22.1.194-202.1977
Bramucci MG, Keggins KM, Lovett PS (1977b) Bacteriophage PMB12 conversion of the sporulation defect in RNA polymerase mutants of *Bacillus subtilis*. J Virol 24(1):194–200. https://doi.org/10.1128/JVI.24.1.194-200.1977
Brandt JW, Chevignon G, Oliver KM, Strand MR (2017) Culture of an aphid heritable symbiont demonstrates its direct role in defence against parasitoids. Proc Biol Sci 284(1866). https://doi.org/10.1098/rspb.2017.1925
Breitbart M, Hewson I, Felts B, Mahaffy JM, Nulton J, Salamon P, Rohwer F (2003) Metagenomic analyses of an uncultured viral community from human feces. J Bacteriol 185(20):6220–6223
Breitbart M, Haynes M, Kelley S, Angly F, Edwards RA, Felts B, Mahaffy JM, Mueller J, Nulton J, Rayhawk S, Rodriguez-Brito B, Salamon P, Rohwer F (2008) Viral diversity and dynamics in an infant gut. Res Microbiol 159(5):367–373. https://doi.org/10.1016/j.resmic.2008.04.006
Breitbart M, Bonnain C, Malki K, Sawaya NA (2018) Phage puppet masters of the marine microbial realm. Nat Microbiol 3(7):754–766. https://doi.org/10.1038/s41564-018-0166-y
Brigotti M, Tazzari PL, Ravanelli E, Carnicelli D, Barbieri S, Rocchi L, Arfilli V, Scavia G, Ricci F, Bontadini A, Alfieri RR, Petronini PG, Pecoraro C, Tozzi AE, Caprioli A (2010) Endothelial damage induced by Shiga toxins delivered by neutrophils during transmigration. J Leukoc Biol 88(1):201–210. https://doi.org/10.1189/jlb.0709475
Brussow H, Canchaya C, Hardt WD (2004) Phages and the evolution of bacterial pathogens: from genomic rearrangements to lysogenic conversion. Microbiol Mol Biol Rev MMBR 68 (3):560–602, table of contents. https://doi.org/10.1128/MMBR.68.3.560-602.2004
Buck GA, Groman NB (1981a) Genetic elements novel for *Corynebacterium diphtheriae*: specialized transducing elements and transposons. J Bacteriol 148(1):143–152
Buck GA, Groman NB (1981b) Physical mapping of beta-converting and gamma-nonconverting corynebacteriophage genomes. J Bacteriol 148(1):131–142
Buck G, Groman N, Falkow S (1978) Relationship between beta converting and gamma non-converting corynebacteriophage DNA. Nature 271(5646):683–685
Buck GA, Cross RE, Wong TP, Loera J, Groman N (1985) DNA relationships among some *tox*-bearing corynebacteriophages. Infect Immun 49(3):679–684
Bushman F (2002) Lateral DNA transfer: mechanisms and consequences. Cold Spring Harbor Laboratory Press, Cold Spring Harbor, NY
Campbell AM (1992) Chromosomal insertion sites for phages and plasmids. J Bacteriol 174:7495–7499
Canchaya C, Proux C, Fournous G, Bruttin A, Brussow H (2003) Prophage genomics. Microbiol Mol Biol Rev MMBR 67(2):238–276, table of contents

Carman RJ, Stevens AL, Lyerly MW, Hiltonsmith MF, Stiles BG, Wilkins TD (2011) *Clostridium difficile* binary toxin (CDT) and diarrhea. Anaerobe 17(4):161–165. https://doi.org/10.1016/j.anaerobe.2011.02.005

Caron DA (1994) Inorganic nutrients, bacteria, and the microbial loop. Microb Ecol 28(2):295–298. https://doi.org/10.1007/BF00166820

Carrolo M, Frias MJ, Pinto FR, Melo-Cristino J, Ramirez M (2010) Prophage spontaneous activation promotes DNA release enhancing biofilm formation in Streptococcus pneumoniae. PLoS One 5(12):e15678. https://doi.org/10.1371/journal.pone.0015678

Casas V, Maloy S (2011) Role of bacteriophage-encoded exotoxins in the evolution of bacterial pathogens. Future Microbiol 6(12):1461–1473. https://doi.org/10.2217/fmb.11.124

Casas V, Miyake J, Balsley H, Roark J, Telles S, Leeds S, Zurita I, Breitbart M, Bartlett D, Azam F, Rohwer F (2006) Widespread occurrence of phage-encoded exotoxin genes in terrestrial and aquatic environments in Southern California. FEMS Microbiol Lett 261(1):141–149. https://doi.org/10.1111/j.1574-6968.2006.00345.x

Casjens S (2003) Prophages and bacterial genomics: what have we learned so far? Mol Microbiol 49(2):277–300. https://doi.org/10.1046/j.1365-2958.2003.03580.x

Castillo D, Kauffman K, Hussain F, Kalatzis P, Rorbo N, Polz MF, Middelboe M (2018) Widespread distribution of prophage-encoded virulence factors in marine *Vibrio* communities. Sci Rep 8(1):9973. https://doi.org/10.1038/s41598-018-28326-9

Chandler WL, Jelacic S, Boster DR, Ciol MA, Williams GD, Watkins SL, Igarashi T, Tarr PI (2002) Prothrombotic coagulation abnormalities preceding the hemolytic-uremic syndrome. N Engl J Med 346(1):23–32. https://doi.org/10.1056/NEJMoa011033

Chaturongakul S, Ounjai P (2014) Phage-host interplay: examples from tailed phages and Gram-negative bacterial pathogens. Front Microbiol 5:442. https://doi.org/10.3389/fmicb.2014.00442

Chung DW, Collier RJ (1977a) Enzymatically active peptide from the adenosine diphosphate-ribosylating toxin of *Pseudomonas aeruginosa*. Infect Immun 16(3):832–841. https://doi.org/10.1128/IAI.16.3.832-841.1977

Chung DW, Collier RJ (1977b) The mechanism of ADP-ribosylation of elongation factor 2 catalyzed by fragment A from diphtheria toxin. Biochim Biophys Acta 483(2):248–257. https://doi.org/10.1016/0005-2744(77)90053-5

Cianciotto N, Rappuoli R, Groman N (1986) Detection of homology to the beta bacteriophage integration site in a wide variety of *Corynebacterium spp*. J Bacteriol 168(1):103–108

Clark MA, Baumann L, Munson MA, Baumann P, Campbell C, Duffus JE, Osborne LS, Moran NA (1992) The eubacterial endosymbionts of whiteflies (Homoptera: Aleyrodoidea) constitute a lineage distinct from the endosymbionts of aphids and mealybugs. Curr Microbiol 25:119–123

Cleary PP, LaPenta D, Vessela R, Lam H, Cue D (1998) A globally disseminated M1 subclone of group A streptococci differs from other subclones by 70 kilobases of prophage DNA and capacity for high-frequency intracellular invasion. Infect Immun 66(11):5592–5597

Coleman D, Knights J, Russell R, Shanley D, Birkbeck TH, Dougan G, Charles I (1991) Insertional inactivation of the *Staphylococcus aureus* beta-toxin by bacteriophage phi 13 occurs by site- and orientation-specific integration of the phi 13 genome. Mol Microbiol 5(4):933–939

Collier RJ, Kandel J (1971) Structure and activity of diphtheria toxin. I. Thiol-dependent dissociation of a fraction of toxin into enzymically active and inactive fragments. J Biol Chem 246 (5):1496–1503

Comeau AM, Bertrand C, Letarov A, Tetart F, Krisch HM (2007) Modular architecture of the T4 phage superfamily: a conserved core genome and a plastic periphery. Virology 362(2):384–396. https://doi.org/10.1016/j.virol.2006.12.031

Cooke FJ, Wain J, Fookes M, Ivens A, Thomson N, Brown DJ, Threlfall EJ, Gunn G, Foster G, Dougan G (2007) Prophage sequences defining hot spots of genome variation in *Salmonella enterica* serovar Typhimurium can be used to discriminate between field isolates. J Clin Microbiol 45(8):2590–2598. https://doi.org/10.1128/JCM.00729-07

Cooper KK, Mandrell RE, Louie JW, Korlach J, Clark TA, Parker CT, Huynh S, Chain PS, Ahmed S, Carter MQ (2014) Comparative genomics of enterohemorrhagic *Escherichia coli*

O145:H28 demonstrates a common evolutionary lineage with Escherichia coli O157:H7. BMC Genomics 15:17. https://doi.org/10.1186/1471-2164-15-17

Cumby N, Edwards AM, Davidson AR, Maxwell KL (2012) The bacteriophage HK97 gp15 moron element encodes a novel superinfection exclusion protein. J Bacteriol 194(18):5012–5019. https://doi.org/10.1128/JB.00843-12

Dannheim H, Riedel T, Neumann-Schaal M, Bunk B, Schober I, Sproer C, Chibani CM, Gronow S, Liesegang H, Overmann J, Schomburg D (2017) Manual curation and reannotation of the genomes of *Clostridium difficile* 630Deltaerm and C. difficile 630. J Med Microbiol 66 (3):286–293. https://doi.org/10.1099/jmm.0.000427

Davis BM, Waldor MK (2003) Filamentous phages linked to virulence of *Vibrio cholerae*. Curr Opin Microbiol 6(1):35–42. https://doi.org/10.1016/s1369-5274(02)00005-x

Davis BM, Moyer KE, Boyd EF, Waldor MK (2000) CTX prophages in classical biotype *Vibrio cholerae*: functional phage genes but dysfunctional phage genomes. J Bacteriol 182 (24):6992–6998

de Carpentier JP, Flanagan PM, Singh IP, Timms MS, Nassar WY (1992) Nasopharyngeal *Corynebacterium ulcerans*: a different diphtheria. J Laryngol Otol 106(9):824–826

Degnan PH, Moran NA (2008a) Diverse phage-encoded toxins in a protective insect endosymbiont. Appl Environ Microbiol 74(21):6782–6791. https://doi.org/10.1128/AEM.01285-08

Degnan PH, Moran NA (2008b) Evolutionary genetics of a defensive facultative symbiont of insects: exchange of toxin-encoding bacteriophage. Mol Ecol 17(3):916–929. https://doi.org/10.1111/j.1365-294X.2007.03616.x

Degnan PH, Lazarus AB, Wernegreen JJ (2005) Genome sequence of *Blochmannia pennsylvanicus* indicates parallel evolutionary trends among bacterial mutualists of insects. Genome Res 15 (8):1023–1033. https://doi.org/10.1101/gr.3771305

De Greve H, Qizhi C, Deboeck F, Hernalsteens JP (2002) The Shiga-toxin VT2-encoding bacteriophage varphi297 integrates at a distinct position in the *Escherichia coli* genome. Biochim Biophys Acta 1579(2-3):196–202. https://doi.org/10.1016/s0167-4781(02)00539-0

Dennis AB, Patel V, Oliver KM, Vorburger C (2017) Parasitoid gene expression changes after adaptation to symbiont-protected hosts. Evolution 71(11):2599–2617. https://doi.org/10.1111/evo.13333

Dewinter LM, Bernard KA, Romney MG (2005) Human clinical isolates of *Corynebacterium diphtheriae* and *Corynebacterium ulcerans* collected in Canada from 1999 to 2003 but not fitting reporting criteria for cases of diphtheria. J Clin Microbiol 43(7):3447–3449

Dong M, Richards DA, Goodnough MC, Tepp WH, Johnson EA, Chapman ER (2003) Synaptotagmins I and II mediate entry of botulinum neurotoxin B into cells. J Cell Biol 162 (7):1293–1303. https://doi.org/10.1083/jcb.200305098

Dorland RB, Middlebrook JL, Leppla SH (1979) Receptor-mediated internalization and degradation of diphtheria toxin by monkey kidney cells. J Biol Chem 254(22):11337–11342

Draper RK, Simon MI (1980) The entry of diphtheria toxin into the mammalian cell cytoplasm: evidence for lysosomal involvement. J Cell Biol 87(3 Pt 1):849–854

Edlin G, Lin L, Kudrna R (1975) Lambda lysogens of *E. coli* reproduce more rapidly than non-lysogens. Nature 255(5511):735–737. https://doi.org/10.1038/255735a0

Edlin G, Lin L, Bitner R (1977) Reproductive fitness of P1, P2, and Mu lysogens of *Escherichia coli*. J Virol 21(2):560–564. https://doi.org/10.1128/JVI.21.2.560-564.1977

Eklund MW, Poysky FT, Reed SM, Smith CA (1971) Bacteriophage and the toxigenicity of *Clostridium botulinum* type C. Science 172(3982):480–482. https://doi.org/10.1126/science.172.3982.480

Eklund MW, Poysky FT, Reed SM (1972) Bacteriophage and the toxigenicity of *Clostridium botulinum* type D. Nat New Biol 235(53):16–17. https://doi.org/10.1038/newbio235016a0

Eklund MW, Poysky FT, Meyers JA, Pelroy GA (1974) Interspecies conversion of *Clostridium botulinum* type C to *Clostridium novyi* type A by bacteriophage. Science 186(4162):456–458. https://doi.org/10.1126/science.186.4162.456

Eklund MW, Poysky FT, Peterson ME, Meyers JA (1976) Relationship of bacteriophages to alpha toxin production in *Clostridium novyi* types A and B. Infect Immun 14(3):793–803. https://doi.org/10.1128/IAI.14.3.793-803.1976

Eklund M, Leino K, Siitonen A (2002) Clinical *Escherichia coli* strains carrying stx genes: stx variants and stx-positive virulence profiles. J Clin Microbiol 40(12):4585–4593. https://doi.org/10.1128/jcm.40.12.4585-4593.2002

Endo Y, Tsurugi K, Yutsudo T, Takeda Y, Ogasawara T, Igarashi K (1988) Site of action of a Vero toxin (VT2) from *Escherichia coli* O157:H7 and of Shiga toxin on eukaryotic ribosomes. RNA N-glycosidase activity of the toxins. Eur J Biochem 171(1-2):45–50

Erez Z, Steinberger-Levy I, Shamir M, Doron S, Stokar-Avihail A, Peleg Y, Melamed S, Leavitt A, Savidor A, Albeck S, Amitai G, Sorek R (2017) Communication between viruses guides lysis-lysogeny decisions. Nature 541(7638):488–493. https://doi.org/10.1038/nature21049

Fancello L, Trape S, Robert C, Boyer M, Popgeorgiev N, Raoult D, Desnues C (2013) Viruses in the desert: a metagenomic survey of viral communities in four perennial ponds of the Mauritanian Sahara. ISME J 7(2):359–369. https://doi.org/10.1038/ismej.2012.101

Faruque SM (2013) Role of phages in the epidemiology of cholera. Curr Topics Microbiol Immunol. https://doi.org/10.1007/82_2013_358

Faruque SM, Albert MJ, Mekalanos JJ (1998) Epidemiology, genetics, and ecology of toxigenic *Vibrio cholerae*. Microbiol Mol Biol Rev 62(4):1301–1314

Fasano A, Baudry B, Pumplin DW, Wasserman SS, Tall BD, Ketley JM, Kaper JB (1991) *Vibrio cholerae* produces a second enterotoxin, which affects intestinal tight junctions. Proc Natl Acad Sci U S A 88(12):5242–5246. https://doi.org/10.1073/pnas.88.12.5242

Field M, Fromm D, al-Awqati Q, Greenough WB 3rd (1972) Effect of cholera enterotoxin on ion transport across isolated ileal mucosa. J Clin Invest 51(4):796–804. https://doi.org/10.1172/JCI106874

Figueroa-Bossi N, Bossi L (1999) Inducible prophages contribute to *Salmonella* virulence in mice. Mol Microbiol 33(1):167–176. https://doi.org/10.1046/j.1365-2958.1999.01461.x

Figueroa-Bossi N, Uzzau S, Maloriol D, Bossi L (2001) Variable assortment of prophages provides a transferable repertoire of pathogenic determinants in *Salmonella*. Mol Microbiol 39 (2):260–271. https://doi.org/10.1046/j.1365-2958.2001.02234.x

Fischer A (2013) Synchronized chaperone function of botulinum neurotoxin domains mediates light chain translocation into neurons. Curr Top Microbiol Immunol 364:115–137. https://doi.org/10.1007/978-3-642-33570-9_6

Fischer A, Montal M (2007) Crucial role of the disulfide bridge between botulinum neurotoxin light and heavy chains in protease translocation across membranes. J Biol Chem 282 (40):29604–29611. https://doi.org/10.1074/jbc.M703619200

Fishmann PH (1990) Mechanism of action of cholera toxin. In: Moss J, Vaughan M (eds) ADP-ribosylating toxins and G proteins. American Society for Microbiology, Washington, DC, pp 127–137

Fortier LC, Sekulovic O (2013) Importance of prophages to evolution and virulence of bacterial pathogens. Virulence 4(5):354–365. https://doi.org/10.4161/viru.24498

Freeman VJ (1951) Studies on the virulence of bacteriophage-infected strains of *Corynebacterium diphtheriae*. J Bacteriol 61(6):675–688. https://doi.org/10.1128/JB.61.6.675-688.1951

Freeman VJ, Morse IU (1952) Further observations on the change to virulence of bacteriophage-infected a virulent strains of *Corynebacterium diphtheriae*. J Bacteriol 63(3):407–414

Furutani M, Kashiwagi K, Ito K, Endo Y, Igarashi K (1992) Comparison of the modes of action of a Vero toxin (a Shiga-like toxin) from *Escherichia coli*, of ricin, and of alpha-sarcin. Arch Biochem Biophys 293(1):140–146

Gamage SD, Patton AK, Hanson JF, Weiss AA (2004) Diversity and host range of Shiga toxin-encoding phage. Infect Immun 72(12):7131–7139. https://doi.org/10.1128/IAI.72.12.7131-7139.2004

Gerding DN, Johnson S, Rupnik M, Aktories K (2014) *Clostridium difficile* binary toxin CDT: mechanism, epidemiology, and potential clinical importance. Gut Microbes 5(1):15–27. https://doi.org/10.4161/gmic.26854

Gill DM, Dinius LL (1971) Observations on the structure of diphtheria toxin. J Biol Chem 246 (5):1485–1491

Gill DM, Pappenheimer AM Jr (1971) Structure-activity relationships in diphtheria toxin. J Biol Chem 246(5):1492–1495

Goerke C, Pantucek R, Holtfreter S, Schulte B, Zink M, Grumann D, Broker BM, Doskar J, Wolz C (2009) Diversity of prophages in dominant *Staphylococcus aureus* clonal lineages. J Bacteriol 191(11):3462–3468. https://doi.org/10.1128/JB.01804-08

Goh S, Ong PF, Song KP, Riley TV, Chang BJ (2007) The complete genome sequence of *Clostridium difficile* phage phiC2 and comparisons to phiCD119 and inducible prophages of CD630. Microbiology 153(Pt 3):676–685. https://doi.org/10.1099/mic.0.2006/002436-0

Groman NB (1953) The relation of bacteriophage to the change of *Corynebacterium diphtheriae* from avirulence to virulence. Science 117(3038):297–299

Groman NB (1984) Conversion by corynephages and its role in the natural history of diphtheria. J Hyg (Lond) 93(3):405–417

Groman NB, Eaton M (1955) Genetic factors in *Corynebacterium diphtheriae* conversion. J Bacteriol 70(6):637–640

Groman N, Cianciotto N, Bjorn M, Rabin M (1983) Detection and expression of DNA homologous to the tox gene in nontoxinogenic isolates of *Corynebacterium diphtheriae*. Infect Immun 42 (1):48–56

Guinee PA, Jansen WH, Peters PW (1985) *Vibrio cholerae* infection and acquired immunity in an adult rabbit model. Zentralbl Bakteriol Mikrobiol Hyg A Med Microbiol Infect Dis Virol Parasitol 259(1):118–131

Guinee PA, Jansen WH, Gielen H, Rijpkema SG, Peters PW (1987) Protective immunity against Vibrio cholerae infection in the rabbit. Zentralbl Bakteriol Mikrobiol Hyg A Med Microbiol Infect Dis Virol Parasitol 266(3–4):552–562

Guinee PA, Jansen WH, Rijpkema SG (1988) Infection and immunity to *Vibrio cholerae*, *Salmonella typhimurium* and *Escherichia coli* in a rabbit model. Zentralbl Bakteriol Mikrobiol Hyg A Med Microbiol Infect Dis Virol Parasitol 270(1-2):260–269

Guttman B, Raya R, Kutter E (2005) Basic phage biology. In: Kutter E, Sulakvelidze A (eds) Bacteriophages: biology and applications. CRC, Boca Raton, FL, pp 29-66

Hadfield TL, McEvoy P, Polotsky Y, Tzinserling VA, Yakovlev AA (2000) The pathology of diphtheria. J Infect Dis 181(Suppl 1):S116–S120

Hambleton P (1992) *Clostridium botulinum* toxins: a general review of involvement in disease, structure, mode of action and preparation for clinical use. J Neurol 239(1):16–20. https://doi.org/10.1007/BF00839205

Hargreaves KR, Kropinski AM, Clokie MR (2014) Bacteriophage behavioral ecology: how phages alter their bacterial host's habits. Bacteriophage 4:e29866. https://doi.org/10.4161/bact.29866

Hatanaka A, Tsunoda A, Okamoto M, Ooe K, Nakamura A, Miyakoshi M, Komiya T, Takahashi M (2003) *Corynebacterium ulcerans* Diphtheria in Japan. Emerg Infect Dis 9(6):752–753. https://doi.org/10.3201/eid0906.020645

Hauser D, Gibert M, Boquet P, Popoff MR (1992) Plasmid localization of a type E botulinal neurotoxin gene homologue in toxigenic *Clostridium butyricum* strains, and absence of this gene in non-toxigenic *C. butyricum* strains. FEMS Microbiol Lett 78(2-3):251–255

Hayashi T, Makino K, Ohnishi M, Kurokawa K, Ishii K, Yokoyama K, Han CG, Ohtsubo E, Nakayama K, Murata T, Tanaka M, Tobe T, Iida T, Takami H, Honda T, Sasakawa C, Ogasawara N, Yasunaga T, Kuhara S, Shiba T, Hattori M, Shinagawa H (2001) Complete genome sequence of enterohemorrhagic *Escherichia coli* O157:H7 and genomic comparison with a laboratory strain K-12. DNA Res 8(1):11–22

Hemphill HE, Whiteley HR (1975) Bacteriophages of *Bacillus subtilis*. Bacteriol Rev 39 (3):257–315

Henkin TM, Grundy FJ, Nicholson WL, Chambliss GH (1991) Catabolite repression of alpha-amylase gene expression in *Bacillus subtilis* involves a trans-acting gene product homologous to the Escherichia coli lacl and galR repressors. Mol Microbiol 5(3):575–584. https://doi.org/10.1111/j.1365-2958.1991.tb00728.x
Hermans AP, Abee T, Zwietering MH, Aarts HJ (2005) Identification of novel *Salmonella enterica* serovar Typhimurium DT104-specific prophage and nonprophage chromosomal sequences among serovar Typhimurium isolates by genomic subtractive hybridization. Appl Environ Microbiol 71(9):4979–4985. https://doi.org/10.1128/AEM.71.9.4979-4985.2005
Hermans AP, Beuling AM, van Hoek AH, Aarts HJ, Abee T, Zwietering MH (2006) Distribution of prophages and SGI-1 antibiotic-resistance genes among different *Salmonella enterica* serovar Typhimurium isolates. Microbiology 152(Pt 7):2137–2147. https://doi.org/10.1099/mic.0.28850-0
Higgins D, Dworkin J (2012) Recent progress in *Bacillus subtilis* sporulation. FEMS Microbiol Rev 36(1):131–148. https://doi.org/10.1111/j.1574-6976.2011.00310.x
Hu VW, Holmes RK (1984) Evidence for direct insertion of fragments A and B of diphtheria toxin into model membranes. J Biol Chem 259(19):12226–12233
Huber KE, Waldor MK (2002) Filamentous phage integration requires the host recombinases XerC and XerD. Nature 417(6889):656–659. https://doi.org/10.1038/nature00782
Hunt JM (2010) Shiga toxin-producing *Escherichia coli* (STEC). Clin Lab Med 30(1):21–45. https://doi.org/10.1016/j.cll.2009.11.001
Hutson RA, Zhou Y, Collins MD, Johnson EA, Hatheway CL, Sugiyama H (1996) Genetic characterization of *Clostridium botulinum* type A containing silent type B neurotoxin gene sequences. J Biol Chem 271(18):10786–10792
Hypsa V, Dale C (1997) In vitro culture and phylogenetic analysis of "*Candidatus Arsenophonus triatominarum*," an intracellular bacterium from the triatomine bug, Triatoma infestans. Int J Syst Bacteriol 47(4):1140–1144. https://doi.org/10.1099/00207713-47-4-1140
Inoue K, Iida H (1970) Conversion of toxigenicity in *Clostridium botulinum* type C. Jpn J Microbiol 14(1):87–89. https://doi.org/10.1111/j.1348-0421.1970.tb00495.x
Inoue K, Iida H (1971) Phage-conversion of toxigenicity in *Clostridium botulinum* types C and D. Jpn J Med Sci Biol 24(1):53–56
Iwasaki M, Ohishi I, Sakaguchi G (1980) Evidence that botulinum C2 toxin has two dissimilar components. Infect Immun 29(2):390–394
Kagan BL, Finkelstein A, Colombini M (1981) Diphtheria toxin fragment forms large pores in phospholipid bilayer membranes. Proc Natl Acad Sci U S A 78(8):4950–4954
Kalatzis PG, Rorbo NI, Castillo D, Mauritzen JJ, Jorgensen J, Kokkari C, Zhang F, Katharios P, Middelboe M (2017) Stumbling across the same phage: comparative genomics of widespread temperate phages infecting the fish pathogen *Vibrio anguillarum*. Viruses 9(5). https://doi.org/10.3390/v9050122
Kaper JB, Fasano A, Trucksis M (1994) Toxins of *Vibrio cholerae*. In: Wachsmuth KI, Blake PA, Olsvik O (eds) *Vibrio cholerae* and cholera: molecular to global perspectives. American Society for Microbiology, Washington, DC, pp 145–176
Kaper JB, Morris JG Jr, Levine MM (1995) Cholera. Clin Microbiol Rev 8(1):48–86
Karama M, Gyles CL (2008) Characterization of verotoxin-encoding phages from *Escherichia coli* O103:H2 strains of bovine and human origins. Appl Environ Microbiol 74(16):5153–5158. https://doi.org/10.1128/AEM.00723-08
Karama M, Johnson RP, Holtslander R, Gyles CL (2008) Phenotypic and genotypic characterization of verotoxin-producing *Escherichia coli* O103:H2 isolates from cattle and humans. J Clin Microbiol 46(11):3569–3575. https://doi.org/10.1128/JCM.01095-08
Karaolis DK, Somara S, Maneval DR Jr, Johnson JA, Kaper JB (1999) A bacteriophage encoding a pathogenicity island, a type-IV pilus and a phage receptor in cholera bacteria. Nature 399 (6734):375–379. https://doi.org/10.1038/20715
Karch H, Denamur E, Dobrindt U, Finlay BB, Hengge R, Johannes L, Ron EZ, Tonjum T, Sansonetti PJ, Vicente M (2012) The enemy within us: lessons from the 2011 European

Escherichia coli O104:H4 outbreak. EMBO Mol Med 4(9):841–848. https://doi.org/10.1002/emmm.201201662
Kaufmann D, Ott P, Ruegg C (2002) Laryngopharyngitis by *Corynebacterium ulcerans*. Infection 30(3):168–170
Keggins KM, Nauman RK, Lovett PS (1978) Sporulation-converting bacteriophages for *Bacillus pumilus*. J Virol 27(3):819–822. https://doi.org/10.1128/JVI.27.3.819-822.1978
King CA, Van Heyningen WE (1973) Deactivation of cholera toxin by a sialidase-resistant monosialosylganglioside. J Infect Dis 127(6):639–647. https://doi.org/10.1093/infdis/127.6.639
Kinney DM, Bramucci MG (1981) Analysis of *Bacillus subtilis* sporulation with spore-converting bacteriophage PMB12. J Bacteriol 145(3):1281–1285. https://doi.org/10.1128/JB.145.3.1281-1285.1981
Koch C, Hertwig S, Appel B (2003) Nucleotide sequence of the integration site of the temperate bacteriophage 6220, which carries the Shiga toxin gene stx(1ox3). J Bacteriol 185 (21):6463–6466. https://doi.org/10.1128/jb.185.21.6463-6466.2003
Komiya T, Seto Y, De Zoysa A, Iwaki M, Hatanaka A, Tsunoda A, Arakawa Y, Kozaki S, Takahashi M (2010) Two Japanese *Corynebacterium ulcerans* isolates from the same hospital: ribotype, toxigenicity and serum antitoxin titre. J Med Microbiol 59(Pt 12):1497–1504. https://doi.org/10.1099/jmm.0.022491-0
Koskella B, Thompson JN, Preston GM, Buckling A (2011) Local biotic environment shapes the spatial scale of bacteriophage adaptation to bacteria. Am Nat 177(4):440–451. https://doi.org/10.1086/658991
Koudelka GB, Arnold JW, Chakraborty D (2018) Evolution of STEC virulence: insights from the antipredator activities of Shiga toxin producing *E. coli*. Int J Med Microbiol 308(7):956–961. https://doi.org/10.1016/j.ijmm.2018.07.001
Krisch HM (2003) The view from Les Treilles on the origins, evolution and diversity of viruses. Res Microbiol 154(4):227–229. https://doi.org/10.1016/s0923-2508(03)00110-4
Kruger A, Lucchesi PM, Parma AE (2011) Verotoxins in bovine and meat verotoxin-producing *Escherichia coli* isolates: type, number of variants, and relationship to cytotoxicity. Appl Environ Microbiol 77(1):73–79. https://doi.org/10.1128/AEM.01445-10
Lacy DB, Stevens RC (1999) Sequence homology and structural analysis of the clostridial neurotoxins. J Mol Biol 291(5):1091–1104. https://doi.org/10.1006/jmbi.1999.2945
Laing CR, Zhang Y, Gilmour MW, Allen V, Johnson R, Thomas JE, Gannon VP (2012) A comparison of Shiga-toxin 2 bacteriophage from classical enterohemorrhagic *Escherichia coli* serotypes and the German E. coli O104:H4 outbreak strain. PLoS One 7(5):e37362. https://doi.org/10.1371/journal.pone.0037362
Laird W, Groman N (1976) Prophage map of converting corynebacteriophage beta. J Virol 19 (1):208–219
Lartigue MF, Monnet X, Le Fleche A, Grimont PA, Benet JJ, Durrbach A, Fabre M, Nordmann P (2005) *Corynebacterium ulcerans* in an immunocompromised patient with diphtheria and her dog. J Clin Microbiol 43(2):999–1001. https://doi.org/10.1128/JCM.43.2.999-1001.2005
Law D (2000) Virulence factors of *Escherichia coli* O157 and other Shiga toxin-producing *E. coli*. J Appl Microbiol 88(5):729–745
Leal J, Gregson DB, Ross T, Church DL, Laupland KB (2008) Epidemiology of *Clostridium* species bacteremia in Calgary, Canada, 2000-2006. J Infect 57(3):198–203. https://doi.org/10.1016/j.jinf.2008.06.018
Leblanc C, Caumont-Sarcos A, Comeau AM, Krisch HM (2009) Isolation and genomic characterization of the first phage infecting Iodobacteria: varphiPLPE, a myovirus having a novel set of features. Environ Microbiol Rep 1(6):499–509. https://doi.org/10.1111/j.1758-2229.2009.00055.x
Lee CY, Iandolo JJ (1986) Lysogenic conversion of staphylococcal lipase is caused by insertion of the bacteriophage L54a genome into the lipase structural gene. J Bacteriol 166(2):385–391

Lee SH, Hava DL, Waldor MK, Camilli A (1999) Regulation and temporal expression patterns of *Vibrio cholerae* virulence genes during infection. Cell 99(6):625–634. https://doi.org/10.1016/s0092-8674(00)81551-2

Lee SH, Butler SM, Camilli A (2001) Selection for in vivo regulators of bacterial virulence. Proc Natl Acad Sci U S A 98(12):6889–6894. https://doi.org/10.1073/pnas.111581598

Lehane AM, Korres H, Verma NK (2005) Bacteriophage-encoded glucosyltransferase GtrII of *Shigella flexneri*: membrane topology and identification of critical residues. Biochem J 389 (Pt 1):137–143. https://doi.org/10.1042/BJ20050102

Levine MM, Kaper JB, Black RE, Clements ML (1983) New knowledge on pathogenesis of bacterial enteric infections as applied to vaccine development. Microbiol Rev 47(4):510–550

Lopez CA, Winter SE, Rivera-Chavez F, Xavier MN, Poon V, Nuccio SP, Tsolis RM, Baumler AJ (2012) Phage-mediated acquisition of a type III secreted effector protein boosts growth of *Salmonella* by nitrate respiration. mBio 3 (3). https://doi.org/10.1128/mBio.00143-12

Lynch KH, Stothard P, Dennis JJ (2010) Genomic analysis and relatedness of P2-like phages of the *Burkholderia cepacia* complex. BMC Genomics 11:599. https://doi.org/10.1186/1471-2164-11-599

Lyon SA, Hutton ML, Rood JI, Cheung JK, Lyras D (2016) CdtR regulates TcdA and TcdB production in *Clostridium difficile*. PLoS Pathog 12(7):e1005758. https://doi.org/10.1371/journal.ppat.1005758

Maiti D, Das B, Saha A, Nandy RK, Nair GB, Bhadra RK (2006) Genetic organization of pre-CTX and CTX prophages in the genome of an environmental *Vibrio cholerae* non-O1, non-O139 strain. Microbiology (Reading) 152(Pt 12):3633–3641. https://doi.org/10.1099/mic.0.2006/000117-0

Marinus MG, Poteete AR (2013) High efficiency generalized transduction in *Escherichia coli* O157:H7. F1000Res 2:7. https://doi.org/10.12688/f1000research.2-7.v1

Martinez AJ, Weldon SR, Oliver KM (2014) Effects of parasitism on aphid nutritional and protective symbioses. Mol Ecol 23(6):1594–1607. https://doi.org/10.1111/mec.12550

Masui S, Kamoda S, Sasaki T, Ishikawa H (2000) Distribution and evolution of bacteriophage WO in *Wolbachia*, the endosymbiont causing sexual alterations in arthropods. J Mol Evol 51 (5):491–497. https://doi.org/10.1007/s002390010112

Mathews MM, Miller PA, Pappenheimer AM Jr (1966) Morphological observations on some diphtherial phages. Virology 29(3):402–409

Matos RC, Lapaque N, Rigottier-Gois L, Debarbieux L, Meylheuc T, Gonzalez-Zorn B, Repoila F, Lopes Mde F, Serror P (2013) *Enterococcus faecalis* prophage dynamics and contributions to pathogenic traits. PLoS Genet 9(6):e1003539. https://doi.org/10.1371/journal.pgen.1003539

Maximescu P (1968) New host-strains for the lysogenic *Corynebacterium diphtheriae* Park Williams No. 8 strain. J Gen Microbiol 53(1):125–133

Maximescu P, Oprisan A, Pop A, Potorac E (1974a) Further studies on *Corynebacterium* species capable of producing diphtheria toxin (*C. diphtheriae*, *C. ulcerans*, *C. ovis*). J Gen Microbiol 82 (1):49–56

Maximescu P, Pop A, Oprisan A, Potorac E (1974b) Diphtheria tox+ gene expressed in *Corynebacterium* species other than *C. diphtheriae*. J Hyg Epidemiol Microbiol Immunol 18 (3):324–326

McClelland M, Sanderson KE, Spieth J, Clifton SW, Latreille P, Courtney L, Porwollik S, Ali J, Dante M, Du F, Hou S, Layman D, Leonard S, Nguyen C, Scott K, Holmes A, Grewal N, Mulvaney E, Ryan E, Sun H, Florea L, Miller W, Stoneking T, Nhan M, Waterston R, Wilson RK (2001) Complete genome sequence of *Salmonella enterica* serovar Typhimurium LT2. Nature 413(6858):852–856. https://doi.org/10.1038/35101614

McLeod SM, Waldor MK (2004) Characterization of XerC- and XerD-dependent CTX phage integration in Vibrio cholerae. Mol Microbiol 54(4):935–947. https://doi.org/10.1111/j.1365-2958.2004.04309.x

Mead PS, Griffin PM (1998) *Escherichia coli* O157:H7. Lancet 352(9135):1207–1212. https://doi.org/10.1016/S0140-6736(98)01267-7

Mekalanos JJ, Rubin EJ, Waldor MK (1997) Cholera: molecular basis for emergence and pathogenesis. FEMS Immunol Med Microbiol 18(4):241–248. https://doi.org/10.1111/j.1574-695X.1997.tb01052.x

Mellmann A, Harmsen D, Cummings CA, Zentz EB, Leopold SR, Rico A, Prior K, Szczepanowski R, Ji Y, Zhang W, McLaughlin SF, Henkhaus JK, Leopold B, Bielaszewska M, Prager R, Brzoska PM, Moore RL, Guenther S, Rothberg JM, Karch H (2011) Prospective genomic characterization of the German enterohemorrhagic *Escherichia coli* O104:H4 outbreak by rapid next generation sequencing technology. PLoS One 6(7): e22751. https://doi.org/10.1371/journal.pone.0022751

Mellor GE, Sim EM, Barlow RS, D'Astek BA, Galli L, Chinen I, Rivas M, Gobius KS (2012) Phylogenetically related Argentinean and Australian *Escherichia coli* O157 isolates are distinguished by virulence clades and alternative Shiga toxin 1 and 2 prophages. Appl Environ Microbiol 78(13):4724–4731. https://doi.org/10.1128/AEM.00365-12

Michel JL, Rappuoli R, Murphy JR, Pappenheimer AM Jr (1982) Restriction endonuclease map of the nontoxigenic corynephage gamma c and its relationship to the toxigenic corynephage beta c. J Virol 42(2):510–518

Middlebrook JL, Dorland RB (1977) Response of cultured mammalian cells to the exotoxins of Pseudomonas aeruginosa and *Corynebacterium diphtheriae:* differential cytotoxicity. Can J Microbiol 23(2):183–189

Middlebrook JL, Dorland RB, Leppla SH (1978) Association of diphtheria toxin with Vero cells. Demonstration of a receptor. J Biol Chem 253(20):7325–7330

Mirold S, Rabsch W, Rohde M, Stender S, Tschape H, Russmann H, Igwe E, Hardt WD (1999) Isolation of a temperate bacteriophage encoding the type III effector protein SopE from an epidemic *Salmonella typhimurium* strain. Proc Natl Acad Sci U S A 96(17):9845–9850. https://doi.org/10.1073/pnas.96.17.9845

Moldave K (1985) Eukaryotic protein synthesis. Annu Rev Biochem 54:1109–1149. https://doi.org/10.1146/annurev.bi.54.070185.005333

Montal M (2010) Botulinum neurotoxin: a marvel of protein design. Annu Rev Biochem 79:591–617. https://doi.org/10.1146/annurev.biochem.051908.125345

Montecucco C, Rossetto O, Schiavo G (2004) Presynaptic receptor arrays for clostridial neurotoxins. Trends Microbiol 12(10):442–446. https://doi.org/10.1016/j.tim.2004.08.002

Mora A, Herrrera A, Lopez C, Dahbi G, Mamani R, Pita JM, Alonso MP, Llovo J, Bernardez MI, Blanco JE, Blanco M, Blanco J (2011) Characteristics of the Shiga-toxin-producing enteroaggregative *Escherichia coli* O104:H4 German outbreak strain and of STEC strains isolated in Spain. Int Microbiol 14(3):121–141. https://doi.org/10.2436/20.1501.01.142

Moran NA (2007) Symbiosis as an adaptive process and source of phenotypic complexity. Proc Natl Acad Sci U S A 104(Suppl 1):8627–8633. https://doi.org/10.1073/pnas.0611659104

Moran NA, Degnan PH, Santos SR, Dunbar HE, Ochman H (2005a) The players in a mutualistic symbiosis: insects, bacteria, viruses, and virulence genes. Proc Natl Acad Sci U S A 102 (47):16919–16926. https://doi.org/10.1073/pnas.0507029102

Moran NA, Russell JA, Koga R, Fukatsu T (2005b) Evolutionary relationships of three new species of *Enterobacteriaceae* living as symbionts of aphids and other insects. Appl Environ Microbiol 71(6):3302–3310. https://doi.org/10.1128/AEM.71.6.3302-3310.2005

Morris RE, Gerstein AS, Bonventre PF, Saelinger CB (1985) Receptor-mediated entry of diphtheria toxin into monkey kidney (Vero) cells: electron microscopic evaluation. Infect Immun 50 (3):721–727

Moskaug JO, Sandvig K, Olsnes S (1988) Low pH-induced release of diphtheria toxin A-fragment in Vero cells. Biochemical evidence for transfer to the cytosol. J Biol Chem 263(5):2518–2525

Muniesa M, Recktenwald J, Bielaszewska M, Karch H, Schmidt H (2000) Characterization of a shiga toxin 2e-converting bacteriophage from an *Escherichia coli* strain of human origin. Infect Immun 68(9):4850–4855. https://doi.org/10.1128/iai.68.9.4850-4855.2000

Muniesa M, Blanco JE, de Simon M, Serra-Moreno R, Blanch AR, Jofre J (2004) Diversity of stx2 converting bacteriophages induced from Shiga-toxin-producing *Escherichia coli* strains isolated from cattle. Microbiology (Reading) 150(Pt 9):2959–2971. https://doi.org/10.1099/mic.0.27188-0

Munro J, Oakey J, Bromage E, Owens L (2003) Experimental bacteriophage-mediated virulence in strains of *Vibrio harveyi*. Dis Aquat Organ 54(3):187–194. https://doi.org/10.3354/dao054187

Murphy JR, Skiver J, McBride G (1976) Isolation and partial characterization of a corynebacteriophage beta, tox operator constitutive-like mutant lysogen of *Corynebacterium diphtheriae*. J Virol 18(1):235–244

Murphy JR, Michel JL, Teng M (1978) Evidence that the regulation of diphtheria toxin production is directed at the level of transcription. J Bacteriol 135(2):511–516

Nakabachi A, Yamashita A, Toh H, Ishikawa H, Dunbar HE, Moran NA, Hattori M (2006) The 160-kilobase genome of the bacterial endosymbiont *Carsonella*. Science 314(5797):267. https://doi.org/10.1126/science.1134196

Nasu H, Iida T, Sugahara T, Yamaichi Y, Park KS, Yokoyama K, Makino K, Shinagawa H, Honda T (2000) A filamentous phage associated with recent pandemic *Vibrio parahaemolyticus* O3:K6 strains. J Clin Microbiol 38(6):2156–2161

Newland JW, Strockbine NA, Miller SF, O'Brien AD, Holmes RK (1985) Cloning of Shiga-like toxin structural genes from a toxin converting phage of *Escherichia coli*. Science 230 (4722):179–181. https://doi.org/10.1126/science.2994228

Newton GJ, Daniels C, Burrows LL, Kropinski AM, Clarke AJ, Lam JS (2001) Three-component-mediated serotype conversion in *Pseudomonas aeruginosa* by bacteriophage D3. Mol Microbiol 39(5):1237–1247. https://doi.org/10.1111/j.1365-2958.2001.02311.x

Nicholson WL, Park YK, Henkin TM, Won M, Weickert MJ, Gaskell JA, Chambliss GH (1987) Catabolite repression-resistant mutations of the *Bacillus subtilis* alpha-amylase promoter affect transcription levels and are in an operator-like sequence. J Mol Biol 198(4):609–618. https://doi.org/10.1016/0022-2836(87)90204-x

Nicholson WL, Munakata N, Horneck G, Melosh HJ, Setlow P (2000) Resistance of *Bacillus* endospores to extreme terrestrial and extraterrestrial environments. Microbiol Mol Biol Rev 64 (3):548–572. https://doi.org/10.1128/mmbr.64.3.548-572.2000

Nishiki T, Kamata Y, Nemoto Y, Omori A, Ito T, Takahashi M, Kozaki S (1994) Identification of protein receptor for *Clostridium botulinum* type B neurotoxin in rat brain synaptosomes. J Biol Chem 269(14):10498–10503

Obeng N, Pratama AA, Elsas JDV (2016) The significance of mutualistic phages for bacterial ecology and evolution. Trends Microbiol 24(6):440–449. https://doi.org/10.1016/j.tim.2015.12.009

O'Brien AD, Newland JW, Miller SF, Holmes RK, Smith HW, Formal SB (1984) Shiga-like toxin-converting phages from *Escherichia coli* strains that cause hemorrhagic colitis or infantile diarrhea. Science 226(4675):694–696. https://doi.org/10.1126/science.6387911

Oguma K, Iida H, Inoue K (1973) Bacteriophage and toxigenicity in *Clostridium botulinum*: an additional evidence for phage conversion. Jpn J Microbiol 17(5):425–426. https://doi.org/10.1111/j.1348-0421.1973.tb00794.x

Ogura Y, Kurokawa K, Ooka T, Tashiro K, Tobe T, Ohnishi M, Nakayama K, Morimoto T, Terajima J, Watanabe H, Kuhara S, Hayashi T (2006) Complexity of the genomic diversity in enterohemorrhagic *Escherichia coli* O157 revealed by the combinational use of the O157 Sakai OligoDNA microarray and the Whole Genome PCR scanning. DNA Res 13(1):3–14. https://doi.org/10.1093/dnares/dsi026

Ohishi I, Iwasaki M, Sakaguchi G (1980) Purification and characterization of two components of botulinum C2 toxin. Infect Immun 30(3):668–673

Ohnishi M, Tanaka C, Kuhara S, Ishii K, Hattori M, Kurokawa K, Yasunaga T, Makino K, Shinagawa H, Murata T, Nakayama K, Terawaki Y, Hayashi T (1999) Chromosome of the enterohemorrhagic *Escherichia coli* O157:H7; comparative analysis with K-12 MG1655 revealed the acquisition of a large amount of foreign DNAs. DNA Res 6(6):361–368

Ohnishi M, Kurokawa K, Hayashi T (2001) Diversification of *Escherichia coli* genomes: are bacteriophages the major contributors? Trends Microbiol 9(10):481–485

Ohnishi M, Terajima J, Kurokawa K, Nakayama K, Murata T, Tamura K, Ogura Y, Watanabe H, Hayashi T (2002) Genomic diversity of enterohemorrhagic *Escherichia coli* O157 revealed by whole genome PCR scanning. Proc Natl Acad Sci U S A 99(26):17043–17048. https://doi.org/10.1073/pnas.262441699

Oliver KM, Higashi CH (2019) Variations on a protective theme: *Hamiltonella defensa* infections in aphids variably impact parasitoid success. Curr Opin Insect Sci 32:1–7. https://doi.org/10.1016/j.cois.2018.08.009

Oliver KM, Russell JA, Moran NA, Hunter MS (2003) Facultative bacterial symbionts in aphids confer resistance to parasitic wasps. Proc Natl Acad Sci U S A 100(4):1803–1807. https://doi.org/10.1073/pnas.0335320100

Oliver KM, Moran NA, Hunter MS (2005) Variation in resistance to parasitism in aphids is due to symbionts not host genotype. Proc Natl Acad Sci U S A 102(36):12795–12800. https://doi.org/10.1073/pnas.0506131102

Oliver KM, Campos J, Moran NA, Hunter MS (2008) Population dynamics of defensive symbionts in aphids. Proc Biol Sci 275(1632):293–299. https://doi.org/10.1098/rspb.2007.1192

Oliver KM, Degnan PH, Hunter MS, Moran NA (2009) Bacteriophages encode factors required for protection in a symbiotic mutualism. Science 325(5943):992–994. https://doi.org/10.1126/science.1174463

Oliver KM, Degnan PH, Burke GR, Moran NA (2010) Facultative symbionts in aphids and the horizontal transfer of ecologically important traits. Annu Rev Entomol 55:247–266. https://doi.org/10.1146/annurev-ento-112408-085305

Olsnes S, Moskaug JO, Stenmark H, Sandvig K (1988) Diphtheria toxin entry: protein translocation in the reverse direction. Trends Biochem Sci 13(9):348–351

Osawa RO, Iyoda S, Nakayama SI, Wada A, Yamai S, Watanabe H (2000) Genotypic variations of Shiga toxin-converting phages from enterohaemorrhagic *Escherichia coli* O157: H7 isolates. J Med Microbiol 49(6):565–574. https://doi.org/10.1099/0022-1317-49-6-565

Pappenheimer AM Jr (1984) The diphtheria bacillus and its toxin: a model system. J Hyg (Lond) 93 (3):397–404

Paredes CJ, Alsaker KV, Papoutsakis ET (2005) A comparative genomic view of clostridial sporulation and physiology. Nat Rev Microbiol 3(12):969–978. https://doi.org/10.1038/nrmicro1288

Park D, Stanton E, Ciezki K, Parrell D, Bozile M, Pike D, Forst SA, Jeong KC, Ivanek R, Dopfer D, Kaspar CW (2013) Evolution of the Stx2-encoding prophage in persistent bovine *Escherichia coli* O157:H7 strains. Appl Environ Microbiol 79(5):1563–1572. https://doi.org/10.1128/AEM.03158-12

Pawluk A, Shah M, Mejdani M, Calmettes C, Moraes TF, Davidson AR, Maxwell KL (2017) Disabling a type I-E CRISPR-Cas nuclease with a bacteriophage-encoded anti-CRISPR protein. mBio 8 (6). https://doi.org/10.1128/mBio.01751-17

Peng L, Berntsson RP, Tepp WH, Pitkin RM, Johnson EA, Stenmark P, Dong M (2012) Botulinum neurotoxin D-C uses synaptotagmin I and II as receptors, and human synaptotagmin II is not an effective receptor for type B, D-C and G toxins. J Cell Sci 125(Pt 13):3233–3242. https://doi.org/10.1242/jcs.103564

Perlak FJ, Mendelsohn CL, Thorne CB (1979) Converting bacteriophage for sporulation and crystal formation in *Bacillus thuringiensis*. J Bacteriol 140(2):699–706. https://doi.org/10.1128/JB.140.2.699-706.1979

Perry LL, SanMiguel P, Minocha U, Terekhov AI, Shroyer ML, Farris LA, Bright N, Reuhs BL, Applegate BM (2009) Sequence analysis of *Escherichia coli* O157:H7 bacteriophage PhiV10 and identification of a phage-encoded immunity protein that modifies the O157 antigen. FEMS Microbiol Lett 292(2):182–186. https://doi.org/10.1111/j.1574-6968.2009.01511.x

Petrov VM, Ratnayaka S, Nolan JM, Miller ES, Karam JD (2010) Genomes of the T4-related bacteriophages as windows on microbial genome evolution. Virol J 7:292. https://doi.org/10.1186/1743-422X-7-292

Pierce NF (1973) Differential inhibitory effects of cholera toxoids and ganglioside on the enterotoxins of *Vibrio cholerae* and *Escherichia coli*. J Exp Med 137(4):1009–1023. https://doi.org/10.1084/jem.137.4.1009

Pilar AV, Reid-Yu SA, Cooper CA, Mulder DT, Coombes BK (2012) GogB is an anti-inflammatory effector that limits tissue damage during *Salmonella* infection through interaction with human FBXO22 and Skp1. PLoS Pathog 8(6):e1002773. https://doi.org/10.1371/journal.ppat.1002773

Popoff MR, Bouvet P (2009) Clostridial toxins. Future Microbiol 4(8):1021–1064. https://doi.org/10.2217/fmb.09.72

Popoff MR, Rubin EJ, Gill DM, Boquet P (1988) Actin-specific ADP-ribosyltransferase produced by a *Clostridium difficile* strain. Infect Immun 56(9):2299–2306. https://doi.org/10.1128/IAI.56.9.2299-2306.1988

Rahimi F, Bouzari M, Katouli M, Pourshafie MR (2012) Prophage and antibiotic resistance profiles of methicillin-resistant *Staphylococcus aureus* strains in Iran. Arch Virol 157(9):1807–1811. https://doi.org/10.1007/s00705-012-1361-4

Rappuoli R, Ratti G (1984) Physical map of the chromosomal region of *Corynebacterium diphtheriae* containing corynephage attachment sites attB1 and attB2. J Bacteriol 158(1):325–330

Ratti G, Covacci A, Rappuoli R (1997) A tRNA(2Arg) gene of *Corynebacterium diphtheriae* is the chromosomal integration site for toxinogenic bacteriophages. Mol Microbiol 25(6):1179–1181

Recktenwald J, Schmidt H (2002) The nucleotide sequence of Shiga toxin (Stx) 2e-encoding phage phiP27 is not related to other Stx phage genomes, but the modular genetic structure is conserved. Infect Immun 70(4):1896–1908. https://doi.org/10.1128/iai.70.4.1896-1908.2002

Rice SA, Tan CH, Mikkelsen PJ, Kung V, Woo J, Tay M, Hauser A, McDougald D, Webb JS, Kjelleberg S (2009) The biofilm life cycle and virulence of *Pseudomonas aeruginosa* are dependent on a filamentous prophage. ISME J 3(3):271–282. https://doi.org/10.1038/ismej.2008.109

Riedel T, Wittmann J, Bunk B, Schober I, Sproer C, Gronow S, Overmann J (2017) A *Clostridioides difficile* bacteriophage genome encodes functional binary toxin-associated genes. J Biotechnol 250:23–28. https://doi.org/10.1016/j.jbiotec.2017.02.017

Rietra PJ, Willshaw GA, Smith HR, Field AM, Scotland SM, Rowe B (1989) Comparison of Vero-cytotoxin-encoding phages from *Escherichia coli* of human and bovine origin. J Gen Microbiol 135(8):2307–2318. https://doi.org/10.1099/00221287-135-8-2307

Rivera-Chavez F, Mekalanos JJ (2019) Cholera toxin promotes pathogen acquisition of host-derived nutrients. Nature 572(7768):244–248. https://doi.org/10.1038/s41586-019-1453-3

Rouil J, Jousselin E, Coeur d'acier A, Cruaud C, Manzano-Marin A (2020) The protector within: comparative genomics of APSE phages across aphids reveals rampant recombination and diverse toxin arsenals. Genome Biol Evol 12(6):878–889. https://doi.org/10.1093/gbe/evaa089

Rummel A (2015) The long journey of botulinum neurotoxins into the synapse. Toxicon 107 (Pt A):9–24. https://doi.org/10.1016/j.toxicon.2015.09.009

Rummel A, Eichner T, Weil T, Karnath T, Gutcaits A, Mahrhold S, Sandhoff K, Proia RL, Acharya KR, Bigalke H, Binz T (2007) Identification of the protein receptor binding site of botulinum neurotoxins B and G proves the double-receptor concept. Proc Natl Acad Sci U S A 104(1):359–364. https://doi.org/10.1073/pnas.0609713104

Russell JA, Latorre A, Sabater-Munoz B, Moya A, Moran NA (2003) Side-stepping secondary symbionts: widespread horizontal transfer across and beyond the Aphidoidea. Mol Ecol 12(4):1061–1075. https://doi.org/10.1046/j.1365-294x.2003.01780.x

Sandman K, Losick R, Youngman P (1987) Genetic analysis of *Bacillus subtilis spo* mutations generated by Tn917-mediated insertional mutagenesis. Genetics 117(4):603–617

Sandstrom JP, Russell JA, White JP, Moran NA (2001) Independent origins and horizontal transfer of bacterial symbionts of aphids. Mol Ecol 10(1):217–228. https://doi.org/10.1046/j.1365-294x.2001.01189.x

Sandvig K, Olsnes S (1980) Diphtheria toxin entry into cells is facilitated by low pH. J Cell Biol 87 (3 Pt 1):828–832

Saxena SK, O'Brien AD, Ackerman EJ (1989) Shiga toxin, Shiga-like toxin II variant, and ricin are all single-site RNA N-glycosidases of 28 S RNA when microinjected into *Xenopus* oocytes. J Biol Chem 264(1):596–601

Scheutz F, Teel LD, Beutin L, Pierard D, Buvens G, Karch H, Mellmann A, Caprioli A, Tozzoli R, Morabito S, Strockbine NA, Melton-Celsa AR, Sanchez M, Persson S, O'Brien AD (2012) Multicenter evaluation of a sequence-based protocol for subtyping Shiga toxins and standardizing Stx nomenclature. J Clin Microbiol 50(9):2951–2963. https://doi.org/10.1128/JCM.00860-12

Schiavo G, Rossetto O, Catsicas S, Polverino de Laureto P, DasGupta BR, Benfenati F, Montecucco C (1993a) Identification of the nerve terminal targets of botulinum neurotoxin serotypes A, D, and E. J Biol Chem 268(32):23784–23787

Schiavo G, Shone CC, Rossetto O, Alexander FC, Montecucco C (1993b) Botulinum neurotoxin serotype F is a zinc endopeptidase specific for VAMP/synaptobrevin. J Biol Chem 268 (16):11516–11519

Schiavo G, Malizio C, Trimble WS, Polverino de Laureto P, Milan G, Sugiyama H, Johnson EA, Montecucco C (1994) Botulinum G neurotoxin cleaves VAMP/synaptobrevin at a single Ala-Ala peptide bond. J Biol Chem 269(32):20213–20216

Schiavo G, Shone CC, Bennett MK, Scheller RH, Montecucco C (1995) Botulinum neurotoxin type C cleaves a single Lys-Ala bond within the carboxyl-terminal region of syntaxins. J Biol Chem 270(18):10566–10570

Schuch R, Fischetti VA (2009) The secret life of the anthrax agent *Bacillus anthracis*: bacteriophage-mediated ecological adaptations. PLoS One 4(8):e6532. https://doi.org/10.1371/journal.pone.0006532

Schuller S (2011) Shiga toxin interaction with human intestinal epithelium. Toxins (Basel) 3 (6):626–639. https://doi.org/10.3390/toxins3060626

Sekizuka T, Yamamoto A, Komiya T, Kenri T, Takeuchi F, Shibayama K, Takahashi M, Kuroda M, Iwaki M (2012) *Corynebacterium ulcerans* 0102 carries the gene encoding diphtheria toxin on a prophage different from the *C. diphtheriae* NCTC 13129 prophage. BMC Microbiol 12:72. https://doi.org/10.1186/1471-2180-12-72

Sekulovic O, Fortier LC (2015) Global transcriptional response of *Clostridium difficile* carrying the CD38 prophage. Appl Environ Microbiol 81(4):1364–1374. https://doi.org/10.1128/AEM.03656-14

Shaikh N, Tarr PI (2003) *Escherichia coli* O157:H7 Shiga toxin-encoding bacteriophages: integrations, excisions, truncations, and evolutionary implications. J Bacteriol 185(12):3596–3605. https://doi.org/10.1128/jb.185.12.3596-3605.2003

Shelford E, Middelboe M, Moller E, Suttle CA (2012) Virus-driven nitrogen cycling enhances phytoplankton growth. Aquat Microb Ecol 66(1):41–46

Shringi S, Schmidt C, Katherine K, Brayton KA, Hancock DD, Besser TE (2012) Carriage of stx2a differentiates clinical and bovine-biased strains of *Escherichia coli* O157. PLoS One 7(12): e51572. https://doi.org/10.1371/journal.pone.0051572

Silver-Mysliwiec TH, Bramucci MG (1990) Bacteriophage-enhanced sporulation: comparison of spore-converting bacteriophages PMB12 and SP10. J Bacteriol 172(4):1948–1953. https://doi.org/10.1128/jb.172.4.1948-1953.1990

Simpson LL (1979) Studies on the mechanism of action of botulinum toxin. Adv Cytopharmacol 3:27–34

Simpson LL (2004) Identification of the major steps in botulinum toxin action. Annu Rev Pharmacol Toxicol 44:167–193. https://doi.org/10.1146/annurev.pharmtox.44.101802.121554

Simpson LL, Rapport MM (1971) The binding of botulinum toxin to membrane lipids: sphingolipids, steroids and fatty acids. J Neurochem 18(9):1751–1759. https://doi.org/10.1111/j.1471-4159.1971.tb03750.x

Sing A, Hogardt M, Bierschenk S, Heesemann J (2003) Detection of differences in the nucleotide and amino acid sequences of diphtheria toxin from *Corynebacterium diphtheriae* and *Corynebacterium ulcerans* causing extrapharyngeal infections. J Clin Microbiol 41(10):4848–4851

Sing A, Bierschenk S, Heesemann J (2005) Classical diphtheria caused by *Corynebacterium ulcerans* in Germany: amino acid sequence differences between diphtheria toxins from *Corynebacterium diphtheriae* and C. ulcerans. Clin Infect Dis 40(2):325–326

Smith TJ, Hill KK, Xie G, Foley BT, Williamson CHD, Foster JT, Johnson SL, Chertkov O, Teshima H, Gibbons HS, Johnsky LA, Karavis MA, Smith LA (2015) Genomic sequences of six botulinum neurotoxin-producing strains representing three clostridial species illustrate the mobility and diversity of botulinum neurotoxin genes. Infect Genet Evol 30:102–113. https://doi.org/10.1016/j.meegid.2014.12.002

Snydman DR, McDermott LA, Jacobus NV, Thorpe C, Stone S, Jenkins SG, Goldstein EJ, Patel R, Forbes BA, Mirrett S, Johnson S, Gerding DN (2015) U.S.-based national sentinel surveillance study for the epidemiology of *Clostridium difficile*-associated diarrheal isolates and their susceptibility to fidaxomicin. Antimicrob Agents Chemother 59(10):6437–6443. https://doi.org/10.1128/AAC.00845-15

Sobel J (2005) Botulism. Clin Infect Dis 41(8):1167–1173. https://doi.org/10.1086/444507

Srinivasiah S, Bhavsar J, Thapar K, Liles M, Schoenfeld T, Wommack KE (2008) Phages across the biosphere: contrasts of viruses in soil and aquatic environments. Res Microbiol 159(5):349–357

Stanley TL, Ellermeier CD, Slauch JM (2000) Tissue-specific gene expression identifies a gene in the lysogenic phage Gifsy-1 that affects Salmonella enterica serovar typhimurium survival in Peyer's patches. J Bacteriol 182(16):4406–4413. https://doi.org/10.1128/jb.182.16.4406-4413.2000

Stewart AW, Johnson MG (1977) Increased numbers of heat-resistant spores produced by two strains of *Clostridium perfringens* bearing temperate phage s9. J Gen Microbiol 103(1):45–50. https://doi.org/10.1099/00221287-103-1-45

Steyert SR, Sahl JW, Fraser CM, Teel LD, Scheutz F, Rasko DA (2012) Comparative genomics and stx phage characterization of LEE-negative Shiga toxin-producing *Escherichia coli*. Front Cell Infect Microbiol 2:133. https://doi.org/10.3389/fcimb.2012.00133

Suttle CA (2005) Viruses in the sea. Nature 437(7057):356–361. https://doi.org/10.1038/nature04160

Suttle CA (2007) Marine viruses—major players in the global ecosystem. Nat Rev Microbiol 5(10):801–812. https://doi.org/10.1038/nrmicro1750

Takahashi I (1979) Catabolite repression-resistant mutants of *Bacillus subtilis*. Can J Microbiol 25(11):1283–1287. https://doi.org/10.1139/m79-202

te Loo DM, Monnens LA, van Der Velden TJ, Vermeer MA, Preyers F, Demacker PN, van Den Heuvel LP, van Hinsbergh VW (2000) Binding and transfer of verocytotoxin by polymorphonuclear leukocytes in hemolytic uremic syndrome. Blood 95(11):3396–3402

Tesh VL (2012) Activation of cell stress response pathways by Shiga toxins. Cell Microbiol 14(1):1–9. https://doi.org/10.1111/j.1462-5822.2011.01684.x

Thomson N, Baker S, Pickard D, Fookes M, Anjum M, Hamlin N, Wain J, House D, Bhutta Z, Chan K, Falkow S, Parkhill J, Woodward M, Ivens A, Dougan G (2004) The role of prophage-like elements in the diversity of *Salmonella enterica* serovars. J Mol Biol 339(2):279–300. https://doi.org/10.1016/j.jmb.2004.03.058

Toh H, Weiss BL, Perkin SA, Yamashita A, Oshima K, Hattori M, Aksoy S (2006) Massive genome erosion and functional adaptations provide insights into the symbiotic lifestyle of *Sodalis glossinidius* in the tsetse host. Genome Res 16(2):149–156. https://doi.org/10.1101/gr.4106106

Touchon M, Hoede C, Tenaillon O, Barbe V, Baeriswyl S, Bidet P, Bingen E, Bonacorsi S, Bouchier C, Bouvet O, Calteau A, Chiapello H, Clermont O, Cruveiller S, Danchin A, Diard M, Dossat C, Karoui ME, Frapy E, Garry L, Ghigo JM, Gilles AM, Johnson J, Le Bouguenec C, Lescat M, Mangenot S, Martinez-Jehanne V, Matic I, Nassif X, Oztas S, Petit MA, Pichon C, Rouy Z, Ruf CS, Schneider D, Tourret J, Vacherie B, Vallenet D, Medigue C,

Rocha EP, Denamur E (2009) Organised genome dynamics in the Escherichia coli species results in highly diverse adaptive paths. PLoS Genet 5(1):e1000344. https://doi.org/10.1371/journal.pgen.1000344

Trost E, Blom J, Soares Sde C, Huang IH, Al-Dilaimi A, Schroder J, Jaenicke S, Dorella FA, Rocha FS, Miyoshi A, Azevedo V, Schneider MP, Silva A, Camello TC, Sabbadini PS, Santos CS, Santos LS, Hirata R Jr, Mattos-Guaraldi AL, Efstratiou A, Schmitt MP, Ton-That H, Tauch A (2012) Pangenomic study of *Corynebacterium diphtheriae* that provides insights into the genomic diversity of pathogenic isolates from cases of classical diphtheria, endocarditis, and pneumonia. J Bacteriol 194(12):3199–3215. https://doi.org/10.1128/JB.00183-12

Trucksis M, Galen JE, Michalski J, Fasano A, Kaper JB (1993) Accessory cholera enterotoxin (Ace), the third toxin of a Vibrio cholerae virulence cassette. Proc Natl Acad Sci U S A 90 (11):5267–5271. https://doi.org/10.1073/pnas.90.11.5267

Tsuzuki K, Kimura K, Fujii N, Yokosawa N, Indoh T, Murakami T, Oguma K (1990) Cloning and complete nucleotide sequence of the gene for the main component of hemagglutinin produced by *Clostridium botulinum* type C. Infect Immun 58(10):3173–3177

Val ME, Bouvier M, Campos J, Sherratt D, Cornet F, Mazel D, Barre FX (2005) The single-stranded genome of phage CTX is the form used for integration into the genome of *Vibrio cholerae*. Mol Cell 19(4):559–566. https://doi.org/10.1016/j.molcel.2005.07.002

van der Wilk F, Dullemans AM, Verbeek M, van den Heuvel JF (1999) Isolation and characterization of APSE-1, a bacteriophage infecting the secondary endosymbiont of *Acyrthosiphon pisum*. Virology 262(1):104–113. https://doi.org/10.1006/viro.1999.9902

Van Ness BG, Howard JB, Bodley JW (1980) ADP-ribosylation of elongation factor 2 by diphtheria toxin. Isolation and properties of the novel ribosyl-amino acid and its hydrolysis products. J Biol Chem 255(22):10717–10720

Ventura M, Canchaya C, Pridmore D, Berger B, Brussow H (2003) Integration and distribution of *Lactobacillus johnsonii* prophages. J Bacteriol 185:4603–4608

Verweyen HM, Karch H, Allerberger F, Zimmerhackl LB (1999) Enterohemorrhagic *Escherichia coli* (EHEC) in pediatric hemolytic-uremic syndrome: a prospective study in Germany and Austria. Infection 27(6):341–347. https://doi.org/10.1007/s150100050040

von Graevenitz A, Bernard K (2006) The genus *Corynebacterium*—medical. In: Dworkin M, Falkow F, Rosenberg E, Schleifer K, Stackebrandt E (eds) The prokaryotes, 3rd edn, vol 3. Springer, New York, pp 819–842

von Hunolstein C, Alfarone G, Scopetti F, Pataracchia M, La Valle R, Franchi F, Pacciani L, Manera A, Giammanco A, Farinelli S, Engler K, De Zoysa A, Efstratiou A (2003) Molecular epidemiology and characteristics of *Corynebacterium diphtheriae* and *Corynebacterium ulcerans* strains isolated in Italy during the 1990s. J Med Microbiol 52(Pt 2):181–188

Wagner J, Ignatius R, Voss S, Hopfner V, Ehlers S, Funke G, Weber U, Hahn H (2001) Infection of the skin caused by *Corynebacterium ulcerans* and mimicking classical cutaneous diphtheria. Clin Infect Dis 33(9):1598–1600. https://doi.org/10.1086/322969

Wagner PL, Livny J, Neely MN, Acheson DW, Friedman DI, Waldor MK (2002) Bacteriophage control of Shiga toxin 1 production and release by *Escherichia coli*. Mol Microbiol 44 (4):957–970. https://doi.org/10.1046/j.1365-2958.2002.02950.x

Wagner KS, White JM, Crowcroft NS, De Martin S, Mann G, Efstratiou A (2010) Diphtheria in the United Kingdom, 1986-2008: the increasing role of *Corynebacterium ulcerans*. Epidemiol Infect 138(11):1519–1530. https://doi.org/10.1017/S0950268810001895

Waldor MK, Mekalanos JJ (1994) Emergence of a new cholera pandemic: molecular analysis of virulence determinants in *Vibrio cholerae* O139 and development of a live vaccine prototype. J Infect Dis 170(2):278–283

Waldor MK, Mekalanos JJ (1996) Lysogenic conversion by a filamentous phage encoding cholera toxin. Science 272(5270):1910–1914. https://doi.org/10.1126/science.272.5270.1910

Waldor MK, Rubin EJ, Pearson GD, Kimsey H, Mekalanos JJ (1997) Regulation, replication, and integration functions of the *Vibrio cholerae* CTXphi are encoded by region RS2. Mol Microbiol 24(5):917–926. https://doi.org/10.1046/j.1365-2958.1997.3911758.x

Wang X, Kim Y, Ma Q, Hong SH, Pokusaeva K, Sturino JM, Wood TK (2010) Cryptic prophages help bacteria cope with adverse environments. Nat Commun 1:147. https://doi.org/10.1038/ncomms1146

Weldon SR, Strand MR, Oliver KM (2013) Phage loss and the breakdown of a defensive symbiosis in aphids. Proc Biol Sci 280(1751):20122103. https://doi.org/10.1098/rspb.2012.2103

Welkos SL, Holmes RK (1981a) Regulation of toxinogenesis in *Corynebacterium diphtheriae.* I. Mutations in bacteriophage beta that alter the effects of iron on toxin production. J Virol 37 (3):936–945

Welkos SL, Holmes RK (1981b) Regulation of toxinogenesis in *Corynebacterium diphtheriae.* II. Genetic mapping of a tox regulatory mutation in bacteriophage beta. J Virol 37(3):946–954

Weynberg KD, Voolstra CR, Neave MJ, Buerger P, van Oppen MJ (2015) From cholera to corals: viruses as drivers of virulence in a major coral bacterial pathogen. Sci Rep 5:17889. https://doi.org/10.1038/srep17889

Whiteley M, Bangera MG, Bumgarner RE, Parsek MR, Teitzel GM, Lory S, Greenberg EP (2001) Gene expression in *Pseudomonas aeruginosa* biofilms. Nature 413(6858):860–864. https://doi.org/10.1038/35101627

Wilcox MH, Fawley WN (2000) Hospital disinfectants and spore formation by *Clostridium difficile*. Lancet 356(9238):1324. https://doi.org/10.1016/S0140-6736(00)02819-1

Wilhelm SW, Suttle CA (1999) Viruses and nutrient cycles in the sea. Bioscience 49(10):781

Wipf JR, Schwendener S, Perreten V (2014) The novel macrolide-Lincosamide-Streptogramin B resistance gene erm(44) is associated with a prophage in *Staphylococcus xylosus*. Antimicrob Agents Chemother 58(10):6133–6138. https://doi.org/10.1128/AAC.02949-14

Wirtz C, Witte W, Wolz C, Goerke C (2009) Transcription of the phage-encoded Panton-Valentine leukocidin of *Staphylococcus aureus* is dependent on the phage life-cycle and on the host background. Microbiology (Reading) 155(Pt 11):3491–3499. https://doi.org/10.1099/mic.0.032466-0

Wommack KE, Colwell RR (2000) Virioplankton: viruses in aquatic ecosystems. Microbiol Mol Biol Rev MMBR 64(1):69–114

Yamaguchi T, Hayashi T, Takami H, Nakasone K, Ohnishi M, Nakayama K, Yamada S, Komatsuzawa H, Sugai M (2000) Phage conversion of exfoliative toxin A production in *Staphylococcus aureus*. Mol Microbiol 38(4):694–705. https://doi.org/10.1046/j.1365-2958.2000.02169.x

Yamaizumi M, Mekada E, Uchida T, Okada Y (1978) One molecule of diphtheria toxin fragment A introduced into a cell can kill the cell. Cell 15(1):245–250

Yang L, Li W, Jiang GZ, Zhang WH, Ding HZ, Liu YH, Zeng ZL, Jiang HX (2017) Characterization of a P1-like bacteriophage carrying CTX-M-27 in *Salmonella spp*. resistant to third generation cephalosporins isolated from pork in China. Sci Rep 7:40710. https://doi.org/10.1038/srep40710

Yokoyama K, Makino K, Kubota Y, Watanabe M, Kimura S, Yutsudo CH, Kurokawa K, Ishii K, Hattori M, Tatsuno I, Abe H, Yoh M, Iida T, Ohnishi M, Hayashi T, Yasunaga T, Honda T, Sasakawa C, Shinagawa H (2000) Complete nucleotide sequence of the prophage VT1-Sakai carrying the Shiga toxin 1 genes of the enterohemorrhagic *Escherichia coli* O157:H7 strain derived from the Sakai outbreak. Gene 258(1-2):127–139

Zhou Y, Sugiyama H, Johnson EA (1993) Transfer of neurotoxigenicity from *Clostridium butyricum* to a nontoxigenic *Clostridium botulinum* type E-like strain. Appl Environ Microbiol 59(11):3825–3831

Zhou Y, Sugiyama H, Nakano H, Johnson EA (1995) The genes for the *Clostridium botulinum* type G toxin complex are on a plasmid. Infect Immun 63(5):2087–2091

Chapter 7
Mycoviruses as Antivirulence Elements of Fungal Pathogens

Mirna Ćurković-Perica, Marin Ježić, and Daniel Rigling

Abstract Plant diseases caused by fungal pathogens are a major cause of yield loss in agriculture and forestry. Mycoviruses have been discovered in many important plant pathogenic fungi. The majority are RNA viruses, that cause persistent infections of their fungal hosts without having an extracellular phase. Virus infection occurs via hyphal anastomosis allowing the virus to move from a virus-infected to uninfected mycelium. While most virus infections remain asymptomatic, some are known to attenuate virulence and sporulation ability of their fungal hosts and thus can act as antivirulence elements against these pathogens. Biological control strategies against plant diseases have been developed using such mycoviruses, the best-known example being hypovirulence in the chestnut blight fungus, *Cryphonectria parasitica*. This chapter gives an overview about the basic concepts of mycovirus-mediated biological control of plant diseases and reviews examples of practical and potential applications of mycoviruses for disease control.

7.1 Introduction

Viruses are infectious agents that contain either DNA or RNA as their genetic material which is usually enclosed within a capsid—a protein coat which contains many copies of one or more different types of polypeptide subunits (Flint et al. 2009; Willey et al. 2010). Some virus species also have envelopes consisting of proteins, lipids and carbohydrates that enclose the nucleocapsid. Viruses are able to transfer their genetic material from one cell to another, imposing their genetic information upon the infected cell and utilising the cell's resources for their own reproduction. They are studied mostly because of their role in causing diseases of all cellular life

M. Ćurković-Perica · M. Ježić
Division of Microbiology, Department of Biology, Faculty of Science, University of Zagreb, Zagreb, Croatia

D. Rigling (✉)
Swiss Federal Research Institute WSL, Birmensdorf, Switzerland
e-mail: daniel.rigling@wsl.ch

C. J. Hurst (ed.), *The Biological Role of a Virus*, Advances in Environmental Microbiology 9, https://doi.org/10.1007/978-3-030-85395-2_7

forms, both prokaryotic and eukaryotic. Some viruses can be utilised for scientific, biotechnological and medicinal purposes to deliver their genetic payload to targeted cells (Roldão et al. 2017). Others are used as biological control agents of pests and pathogens by utilizing their debilitating effect on their hosts (Köhl et al. 2019) which has been exploited by humans.

The suitability and selection of mycoviruses for use as biological tools depends in part upon both the traits of a virus strain and the situation for which a virus will be needed. The appropriate factors to be considered when selecting viral attributes will be discussed later in this chapter. One example of those attributes would be that viral presence in the actively growing hyphae is important for horizontal transmission of the virus, and yet viral presence in the fungal reproductive structures is important for achieving vertical transmission of a virus. Another example is recognizing situations where achieving rapid effectiveness of virus transmission is more important than the extent to which a virus affects fungal virulence.

7.2 Mycoviruses

The term mycovirus refers to a large group of taxonomically unrelated viruses that infect fungi, also called fungal viruses. Since the first description of a virus in the cultivated mushroom *Agaricus bisporus* (J.E. Lange) Imbach more than 50 years ago (Hollings 1962), the knowledge of mycoviruses has expanded extensively (Buck 1986; Ghabrial et al. 2015; Hillman and Suzuki 2004; Son et al. 2015). The vast majority of mycoviruses do not have an extracellular phase, and some of them do not even form true virions due to the absence of coat proteins, i.e., they only consist of nucleic acid. Most mycoviruses have a single-stranded RNA (ssRNA) or double-stranded RNA (dsRNA) genome. Only a few mycoviruses with a DNA genome are known thus far. The RNA genomes always encode an RNA-dependent RNA polymerase (RdRP), which is responsible for replication and transcription of the virus. The genome of encapsidated mycoviruses contain additional genes for the coat proteins (CP genes) (Ghabrial et al. 2015; Pearson et al. 2009).

Mycoviruses are typically found in the cytoplasm of their fungal hosts. However, some mycoviruses (e.g., mitoviruses) are associated with the mitochondria of the fungus utilising the mitochondrial genetic coding. Due to the general lack of an extracellular phase, mycoviruses are usually limited to the inside of fungal host cells. They can be transmitted vertically into spores produced by a virus-infected fungal host during its sexual or asexual reproduction (Abbas 2016). Virus transmission into asexual spores (conidia) is more common than into sexual spores. Horizontal virus transmission can occur between two fungal strains when they come in contact and form hyphal anastomoses, allowing transmission of a mycovirus by cytoplasmic exchange. Likewise, virus-infected, germinating spores can transfer their virus load to a new host via hyphal anastomosis. Horizontal virus transmission is restricted by a vegetative incompatibility (also known as heterokaryon incompatibility) system in the fungal host (Cortesi et al. 2001; Debets et al. 2014; Saupe 2000). This is a self versus non-self recognition system, which controls the formation of hyphal anastomosis and cytoplasmic exchange between fungal strains (Paoletti 2016).

7.3 Plant Pathogenic Fungi as Hosts of Mycoviruses

Fungi have a close association with plants, i.e., most fungi live from nutrients produced by plants. In this dependence, fungi have developed different life strategies in the course of evolution (Heitman et al. 2017). Pathogenic fungi can infect living plants while saprophytic fungi can only colonise dead plant material. There are also fungi that live in a mutualistic interaction with the host plant (e.g., mycorrhizal fungi) where both partners benefit from each other (van der Heijden et al. 2015). Among the pathogenic fungi, necrotrophic and biotrophic pathogens are usually distinguished. Necrotrophic pathogens first kill the infected plant tissue and then invade it and exploit its nutrients. In this type of interaction, the pathogen often uses toxins and cell wall degrading enzymes to kill the host cells. Necrotrophic pathogens typically have the capability to survive saprophytically on dead hosts and are therefore also called nonobligate pathogens. Biotrophic pathogens grow and reproduce only on living plants. These obligate pathogens derive nutrients from living cells without killing them rapidly (Agrios 2005).

The life cycle of a plant pathogenic fungus typically begins with fungal spores that germinate on the surface of the plants. This is followed by penetration into host tissue and parasitic growth inside the host. The severity of the infection depends on which type of host tissue is attacked. For example, leaf infections can have little effects on the host while root or stem infections can kill the entire plant. The fungal life cycle closes with the formation of new fungal spores, which are produced on infected plant tissue and spread to new hosts by wind, rain, or insect vectors (Doehlemann et al. 2017). Fungal pathogens can have both a sexual and an asexual reproductive cycle. In heterothallic fungi, sexual reproduction usually requires mating of two strains with opposite mating types, which results in the production of genetically recombinant sexual spores. Self-fertilization occurs in homothallic fungi, i.e. sexual fruiting bodies are produced without mating type partner (Lee et al. 2010). The asexual spores are always produced by single strains and are all genetically identical. They often act as spermatia during sexual reproduction but can also directly cause new infections.

7.4 Mycoviruses as Biological Control Agents

Biological control is a method to manage pests and pathogens by utilising their natural enemies like predators, parasitoids and pathogens, instead of relying on chemical pesticides. There is an increasing amount of scientific evidence about the effective use of viruses in controlling pathogens, including plant pathogenic fungi

(Muñoz-Adalia et al. 2016; Nuss 2005; Pearson et al. 2009; Van De Sande et al. 2010; Xie and Jiang 2014). Plant pathogens are a major cause of yield loss in agriculture and forestry due to the diseases they cause on many different commercially important plant species. Although mycoviruses have been discovered in almost all groups of fungi, a relatively small number are known to attenuate the growth and virulence of their fungal hosts (García-Pedrajas et al. 2019; Muñoz-Adalia et al. 2016). Biological control by mycoviruses is a relatively new disease management option that offers alternatives to chemical control with fungicides, which can have undesirable side effects on the environment. To develop an efficient biological control tool utilising a mycovirus, a few basic criteria must be fulfilled:

1. A mycovirus candidate, which is considered as a biocontrol agent of a pathogenic fungus, should cause hypovirulence (i.e., reduce the virulence of the pathogen). This is achieved if a virus infection has a negative impact on the parasitic growth or reproduction of the pathogen. If either, or both, processes are suppressed, then the disease epidemic will be disrupted and the severity of the disease reduced.
2. The mycovirus should be able to infect the fungal pathogen that causes the disease. As most mycoviruses do not have an extracellular phase a virus infection can only occur after hyphal anastomosis and cytoplasmic exchange between a virus-infected and a virus-free fungal strain.
3. A mycovirus candidate should be able to overcome fungal cellular defence mechanisms against virus infections. The mechanisms of RNA silencing, which are used by many eukaryotes, including fungi, as an antiviral defence pathway can be suppressed by certain mycoviruses.
4. After an initial infection, a mycovirus must be able to replicate and spread through the thallus (vegetative body) of the fungus. A fungal thallus typically consists of a network of filamentous hyphae which grow actively at the edge of the expanding colony, while reproductive structures (e.g., fruiting bodies) develop in the older parts. A mycovirus needs to spread into actively growing hyphae in order to be able to interfere with the infection process during plant-pathogen interaction. Spread to actively growing hyphae is also important for horizontal virus transmission, while spread into reproductive structures is required for vertical virus transmission into sexual or asexual spores.
5. If the aim is a self-sustainable biocontrol, the mycovirus should spread and persist in a fungal pathogen population. This may require both vertical and horizontal virus transmission. Ideally, the mycovirus is transmitted at high frequency into spores, which disseminate the virus. Horizontal virus spread among individuals is affected by barriers imposed by the vegetative incompatibility system in fungi. Successful virus spread is typically correlated with a low diversity of vegetative compatibility (vc) types in a pathogen population.

Mycoviruses that induce a hypovirulent phenotype upon infecting their host have a potential to be used as biological control agents. The best-known and studied virus used for biocontrol is *Cryphonectria hypovirus 1* (CHV1) which causes hypovirulence in the chestnut blight fungus *Cryphonectria parasitica*. This biological control system is described in detail in the following sections, supplemented by other well-known examples. A summary of the examples is shown in Table 7.1.

7.4.1 *Hypovirulence in the Chestnut Blight Fungus,* Cryphonectria parasitica

The ascomycete fungus *Cryphonectria parasitica* (Murr.) Barr which causes chestnut blight, was accidentally introduced into the USA and Europe from eastern Asia (Anagnostakis 1987; Rigling and Prospero 2018). This pathogen is considered to be one of the most aggressive invasive species. Two chestnut species are highly susceptible to destructive attack by this fungus—American chestnut, *Castanea dentata* (Marshall) Borkh., which is native to eastern North America (Fralish and Franklin 2002; Russell 1987) and European chestnut, *Castanea sativa* Mill., which is found mainly in southern Europe, Turkey and the Caucasus mountain range (Conedera et al. 2016; Mattioni et al. 2017; Poljak et al. 2017). The native hosts in Asia, the Chinese chestnut, *C. mollissima* Blume, and the Japanese chestnut, *C. crenata* Sibold & Zucc. are tolerant to *C. parasitica*, presumably because of their long co-evolution with the pathogen (Anagnostakis 1992).

The pathogen enters the host through wounds in the bark and causes bark lesions (so-called 'cankers') on stems and branches. These cankers lead to the wilting of plant parts that are distal to the point of infection. In North America, the chestnut blight epidemic resulted in the ecological extinction of almost 4 billion American chestnut trees. In Europe, after an initial epidemic that caused severe damage, chestnut trees showed atypical, superficial, non-lethal bark cankers, that were first observed in Italy and France (Grente 1965; Heiniger and Rigling 1994). *Cryphonectria parasitica* strains isolated from these cankers exhibited a different phenotype than those recovered from deep, lethal cankers. As opposed to strains usually obtained from deep, lethal cankers, which exhibit an orange phenotype and abundant conidia formation in culture, strains isolated from superficial cankers showed a white culture morphology with low conidial production (Fig. 7.1). Subsequently, it was demonstrated that the atypical strains exhibited a reduced virulence towards the chestnut and were therefore called hypovirulent strains. Hypovirulence was shown to be associated with the presence of a double-stranded (ds) RNA element in the *C. parasitica* mycelium. Later, it was determined that this dsRNA present in hypovirulent *C. parasitica* strains represents the replicative form of *Cryphonectria hypovirus 1* (CHV1), the first described hypovirus (Choi and Nuss 1992; Suzuki et al. 2018).

Table 7.1 Examples of mycoviruses that induce hypovirulence in plant pathogenic fungi

Plant disease	Fungal pathogen	Mycovirus (family)	Characteristics of mycovirus-fungus interaction: Virus transmission (T), effect on host (E), biocontrol potential (B)
Chestnut blight on *Castanea* species	*Cryphonectria parasitica*	CHV1 *Cryphonectria hypovirus 1* (*Hypoviridae*)	T: via hyphal anastomosis and conidia E: Reduced mycelial growth, sporulation and pigmentation, female sterility B: Natural dissemination and therapeutic treatments of chestnut blight in Europe
		CHV2 *Cryphonectria hypovirus 2* (*Hypoviridae*)	T: via hyphal anastomosis and rarely via conidia E: Severely reduced mycelial growth, moderately reduced fungal sporulation and pigmentation B: Laboratory experiments
		CHV3 *Cryphonectria hypovirus 3* (*Hypoviridae*)	T: via hyphal anastomosis and rarely via conidia E: Slow growth in culture B: Therapeutic treatments of chestnut blight cankers in the USA
		CpMyRV1 *Cryphonectria mycoreovirus 1* (*Reoviridae*)	T: Infectious as virus particles in transfection assays E: Altered culture morphology B: Laboratory experiments
		CpMV1 *Cryphonectria parasitica mitovirus 1* (*Narnaviridae*)	T: via hyphal anastomosis, vertical transmission into conidia, maternal inheritance in sexual crosses E: Mildly reduced mycelial growth B: Laboratory experiments using excised chestnut wood and bark
Rice blast	*Magnaporthe oryzae*	MoCV1 *Magnaporthe oryzae chrysovirus 1* (*Chrysoviridae*)	T: via hyphal anastomosis; transfection of yeast possible and impairs cell growth E: Reduced mycelial growth, altered colony morphology and reduced pigmentation B: Laboratory experiments

(continued)

Table 7.1 (continued)

Plant disease	Fungal pathogen	Mycovirus (family)	Characteristics of mycovirus-fungus interaction: Virus transmission (T), effect on host (E), biocontrol potential (B)
Leaf spot diseases on various crops	*Alternaria alternata*	AaHV1 *Alternaria alternata hypovirus 1* (unclassified hypovirus-related virus)	T: via hyphal anastomosis and via conidia; transfection of *Botryosphaeria dothidea* possible and resulting in a hypovirulent phenotype E: Reduced mycelial growth B: Laboratory experiments
Rice sheath blight	*Rhizoctonia solani*	RsEV1 *Rhizoctonia solani endornavirus 1* (unclassified *Endornaviridae*-related virus)	T: via hyphal anastomosis E: Change in mycelial pigmentation, growth rate and morphology, smaller size of sclerotia, metabolic perturbance B: Laboratory experiments
Fusarium head blight of wheat, barley and other small-grain cereals	*Fusarium graminearum*	FgV1 *Fusarium graminearum virus 1* (*Chrysoviridae*)	T: via hyphal anastomosis and conidia E: Reduced mycelial growth, increased pigmentation, inhibition of mycotoxin production B: Laboratory experiments
		FgV-ch9 *Fusarium graminearum mycovirus China 9* (unclassified hypovirus-related virus)	T: via conidia E: Reduced mycelial growth and sporulation, abnormal colony morphology, disorganized cytoplasm B: Laboratory and greenhouse experiments
		FgHV2 *Fusarium graminearum hypovirus 2* (unclassified hypovirus-related virus)	T: Not known E: Reduced mycelial growth rate, conidiation and mycotoxin production B: Laboratory experiments with wheat spikes
Root disease in many coniferous and some broadleaf species	*Heterobasidion annosum* and *H. parviporum*	HetPV13-an1 *Heterobasidion partitivirus 13* (*Partitiviridae*)	T: via hyphal anastomosis E: Reduced mycelial growth rate, metabolic perturbance B: Laboratory experiments and field experiment on Norway spruce

(continued)

Table 7.1 (continued)

Plant disease	Fungal pathogen	Mycovirus (family)	Characteristics of mycovirus-fungus interaction: Virus transmission (T), effect on host (E), biocontrol potential (B)
Dutch elm disease	*Ophiostoma novo-ulmi*	OnuMV Ophiostoma novo-ulmi mitoviruses (*Narnaviridae*)	T: via hyphal anastomosis and conidia E: Debilitated growth, reduced conidial viability, mitochondrial malfunction B: Inoculations studies with elm trees
Grey mold or stem rot on many plant species	*Botrytis cinerea*	BcMV1 *Botrytis cinerea mitovirus 1* (*Narnaviridae*)	T: via hyphal anastomosis and conidia E: Slow mycelial growth, abnormal colony sectors, decreased formation of infection cushions B: Laboratory experiments with detached leaves
		BcRV1 *Botrytis cinerea RNA virus 1* (unclassified)	T: via hyphal anastomosis and vertical transmission to macroconidia E: Reduced mycelial growth, sectored colony margin B: Laboratory experiments with detached leaves
		BcHV1 *Botrytis cinerea hypovirus 1* (unclassified hypovirus-related virus)	T: via hyphal anastomosis E: Decreased formation of infection cushions B: Laboratory experiments with detached leaves
		BcPV2 *Botrytis cinerea partitivirus 2* (*Partitiviridae*)	T: via hyphal anastomosis E: Altered culture pigmentation, reduced production of conidia and sclerotia B: Laboratory experiments with detached leaves
		Bc378V1 *Botrytis cinerea CCg378 virus 1* (*Partitiviridae*)	T: Not known E: Reduction of sporulation and laccase activity B: Laboratory experiments with detached leaves

(continued)

Table 7.1 (continued)

Plant disease	Fungal pathogen	Mycovirus (family)	Characteristics of mycovirus-fungus interaction: Virus transmission (T), effect on host (E), biocontrol potential (B)
Root rot in many woody plant species	*Rosellinia necatrix*	RnMBV1 *Rosellinia necatrix megabirnavirus 1* (*Megabirnaviridae*)	T: Infectious as virus particles in transfection assays E: Reduced mycelial growth B: Greenhouse experiments with apple seedlings
		MyRV3 *Rosellinia necatrix Mycoreovirus 3* (*Reoviridae*)	T: Infectious as virus particles in transfection assays E: Reduced mycelial growth B: Greenhouse experiments with apple seedlings
Victoria blight of oats	*Helminthosporium victoriae*	HvV190S *Helminthosporium victoriae virus 190S* (*Totiviridae*)	T: Not known, transfection of *C. parasitica* possible resulting in a hypovirulent phenotype E: Reduced growth, production of abundant white aerial mycelium, excessive sectoring of colonies B: Laboratory experiments with apples and branches
White mold or stem rot on many plant species	*Sclerotinia sclerotiorum*	SsHADV-1 *Sclerotinia sclerotiorum hypovirulence-associated DNA virus 1* (*Genomoviridae*)	T: via hyphal anastomosis, extracellular transmission of virus particles E: Abnormal culture morphology, reduced sclerotia size B: Preventive application, growth chamber experiments and field trials
		SsPV1 Sclerotinia sclerotiorum partitivirus 1 (Partitiviridae)	T: via hyphal anastomosis, transmission into sclerotia, interspecific transmission to other *Sclerotinia* species. E: Reduced mycelial growth, reduced sclerotia production B: Laboratory experiments
		SsMV1 *Sclerotinia sclerotiorum mitovirus 1* (*Narnaviridae*)	T: via hyphal anastomosis E: Reduced mycelial growth, mitochondrial malformation B: Laboratory experiments with detached leaves

(continued)

Table 7.1 (continued)

Plant disease	Fungal pathogen	Mycovirus (family)	Characteristics of mycovirus-fungus interaction: Virus transmission (T), effect on host (E), biocontrol potential (B)
		SsHV2 *Sclerotinia sclerotiorum hypovirus 2* (unclassified hypovirus-related virus)	T: not known E: Reduced mycelial growth B: Laboratory experiments with detached leaves

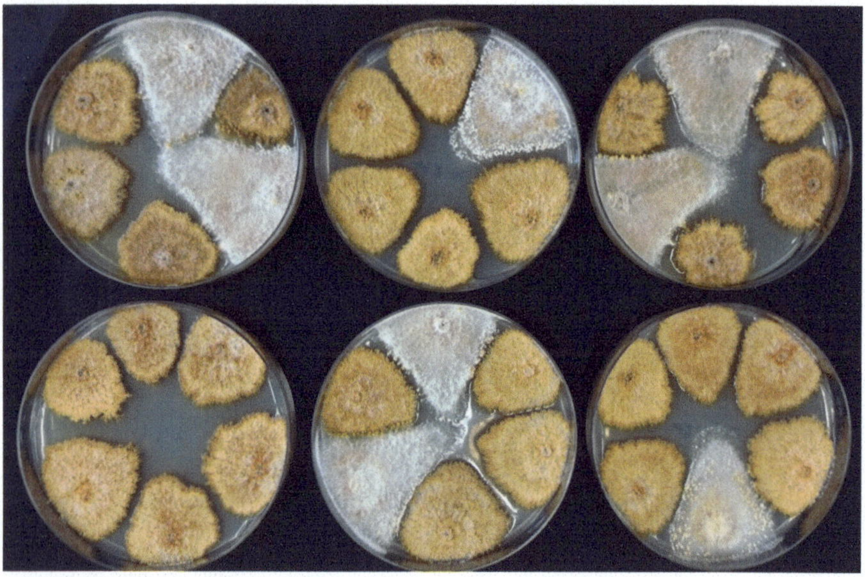

Fig. 7.1 Cultures of *Cryphonectria parasitica* that were isolated from chestnut blight cankers in southern Switzerland. Normal virus-free cultures produce an orange pigmentation when grown on potato dextrose agar, whereas cultures infected by Cryphonectria hypovirus 1 (CHV1) remain white. Six cultures were inoculated at the edge of each plate. Note that CHV1 infected strains often grow better in vitro than virus-free strains

7.4.1.1 Mycovirus Species Belonging to the Genus *Hypovirus*

Beside the already mentioned CHV1, other unencapsidated mycoviruses from *C. parasitica*, have been characterised (Hillman et al. 1992; Linder-Basso et al. 2005; Tartaglia et al. 1986), and placed within the genus *Hypovirus* (Suzuki et al. 2018). All hypoviruses have a positive sense single stranded (+ss) RNA genome and are located in the cytoplasm of their fungal host. They do not encode a capsid protein (i.e. are unencapsidated, cannot produce virions and hence possess no extracellular phase) and are associated with fungal membrane vesicles (Hansen et al. 1985).

The genus *Hypovirus* comprises of four species: CHV1, CHV2, CHV3, and CHV4. While the species are taxonomically related, they have vastly different effects on the fungal host (Hillman and Suzuki 2004; Suzuki et al. 2021). Of these, CHV2 and some strains of CHV3 also induce a hypovirulent phenotype in *C. parasitica*, whereas CHV4 does not affect the virulence of its host. Thus far, CHV1 is the only member of the genus found in Europe and it has also been detected in the native *C. parasitica* populations in Asia—Japan, China and Korea (Park et al. 2004; Peever et al. 1997), from where it was most likely introduced into Europe. The latter three hypovirus species have been found in North America, with CHV4 occurring throughout the natural chestnut range of *C. dentata*, while CHV2 and CHV3 have a more restricted distribution (Peever et al. 1997).

The best characterised and most studied hypovirus is CHV1 (reviewed by Dawe and Nuss 2001; Nuss 2005). The genome sequence of CHV1 consist of 12712 nucleotides (nts) and contains two open reading frames designated ORFA and ORFB (Shapira et al. 1991). ORFA (1869 nts) encodes the polyprotein p69, which is autoproteolytically cleaved into two polypeptides (p29 and p40) during translation. ORFB (9498 nts) also encodes a polyprotein, which autoproteolytically releases the polypeptide p48 from its N-terminal end. The remaining polypeptide of ORFB contains RNA-dependent RNA polymerase (pol) and RNA helicase (hel) domains (Suzuki et al. 2018).

7.4.1.2 *Cryphonectria hypovirus 1* Subtypes

Several subtype of CHV1 have been found in Europe. Their current geographic distribution reflects the invasion history of *C. parasitica*, which is characterised by several introduction events. The Italian subtype (I) is widespread in southern and south-eastern Europe including Switzerland, Italy, south-eastern France, Bosnia and Herzegovina, Croatia, Slovenia, North Macedonia, Turkey and Greece (Allemann et al. 1999; Gobbin et al. 2003; Krstin et al. 2008, 2011; Robin et al. 2010; Sotirovski et al. 2004). Other CHV1 subtypes have a more restricted distribution in Europe. Subtypes F1 and F2 have been found in France and Spain (Feau et al. 2014), while subtype D/E was reported in Germany and Spain (Peters et al. 2014; Trapiello et al. 2017). In Eurasian Georgia a unique subtype (subtype G) partially related to F2 was detected (Rigling et al. 2018). Recombination events apparently have played an important role in the evolution of the different CHV1 subtypes (Feau et al. 2014; Mlinarec et al. 2018a). The different subtypes usually vary in their virulence against *C. parasitica*. Specifically, subtype I is considered to have a mild effect on fungal virulence and sporulation and is commonly associated with a high natural prevalence of hypovirulence in *C. parasitica* populations (Rigling and Prospero 2018). In contrast, subtypes F1 and F2 usually have a severe effect on their fungal host and almost completely inhibit its parasitic growth and sporulation (Robin et al. 2010).

These differences between subtypes have important consequences for biological control of the chestnut blight. CHV1 isolates of French subtypes F1 and F2 may be particularly suited for therapeutic treatment of individual cankers on high-value chestnut trees (e.g. plantations) as they enable rapid healing of the cankers (Prospero and Rigling 2013). By contrast, CHV1 isolates of the Italian subtype, which show a higher potential for natural dissemination, should be used when the goal of the treatments is the establishment and maintenance of hypovirulence in chestnut stands (Robin et al. 2010).

7.4.1.3 Phenotypic Effects of *Cryphonectria hypovirus 1* on Its Fungal Host

CHV1 has pronounced phenotypic effects on *C. parasitica*. Besides reducing pigmentation, CHV1 inhibits sexual reproduction, strongly reduces asexual sporulation, and attenuates parasitic growth of CHV1-infected strains. The severity of the observed effects depends on several factors, including the CHV1 subtype and the particular virus isolate, the fungal strain, and the combination of these factors (Bryner and Rigling 2011; Krstin et al. 2017; Nuskern et al. 2017a, b; Rigling et al. 1989). Hypovirulent *C. parasitica* strains are not able to colonize and kill the cambium of infected chestnut trees (Hebard et al. 1984). As a result, the tree can defend itself against the pathogen either by callusing the cankers or restricting the infection to the outer part of the bark (Fig. 7.2) (Bryner et al. 2014; Ježić et al. 2019). This significantly reduces lethality of the infection, allowing the parts of the plant distal to the infection site to survive.

Brusini et al. (2017) used isogenic *C. parasitica* strains infected with six hypovirus isolates to investigate the effects of CHV1 on life-history traits (mycelial growth and asexual sporulation) of its fungal host *C. parasitica*. A significant negative correlation was found between somatic growth of the fungus and asexual reproduction capability for the virulent *C. parasitica* strains, indicating a trade-off between these two phenotypic traits. This was not observed in hypovirulent strains, illustrating the profound changes in host resource allocation induced by CHV1 infection. In hypovirulent strains, a significant and positive correlation was found between somatic growth and vertical virus transmission into conidia (Brusini et al. 2017). Since CHV1 is transmitted only into asexual conidia (Peever et al. 2000), and not sexual ascospores (Anagnostakis 1988) this alteration of *C. parasitica* physiology is not surprising. Furthermore, Prospero et al. (2006) demonstrated that CHV1-infected *C. parasitica* strains can successfully propagate asexually on dead chestnut wood and produce virus-infected conidia.

7.4.1.4 Effects of *Cryphonectria hypovirus 1* at the Molecular Level

Gene expression studies reveal that CHV1 infection deregulates (either downregulates or upregulates) expression patterns of a number of genes in the

Fig. 7.2 Virulent (left image) and hypovirulent (right image) chestnut blight cankers on European chestnut (*Castanea sativa*) trees. Virulent cankers induced by hypovirus-free strains of *Cryphonectria parasitica* exhibit a reddish discoloration and are actively expanding. The chestnut tree reacts by producing epicormic shoots below the cankers. Hypovirulent cankers became infected by the hypovirus and the chestnut trees are able to fight off the cankers, which eventually stop expanding

fungus *C. parasitica* (e.g. Rigling and Van Alfen 1991). Allen et al. (2003) reported that 13.4% of approximately 2200 studied genes were responsive to CHV1 infection, showing either increased or decreased mRNA accumulation. Transcription activity of the genes that were responsive to CHV1 infection revealed their involvement in a broad spectrum of biological processes.

One cellular process affected by CHV1 is the fungal signal transduction pathway (Choi et al. 1995; Larson et al. 1992; Park et al. 2004). If this pathway is disturbed by the virus, the fungus is no longer able to respond adequately to external signals. Such reaction can be observed when the fungus is exposed to light while cultivated in vitro on artificial medium (e.g., potato dextrose agar). Under this condition, virus-free strains react to light stimulus and produce the typical orange pigmentation and abundant asexual sporulation. In contrast, CHV1-infected strains largely remain white and only scarcely sporulate (Fig. 7.1). Signal transduction processes are considered important for the pathogen to infect a chestnut tree, and CHV1-induced hypovirulence is likely the result of a disruption of this pathway (Nuss 2005).

Comparative transcriptomic analyses (Allen et al. 2003; Allen and Nuss 2004) showed that CHV1 infection up-regulated the expression of *C. parasitica* genes involved in cellular methylation reactions, namely genes for putative S-adenosylmethionine synthase (SAMS) and putative S-adenosylhomocysteine hydrolase (SAHH). SAMS catalyses the generation of the primary methyl group donor S-adenosyl methionine (SAM), while SAHH hydrolyses S-adenosyl homocysteine (SAH) to adenosine and homocysteine (Liao et al. 2012). Infection with

CHV1 affects the general methylation pattern of *C. parasitica* genome; usually it increases the number and diversity of methylation-sensitive amplification polymorphism (MSAP) markers compared to isogenic uninfected controls, especially methylated markers, while the effect on hemi-methylated and unmethylated markers was inconsistent (Nuskern et al. 2018). The increase in genomic methylation levels correlates well with the CHV1-induced reduction of fungal growth in vitro, indicating that *C. parasitica* genome methylation is likely directly affected by the virus itself rather than being a defensive mechanism of the fungus. The CHV1 effect on the methylation levels in *C. parasitica* depends mostly on individual CHV1 isolates and on the combination of host and virus genotypes, a trend observed on phenotypic level as well (Nuskern et al. 2018).

Different virus isolates, which have severe or mild effects on *C. parasitica* were genetically characterised by Dawe and Nuss (2001). By using chimaera between mild and severe hypoviruses and deletion mutants, Donald Nuss and co-workers were able to map specific regions of the CHV1 genome to different host phenotypes affected by the hypovirus (Nuss 2005). These studies indicate that the fungal pigmentation and asexual sporulation is suppressed by a specific domain in the viral ORFA, while reduced colony and canker growth is determined by part of ORFB (Chen et al. 2000; Suzuki and Nuss 2002). Furthermore, (Suzuki and Nuss 2002) demonstrated that the first 24 amino acids of ORFA are crucial for the virus replication. However, the attempts to establish an association between specific region(s) of the CHV1 genome and phenotypic effects are often hampered because even isolates which belong to the same CHV1 subtype (and have a very high amino acid sequence identity) differ in their impact on severity of symptoms such as virulence attenuation (hypovirulence) and viral RNA accumulation. Studies by Zhang et al. (2012) imply the importance of the central region of the CHV1 genome in attenuation of *C. parasitica* virulence. Xiong et al. (2019) noted highly conserved cysteine and histidine residues in papain-like protease p29 coding region of several sequenced CHV1 genomes—CN280 from China, and CHV1-EP713, CHV1-Euro7 and CHV1-EP721 from Europe. However, the authors also found some instances of profound sequence dissimilarity within the same genomic region, and suggested a possible link between this dissimilarity and the phenotypic differences induced by the aforementioned CHV1 strains.

In most cases, findings from proteomic studies (Kim et al. 2012) have proven congruent with the results obtained by transcriptomic studies. Proteins responsive to CHV1 infection include metabolic enzymes, stress-related and heat-shock proteins, signalling and cellular process-related proteins, some structural proteins and glutathione S-transferase. Transcriptomic and proteomic analysis, however, did not always give the same quantitative results, and differences were noted between the accumulation of mRNAs and the quantity of the corresponding proteins (Kim et al. 2012).

Cryphonectria parasitica RNA silencing interferes with viral replication and acts as an important defence response of the fungus against viral infection (Segers et al. 2007). On the other hand, CHV1 may counteract this defence process by suppressing RNA silencing through the production a specific viral protease (Segers et al. 2006).

The RNA silencing pathway apparently also contributes to viral RNA recombination in *C. parasitica* (Dawe and Nuss 2013).

7.4.1.5 Effect of *Cryphonectria hypovirus 1* on the *Cryphonectria parasitica* Secretome and Enzyme Activities

Through changes in the secretome, CHV1 probably affects several biochemical pathways and physiological changes in infected *C. parasitica* (Wang et al. 2016). These authors identified and quantified 403 unique proteins in the secretome of the virus-free strain EP155. Of these proteins, 329 were predicted to be involved in known secretory pathways and are primarily composed of metabolic enzymes, biological regulators, responders to stimuli and components involved in plant-pathogen interactions. When infected with CHV1, 99 of those proteins, mainly related to plant cell wall degradation, response to host defence, fungal virulence and intracellular structure, were found to be differently expressed. One of the proteins whose quantity changed was ubiquitin, which was upregulated in the virus-infected fungal strain. Ubiquitin in *C. parasitica* is encoded by the polyubiquitin gene, *cpubi4* (Chen et al. 2018) and is responsible for ubiquitination of a wide range of proteins, some of which are critically important in biological and metabolic processes.

Furthermore, knowledge about the impact of CHV1 on *C. parasitica* is complemented by measurements of the activity of various enzymes. CHV1 induced a reduction of the enzyme activity of some plant cell wall degrading and pathogenicity related enzymes, including laccase, cutinase and polygalacturonase (Gao and Shain 1995; Kim 1995; Rigling et al. 1989; Varley et al. 1992). However, recent measurements of intracellular laccase activity of different *C. parasitica* strains infected by various CHV1 isolates, were unable to support the ubiquity of this pattern (Nuskern et al. 2017a). Extracellular laccase activity was only reduced in some of the aforementioned combinations, while in other situations laccase activity either remained unchanged, or even increased. Nuskern et al. (2017a) speculate that both host genome and characteristics specific to a viral isolate significantly contribute to the outcome of this interaction and both should be considered for meaningful inference about this pathosystem. Alteration in the activity of antioxidative enzymes such as catalase, superoxide dismutase and glutathione S-transferase as a consequence of CHV1 infection, has also been reported (Nuskern et al. 2017b).

7.4.1.6 Trilateral Interaction Between Chestnut Tree, *Cryphonectria parasitica* and *Cryphonectria hypovirus 1*

It is a well-known fact that *C. parasitica* hosts belonging to different tree species show different levels of susceptibility or tolerance toward *C. parasitica* infection (Anagnostakis 1992). The most susceptible species is *C. dentata*, followed by *C. sativa*. On the other hand, the native Asian chestnut hosts, *C. crenata* and

C. mollissima, are mostly tolerant to *C. parasitica* infection and show only moderate disease symptoms, presumably because of their long co-evolution with the pathogen (Rigling and Prospero 2018). The lower susceptibility of European chestnut compared to American chestnut is thought to be an important factor for the success of hypovirulence in Europe, as a slower canker expansion allows more time for hypovirus infection of the canker-causing fungi (Griffin 1986). Beyond that, intraspecific variation in susceptibility of *C. sativa* to *C. parasitica* has been observed as well (Anagnostakis 1992; Bolvanský et al. 2018; Graves 1950; Huang 1996). Trilateral interaction between the virus, the fungus and the tree host, plays an important role in the severity of the chestnut blight disease. Ježić et al. (2014) and Krstin et al. (2017) have explored these differences in susceptibility primarily between different sweet chestnut genotypes. Ježić et al. (2014) showed that despite the presence of naturally occurring CHV1 evenly distributed among the resident *C. parasitica* population in mixed chestnut forests, recovery of the marron-producing chestnut cultivar is slower than that of wild, naturally growing chestnut trees. The results imply that this cultured chestnut genotype is especially vulnerable to *C. parasitica* infection, and its ability to recover is limited, even in the presence of naturally occurring hypovirulence. Krstin et al. (2017) inoculated different virulent *C. parasitica* strains and isogenic strains infected with various CHV1 isolates belonging to subtypes I and F1 on different *C. sativa* genotypes. The lesion development depended on the combination of hypovirus isolate and *C. parasitica* strain but was affected by the genotype of the inoculated chestnut as well, implying that naturally growing chestnut trees vary in their susceptibility to hypovirulent and virulent *C. parasitica* strains. These findings imply that chestnut susceptibility and recovery depend on a particular genotypic combination of chestnut and *C. parasitica* as well as on a particular CHV1 strain. Furthermore, Krstin et al. (2017), found that one particular CHV1 subtype I isolate had a very similar and severe effect on *C. parasitica* as did subtype F1 isolate CHV1-EP713, further indicating the importance of a particular CHV1 genotype on its hypovirulent effect, rather than the subtype alone.

Beyond genetic effects attributed to the chestnut trees, plus the specificity of interactions between *C. parasitica* strains and particular CHV1 isolates, environmental factors must be considered as well. For example, Bryner and Rigling (2011) pointed out the importance of the temperature in this interplay, showing that it can affect the interaction between CHV1 and *C. parasitica*. Their results suggest that different host and virus genotypes would be selected under different climatic conditions, affecting the coevolutionary dynamics of the host-parasite interaction and the course of chestnut blight epidemic as it spreads over climatically diverse regions.

7.4.1.7 *Cryphonectria hypovirus 1* Transmission and Spread

A successful biological control of chestnut blight depends on efficient transmission of the hypovirus from infected to non-infected *C. parasitica* strains via hyphal anastomosis (horizontal transmission) (Fig. 7.3), which is limited by a vegetative incompatibility (*vic*) system (Cortesi and Milgroom 1998; Cortesi et al. 2001). This allorecognition system is able to distinguish between self and nonself within a species and controls the formation of hyphal anastomosis and cytoplasmic exchange between fungal strains. Vegetatively incompatible strains contain nuclei with different alleles at one or more *vic* loci that are co-expressed during hyphal contact and induce localized cell death (Paoletti 2016).

In *C. parasitica*, six diallelic *vic* loci have been identified in Europe to date, defining $2^6 = 64$ *vic* genotypes or EU vc types (Cortesi and Milgroom 1998). Additional vc types have been detected in the first decade of the twenty-first century. An additional allele on one of the already known six loci or an entirely new locus has been suggested as a possible explanation for this (Robin et al. 2009; Zamora et al. 2008). Two *C. parasitica* strains are compatible if they have the same alleles at all *vic* loci. The vc type of a fungal strain can be determined either by co-culturing it with defined EU vc type tester strains or by genotyping the *vic* loci directly (Cornejo et al. 2019; Mlinarec et al. 2018b; Short et al. 2015). When co-culturing, compatible strains (i.e., belonging to the same vc type) merge into a single culture, whereas a barrage line is formed between incompatible strains (i.e., belonging to different vc types) as the result of the induced cell death. CHV1 is readily transmitted between fungal strains belonging to the same vc type, whereas transmission rates between different vc types are in most cases lower (Cortesi et al. 2001; Liu and Milgroom 1996). The diversity of vc types in local *C. parasitica* populations is higher in North America than in Europe, limiting natural dissemination of the CHV1, which, along with the lower resistance of American chestnut to *C. parasitica*, is considered a major reason why biological control of chestnut blight was not successful in North America (Milgroom and Cortesi 2004).

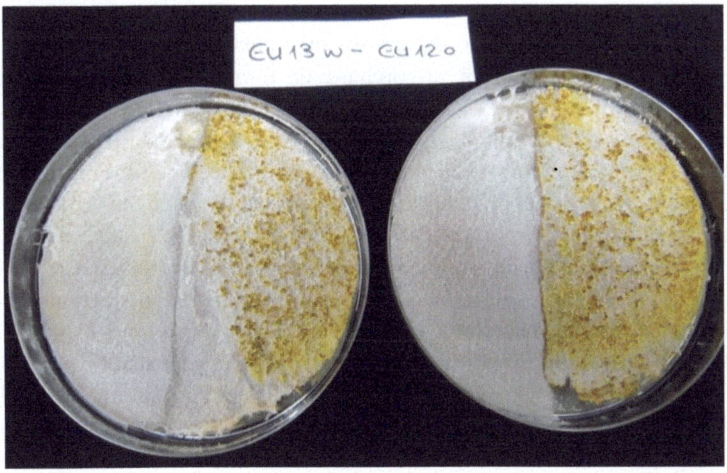

Fig. 7.3 Laboratory transmission of CHV1 between strains of *Cryphonectria parasitica*. Pairs of hypovirus-infected (white, vc type EU13) and hypovirus-free (orange, vc type EU12) strains were co-cultured on potato dextrose agar. Vegetative incompatibility between EU12 and EU13 strains prevented hypovirus transmission in the right plate, but not in the left plate

CHV1 is transmitted with variable frequency into asexual conidia (vertical transmission), which are supposed to play an important role for virus dissemination (Peever et al. 2000). Conidia are mainly splash dispersed over short distances, but can also spread over longer distances by wind-driven rain, insects or birds (Griffin 1986). In contrast to conidia, hypovirus-infected sexual ascospores have never been observed (Anagnostakis 1988; Prospero et al. 2006). Sexual reproduction in *C. parasitica*, therefore, obstructs hypovirus dissemination in two ways: (1) by contributing to the spread of only the virulent, hypovirus-free strains; and (2) by maintaining or even increasing the diversity of vc types through recombination of *vic* genes.

7.4.1.8 Human-Mediated Biological Control of Chestnut Blight

Biological control of chestnut blight using CHV1 started decades ago and today represents the best-known and most successful example of biocontrol of a tree disease. The method relies on inoculation of virulent chestnut blight cankers with hypovirus-infected *C. parasitica* strains (Heiniger and Rigling 1994). Inoculated cankers heal following the treatment, and even more, the applied hypovirus spreads from treated cankers to untreated cankers on surrounding trees (Hoegger et al. 2003; Prospero and Rigling 2016). Biological control using different CHV1 subtypes and strains was applied in France, Greece, Switzerland, Croatia and USA (Diamandis 2018; Halambek and Novak Agbaba 1989; Heiniger and Rigling 2009; Milgroom and Cortesi 2004; Robin et al. 2000). The biocontrol treatment protocol consisted of making holes in the bark of infected chestnut with a cork borer around the margins of the cankers and filling them with a paste consisting of a hypovirus-infected *C. parasitica* strain (Fig. 7.4).

Before any field application of hypovirulent strains, it is preferable to determine which vc types are present in an orchard or forest stand, especially if local *C. parasitica* populations are known or suspected to have a high vc type diversity. Based on this information, suitable hypovirulent strains, adapted to the local vc types present in a population (or even canker) can be selected. In France, mixtures of hypovirulent strains with different vc types are used to match the vc type diversity of the target *C. parasitica* populations (Robin et al. 2000). Periodic monitoring of *C. parasitica* populations is recommend, especially in populations where several vc types and both mating types have been observed, since the composition and diversity of vc types can change over time (Ježić et al. 2018). Biocontrol using CHV1 is generally quite successful in Europe. Therapeutic treatment of individual cankers usually stops canker expansion (Fig. 7.4) and the tree can fend off the infection (Diamandis et al. 2015; Heiniger and Rigling 2009). The success of treatments with hypovirulent paste has also been verified by molecular identification of the applied CHV1 isolates in treated and untreated cankers (Hoegger et al. 2003; Prospero and Rigling 2016).

However, biocontrol of chestnut blight in the USA generally remains less successful than in Europe. This has been attributed to the higher susceptibility of

Fig. 7.4 Biological control treatment of a chestnut blight canker. Holes at the margin of an actively growing canker are filled with the mycelium of a hypovirus-infected *Cryphonectria parasitica* strain (left image). The hypovirus is transmitted into the canker-causing strain via hyphal anastomosis, converting the infecting *C. parasitica* strain to a hypovirulent strain, thereby healing the canker (right image)

American chestnut to *C. parasitica*, compared to European chestnut and higher vc type diversity of the pathogen in the USA than in Europe (Ježić et al. 2019; Milgroom and Cortesi 2004). Recent approaches in biocontrol of chestnut blight in the USA employed genetically modified *C. parasitica* strains. The first approach encompasses application of transgenic strains carrying an infectious cDNA copy of CHV1 that is integrated into the fungal genome (Dawe and Nuss 2001). These strains provide enhanced dissemination potential for the hypovirus by transmitting the virus into sexual spores (via the nuclear copy of the viral cDNA) and thereby into all vc types produced in sexual crosses (Chen et al. 1993). The other approach utilizes application of the so called "hypovirus super donor strains", in which deletions have been introduced into *vic* loci, thereby disrupting their function (Zhang and Nuss 2016). Hence, such genetically modified fungal strains are able to transmit any cytoplasmic genetic elements, including the hypoviruses, to recipient strains that are otherwise vegetatively incompatible. In a recent field trial in the US, the use of the super donor strains resulted in increased CHV1 transmission rate into the treated chestnut blight cankers (Stauder et al. 2019). Super donor strains are expected to be particularly suitable for biocontrol of *C. parasitica* populations with high vc type diversity.

7.4.1.9 Other Viruses Belonging to the Genus *Hypovirus*

Cryphonectria hypovirus 2 (CHV2) was discovered in 1988 in several chestnut cankers in New Jersey. Some of the *C. parasitica* cultures isolated from those cankers expressed brown, rather than orange colony morphology and exhibited a hypovirulent phenotype (Chung 1994; Hillman and Suzuki 2004; Hillman et al.

1992). CHV2 was found in eastern Asia as well (Peever et al. 1998). Beyond changes in colony morphology, infection with CHV2 also moderately reduces fungal sporulation (Hillman and Suzuki 2004). The biocontrol potential of CHV2, however, is weaker than that of CHV1 and CHV3. It is rarely transmitted into conidia (Hillman and Suzuki 2004; Smart et al. 1999) and its extremely debilitating effect on *C. parasitica* significantly reduces fitness and survivability of the infected mycelium in natural environment, hence it has never been used in biocontrol trials.

Cryphonectria hypovirus 3 (CHV3) was detected in *C. parasitica* isolates collected from trees recovering from chestnut blight in Michigan, outside the natural distribution range of American chestnut (Fulbright et al. 1983; Milgroom and Cortesi 2004). In contrast to CHV1 and CHV2, this hypovirus has only one open reading frame and significantly smaller genome (Hillman and Suzuki 2004). There are several stands of American chestnut in Michigan and Ontario where CHV3 apparently successfully controls chestnut blight without human intervention. CHV3 was used initially in biocontrol trials (1992–1995) in the West Salem, Wisconsin (Double et al. 2018). It performed rather poorly, compared to CHV1-Euro7 strain, meaning that although directly treated cankers mostly healed, hypovirulence largely failed to spread through the *C. parasitica* population (Double et al. 2018). Biocontrol attempts using CHV3 were discontinued and since then the strain CHV1-Euro7 has been used.

Cryphonectria hypovirus 4 (CHV4) is widespread in the Appalachian Mountains (Peever et al. 1997), but infected *C. parasitica* colonies are difficult to identify (Enebak et al. 1994) because infection is asymptomatic and does not cause any discernible hypovirulent effect on its host (Enebak et al. 1994; MacDonald and Double 2005). Therefore, this virus has no potential as biocontrol agent against chestnut blight. However, CHV4 was shown to facilitate stable infection of *C. parasitica* with dsRNA virus *Mycoreovirus 2* (MyRV2), likely through suppression of antiviral RNA silencing (Aulia et al. 2019).

7.4.1.10 Other Viruses Infecting *Cryphonectria parasitica*

Besides viruses belonging to the family *Hypoviridae*, other mycoviruses that induce hypovirulent phenotype have been identified in *C. parasitica*: two mycoreoviruses: *Cryphonectria parasitica mycoreovirus 1* (CpMyRV1) (Suzuki et al. 2004), and *Cryphonectria parasitica mycoreovirus 2* (CpMyRV2) and one mitovirus (Hillman and Suzuki 2004). Both mycoreoviruses have segmented genomes comprising of 11 dsRNA segments and induce a strong hypovirulent phenotype in *C. parasitica*. It is worth noting that these viruses, despite having dsRNA genomes, interact with CHV1 p29, which may facilitate mycoreovirus genome rearrangements by supressing *C. parasitica* RNA silencing pathways (Eusebio-Cope and Suzuki 2015). Unlike hypoviruses, which lack genes encoding capsid proteins and an extracellular phase, CpMyRV2 produces infectious particles (virions) and can be easily transmitted between different *C. parasitica* vc types. *Cryphonectria parasitica mitovirus 1* strain CpMV1/NB631, which has been mostly studied thus far, has only a mild effect on *C. parasitica* (Hillman and Suzuki 2004). This virus is transmitted into asexual spores at high frequency and, as expected for a mitochondrial mycovirus, is maternally inherited as well. A recent study on chestnut trees confirmed the biocontrol potential of CpMV1 and CpMyRV1, although the performance was less robust than that of CHV1 or CHV2 (Suzuki et al. 2021).

7.4.2 Hypoviruses in Other Fungal Genera

As mentioned previously, all hypoviruses are characterized by +ssRNA genomes, similar genomic organisation and lack of a capsid. In addition to those infecting the genus *Cryphonectria*, several other hypoviruses have been identified. Except for one virus found in the fungal genus *Sclerotinia*, which belongs to the class Leotiomycetes, all other viruses are associated with fungal hosts belonging to the class Sordariomycetes, and many of those fungi are phytopathogenic. All hypoviruses are phylogenetically related and distinctly different from all other known mycoviruses. The family *Hypoviridae* currently includes only a single recognized genus, *Hypovirus*, with four accepted species: CHV1, CHV2, CHV3 and CHV4 (Suzuki et al. 2018). *Fusarium graminearum hypovirus 1* and *2* (FgHV1 and FgHV2) as well as *Sclerotinia sclerotiorum hypovirus 2* (SsHV2) grouped together in the same clade with CHV1 and CHV2 and were placed into a proposed genus *Alphahypovirus* by Li et al. (2015). *Sclerotinia sclerotiorum hypovirus 1* (SsHV1), *Valsa ceratosperma hypovirus 1* (VcHV1) and *Phomopsis longicolla hypovirus 1* (PlHV1) show a significant sequence similarity with CHV3 and CHV4. Therefore, it has been suggested that they should be placed inside the second genus *Betahypovirus* (Li et al. 2015). This taxonomical solution is, however, by the time of writing this text, only a proposed solution and ICTV (International Committee on Taxonomy of Viruses) does not yet recognize the aforementioned genera.

It is worth noting that this grouping does not necessarily predict viruses capable of inducing a hypovirulent phenotype. For example, CHV1, CHV2, FgHV2 and SsHV2 reduce their host virulence to various degrees, while FgHV1 does not reduce host virulence despite belonging to the same phylogenetic cluster (*Alphahypovirus*). Of the viruses belonging to the second cluster (*Betahypovirus*), CHV3 induces hypovirulence, whereas SsHV1 is only able to do so if a satellite-like dsRNA element is present as well. PlHV1 has not been investigated in detail, but preliminary analysis showed that infection with this virus can be either completely asymptomatic, severely debilitating to the host, or somewhere in between (Koloniuk et al. 2014). Infection with VcHV1 had no effect on its host, as suggested by observation of isogenic virus-free and virus-infected fungal strains in vitro. While the hypothetical hypovirulent effect might not always be apparent in such in vitro experiments, even this preliminary finding gives further credence to the impression that not all hypoviruses can induce a hypovirulent phenotype in their hosts, regardless of phylogenetic relatedness (Yaegashi et al. 2012).

7.4.3 *Hypovirulence in* Magnaporthe oryzae *(*Magnaporthe grisea*)*

Rice blast, caused by fungus *Magnaporthe oryzae* B.C. Couch or *Magnaporthe grisea* (T.T. Hebert) M.E. Barr, is one of the most important rice diseases in the world, causing significant economic losses (Zhang et al. 2016). *Magnaporthe oryzae* is a hemiobiotrophic pathogen meaning that initially the fungus establishes a biotrophic relationship with its host, and later switches to necrotrophic lifestyle in which the infected plant tissue is destroyed.

Magnaporthe oryzae chrysovirus 1-D (MoCV1-D) was found in a Japanese isolate of the rice blast fungus and subsequently characterized (Urayama et al. 2010). Infection with this virus causes growth inhibition of its host. The genome of MoCV1-D consists of five dsRNA segments, one of which, dsRNA4, is responsible for inducing the hypovirulent phenotype. This was concluded since the protein encoded by dsRNA4 impairs cell growth when expressed in yeast cells. Beyond that, studies in yeast demonstrated that it causes abnormal pigmentation and colony albinization, a phenotypic change associated with reduced accumulation of the melanin biosynthesis intermediate scylatone. MoCV1-D is readily transmissible via hyphal anastomosis. These results suggest that MoCV1-D might be a potential candidate for biological control of the rice blast fungus (Higashiura et al. 2019). Two other mycoviruses were found in the original *M. oryze* isolate, but their effects on the fungus are currently unknown.

7.4.4 *Hypovirulence in* Alternaria alternata

Alternaria alternata (Fr.) Keissl. is an anamorphic ascomycete fungus that causes leaf spot diseases on various crops (Troncoso-Rojas and Tiznado-Hernández 2014). A novel ssRNA mycovirus named *Alternaria alternata hypovirus 1* (AaHV1) closely related to the genus *Hypovirus*, was recently identified in a fungal strain isolated from apple leaves (Li et al. 2019a). The genome of AaHV1 contains a single large open-reading frame encoding a putative polyprotein with cysteine proteinase-like domain and an RNA-dependant RNA polymerase domain. The virus causes reduced growth and attenuated virulence of the host *A. alternata*. Total RNA extracts from *A. alternata* (containing viral RNAs) were successfully used to transfect freshly prepared protoplasts of *Botryosphaeria dothidea* (Moug.) Ces. & De Not. The virus was also able to replicate and confer hypovirulence in this new host (Li et al. 2019a). This indicates a potential of AaHV1 to induce a hypovirulent effect in other plant pathogenic fungi.

7.4.5 *Hypovirulence in* Rhizoctonia solani

Rhizoctonia solani J.G. Kühn (teleomorph *Thanatephorus cucumeris* (A.B. Frank) Donk) represents a species complex of soil borne fungi which can infect many plant species (Lübeck 2004). Especially important is strain *R. solani* AG1-1A, which is the causal agent of rice sheath blight. A new member of *Endornaviridae*, provisionally named as *Rhizoctonia solani endornavirus 1* (RsEV1) was recently isolated from a hypovirulent strain of *R. solani* AG-1 IA GD2 (Zheng et al. 2019). RsEV1 can be transmitted horizontally into a virulent strain of the pathogen, altering its phenotype. In *R. solani* strains infected with RsEV1 a metabolic perturbance was noted, changing several metabolic pathways, including pentose and glucuronate interconversions and glyoxylate, dicarboxylate, starch, and sucrose metabolism. Beyond this metabolic disorder, a hypovirulent phenotype was noticeable by changes in mycelial pigmentation, growth rate and morphology as well as smaller size of sclerotia, indicating a potential of RsEV1 in biocontrol of rice sheath blight.

7.4.6 *Hypovirulence in* Fusarium *Species*

Fusarium graminearum Schwabe (teleomorph *Gibberella zeae* (Schwein.) Petch) (Gräfenhan et al. 2011) causes Fusarium head blight, a severe disease of wheat, barley and other small-grain cereal crops around the world. Many mycoviruses have been detected in this pathogen but only four of them are associated with a significant hypovirulent effect. Based on their genomes the *Fusarium* mycoviruses can be placed into two groups—dsRNA viruses, most of which have segmented genomes, and ssRNA viruses that possess a single RNA molecule as a genome. Most of ssRNA viruses have a positive sense RNA as a genome, with the only exception being *Fusarium graminearum negative-stranded RNA virus 1* (FgNSRV1) (Li et al. 2019b).

Almost 20 years ago a dsRNA element associated with a hypovirulent phenotype was detected in *F. graminearum* isolates from infected maize in Korea. Subsequent studies demonstrated various peculiarities of dsRNA-infected strains, including reduction in growth, increased pigmentation, reduced virulence towards wheat, and decreased production of certain mycotoxins. The hypovirulent phenotype was horizontally transmissible to other strains via hyphal anastomosis and vertically into approximately 50% of conidia (Chu et al. 2002). This virus was eventually named *Fusarium graminearum virus 1* strain DK21 (FgV1-DK21) and placed into the new family *Fusariviridae* (Li et al. 2019b). Subsequent proteomic analyses uncovered 148 differentially represented proteins, out of which 33 exhibit consistent differences in expression, when comparing virus-infected with virus-free fungal strains. Upregulated proteins had various functions, regulating sporulation, sugar metabolism, protein synthesis and differentiation, while downregulated genes included mainly metabolic and defence-response genes involved in ROS detoxification (Kwon et al. 2009).

In addition to FgV1, three members of the family *Chrysoviridae, Fusarium graminearum virus* strain China 9 (FgV-ch9), *Fusarium graminearum virus 2* FgV2 and *Fusarium oxysporum f. sp. dianthi mycovirus 1* (FodV1) were found to induce a hypovirulent phenotype (Li et al. 2019b). The mycovirus FgV-ch9 was detected in *F. graminearum* strains recovered from cereals in China. Like FgV1, this virus is vertically transmitted to asexual conidia. Symptom expression by FgV-ch9 depends on the virus load in *F. graminearum* infected fungal strains. At high and medium virus concentrations, the most noticeable symptoms are reduced mycelial growth, reduced sporulation (both sexual and asexual), abnormal colony morphology, disorganized cytoplasm and attenuated virulence of the infected fungus on wheat and maize. At low virus concentrations the infection of the fungus remains asymptomatic.

Fusarium oxysporum Schltdl. is an anamorphic species complex, a member of which, *F. oxysporum* f.sp. *dianthi* (Prill. & Delacr.) W.C. Snyder & H.N. Hansen is a plant pathogen infecting carnations (Fourie et al. 2011; Lemus-Minor et al. 2018). *Fusarium oxysporum f. sp. dianthi virus 1* (FodV1) was the first mycovirus detected in the *Fusarium oxysporum* species complex with the ability to induce hypovirulence (Lemus-Minor et al. 2018). Its genome consists of four dsRNA segments. Infection with FodV1 causes significant phenotypic alterations in vegetative growth and virulence of its host. This characteristic makes this mycovirus an interesting candidate as a potential biocontrol agent for *Fusarium* wilt of carnations.

Fusarium graminearum virus 2 (FgV2) has a similar effect on its host as FgV1. In a study by Lee et al. (2014) the hypovirulence effect of four virus species on several *Fusarium* spp. strains were examined. The results showed that both, FgV1 and FgV2, induce a hypovirulent phenotype, but they differ in their effect on physiology and gene expression of its respective host. Additional viruses detected in *F. graminearum* include FgV3 and FgV4, but both do not seem to alter the virulence of infected fungal strains, despite affecting gene expressions (Lee et al. 2014).

Two unclassified hypovirus-related viruses of the family *Hypoviridae—Fusarium graminearum hypovirus 1* and 2 have been identified more recently. Their +ssRNA unsegmented genomes contain two (FgHV1) or one (FgHV2) ORFs, with highly conserved protease, RNA-dependent RNA polymerase, and helicase domains, characteristic for all members of *Hypoviridae*. Interestingly, only FgHV2, a member of the proposed genus *Alphahypovirus* induces hypovirulence, by negatively affecting mycelial growth, conidiation, and mycotoxin production of infected *F. graminearum* strains (Chu et al. 2004; Li et al. 2019b).

7.4.7 Hypovirulence in Heterobasidion *Species*

The fungal genus *Heterobasidion* Bref. includes some of the most devastating forest pathogens that cause root disease in many coniferous and even some broadleaf species (Garbelotto and Gonthier 2013; Gonthier et al. 2014). Because of the economic losses caused by these tree pathogens, viruses with the ability to cause a hypovirulent phenotype are intensively searched for in different *Heterobasidion* species. A first study by Ihrmark et al. (2001) revealed that approximately 15% of *H. annosum* s.l. (Fr.) Bref. isolates in Europe and Western Asia harbour dsRNA viruses. Vainio and Hantula (2016) reported that the genus *Heterobasidion* hosts a widespread and diverse mycovirus community composed of more than 16 species of *Partitiviridae*, a species of *Narnaviridae* and one taxonomically unassigned virus related to the *Curvularia thermal tolerance virus*. Of these, *Heterobasidion partitivirus 13* (HetPV13-an1) from *H. annosum* shows the highest potential as a biocontrol agent. HetPV13-an1 causes severe phenotypic debilitation in the host fungus (Vainio et al. 2017), affecting the transcription of 683 genes, of which 60% are downregulated and 40% upregulated. Alterations observed in carbohydrate and amino acid metabolism suggest that the virus causes a state of starvation, which is compensated for by alternative synthesis routes. HetPV13-an1 was transferred into different *H. annosum* and *H. parviporum* Niemelä & Korhonen strains. While the virus caused growth reduction in all three newly infected *H. parviporum* strains, only two of the six infected *H. annosum* strains showed significant debilitation. In a field experiment on Norway spruce, a HetPV13-an1-infected *H. parviporum* strain showed considerably less growth within living trees than did the isogenic virus-free strain of the fungus (Vainio et al. 2017).

7.4.8 Hypovirulence in Ophiostoma novo-ulmi

Dutch elm disease (DED) is a wilting disease of elm trees caused by an ascomycete fungus of the genus *Ophiostoma* Syd. & P. Syd. The first DED pandemic was caused by a less aggressive species, *O. ulmi* (Buisman) Nannf., while the second (and current) pandemic is caused by a more aggressive species, *O. novo-ulmi* Brasier (Brasier and Buck 2001; Katanić et al. 2020). Elm bark beetles play an important role in the spread of the disease by carrying fungal spores from infected to healthy elms (Santini and Faccoli 2014).

The first mycoviruses that were shown to induce severe debilitation of *O. novo-ulmi* were originally described as d-factors that cause a reduction in fungal persistence, xylem infection levels, growth rates (Sutherland and Brasier 1995) and conidial viability (Sutherland and Brasier 1997). Later, several d-factors were characterised, and based on phylogenetic analysis, assigned to the genus *Mitovirus* (Doherty et al. 2006).

A proposed approach for biological control of DED is to use elm bark beetles as carriers for virus-infected fungal conidia, thereby introducing the virus into local *O. novo-ulmi* populations.

7.4.9 *Hypovirulence in* Botrytis *Species*

Botrytis cinerea Pers. (teleomorph *Botryotinia fuckeliana* (de Bary) Whetzel) is a necrotrophic pathogen that infects many plant species in temperate and subtropical climates. Many of these host plants are economically important (e.g., grapes, strawberries, solanaceous vegetables). Symptoms caused by *B. cinerea* vary depending on host plants and plant tissues affected and include leaf blight, blossom blight, and post-harvest fruit rots (Williamson et al. 2007). This pathogen is mainly controlled using chemical fungicides (so called "botryticides"), which increasingly carries the risk of inducing fungicide resistance (Jacometti et al. 2010). This fungus is also responsible for noble rot of grapes as well, producing highly valuable and sweet botrytised vines (Fournier et al. 2013; Negri et al. 2017).

Numerous mycoviruses belonging to different virus families have been reported in *B. cinerea* and several of them were associated with hypovirulence of its host, namely *Botrytis cinerea mitovirus 1* (BcMV1, previously named BcDRV), *Botrytis cinerea RNA virus 1* (BcRV1), *Botrytis cinerea hypovirus 1* (BcHV1), *Botrytis cinerea partitivirus 2* (BcPV2) and *Botrytis cinerea CCg378 virus 1* (Bc378V1). BcMV1 was found in a hypovirulent *B. cinerea* isolate from oilseed rape in China and appears to debilitate mitochondria in the infected hyphae. Laboratory experiments showed that BcMV1 can be transmitted both vertically to asexual conidia and horizontally to other fungal strains (Wu et al. 2010). BcRV1 was detected in a hypovirulent isolate of *B. cinerea* from *Berberis* sp. in China. The degree of hypovirulence is greatly affected by the accumulation level of BcRV1 in *B. cinerea* strains. This mycovirus can spread vertically via macroconidia, as well as horizontally through hyphal anastomosis (Yu et al. 2015). As in other fungi, horizontal virus transmission of BcRV1 and BcMV1seems to be limited by a vegetative incompatibility system of *B. cinerea* (Wu et al. 2010; Yu et al. 2015).

In contrast to BcMV1 and BcRV1, which reduce mycelial growth of the pathogen, infection with BcHV1 causes hypovirulent effect without affecting the growth of the infected fungus. In this case, reduced virulence seems to arise from specific changes in the expression of genes associated with the formation of infection cushions, which are important because they facilitate penetration of plant tissues during the infection process. BcHV1 can be transmitted horizontally, while vertical transmission has not been demonstrated thus far (Hao et al. 2018). BcPV2 is a partitivirus that belongs to the genus *Alphapartitivirus* and was detected in an atypically pink strain of *B. cinerea* (Kamaruzzaman et al. 2019). Strains infected with this virus showed normal vegetative growth, but reduced production of conidia and sclerotia. Importantly, BcPV2 was found to attenuate virulence of *B. cinerea* on several crops, including apples, tomatoes, and potatoes. This virus was successfully transmitted to several virulent strains of *B. cinerea*, inducing a hypovirulent phenotype. Another partitivirus, Bc378V1, was detected in a wild-type *B. cinerea* strain that was co-infected with another mycovirus. While the second mycovirus has not been further characterized, the presence of both mycoviruses appears to be necessary for inducing a hypovirulent phenotype (Potgieter et al. 2013).

Botrytis spp. also cause grey mold disease on several *Allium* L. species, mostly on garlic (*A. sativum* L.), garlic chives (*A. tuberosum* Rottler ex Spreng.) and bulb onions (*A. cepa* L.) (Wu et al. 2012). Beyond that, *Botrytis porri* N.F. Buchw. (teleomorph *Botryotinia porri* (H.J.F. Beyma) Whetzel) can cause garlic clove rot (Dugan et al. 2007), garlic leaf blight (Zhang et al. 2009), and leek leaf rot (Asiedu 1986). In 2012, Wu et al. detected dsRNA elements in hypovirulent *B. porri* strains obtained from garlic and onion plants. They identified a bipartite segmented virus associated with ~35 nm particles and called the virus *Botrytis porri RNA virus 1* (BpRV1). The particles are comprised of three structural proteins (SP) and two dsRNA elements ~6 kb long, coding for three SPs and the RdRP. Fungal strains harbouring this virus grew much slower in vitro and were only able to induce small lesions on inoculated garlic leaves. Besides that, the infected strains were not able to produce sclerotia and the hyphae showed ultrastructural aberrations. Phylogenetically this BpRV1 seemed to be related to members of the genera *Totivirius* and *Victorivirus*, although the support for this classification is weak. The authors were also able to demonstrate vertical transmission of the virus into conidia, while the success of horizontal transmission appeared to be dependent on the donor and recipient fungal strain (Wu et al. 2012).

The detection of several mycoviruses that reduce the virulence of *Botrytis* spp. gives hope for the development of a biocontrol system against this important plant pathogen. However, two major limiting factors still need to be overcome: the low competitive ability of mycovirus-infected strains relative to mycovirus-free strains and the limited horizontal transmissibility of the mycoviruses, which is mainly due to the presence of vegetative incompatibility barriers in the fungus.

7.4.10 Hypovirulence in Rosellinia necatrix

Rossellinia necatrix Berl. ex Prill. is a soil borne fungi that causes white root rot disease in many plant species including grapevine and several fruit trees. As a necrotrophic pathogen it can also survive as a saprophyte in the soil. Due to its soil-borne lifestyle and the ability to remain dormant and infectious for many years (Kulshrestha et al. 2014), the pathogen is difficult to control. Several mycoviruses were detected in *R. necatrix* but most of them cause asymptomatic infections. Two mycoviruses that induce a hypovirulent phenotype were reported: *Rosellinia necatrix megabirnavirus 1* (RnMBV1) and *Mycoreovirus 3* (MyRV3). The first virus is a bipartite dsRNA virus consisting of ~7 and ~9 kb genome segments and viral particles ~50 nm in diameter with a single major capsid protein of 135 kDa. Purified viral particles were able to infect fungal protoplasts, which developed an altered phenotype after transfection. Fungal strains infected with this virus showed attenuated virulence on inoculated plants, causing only superficial rather than lethal lesions (Chiba et al. 2009). Similarly, MyRV3 particles were used for transfection of virulent *R. neacatrix* protoplasts, in which a hypovirulent phenotype was induced. The hypovirulent phenotype of the fungus included reduced growth of the infected colony and smaller lesions on inoculated apple fruits. Kanematsu et al. (2010) successfully transfected several other phytopathogenic fungi with MyRV3 particles. After transfection, *Diaporthe* Nitschke sp., *C. parasitica* and *Valsa ceratosperma* (Tode) Maire showed reduced colony growth and smaller lesion on inoculated apples. These experiments showed that purified particles of both mycoviruses can be used to transfect virus free *R. necatrix* strains. Furthermore, other pathogenic fungi transfected with MyRV3 subsequently become hypovirulent. These experiments were conducted in the laboratory, and the direct use of virus particles for biocontrol in the field remains to be demonstrated.

7.4.11 Hypovirulence in Helminthosporium victoriae

The plant pathogenic fungus *Helminthosporium victoriae* Meehan & H.C. Murphy, (teleomorph: *Cochliobolus victoriae* R.R. Nelson), is the causal agent of Victoria blight of oats (*Avena sativa* L.). In the late 1940s, this disease caused considerable yield losses in the oat-growing regions of the USA (Meehan and Murphy 1946). In the 1950s, only minor damage was observed in some oat fields in Louisiana, despite widespread infection with *H. victoriae*. Isolates recovered from infected plants produced highly sectored colonies and their growth was stunted (Lindberg 1960). By means of co-culturing, this phenotype was transmitted from a diseased isolate to 'normally' growing isolates. It was later demonstrated that the diseased *H. victoriae* isolates were infected with two viruses, the *Helminthosporium victoriae virus 190S* (HvV190S) and the *Helminthosporium victoriae chrysovirus 145S* (HvV145S). Recent studies have suggested that infection with HvV190S alone is responsible for hypovirulence in *H. victoriae*. Interestingly, after artificial infection, HvV190S also induces a hypovirulent phenotype in the chestnut blight fungus *C. parasitica* (Xie et al. 2016). Furthermore, Wu et al. (2020) revealed that another species, *Bipolaris maydis* (Y. Nisik. & C. Miyake) Shoemaker (teleomorph: *Cochliobolus heterostrophus* (Drechsler) Drechsler) is also a natural host of HvV190S. This could indicate the potential of HvV190S-mediated hypovirulence in other plant pathogenic fungi.

7.4.12 Hypovirulence in Sclerotinia sclerotiorum

The ascomycete fungus *Sclerotinia sclerotiorum* (Lib.) de Bary is a necrotrophic plant pathogen that causes white mold or stem rot on many different plant species. It has a broad host range that includes important agricultural host species such as bean, rapeseed, soybean, and lettuce. The pathogen is found worldwide in temperate and subtemperate regions. The pathogen is homothallic and forms sexual fruiting bodies (apothecia) by self-fertilisation. The sexual ascospores are spread by wind and are the main source of new infections. The pathogen produces special survival structures called sclerotia, which also serve as asexual dissemination propagules.

Several mycoviruses are known to cause hypovirulence in *S. sclerotiorum* with *Sclerotinia sclerotiorum hypovirulence-associated DNA virus 1* (SsHADV-1) being the most promising biological control agent (Yu et al. 2010). This virus is a rare example of a DNA mycovirus. To date, it is also the only mycovirus of which purified virus particles can be directly used to infect a fungal host (Yu et al. 2013). This was demonstrated by spraying purified virus particles onto leaves, which resulted in suppression of *S. sclerotiorum* infections. In field trials with rapeseed, the preventive application of hyphal fragments containing SsHAVDV-1virus particles significantly reduced Sclerotinia stem rot and increased crop yield.

Sclerotinia sclerotiorum also hosts a mycovirus related to the family *Hypoviridae*, which causes hypovirulence (Hu et al. 2014; Khalifa and Pearson 2014). *Sclerotinia sclerotiorum hypovirus 2* (SsHV2) induces the hypovirulent phenotype by reducing mycelial growth, sporulation, and the formation of sclerotia. By using a detached leaf assay, a wide variation in symptom expression among SsHV2-infected strains was observed. Several virus-infected strains showed a pronounced virulence reduction, which could indicate potential candidates for biological control of *S. sclerotiorum*.

Sclerotinia sclerotiorum partitivirus 1 (SsPV1) is another mycovirus that reduces virulence of *S. sclerotiorum* and related species (Xiao et al. 2014). Fungal strains infected with SsPV1 exhibit abnormal colony morphology and severe growth reduction. Horizontal transmission of SsPV1 to virus-free, virulent strains, results in their conversion to the hypovirulent phenotypes.

A mitochondrial mycovirus, *Sclerotinia sclerotiorum mitovirus 1* (SsMV1/HC025) also induces hypovirulence in *S. sclerotiorum* (Xu et al. 2015). Infection with this mitovirus causes mitochondrial malformations and greatly reduces mycelial growth and sclerotia production. On inoculated leaves of rapeseed and soybean plants, virus-infected fungal strains produce no or only very small lesions. SsMV1/HC025 can be transmitted horizontally via hyphal anastomosis, which leads to conversion of the recipient virulent strains to the hypovirulent phenotype.

7.5 Mycoviruses Related to Plant Viruses

Fungal viruses are a diverse and phylogenetically heterogenous group of parasitic genetic elements which infect a major eukaryotic lineage. Most mycoviruses characterised so far have RNA (either ssRNA or dsRNA) genomes, while DNA mycoviruses have been identified much less frequently (Abbas 2016). Several general observations can be made, however, regarding their pathogenic effect and organization of their genomes. Many of them have a mild effect on their host, causing usually asymptomatic infections. Secondarily, a common element of their genomes is an RNA-dependant RNA polymerase (RdRP), ubiquitous for their replication. Their polymerases seem to be related indicating a common phylogenetic ancestor, at least for some of the mycoviruses (Ghabrial 1998).

Another important aspect of mycovirus evolution is related to their hosts' close and long-lasting symbiotic relationship with plants, either mutualistic as in mycorrhizal associations (van der Heijden et al. 2015), or parasitic as demonstrated with a large number of fungal plant pathogens (Han 2019; Möller and Stukenbrock 2017). This relationship has existed at least since both groups (i.e. plants and fungi) began to colonise the land in the early Palaeozoic (Field et al. 2015; Morris et al. 2018). This close and long-lasting symbiotic relationship between various fungal and early plant linages might have facilitated the exchange of the viruses between these two kingdoms, and for many mycoviruses their phytovirus counterparts have been found. Furthermore, many plant viruses have +ssRNA genomic organization, similar to several mycovirus lineages, which is especially evident in their conserved RdRP domains (Roossinck 2019).

In 2017 a putative +ssRNA virus was discovered in the plant pathogenic fungus *Phomopsis longicolla* (Hrabáková et al. 2017). The fungal isolate in which this virus was found had a debilitated growth and reduced virulence and was already known to contain *Phomopsis longicolla hypovirus* 1, which is known to induce a wide range of symptoms in its host. This combination of a known hypovirus and a novel RNA element, placed in a proposed *Ourmiavirus* genus which is a sister genus to mitoviruses, is perhaps responsible for the hypovirulent effect. Interestingly sequence analysis of this virus reveal that its RdRP is actually related to the RNA polymerase of plant viruses, rather than other mycoviruses (Hrabáková et al. 2017). This taken with findings of Nerva et al. (2017), who reported that certain fungal viruses can replicate inside plant cells, sheds light on an interesting aspect of mycovirus evolution. Fungal viruses, thus, beyond being interesting as biocontrol agents of plant diseases, are suspected to be evolutionary connected with plant viruses, as deduced by sequence similarities between some plant and fungal viruses (Pearson et al. 2009).

7.6 Conclusion

Recent research in fungal virology has made great progress in our understanding of the biology of mycoviruses and their interactions with fungal hosts. A key driver of this development is the prospect of using mycoviruses as natural enemies against fungal pathogens. Indeed, a large number of new mycoviruses have been discovered in various plant pathogenic fungi in recent years. Several of these mycoviruses attenuate virulence and some affect the reproduction capacity of their fungal hosts, traits that suggest these viruses have the potential to be used as biological control agents. Hypovirulence of the chestnut blight fungus *C. parasitica* still provides the prime example of mycovirus-mediated biocontrol of a plant disease. The recent results in mycovirus research, however, are very promising for further applications of mycoviruses for plant disease control.

Acknowledgements Our chapter is dedicated to the late Hélène Blauenstein, who for many years provided excellent assistance to our work on mycovirus-mediated biological control of chestnut blight. Our research and collaboration have been supported by grants of the Swiss National Science Foundation (SCOPES projects IZ73Z0_127922 and IZ73Z0_152525), the Competence Center Environment and Sustainability of the ETH Domain (CCES-GEDIHAP project), the Swiss Federal Office for the Environment (FOEN), the Croatian-Swiss Research Programme (project IZHRZ0_180651), Croatian National Science Foundation (project IP-2018-01-1295), and by the University of Zagreb.

References

Abbas A (2016) A review paper on mycoviruses. J Plant Pathol Microbiol 7:10–13. https://doi.org/10.4172/2157-7471.1000390

Agrios GN (2005) Plant pathology. Elsevier Academic, Cambridge

Allemann C, Hoegger P, Heiniger U et al (1999) Genetic variation of Cryphonectria hypoviruses (CHV1) in Europe, assessed using restriction fragment length polymorphism (RFLP) markers. Mol Ecol 8:843–854. https://doi.org/10.1046/j.1365-294X.1999.00639.x

Allen TD, Nuss DL (2004) Specific and common alterations in host gene transcript accumulation following infection of the chestnut blight fungus by mild and severe hypoviruses. J Virol 78:4145–4155. https://doi.org/10.1128/JVI.78.8.4145-4155.2004

Allen TD, Dawe AL, Nuss DL (2003) Use of cDNA microarrays to monitor transcriptional responses of the chestnut blight fungus *Cryphonectria parasitica* to infection by virulence-attenuating hypoviruses. Eukaryot Cell 2:1253–1265. https://doi.org/10.1128/EC.2.6.1253-1265.2003

Anagnostakis SL (1987) Chestnut blight: the classical problem of an introduced pathogen. Mycologia 79(1):23–37. https://doi.org/10.2307/3807741

Anagnostakis SL (1988) *Cryphonectria parasitica*, cause of chestnut blight. Adv Plant Pathol 6:123–136. https://doi.org/10.1016/B978-0-12-033706-4.50011-6

Anagnostakis SL (1992) Measuring resistance of chestnut trees to chestnut blight. Can J For Res 22:568–571. https://doi.org/10.1139/x92-075

Asiedu SK (1986) *Botrytis porri* on leek in Canada. Plant Dis 70:259. https://doi.org/10.1094/PD-70-259c

Aulia A, Andika IB, Kondo H et al (2019) A symptomless hypovirus, CHV4, facilitates stable infection of the chestnut blight fungus by a coinfecting reovirus likely through suppression of antiviral RNA silencing. Virology 533:99–107. https://doi.org/10.1016/j.virol.2019.05.004

Bolvanský M, Pažitný J, Adamčíková K (2018) Grading of blight resistance in different chestnut accessions. Acta Hortic 1220:87–94. https://doi.org/10.17660/ActaHortic.2018.1220.13

Brasier CM, Buck KW (2001) Rapid evolutionary changes in a globally invading fungal pathogen (Dutch elm disease). Biol Invasions 3:223–233. https://doi.org/10.1023/A:1015248819864

Brusini J, Wayne ML, Franc A et al (2017) The impact of parasitism on resource allocation in a fungal host: the case of *Cryphonectria parasitica* and its mycovirus, Cryphonectria Hypovirus 1. Ecol Evol 7:5967–5976. https://doi.org/10.1002/ece3.3143

Bryner SFS, Rigling D (2011) Temperature-dependent genotype-by-genotype interaction between a pathogenic fungus and its hyperparasitic virus. Am Nat 177:65–74. https://doi.org/10.1086/657620

Bryner SF, Prospero S, Rigling D (2014) Dynamics of Cryphonectria hypovirus infection in chestnut blight cankers. Biol Control 104:918–925. https://doi.org/10.1094/PHYTO-03-13-0069-R

Buck KW (1986) Fungal virology—an overview. In: Buck KW (ed) Fungal virology, 1st edn. CRC, Boca Raton, pp 1–84

Chen B, Choi GH, Nuss DL (1993) Mitotic stability and nuclear inheritance of integrated viral cDNA in engineered hypovirulent strains of the chestnut blight fungus. EMBO J 12:2991–2998. https://doi.org/10.1002/j.1460-2075.1993.tb05967.x

Chen B, Geletka LM, Nuss DL (2000) Using chimeric hypoviruses to fine-tune the interaction between a pathogenic fungus and its plant host. J Virol 74:7562–7567. https://doi.org/10.1128/JVI.74.16.7562-7567.2000

Chen Q, Li Y, Wang J et al (2018) cpubi4 is essential for development and virulence in chestnut blight fungus. Front Microbiol 9:1–15. https://doi.org/10.3389/fmicb.2018.01286

Chiba S, Salaipeth L, Lin Y-H et al (2009) A novel bipartite double-stranded RNA mycovirus from the white root rot fungus *Rosellinia necatrix*: molecular and biological characterization, taxonomic considerations, and potential for biological control. J Virol 83:12801–12812. https://doi.org/10.1128/JVI.01830-09

Choi G, Nuss D (1992) Hypovirulence of chestnut blight fungus conferred by an infectious viral cDNA. Science 257:800–803. https://doi.org/10.1126/science.1496400

Choi GH, Chen B, Nuss DL (1995) Virus-mediated or transgenic suppression of a G-protein alpha subunit and attenuation of fungal virulence. Proc Natl Acad Sci USA 92:305–309. https://doi.org/10.1073/pnas.92.1.305

Chu YY, Jeon JJ, Yea SS et al (2002) Double-stranded RNA mycovirus from *Fusarium graminearum*. Appl Environ Microbiol 68:2529–2534. https://doi.org/10.1128/AEM.68.5.2529-2534.2002

Chu Y-M, Lim W-S, Yea S-J et al (2004) Complexity of dsRNA mycovirus isolated from *Fusarium graminearum*. Virus Genes 28:135–143. https://doi.org/10.1023/B:VIRU.0000012270.67302.35

Chung P-H (1994) Diversity of *Cryphonectria parasitica* hypovirulence-associated double-stranded RNAs within a chestnut population in New Jersey. Phytopathology 84:984–990. https://doi.org/10.1094/Phyto-84-984

Conedera M, Tinner W, Krebs P et al (2016) *Castanea sativa* in Europe: distribution, habitat, usage and threats. In: San-Miguel-Ayanz J, de Rigo D, Caudullo G, Houston Durrant T, Mauri A (eds) European atlas for tree species. Publ Off EU, Luxembourg, pe0125e0+

Cornejo C, Šever B, Kupper Q et al (2019) A multiplexed genotyping assay to determine vegetative incompatibility and mating type in *Cryphonectria parasitica*. Eur J Plant Pathol 55(1):81–91. https://doi.org/10.1007/s10658-019-01751-w

Cortesi P, Milgroom MG (1998) Genetics of vegetative incompatibility in *Cryphonectria parasitica*. Appl Environ Microbiol 64:2988–2994

Cortesi P, McCulloch CE, Song H et al (2001) Genetic control of horizontal virus transmission in the chestnut blight fungus, *Cryphonectria parasitica*. Genetics 159:107–118

Dawe AL, Nuss DL (2001) Hypoviruses and chestnut blight: exploiting viruses to understand and modulate fungal pathogenesis. Annu Rev Genet 35:1–29

Dawe AL, Nuss DL (2013) Hypovirus molecular biology: from Koch's postulates to host self-recognition genes that restrict virus transmission. Adv Virus Res 86:109–147. https://doi.org/10.1016/B978-0-12-394315-6.00005-2

Debets AJM, Aanen DK, Glass NL et al (2014) Biology and genetics of vegetative incompatibility in Fungi. In: Borkovich KA, Ebbole DJ (eds) Cellular and molecular biology of filamentous fungi. ASM, Washington, pp 274–288. https://doi.org/10.1128/9781555816636.ch20

Diamandis S (2018) Management of chestnut blight in Greece using hypovirulence and silvicultural interventions. Forests 9(8):492. https://doi.org/10.3390/f9080492

Diamandis S, Perlerou C, Tziros GT et al (2015) Establishment and dissemination of hypovirulent strains of *Cryphonectria parasitica* in Greece. For Pathol 45:408–414. https://doi.org/10.1111/efp.12188

Doehlemann G, Ökmen B, Zhu W et al (2017) Plant pathogenic fungi. In: Heitman J, Howlett BJ, Crous PW, Stukenbrock EH, James TY, Gow NAR (eds) The fungal kingdom. ASM Press, Washington, pp 701–726

Doherty M, Coutts RHA, Brasier CM et al (2006) Sequence of RNA-dependent RNA polymerase genes provides evidence for three more distinct mitoviruses in *Ophiostoma novo-ulmi* isolate Ld. Virus Genes 33:41–44. https://doi.org/10.1007/s11262-005-0029-5

Double ML, Jarosz AM, Fulbright DW et al (2018) Evaluation of two decades of *Cryphonectria parasitica* Hypovirus introduction in an American chestnut stand in Wisconsin. Phytopathology 108:702–710. https://doi.org/10.1094/PHYTO-10-17-0354-R

Dugan FM, Hellier BC, Lupien SL (2007) Pathogenic fungi in garlic seed cloves from the United States and China, and efficacy of fungicides against pathogens in garlic germplasm in Washington State. J Phytopathol 155:437–445. https://doi.org/10.1111/j.1439-0434.2007.01255.x

Enebak S, Hillman B, MacDonald W (1994) A hypovirulent isolate of *Cryphonectria parasitica* with multiple, genetically unique dsRNA segments. Mol Plant Microbe Interact 7:590–595

Eusebio-Cope A, Suzuki N (2015) Mycoreovirus genome rearrangements associated with RNA silencing deficiency. Nucleic Acids Res 43:3802–3813. https://doi.org/10.1093/nar/gkv239

Feau N, Dutech C, Brusini J et al (2014) Multiple introductions and recombination in Cryphonectria hypovirus 1: perspective for a sustainable biological control of chestnut blight. Evol Appl 7:580–596. https://doi.org/10.1111/eva.12157

Field KJ, Pressel S, Duckett JG et al (2015) Symbiotic options for the conquest of land. Trends Ecol Evol 30:477–486. https://doi.org/10.1016/j.tree.2015.05.007

Flint SJ, Racaniello VR, Rall GF et al (2009) Principles of virology, 4th edn. ASM Press, Washington

Fourie G, Steenkamp ET, Ploetz RC et al (2011) Current status of the taxonomic position of *Fusarium oxysporum formae specialis cubense* within the *Fusarium oxysporum* complex. Infect Genet Evol 11:533–542. https://doi.org/10.1016/j.meegid.2011.01.012

Fournier E, Gladieux P, Giraud T (2013) The 'Dr Jekyll and Mr Hyde fungus': noble rot versus gray mold symptoms of *Botrytis cinerea* on grapes. Evol Appl 6:960–969. https://doi.org/10.1111/eva.12079

Fralish JS, Franklin SB (2002) Taxonomy and ecology of woody plants in North American forests (Excluding Mexico and subtropical Florida). Wiley, New York

Fulbright DW, Weidlich WH, Haufler KZ et al (1983) Chestnut blight and recovering American chestnut trees in Michigan. Can J Bot 61:3164–3171. https://doi.org/10.1139/b83-354

Gao S, Shain L (1995) Activity of polygalacturonase produced by *Cryphonectria parasitica* in chestnut bark and its inhibition by extracts from American and Chinese chestnut. Physiol Mol Plant Pathol 46:199–213. https://doi.org/10.1006/pmpp.1995.1016

Garbelotto M, Gonthier P (2013) Biology, epidemiology, and control of *Heterobasidion* species worldwide. Annu Rev Phytopathol 51:39–59. https://doi.org/10.1146/annurev-phyto-082712-102225

García-Pedrajas MD, Cañizares MC, Sarmiento-Villamil JL et al (2019) Mycoviruses in biological control: from basic research to field implementation. Phytopathology 109:1828–1839. https://doi.org/10.1094/PHYTO-05-19-0166-RVW

Ghabrial SA (1998) Origin, adaptation and evolutionary pathways of fungal viruses. Virus Genes 16:119–131. https://doi.org/10.1023/A:1007966229595

Ghabrial SA, Castón JR, Jiang D et al (2015) 50-plus years of fungal viruses. Virology 479–480:356–368. https://doi.org/10.1016/j.virol.2015.02.034

Gobbin D, Hoegger P, Heiniger U et al (2003) Sequence variation and evolution of Cryphonectria hypovirus 1 (CHV-1) in Europe. Virus Res 97:39–46. https://doi.org/10.1016/S0168-1702(03)00220-X

Gonthier P, Anselmi N, Capretti P et al (2014) An integrated approach to control the introduced forest pathogen *Heterobasidion irregulare* in Europe. Forestry 87:471–481. https://doi.org/10.1093/forestry/cpu015

Gräfenhan T, Schroers H-J, Nirenberg HI et al (2011) An overview of the taxonomy, phylogeny, and typification of nectriaceous fungi in Cosmospora, Acremonium, Fusarium, Stilbella, and Volutella. Stud Mycol 68:79–113. https://doi.org/10.3114/sim.2011.68.04

Graves A (1950) Relative blight resistance in species and hybrids of *Castanea*. Phytopathology 49:1125–1131

Grente J (1965) Les formes hypovirulentes d'Endothia parasitica et In espories de lutte contre le chancre du chataignier. C R Acad Agr Fr 51:1033–1037

Griffin GJ (1986) Chestnut blight and its control. Hortic Rev *8*:291–336. https://doi.org/10.1002/9781118060810.ch8

Halambek M, Novak Agbaba S (1989) Istraživanje mogućnosti suzbijanja raka kore pitomog kestena pomoću hipovirulentnih sojeva Endothia parasitica. Glasnik zaštite bilja 11:241–252

Han G (2019) Origin and evolution of the plant immune system. New Phytol 222:70–83. https://doi.org/10.1111/nph.15596

Hansen DR, Van Alfen NK, Gillies K et al (1985) Naked dsRNA associated with hypovirulence of *Endothia parasitica* is packaged in fungal vesicles. J Gen Virol 66:2605–2614. https://doi.org/10.1099/0022-1317-66-12-2605

Hao F, Ding T, Wu M et al (2018) Two novel hypovirulence-associated mycoviruses in the phytopathogenic fungus *Botrytis cinerea*: molecular characterization and suppression of infection cushion formation. Viruses 10. https://doi.org/10.3390/v10050254

Hebard F, Griffin G, Elkins J (1984) Developmental histopathology of cankers incited by hypovirulent and virulent isolates of *Endothia parasitica* on susceptible and resistant chestnut trees. Phytopathology 74:140–149

Heiniger U, Rigling D (1994) Biological control of chestnut blight in Europe. Annu Rev Phytopathol 32:581–599. https://doi.org/10.1146/annurev.py.32.090194.003053

Heiniger U, Rigling D (2009) Application of the Cryphonectria hypovirus 1 (CHV-1) to control the chestnut blight, experience from Switzerland. Acta Hortic 815:233–246. https://doi.org/10.17660/ActaHortic.2009.815.31

Heitman J, Howlett BJ, Crous PW et al (eds) (2017) The fungal kingdom. ASM Press, Washington, DC. https://doi.org/10.1128/9781555819583

Higashiura T, Katoh Y, Urayama S et al (2019) Magnaporthe oryzae chrysovirus 1 strain D confers growth inhibition to the host fungus and exhibits multiform viral structural proteins. Virology 535:241–254. https://doi.org/10.1016/j.virol.2019.07.014

Hillman BI, Suzuki N (2004) Viruses of the chestnut blight fungus, *Cryphonectria parasitica*. Adv Virus Res 63:423–472. https://doi.org/10.1016/S0065-3527(04)63007-7

Hillman BI, Tian Y, Bedker PJ et al (1992) A North American hypovirulent isolate of the chestnut blight fungus with European isolate-related dsRNA. J Gen Virol 73:681–686

Hoegger P, Heiniger U, Holdenrieder O et al (2003) Differential transfer and dissemination of hypovirus and nuclear and mitochondrial genomes of a hypovirus-infected *Cryphonectria parasitica* strain after introduction into a natural population. Appl Envion Microbiol 69:3767–3771. https://doi.org/10.1128/AEM.69.7.3767

Hollings M (1962) Viruses associated with a die-back disease of cultivated mushroom. Nature 196:962–965. https://doi.org/10.1038/196962a0

Hrabáková L, Koloniuk I, Petrzik K (2017) Phomopsis longicolla RNA virus 1—novel virus at the edge of myco- and plant viruses. Virology 506:14–18. https://doi.org/10.1016/j.virol.2017.03.003

Hu Z, Wu S, Cheng J et al (2014) Molecular characterization of two positive-strand RNA viruses co-infecting a hypovirulent strain of *Sclerotinia sclerotiorum*. Virology 464:450–459. https://doi.org/10.1016/j.virol.2014.07.007

Huang H (1996) Evaluation of Chinese chestnut cultivars for resistance to *Cryphonectria parasitica*. Plant Dis 80:45–47. https://doi.org/10.1094/PD-80-0045

Ihrmark K, Zheng J, Stenstrom E et al (2001) Presence of double-stranded RNA in *Heterobasidion annosum*. For Pathol 31:387–394. https://doi.org/10.1046/j.1439-0329.2001.00263.x

Jacometti MA, Wratten SD, Walter M (2010) Review: Alternatives to synthetic fungicides for *Botrytis cinerea* management in vineyards. Aust J Grape Wine Res 16:154–172. https://doi.org/10.1111/j.1755-0238.2009.0067.x

Ježić M, Krstin L, Poljak I et al (2014) *Castanea sativa*: genotype-dependent recovery from chestnut blight. Tree Genet Genomes 10:101–110. https://doi.org/10.1007/s11295-013-0667-z

Ježić M, Mlinarec J, Vuković R et al (2018) Changes in *Cryphonectria parasitica* populations affect natural biological control of chestnut blight. Phytopathology 108:870–877. https://doi.org/10.1094/PHYTO-07-17-0252-R

Ježić M, Kolp M, Prospero S et al (2019) Diversity of *Cryphonectria parasitica* in callused chestnut blight cankers on European and American chestnut. For Pathol 49(6):e12566. https://doi.org/10.1111/efp.12566

Kamaruzzaman M, He G, Wu M et al (2019) A novel partitivirus in the hypovirulent isolate qt5-19 of the plant pathogenic fungus *Botrytis cinerea*. Viruses 11(1):24. https://doi.org/10.3390/v11010024

Kanematsu S, Sasaki A, Onoue M et al (2010) Extending the fungal host range of a partitivirus and a mycoreovirus from *Rosellinia necatrix* by inoculation of protoplasts with virus particles. Phytopathology 100:922–930. https://doi.org/10.1094/PHYTO-100-9-0922

Katanić Z, Krstin L, Ježić M et al (2020) Identification and characterization of the causal agent of Dutch elm disease in Croatia. Eur J For Res 139:805–815. https://doi.org/10.1007/s10342-020-01286-4

Khalifa ME, Pearson MN (2014) Characterisation of a novel hypovirus from *Sclerotinia sclerotiorum* potentially representing a new genus within the Hypoviridae. Virology 464:441–449. https://doi.org/10.1016/j.virol.2014.07.005

Kim DH (1995) A new extracellular laccase of *Cryphonectha parasitica* Is revealed by deletion of lac1. Mol Plant Microbe Interact 8:259–266. https://doi.org/10.1094/MPMI-8-0259

Kim J-M, Park J-A, Kim D-H (2012) Comparative proteomic analysis of chestnut blight fungus, *Cryphonectria parasitica*, under tannic-acid-inducing and hypovirus-regulating conditions. Can J Microbiol 58:863–871. https://doi.org/10.1139/w2012-065

Köhl J, Kolnaar R, Ravensberg WJ (2019) Mode of action of microbial biological control agents against plant diseases: relevance beyond efficacy. Front Plant Sci 10:1–19. https://doi.org/10.3389/fpls.2019.00845

Koloniuk I, El-Habbak MH, Petrzik K, Ghabrial SA (2014) Complete genome sequence of a novel hypovirus infecting *Phomopsis longicolla*. Arch Virol 159:1861–1863. https://doi.org/10.1007/s00705-014-1992-8

Krstin L, Novak-Agbaba S, Rigling D et al (2008) Chestnut blight fungus in Croatia: diversity of vegetative compatibility types, mating types and genetic variability of associated Cryphonectria hypovirus 1. Plant Pathol 57:1086–1096. https://doi.org/10.1111/j.1365-3059.2008.01905.x

Krstin L, Novak-Agbaba S, Rigling D et al (2011) Diversity of vegetative compatibility types and mating types of *Cryphonectria parasitica* in Slovenia and occurrence of associated Cryphonectria hypovirus 1. Plant Pathol 60:752–761. https://doi.org/10.1111/j.1365-3059.2011.02438.x

Krstin L, Katanić Z, Ježić M et al (2017) Biological control of chestnut blight in Croatia: an interaction between host sweet chestnut, its pathogen *Cryphonectria parasitica* and the biocontrol agent Cryphonectria hypovirus 1. Pest Manag Sci 73:582–589. https://doi.org/10.1002/ps.4335

Kulshrestha S, Seth CA, Sharma M et al (2014) Biology and control of *Rosellinia necatrix* causing white root rot disease: a review. J Pure Appl Microbiol 8:1803–1814

Kwon SJ, Cho SY, Lee KM et al (2009) Proteomic analysis of fungal host factors differentially expressed by *Fusarium graminearum* infected with Fusarium graminearum virus-DK21. Virus Res 144:96–106. https://doi.org/10.1016/j.virusres.2009.04.004

Larson TG, Choi GH, Nuss DL (1992) Regulatory pathways governing modulation of fungal gene expression by a virulence-attenuating mycovirus. EMBO J 11:4539–4548. https://doi.org/10.1002/j.1460-2075.1992.tb05555.x

Lee SC, Ni M, Li W et al (2010) The evolution of sex: a perspective from the fungal kingdom. Microbiol Mol Biol Rev 74:298–340. https://doi.org/10.1128/MMBR.00005-10

Lee K-M, Cho WK, Yu J et al (2014) A comparison of transcriptional patterns and mycological phenotypes following infection of *Fusarium graminearum* by four mycoviruses. PLoS One 9 (6):e100989. https://doi.org/10.1371/journal.pone.0100989

Lemus-Minor CG, Cañizares MC, García-Pedrajas MD et al (2018) Fusarium oxysporum f. sp. dianthi virus 1 accumulation is correlated with changes in virulence and other phenotypic traits of its fungal host. Phytopathology 108:957–963. https://doi.org/10.1094/PHYTO-06-17-0200-R

Li P, Zhang H, Chen X et al (2015) Molecular characterization of a novel hypovirus from the plant pathogenic fungus *Fusarium graminearum*. Virology 481:151–160. https://doi.org/10.1016/j.virol.2015.02.047

Li H, Bian R, Liu Q et al (2019a) Identification of a novel hypovirulence-inducing hypovirus from *Alternaria alternata*. Front Microbiol 10:1–14. https://doi.org/10.3389/fmicb.2019.01076

Li P, Bhattacharjee P, Wang S et al (2019b) Mycoviruses in Fusarium species: an update. Front Cell Infect Microbiol 9:1–11. https://doi.org/10.3389/fcimb.2019.00257

Liao S, Li R, Shi L et al (2012) Functional analysis of an S-adenosylhomocysteine hydrolase homolog of chestnut blight fungus. FEMS Microbiol Lett 336:64–72. https://doi.org/10.1111/j.1574-6968.2012.02657.x

Lindberg GD (1960) Reduction in pathogenicity and toxin production in diseased *Helminthosporium victoriae*. Phytopathology 50:457–460

Linder-Basso D, Dynek JN, Hillman BI (2005) Genome analysis of *Cryphonectria hypovirus 4*, the most common hypovirus species in North America. Virology 337:192–203. https://doi.org/10.1016/j.virol.2005.03.038

Liu Y, Milgroom M (1996) Correlation between hypovirus transmission and the number of vegetative incompatibility (*vic*) genes different among isolates from a natural population of *Cryphonectria parasitica*. Phytopathology 86:79–86

Lübeck M (2004) Molecular characterization of *Rhizoctonia solani*. In: Arora DK, Khachatourians GG (eds) Applied mycology and biotechnology. Elsevier, Amsterdam, pp 205–244. https://doi.org/10.1016/S1874-5334(04)80011-X

MacDonald WL, Double ML (2005) Hypovirulence: use and limitations as a chestnut blight biological control. In: Steiner KC, Carlson JE (eds) Proceedings of conference on restoration of American chestnut to forest lands, pp 87–95

Mattioni C, Martin MA, Chiocchini F et al (2017) Landscape genetics structure of European sweet chestnut (*Castanea sativa* Mill): indications for conservation priorities. Tree Genet Genomes 13:39. https://doi.org/10.1007/s11295-017-1123-2

Meehan F, Murphy HC (1946) A new helminthosporium blight of oats. Science 104:413–414. https://doi.org/10.1126/science.104.2705.413

Milgroom MG, Cortesi P (2004) Biological control of chestnut blight with hypovirulence: a critical analysis. Annu Rev Phytopathol 42:311–338. https://doi.org/10.1146/annurev.phyto.42.040803.140325

Mlinarec J, Ježić M, Ćosić J et al (2018a) Multilocus PCR assay reveals high diversity of vegetative compatibility types in populations of *Cryphonectria parasitica* in Croatia. Plant Pathol 67:741–749. https://doi.org/10.1111/ppa.12751

Mlinarec J, Nuskern L, Ježić M et al (2018b) Molecular evolution and invasion pattern of Cryphonectria hypovirus 1 in Europe: mutation rate, and selection pressure differ between genome domains. Virology 514:156–164. https://doi.org/10.1016/j.virol.2017.11.011

Möller M, Stukenbrock EH (2017) Evolution and genome architecture in fungal plant pathogens. Nat Rev Microbiol 15:756–771. https://doi.org/10.1038/nrmicro.2017.76

Morris JL, Puttick MN, Clark JW et al (2018) The timescale of early land plant evolution. Proc Natl Acad Sci USA 115:E2274–E2283. https://doi.org/10.1073/pnas.1719588115

Muñoz-Adalia EJ, Fernández MM, Diez JJ (2016) The use of mycoviruses in the control of forest diseases. Biocontrol Sci Technol 26:577–604. https://doi.org/10.1080/09583157.2015.1135877

Negri S, Lovato A, Boscaini F et al (2017) The induction of noble rot (*Botrytis cinerea*) infection during postharvest withering changes the metabolome of grapevine berries (*Vitis vinifera* L., CV. Garganega). Front Plant Sci 8:1–12. https://doi.org/10.3389/fpls.2017.01002

Nerva L, Varese GC, Falk BW et al (2017) Mycoviruses of an endophytic fungus can replicate in plant cells: evolutionary implications. Sci Rep 7:1908. https://doi.org/10.1038/s41598-017-02017-3

Nuskern L, Tkalec M, Gačar M et al (2017a) Laccase activity change in Cryphonectria hypovirus 1-infected *Cryphonectria parasitica* is influenced by culture conditions, fungal isolate and virus strain. In: FEMS 2017 7th congress of European microbiologists, Valencia, 9–13 July 2017

Nuskern L, Tkalec M, Ježić M et al (2017b) Cryphonectria hypovirus 1-induced changes of stress enzyme activity in transfected phytopathogenic fungus *Cryphonectria parasitica*. Microb Ecol 74:302–311. https://doi.org/10.1007/s00248-017-0945-7

Nuskern L, Ježić M, Liber Z et al (2018) Cryphonectria hypovirus 1-induced epigenetic changes in infected phytopathogenic fungus *Cryphonectria parasitica*. Microb Ecol 75:790–798. https://doi.org/10.1007/s00248-017-1064-1

Nuss DL (2005) Hypovirulence: mycoviruses at the fungal–plant interface. Nat Rev Microbiol 3:632–642

Paoletti M (2016) Vegetative incompatibility in fungi: from recognition to cell death, whatever does the trick. Fungal Biol Rev 30:152–162. https://doi.org/10.1016/j.fbr.2016.08.002

Park SM, Choi ES, Kim MJ et al (2004) Characterization of HOG1 homologue, CpMK1 , from *Cryphonectria parasitica* and evidence for hypovirus-mediated perturbation of its phosphorylation in response to hypertonic stress. Mol Microbiol 51:1267–1277. https://doi.org/10.1046/j.1365-2958.2004.03919.x

Pearson MN, Beever RE, Boine B et al (2009) Mycoviruses of filamentous fungi and their relevance to plant pathology. Mol Plant Pathol 10:115–128. https://doi.org/10.1111/j.1364-3703.2008.00503.x

Peever TL, Liu YC, Milgroom MG (1997) Diversity of hypoviruses and other double-stranded RNAs in *Cryphonectria parasitica* in North America. Phytopathology 87:1026–1033. https://doi.org/10.1094/PHYTO.1997.87.10.1026

Peever TL, Liu YC, Wang K et al (1998) Incidence and diversity of double-stranded RNAs occurring in the chestnut blight fungus, *Cryphonectria parasitica*, in China and Japan. Phytopathology 88:811–817. https://doi.org/10.1094/PHYTO.1998.88.8.811

Peever TL, Liu Y-CY-C, Cortesi P et al (2000) Variation in tolerance and virulence in the chestnut blight fungus-hypovirus interaction. Appl Environ Microbiol 66:4863–4869. https://doi.org/10.1128/AEM.66.11.4863-4869.2000

Peters FS, Bußkamp J, Prospero S et al (2014) Genetic diversification of the chestnut blight fungus *Cryphonectria parasitica* and its associated hypovirus in Germany. Fungal Biol 118:193–210. https://doi.org/10.1016/j.funbio.2013.11.009

Poljak I, Idžojtić M, Šatović Z et al (2017) Genetic diversity of the sweet chestnut (*Castanea sativa* Mill.) in Central Europe and the Western part of the Balkan Peninsula and evidence of marron genotype introgression into wild populations. Tree Genet Genomes 13:18. https://doi.org/10.1007/s11295-017-1107-2

Potgieter CA, Castillo A, Castro M et al (2013) A wild-type *Botrytis cinerea* strain co-infected by double-stranded RNA mycoviruses presents hypovirulence-associated traits. Virol J 10:1–9. https://doi.org/10.1186/1743-422X-10-220

Prospero S, Rigling D (2013) Chestnut blight. In: Gonthier P, Nicolotti G (eds) Infectious forest diseases. CABI, Wallingford, pp 318–339

Prospero S, Rigling D (2016) Using molecular markers to assess the establishment and spread of a mycovirus applied as a biological control agent against chestnut blight. BioControl 61:313–323. https://doi.org/10.1007/s10526-015-9713-0

Prospero S, Conedera M, Heiniger U et al (2006) Saprophytic activity and sporulation of *Cryphonectria parasitica* on dead chestnut wood in forests with naturally established hypovirulence. Phytopathology 96(12):1337–1344

Rigling D, Prospero S (2018) *Cryphonectria parasitica*, the causal agent of chestnut blight: invasion history, population biology and disease control. Mol Plant Pathol 19:7–20. https://doi.org/10.1111/mpp.12542

Rigling D, Van Alfen NK (1991) Regulation of laccase biosynthesis of the plant pathogenic fungus *Cryphonectria parasitica* by double-stranded RNA. J Bacteriol 173:8000–8003

Rigling D, Heiniger U, Hohl HR (1989) Reduction of laccase activity in dsRNA-containing hypovirulent strains of *Cryphonectria* (*Endothia*) *parasitica*. Phytopathology 79:219–223

Rigling D, Borst N, Cornejo C et al (2018) Genetic and phenotypic characterization of *Cryphonectria hypovirus 1* from Eurasian Georgia. Viruses 10:687. https://doi.org/10.3390/v10120687

Robin C, Anziani C, Cortesi P (2000) Relationship between biological control, incidence of hypovirulence, and diversity of vegetative compatibility types of *Cryphonectria parasitica* in France. Phytopathology 90:730–737. https://doi.org/10.1094/PHYTO.2000.90.7.730

Robin C, Capdevielle X, Martin M et al (2009) *Cryphonectria parasitica* vegetative compatibility type analysis of populations in South-Western France and Northern Spain. Plant Pathol 58:527–535. https://doi.org/10.1111/j.1365-3059.2008.01986.x

Robin C, Lanz S, Soutrenon A et al (2010) Dominance of natural over released biological control agents of the chestnut blight fungus *Cryphonectria parasitica* in south-eastern France is associated with fitness-related traits. Biol Control 53:55–61. https://doi.org/10.1016/j.biocontrol.2009.10.013

Roldão A, Silva AC, Mellado MCM et al (2017) Viruses and virus-like particles in biotechnology: fundamentals and applications. Compr Biotechnol 633–656. https://doi.org/10.1016/B978-0-12-809633-8.09046-4

Roossinck MJ (2019) Evolutionary and ecological links between plant and fungal viruses. New Phytol 221:86–92. https://doi.org/10.1111/nph.15364

Russell EWB (1987) Pre-blight distribution of *Castanea dentata* (Marsh.) Borkh. Bull Torrey Bot Club 114:183–190. https://doi.org/10.2307/2996129

Santini A, Faccoli M (2014) Dutch elm disease and elm bark beetles: a century of association. IForest 8:126–134. https://doi.org/10.3832/ifor1231-008

Saupe SJ (2000) Molecular genetics of heterokaryon incompatibility in filamentous ascomycetes. Microbiol Mol Biol Rev 64:489–502. https://doi.org/10.1128/MMBR.64.3.489-502.2000

Segers GC, Van Wezel R, Zhang X et al (2006) Hypovirus papain-like protease p29 suppresses RNA silencing in the natural fungal host and in a heterologous plant system. Eukaryot Cell 5:896–904. https://doi.org/10.1128/EC.00373-05

Segers GC, Zhang X, Deng F et al (2007) Evidence that RNA silencing functions as an antiviral defense mechanism in fungi. Proc Natl Acad Sci USA 104:12902–12906. https://doi.org/10.1073/pnas.0702500104

Shapira R, Choi GH, Nuss DL (1991) Virus-like genetic organization and expression strategy for a double-stranded RNA genetic element associated with biological control of chestnut blight. EMBO J 10:731–739

Short DPG, Double M, Nuss DL et al (2015) Multilocus PCR assays elucidate vegetative incompatibility gene profiles of *Cryphonectria parasitica* in the United States. Appl Environ Microbiol 81:5736–5742. https://doi.org/10.1128/AEM.00926-15

Smart CD, Yuan W, Foglia R et al (1999) *Cryphonectria hypovirus 3*, a virus species in the family hypoviridae with a single open reading frame. Virology 265:66–73. https://doi.org/10.1006/viro.1999.0039

Son M, Yu J, Kim K-H (2015) Five questions about mycoviruses. PLOS Pathog 11:e1005172. https://doi.org/10.1371/journal.ppat.1005172

Sotirovski K, Papazova-Anakieva I, Grünwald N et al (2004) Low diversity of vegetative compatibility types and mating type of *Cryphonectria parasitica* in the southern Balkans. Plant Pathol 53:325–333. https://doi.org/10.1111/j.1365-3059.2004.01006.x

Stauder CM, Nuss DL, Zhang DX et al (2019) Enhanced hypovirus transmission by engineered super donor strains of the chestnut blight fungus, *Cryphonectria parasitica*, into a natural population of strains exhibiting diverse vegetative compatibility genotypes. Virology 528:1–6. https://doi.org/10.1016/j.virol.2018.12.007

Sutherland ML, Brasier CM (1995) Effect of d-factors on in vitro cerato-ulmin production by the Dutch elm disease pathogen *Ophiostoma novo-ulmi*. Mycol Res 99:1211–1217. https://doi.org/10.1016/S0953-7562(09)80281-5

Sutherland ML, Brasier CM (1997) A comparison of thirteen d-factors as potential biological control agents of *Ophiostoma novo-ulmi*. Plant Pathol 46:680–693. https://doi.org/10.1046/j.1365-3059.1997.d01-62.x

Suzuki N, Nuss DL (2002) Contribution of protein p40 to hypovirus-mediated modulation of fungal host phenotype and viral RNA accumulation. J Virol 76:7747–7759. https://doi.org/10.1128/JVI.76.15.7747-7759.2002

Suzuki N, Supyani S, Maruyama K et al (2004) Complete genome sequence of Mycoreovirus-1/Cp9B21, a member of a novel genus within the family Reoviridae, isolated from the chestnut blight fungus *Cryphonectria parasitica*. J Gen Virol 85:3437–3448. https://doi.org/10.1099/vir.0.80293-0

Suzuki N, Ghabrial SA, Kim K-H et al (2018) ICTV Virus taxonomy profile: Hypoviridae. J Gen Virol 99:615–616. https://doi.org/10.1099/jgv.0.001055

Suzuki N, Cornejo C, Aulia A et al (2021) In-tree behavior of diverse viruses harbored in the chestnut blight fungus, *Cryphonectria parasitica*. J Virol 95(6):e01962-20 (18 pp.). https://doi.org/10.1128/JVI.01962-20

Tartaglia J, Paul CP, Fulbright DW et al (1986) Structural properties of double-stranded RNAs associated with biological control of chestnut blight fungus. Proc Natl Acad Sci USA 83:9109–9113. https://doi.org/10.1073/pnas.83.23.9109

Trapiello E, Rigling D, González AJ (2017) Occurrence of hypovirus-infected *Cryphonectria parasitica* isolates in northern Spain: an encouraging situation for biological control of chestnut blight in Asturian forests. Eur J Plant Pathol 149:503–514. https://doi.org/10.1007/s10658-017-1199-4

Troncoso-Rojas R, Tiznado-Hernández ME (2014) *Alternaria alternata* (black rot, black spot). In: Bautista-Baños S (ed) Postharvest decay. Elsevier, London, pp 147–187

Urayama S, Kato S, Suzuki Y (2010) Mycoviruses related to Chrysovirus affect vegetative growth in the rice blast fungus *Magnaporthe oryzae*. J Gen Virol 91(12):3085–3094. https://doi.org/10.1099/vir.0.025411-0

Vainio EJ, Hantula J (2016) Taxonomy, biogeography and importance of Heterobasidion viruses. Virus Res 219:2–10. https://doi.org/10.1016/j.virusres.2015.10.014

Vainio EJ, Jurvansuu J, Hyder R et al (2017) Heterobasidion partitivirus 13 mediates severe growth debilitation and major alterations in the gene expression of a fungal forest pathogen. J Virol 92:1–15. https://doi.org/10.1128/jvi.01744-17

van der Heijden MGA, Martin FM, Selosse MA et al (2015) Mycorrhizal ecology and evolution: the past, the present, and the future. New Phytol 205:1406–1423. https://doi.org/10.1111/nph.13288

Van De Sande WWJ, Lo-Ten-Foe JR, Van Belkum A et al (2010) Mycoviruses: future therapeutic agents of invasive fungal infections in humans? Eur J Clin Microbiol Infect Dis 29:755–763. https://doi.org/10.1007/s10096-010-0946-7

Varley DA, Podila GK, Hiremath ST (1992) Cutinase in *Cryphonectria parasitica*, the chestnut blight fungus: suppression of cutinase gene expression in isogenic hypovirulent strains containing double-stranded RNAs. Mol Cell Biol 12:4539–4544. https://doi.org/10.1128/MCB.12.10.4539

Wang J, Shi L, He X et al (2016) Comparative secretome analysis reveals perturbation of host secretion pathways by a hypovirus. Sci Rep 6:34308. https://doi.org/10.1038/srep34308

Willey J, Sherwood L, Woolverton C (2010) Prescott's microbiology. McGraw Hill, New York

Williamson B, Tudzynski B, Tudzynski P et al (2007) *Botrytis cinerea*: the cause of grey mould disease. Mol Plant Pathol 8:561–580. https://doi.org/10.1111/j.1364-3703.2007.00417.x

Wu M, Zhang L, Li G et al (2010) Genome characterization of a debilitation-associated mitovirus infecting the phytopathogenic fungus *Botrytis cinerea*. Virology 406:117–126. https://doi.org/10.1016/j.virol.2010.07.010

Wu M, Jin F, Zhang J et al (2012) Characterization of a novel bipartite double-stranded RNA mycovirus conferring hypovirulence in the phytopathogenic fungus *Botrytis porri*. J Virol 86:6605–6619. https://doi.org/10.1128/jvi.00292-12

Wu R, Yang Y, Duan X et al (2020) Four distinct isolates of Helminthosporium victoriae virus 190S identified from *Bipolaris maydis*. Virus Res 285. https://doi.org/10.1016/j.virusres.2020.197941

Xiao X, Cheng J, Tang J et al (2014) A novel partitivirus that confers hypovirulence on plant pathogenic fungi. J Virol 88(17):10120–10133. https://doi.org/10.1128/JVI.01036-14

Xie J, Jiang D (2014) New insights into mycoviruses and exploration for the biological control of crop fungal diseases. Annu Rev Phytopathol 52:45–68. https://doi.org/10.1146/annurev-phyto-102313-050222

Xie J, Havens WM, Lin Y-H et al (2016) The victorivirus helminthosporium victoriae virus 190S is the primary cause of disease/hypovirulence in its natural host and a heterologous host. Virus Res 213:238–245. https://doi.org/10.1016/j.virusres.2015.12.011

Xiong Q, Zhang L, Waletich J et al (2019) Characterization of the papain-like protease p29 of the hypovirus CHV1-CN280 in its natural host fungus *Cryphonectria parasitica* and nonhost fungus *Magnaporthe oryzae*. Phytopathology 109:736–747. https://doi.org/10.1094/PHYTO-08-18-0318-R

Xu Z, Wu S, Liu L et al (2015) A mitovirus related to plant mitochondrial gene confers hypovirulence on the phytopathogenic fungus *Sclerotinia sclerotiorum*. Virus Res 197:127–136. https://doi.org/10.1016/j.virusres.2014.12.023

Yaegashi H, Kanematsu S, Ito T (2012) Molecular characterization of a new hypovirus infecting a phytopathogenic fungus, *Valsa ceratosperma*. Virus Res 165:143–150. https://doi.org/10.1016/j.virusres.2012.02.008

Yu X, Li B, Fu Y et al (2010) A geminivirus-related DNA mycovirus that confers hypovirulence to a plant pathogenic fungus. Proc Natl Acad Sci USA 107(18):8387–8392. https://doi.org/10.1073/pnas.0913535107

Yu X, Li B, Fu Y et al (2013) Extracellular transmission of a DNA mycovirus and its use as a natural fungicide. Proc Natl Acad Sci USA 110(4):1452–1457. https://doi.org/10.1073/pnas.1213755110

Yu L, Sang W, Wu M-D et al (2015) Novel hypovirulence-associated RNA mycovirus in the plant-pathogenic fungus *Botrytis cinerea*: molecular and biological characterization. Appl Environ Microbiol 81:2299–2310. https://doi.org/10.1128/AEM.03992-14

Zamora P, Martín AB, Arrate J et al (2008) Detection of the vegetative compatibility groups of *Cryphonectria parasitica* in Castilla Y Leon region. Acta Hortic 159–162. https://doi.org/10.17660/ActaHortic.2008.784.24

Zhang D-X, Nuss DL (2016) Engineering super mycovirus donor strains of chestnut blight fungus by systematic disruption of multilocus *vic* genes. Proc Natl Acad Sci USA 113:2062–2067. https://doi.org/10.1073/pnas.1522219113

Zhang J, Li GQ, Jiang DH (2009) First report of garlic leaf blight caused by *Botrytis porri* in China. Plant Dis 93:1216–1216. https://doi.org/10.1094/PDIS-93-11-1216B

Zhang X, Shi D, Nuss DL (2012) Variations in hypovirus interactions with the fungal-host RNA-silencing antiviral-defense response. J Virol 86:12933–12939. https://doi.org/10.1128/JVI.00961-12

Zhang H, Zheng X, Zhang Z (2016) The *Magnaporthe grisea* species complex and plant pathogenesis. Mol Plant Pathol 17:796–804. https://doi.org/10.1111/mpp.12342

Zheng L, Shu C, Zhang M et al (2019) Molecular characterization of a novel *Endornavirus* conferring hypovirulence in rice sheath blight fungus *Rhizoctonia solani* AG-1 IA Strain GD-2. Viruses 11:178. https://doi.org/10.3390/v110201781293

Part III
Defending the Health of Its Hosts

Chapter 8
The Contribution of Viruses to Immune Systems

Felix Broecker

Abstract Cellular organisms have evolved a plethora of immune systems to defend against selfish genetic elements and pathogens, including viruses. Virus-derived and virus-related sequences (such as those of mobile genetic elements) constitute a substantial portion of the genomes of all life forms. Some of these sequences mediate resistance to viruses or virus-like parasites and are integral components of many immune systems acting against various invaders (viral and cellular), such as the Rag1/2 system of vertebrates, which generates antibody and T cell receptor diversity. Recently, intimate evolutionary relationships between viral and virus-like sequences and various cellular immune pathways of both pro- and eukaryotes have been uncovered. Here, I argue that the most basic—and likely evolutionarily the first—immune system may be the superinfection exclusion (SIE) mechanism, a phenomenon where one parasitic element (a virus or virus-like entity) prevents or restricts invasion of a compartment (e.g., a cell) by another parasitic element. The SIE mechanism is still a feature of many extant viruses and the related viroids, which are putative relics of an ancient RNA world that existed before the emergence of cells. During cellular evolution, various more complex immune mechanisms fully or partially derived from viruses and virus-like element have evolved. In this chapter, I summarize the current knowledge on the contribution of viral and virus-like elements to the evolution of various cellular immune systems. The emerging picture is that many of today's cellular immune systems have evolved from simple SIE mechanisms to highly complex defense strategies, frequently involving viral or virus-like sequences.

8.1 Introduction

The idea that immune systems may have evolved from viral sequences is not new, and to my knowledge there are two pioneering scientists who mainly brought forward this idea. First, Eugene V. Koonin, who (amongst many other topics) studies

F. Broecker (✉)
Idorsia Pharmaceuticals Ltd., Allschwil, Switzerland

C. J. Hurst (ed.), *The Biological Role of a Virus*, Advances in Environmental Microbiology 9, https://doi.org/10.1007/978-3-030-85395-2_8

the evolution of viruses and virus-like elements, reported in 2008 that eukaryotic RNA interference (RNAi), a type of immune system that likely originated as an antiviral defense mechanism, "seems to have been pieced together from ancestral archaeal, bacterial and *phage* proteins" (my emphasis) (Shabalina and Koonin 2008), whereby phages are the viruses infecting prokaryotes. Many more scientific articles by Koonin's group on the origin of immune systems from viral sequences have followed. Second, evolutionary scientist Luis P. Villarreal wrote, in 2011, a seminal article on 'Viral Ancestors of Antiviral Systems' (Villarreal 2011). The concepts outlined in this chapter are mainly based on Koonin's and Villarreal's ideas, and I complement these with my own ideas wherever possible. The person who initially stimulated my interest in this question, however, is virologist Karin Moelling, who already in 2006 recognized distinct similarities between the enzymes of the RNAi machinery and those involved in retroviral replication, suggesting a common evolutionary origin (Moelling et al. 2006). Finally, I wish to mention two more names, Emmanuelle Charpentier and Jennifer Doudna, who have been awarded the Nobel Prize in Chemistry "for the development of a method for genome editing" in 2020. This genome editing method is CRISPR-Cas, nowadays widely used for genetic manipulation of all kinds of genomes in laboratories worldwide. CRISPR-Cas, however, is not only a very useful tool in molecular biology, but also an evolutionarily ancient immune system of prokaryotes that acts against genetic parasites such as phages and plasmids. The reason why I mention CRISPR-Cas here is because its evolution involved various virus-like sequences (Koonin and Makarova 2017, explained in more detail in Sect. 8.6.2). Briefly, CRISPR-Cas could only evolve because of the existence of a virus-related genetic parasite, a transposon called *casposon* (Krupovic et al. 2014). CRISPR-Cas is therefore a perfect example of how virus-like sequences have been coopted by a host cell to exert anti-viral functions. But before going into more details, I like to provide a definition of a virus, which brings us to the idea of the *Greater Virus World.*

8.2 The Greater Virus World

What is a virus? Conventionally, it is defined as an infectious agent that is an obligate intracellular parasite. According to this textbook definition, a virus exists in two different states, an extracellular one with surface structures such as capsid or envelope proteins that allows it to move between organisms or cells, and an intracellular one during which new viruses are produced with the help of the cellular protein synthesis machinery. This classical definition has recently been challenged by sequence analyses of viruses and their evolutionary relatives, the mobile genetic elements (MGEs). First, there is evidence that viruses (or their evolutionary ancestors) were already diverse at the time of the last universal cellular ancestor (LUCA), with extensive evolution (likely from RNA to DNA viruses or virus-like entities) even before the existence of LUCA (Krupovic et al. 2020). This means that the ancestors of viruses have not originated as cellular parasites, since they may have predated cells during evolution. Second, MGEs (which include (retro)transposons, plasmids and viroids) and *bona fide* viruses (cell-infecting genetic parasites with an intracellular and an extracellular state) share a number of hallmark genes, e.g., those involved in replication, which suggests that viruses have evolved from MGEs and vice versa multiple times throughout evolution (Koonin and Dolja 2014). 'The Greater Virus World', a term coined by Koonin and his colleague Valerian Dolja, thus includes both bona fide viruses and MGEs. In this article, I will adopt this broad definition of a virus and highlight the contribution of both, bona fide viruses and MGEs, to the evolution of the various immune systems.

8.3 Viruses as Drivers of Evolution

Parasite-host coevolution is a major aspect of the evolution of all (cellular and pre-cellular) life (Koonin 2016). Of note, the genetic diversity of the virosphere, which is the entirety of the viruses on Earth, is substantially higher than that of cellular life forms, and viral evolution typically occurs at much faster pace, especially in the case of RNA viruses that have limited proofreading mechanisms during replication of their genomes (Paez-Espino et al. 2016; Koonin and Dolja 2013). Consequently, large numbers of novel viruses are currently being detected by high-throughput sequencing (Wolf et al. 2020). Viruses frequently hijack cellular genes and vice versa, i.e., there is constant genetic exchange. Rather than simply being understood as solely disease-causing or detrimental agents, viruses and other parasitic MGEs are increasingly being recognized as entities that can provide benefits to the host, e.g., by protecting from superinfection and through exaptation of genetic material of the viral parasite for host functions (Koonin 2016; Broecker and Moelling 2019a). Moreover, a large portion of adaptations of cellular proteins are driven by the action of viruses; e.g., in an estimated 30% of all conserved mammalian proteins (Enard et al. 2016). All known life forms harbor such genetic parasites, and sequences originating from viruses and MGEs constitute large fractions of cellular genomes, up to 90% in some plant species and up to two thirds of the human genome (Koonin 2016; de Koning et al. 2011). Originally dismissed as 'junk DNA', it is now well-established that these sequences are frequently transcribed, provide promoters, enhancers, polyadenylation and splice sites and are thereby substantially involved in host gene regulation (Gogvadze and Buzdin 2009). Moreover, they contribute to the formation of new genes, either directly (e.g., the syncytin genes originating from retroviruses which will be described in Sect. 8.5.1) or indirectly through pseudogene formation mediated by retroelements, whose replication machinery can reverse-transcribe mRNAs of cellular genes and then re-insert the DNA copies into the genome. Some of the viral genes are involved in the various immune systems that have evolved in cellular life forms.

8.4 Immune Systems: An Evolutionary Perspective

The following sections contain a noncomprehensive overview of the involvement of viruses and virus-like elements in the evolution of various immune systems. I will start with an attempt to define the term *immune system*.

8.4.1 What Is an Immune System?

The emergence of identity, i.e., the ability of an entity (e.g., a cell) to discriminate self from non-self, has likely been a crucial step in the evolution from an inanimate predecessor world to the living world in which we exist (Villarreal and Witzany 2013). The immunological self/non-self model, proposed by Frank Macfarlane Burnet in 1949, applies the concept of identity to the immune system, stating that any foreign (*non-self*) element triggers an immune reaction of an organism, whereas any component of the organism itself (*self*) does not (Burnet and Fenner 1949). This is, based on today's knowledge, an oversimplification, since many self-structures are recognized and eliminated by the mammalian immune system (e.g., dead cells are eliminated by phagocytes, and (pre)tumor cells are frequently recognized and eliminated as well) and many non-self-structures are tolerated, e.g., the approximately 4×10^{13} bacterial cells that reside as commensals, with many beneficial functions in digestion and immune function, in the intestinal tract of every human being (Pradeu and Carosella 2006). Viruses are also frequently tolerated by the human immune system, including *bona fide* viruses such as herpesviruses (which in immunocompetent individuals are mostly symptomless), as well as the about 100,000 endogenous retroviruses (ERVs) or fractions thereof in the human genome that could be considered as foreign sequences. The ERVs are typically benign or beneficial, however, they can be abnormally activated in and thus potentially contribute to certain disease states, including cancer, and have consequently been described as 'the enemy within' (Wilkins 2010). These ERVs originated from germline cells infected with *bona fide* retroviruses, mostly genomic introductions that occurred millions of years ago, and these infected germline cells should have been eliminated by the immune system according to the self/non-self model (a virus-infected cell is typically eliminated by the immune system as it presents virus-encoded non-self structures on the surface). However instead, the genetic information of many ERVs has been fixed in the genomes of many species, including humans. The complex immune systems of the various species on earth, pro- and eukaryotic, single- and multicellular, in many cases consequently do not merely distinguish self from non-self, but rather harmless (or even useful) from harmful, which is a much more elaborate process. Moreover, the involvement of microbes in immune systems is increasingly becoming recognized, i.e., the ability of the host to "manage and exploit beneficial microbes to fend off nasty ones" (Travis 2009), which is the subject of this chapter when 'microbes' are specified to 'viruses'.

8.4.2 A Simple Immune System Based on RNA?

It is possible to design RNA molecules that can cleave other RNA molecules in a sequence-specific manner. This type of catalytic RNA is a so-called hammerhead ribozyme that forms Watson-Crick basepairs with the target RNA and then cleaves a specific phosphodiester bond of the target RNA. This idea is being investigated as a potential therapeutic approach against autoimmune diseases and cancer (Citti and Rainaldi 2005). The artificial ribozyme could constitute an immune system as defined in the previous paragraph, if the target RNA is a harmful one, e.g., a parasitic RNA. The ribozyme does not "blindly" discriminate self from non-self. It specifically eliminates those RNAs that have sufficient complementarity and tolerates all other RNAs. It can distinguish harmful from harmless, if the information (harmful vs. harmless) is encoded in the RNA sequence of the parasite.

In nature, hammerhead ribozymes are found in viroids, which are virus-related, protein-free infectious agents consisting of highly structured, circular non-coding RNA. Viroids are possible remnants of the ancient RNA world thought to have existed before the evolution of DNA or proteins (Diener 1989; Flores et al. 2014). In the ancient RNA world, a primordial RNA-based immune system could have been constituted by a viroid that eliminates another viroid via ribozymatic cleavage *in trans* (Table 8.1). Although known natural viroids are generally self-cleaving, they can be modified relatively easily to yield *trans*-cleaving derivatives (Jimenez et al. 2015), suggesting that *trans*-cleaving ribozymes may have existed or may still exist naturally. However, this example is merely a molecule acting against other molecules. If we add a cell-like structure or compartment, perhaps an early primordial cell in the RNA world, such a *trans* cleaving ribozyme might be beneficial to that cell by protecting against other, parasitic RNAs. (As a side note, parasites inevitably occur within evolving life forms (Koonin et al. 2017). No life exists without parasites, and even parasites often have parasites—there will be an example of a virus infecting another virus below in Sect. 8.6.1). Thus, the cell or compartment might benefit from hosting such a parasite-cleaving RNA. The parasite-cleaving RNA might itself be a parasite of the cell (i.e., a viroid-like structure), which however may be tolerated since its presence provides a net benefit to the cell. This hypothetical primordial immune system highlights an important phenomenon that may be at the origin of various immune systems; superinfection exclusion (SIE). This phenomenon is defined as an infection by a virus or viroid that protects against superinfection by the same or a different virus or viroid (Ziebell and Carr 2010). (I designated Sect. 8.5 below for a more detailed discussion of SIE.) If the first virus or viroid is asymptomatic or causing mild symptoms only but protects against superinfection by a more virulent virus or viroid then there is a net benefit for the cell. In plants, SIE has been reported for both *bona fide* viruses as well as viroids. Coming back to the abovementioned RNA-based immunity against RNA infectious agents in the present-day world: The mechanism of action of SIE by extant hammerhead viroids is likely not based on RNA cleavage (since natural hammerhead ribozymes can only cleave *in cis*) but on RNA silencing mechanisms provided by the host cell

Table 8.1 Examples of immune systems whose evolution involved viruses or virus-like elements including: ERVs (endogenous retroviruses); piRNA (Piwi-interacting RNA); SIE (superinfection exclusion); and siRNA (small interfering RNA)

Immune system	Organisms	Function	Involvement of viruses/virus-like sequences
Viroid/ribozyme-based immunity (speculative)	Pre-cellular life or early cells (RNA world?)	A catalytic RNA destroys other (potentially parasitic) RNAs in a sequence-dependent manner	Known ribozymes are self-cleaving but can be modified easily to *trans*-cleaving ones that may exist or may have existed naturally (Jimenez et al. 2015)
Restriction-modification	Prokaryotes	The host genome is methylated at a specific sequence, a restriction endonuclease cuts unmethylated (invading) DNA of that sequence	Phages and MGEs mediate horizontal gene transfer of restriction-modification systems (Furuta et al. 2010; Murphy et al. 2013)
Prophages	Prokaryotes	SIE mediated by prophages or prophage-derived genes against exogenous phages	Prophages or prophage-derived genes mediate resistance to phage infection by various mechanisms (Bondy-Denomy et al. 2016)
CRISPR-Cas	Prokaryotes	Adaptive immune system of prokaryotes; a piece of invading genetic information is integrated into the host genome and is used to guide nucleases to invaders with homologous sequences	At least four mobile genetic elements contributed to the evolution of CRISPR-Cas systems (Koonin and Makarova 2017). The transposon called casposon is the origin of all CRISPR-Cas systems
Argonaute-based targeting	Prokaryotes	Genomes of invaders are cleaved into small fragments that are used to guide Argonaute nucleases to invaders with sequence complementarity, inducing their degradation	Argonaute proteins and the retroviral reverse transcriptase-RNase H proteins share structural and functional properties, suggesting a common evolutionary origin (Moelling et al. 2006)
piRNA	Eukaryotes	piRNAs are transcribed from piRNA clusters and guide Argonaute proteins to invaders with sequence complementarity that are consequently degraded	Most piRNA sequences originate from transposons and mediate transposon suppression, especially in germline cells (Iwasaki et al. 2015). Some piRNA sequences originate from viruses with putative anti-viral activity, mainly in insects (Ophinni et al. 2019)
RNA interference	Eukaryotes	RNA interference has likely evolved as an antiviral mechanism (Shabalina and Koonin 2008). It acts by processing	The RNA interference and retroviral replication machineries share structural and functional homologies that

(continued)

Table 8.1 (continued)

Immune system	Organisms	Function	Involvement of viruses/virus-like sequences
		invading RNAs into small RNAs (siRNAs) that then guide the RNA-induced silencing complex to invading RNAs with sequence complementarity, which induces the degradation of the invading RNA	suggest a common evolutionary origin (Moelling et al. 2006). The 'endo-siRNA' pathway uses siRNAs from sense-antisense pairs of RNAs from transposons to suppress transposon activity in mouse brain and embryonic stem cells (Nandi et al. 2016; Berrens et al. 2017)
Endogenous retroviruses	Eukaryotes	Genes derived from endogenous retroviruses mediate SIE against exogenous retroviruses	Various examples of coopted retroviral *env* and *gag* genes with antiviral activity have been described (see Sects. 8.5.1 and 8.5.2)
Interferon system	Eukaryotes	The interferon system is part of the antiviral innate immune system of animals (metazoans)	ERVs have been coopted to provide transcription factor binding sites to various interferon-stimulated genes and thereby are involved in regulating antiviral responses (Chuong et al. 2016; Ito et al. 2017)
Antibodies and T cell receptors	Eukaryotes	In jawed vertebrates, antibodies and T-cell receptors are the basis of adaptive immunity	The Rag1/2 system that generates the diversity of antibodies and T-cell receptors as well as the recognition sites for recombination originate from a *Transib* transposon (Kapitonov and Koonin 2015; Huang et al. 2016)
Mucosal immunity	Eukaryotes	In animals (metazoans), phages adhering to mucosal surfaces mediate immunity against bacterial infections	The mucus layers of metazoans and bacteriophages may have co-evolved such that phages adhere to mucus via immunoglobulin-like proteins and thus provide a protective barrier against invading bacteria (Barr et al. 2013)

(Kovalskaya and Hammond 2014). However, it is possible to express hammerhead ribozymes targeting a pathogenic viroid to "immunize" plant cells against disease. Thus, the hypothetical primordial immune system described above is not as far-fetched and might exist or have existed in nature.

8.4.3 Innate and Adaptive Immunity

The term "immune system" commonly refers to the eukaryotic, or more specific, the mammalian immune system (prokaryotic immune systems will be described below in Sects. 8.5.7 and 8.6.2). The mammalian immune system can be subdivided into two arms, the innate and the adaptive immune system. Examples for the former include the well-described toll-like receptors (TLRs) which are a group of pattern-recognition receptors (PRRs) that recognize microbial/viral/fungal structures such as lipopolysaccharide, double-stranded RNA, zymosan, etc. (Mahla et al. 2013). From an evolutionary perspective, the TLR protein family is over 700 million years old and found throughout the eumetazoan clade (all multicellular organisms except sponges and placozoa, the simplest known animals), a group which includes diverse animals such as squids, jellyfish, mammals, annelids, as well as insects, to name a few (Leulier and Lemaitre 2008). These types of receptors have been first described in the model organism *Drosophila melanogaster* (fruit fly) in 1985 (Anderson et al. 1985a, b), by the group of Nobel Prize Awardee Christiane Nüsslein-Volhard. (As a side note, the toll receptors of *D. melanogaster* are mainly involved in embryonic development and not in immunity and have thus evolved functional divergence in different species (Kambris et al. 2002)). Other PRRs include RIG-I-like receptors (RLRs), NOD-like receptors (NLRs) and C-type lectin receptors (CLRs) (Mahla et al. 2013). The activation of PRRs leads to complex antimicrobial responses. Like TLRs, RLRs, NLRs and CLRs are found in both invertebrate and vertebrate genomes; homologs of NLRs can also be found in plants, suggesting an even earlier evolutionary origin (Lange et al. 2011; Zou et al. 2009; Sattler et al. 2012; Jones et al. 2016). Another type of innate immune system that has likely evolved as a defense against viruses and MGEs is RNA interference (RNAi), which acts by small interfering RNAs (siRNAs) and is found not only in animals but also in plants and fungi and is therefore evolutionarily more ancient than PRR-based systems (Obbard et al. 2009; Shabalina and Koonin 2008; Ge and Zamore 2013). In animals, a variation of RNAi, using PIWI-interacting RNAs (piRNAs) is used to silence transposons in germline cells to ensure fertility (Ge and Zamore 2013).

Several hundred million years after the emergence of the abovementioned innate immune systems passed until about 450–500 million years ago a new type of immune system evolved, the adaptive immune system. Remarkably, comparable adaptive immune systems evolved in parallel at least twice, in jawless vertebrates (agnathans) as well as in jawed vertebrates (gnathostomes). The prerequisites for such an adaptive immune system are (1) a molecular machinery that allows for the rearrangement of germline-encoded antigen-receptor genes and (2) a dedicated repertoire of cells, each of which expresses a different antigen receptor, e.g., B cells expressing specific immunoglobulins (Igs) (Bayne 2003). The recombination machineries involved in diversifying the antigen-receptors in vertebrates allow for the generation of theoretically over 10^{14} receptors with different specificities, a number which is impossible to be encoded in any genome (e.g., mammalian genomes encode only about 20,000 genes). In jawed vertebrates, these diverse receptors are Igs expressed on B cells and T cell receptors (TCRs) expressed on T cells. Recognition of a specific structure of a pathogen by an antigen-receptor triggers clonal amplification of that cell, its differentiation and, in the case of B cells, production of antibodies with the same antigen binding specificity (Cooper and Alder 2006). As will be discussed below the mechanism that diversifies Igs and TCRs, V(D)J recombination, relies on an ancient transposon that allows for the key step to occur, genomic recombination. Agnathans have immune cells that share similarities with B and T cells but antigen receptor diversity (which is in the same theoretical order of magnitude as the one achieved by V(D)J recombination) is not generated by recombination, but instead by gene conversion (Boehm et al. 2012).

8.5 Viruses Against Viruses: Superinfection Exclusion, a Simple Immune System

Superinfection exclusion (SIE) is the ability of a first viral infection to restrict secondary viral infections of the same cell. In most cases, SIE prevents infections by the same virus that caused the first infection, or closely related ones, but in some cases (e.g., in the case of the virophages described below in Sect. 8.6.1) also restricts infection by genetically non-related viruses. Importantly, SIE can be regarded as a simple type of immune system. This phenomenon was discovered in the 1920s in tobacco. Tobacco plants infected with a non-virulent variant of the Tobacco mosaic virus (TMV) were shown to be protected against a more virulent TMV isolate (McKinney 1929). This example highlights that the host (here, the tobacco plant) can benefit from SIE under certain conditions if: (1) The first virus infecting the cell exerts little or no fitness cost to the host (it is relatively benign); (2) The first virus establishes a latent infection that is not cleared by the host's immune system (otherwise, the benefit would only be transient) and (3) The infection confers protection against one or several viruses that are more virulent than the first virus. If we now imagine that the tobacco plant evolves such that the genome of the benign TMV strain becomes stably inherited to the following plant generations (e.g., by genomic integration of the whole TMV genome, or of the gene(s) that mediate the resistance to virulent TMV strains), then the result would be an inheritable immune system (to my knowledge, endogenous TMV elements have not been reported, so this remains a thought experiment, but the sequences of many other plant viruses are found in the genomes of various plant species, with possible antiviral functions including the generation of small RNAs used for the RNAi machinery (Chu et al. 2014; da Fonseca et al. 2016)).

Superinfection exclusion is not restricted to plant cells but occurs widely in prokaryotic as well as eukaryotic (single-cell or multicellular organisms) systems (Broecker and Moelling 2019b). Prokaryotic SIE mechanisms will be described below and they have a broad presence. SIE has also been described for several human pathogenic viruses, including vaccinia, measles, hepatitis C, West Nile, influenza virus and others (Birukov and Meyers 2018). Another example is human immunodeficiency virus (HIV). One of the essential steps in the replication cycle of HIV (as for all retroviruses) is the integration of the (reverse-transcribed) genome into the genome of the host cell (forming a so-called provirus). This is an important feature, as an integrated retroviral genome (or a part thereof) can be inherited by the next host generation (which, however, requires infection of germline cells, see below). It is important to note that HIV mediates SIE, and it has been shown that the virus does so by expressing an accessory protein, Nef, which downregulates the receptor for HIV, CD4, and one coreceptor, CCR5 (Michel et al. 2005). Thereby, an HIV-infected cell is less likely to become superinfected by other HIV particles.

In 2014, an article was published claiming that HIV was '*en route* to endogenization' (Colson et al. 2014). Peripheral blood mononuclear cells (PBMCs) of one described patient (who tested negative for the protective CCR5-Δ32 genomic mutation) harbored defective HIV-1 proviruses. These PBMCs could not be (super)infected with the identical HIV-1 strain in vitro, which suggests that the HIV-1 proviruses conferred resistance to infection. Although most proviruses had premature stop codons, some of them showed intact open reading frames. The presence of apparently protective HIV-1 proviruses suggested that HIV-1 proviruses could potentially mediate SIE. However, whether the virus is being *endogenized* is a more complex question. Endogenization is a two-step process whereby viral genetic information becomes part of a host's genome. First, the viral genome needs to be integrated into the genome of the host cell (this occurs during the normal retroviral lifecycle but can also occur 'accidentally' for other viruses, see below). In multicellular organisms (such as humans) it is necessary that genomic integration occurs in germline cells, as genomic integrations in somatic cells will not be inherited. Second, the viral genome in its entirety, or parts thereof, become fixed in the population. The human genome, for example, contains about 700,000 endogenous retroviruses (complete or fragments thereof), which constitute up to 8% of the genomic sequence, compared to ~20,000 human genes which account for ~2% of the total coding sequence of the genome (Belshaw et al. 2004; Escalera-Zamudio and Greenwood 2016). In the HIV example described above, the patient had proviruses in their PBMCs (i.e., in somatic cells) but not in germline cells, which would be required for vertical transmission and thus, endogenization. Therefore, only a germline infection with HIV-1 may confer inheritable resistance against HIV-1 induced disease at the population level. It is debated if HIV is able to infect germline cells (spermatozoa and oocytes) and whether integrated proviruses can be transmitted vertically (Baccetti et al. 1994, 1998; Bagasra et al. 1994; Barboza et al. 2004; Cardona-Maya et al. 2009, 2011; Nuovo et al. 1994), and in general the more complex retroviruses (genus lentivirus, of which HIV is an example) appear to enter the germline much less frequently than do simple retroviruses (e.g., all of the 100,000 known HERVs in the human genome are derived from simple retroviruses, and not a single one from a lentivirus). Endogenous lentiviruses have been described, however, in the genomes of other species, such as rabbits, non-human primates, weasels, colugos, ferrets and bats (Katzourakis et al. 2007; Gifford et al. 2008; Gilbert et al. 2009; Han and Worobey 2012, 2015; Cui and Holmes 2012; Jebb et al. 2020). It thus appears to be theoretically possible that HIV-1 may indeed become endogenized at some point in the future. The propensity for endogenization of simple retroviruses, however, appears to be substantially higher, given that the vast majority of known ERVs are derived from simple retroviruses. For these simple ERVs there are several examples for SIE described in the literature, as summarized in the following section.

8.5.1 Endogenous Retrovirus-Mediated Immunity in Eukaryotes: The Envelope Protein

A virus is endogenized when it enters the genome of a germline cell, is transmitted vertically to the next generations of host and becomes fixed in the host population (Fig. 8.1a). In eukaryotes, retroviruses are by far the most frequently endogenized viruses, as during their replication there is an obligatory step during which the retroviral genome is integrated into the host's genome. These ERVs, or parts thereof,

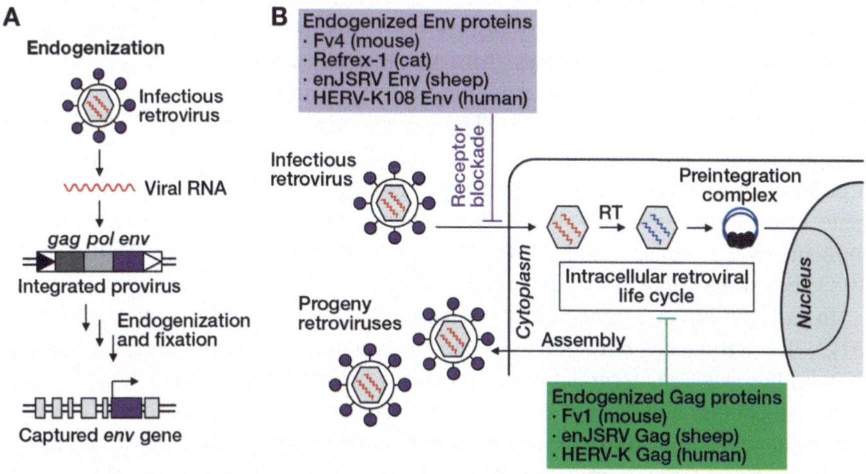

Fig. 8.1 Endogenous retrovirus-mediated antiviral immunity. (**a**) Schematic of the process of endogenization of a retrovirus. A retrovirus infects a germline cell. The viral RNA is reverse-transcribed into a DNA copy and integrated into the cellular genome, forming a provirus. If this provirus is beneficial or at least not detrimental to the survival of the cell and the organism, it can become endogenized and eventually fixed in the population. Most of the provirus can decay over time due to random mutations, but certain beneficial genes (here, the retroviral *env* gene is used as an example) can be captured and exert novel functions, such as serving as an antiviral defense mechanism. (**b**) Examples of endogenous retrovirus-derived genes that serve as immune defense against retroviruses. Env proteins can act by blocking receptors (termed a receptor blockade) to prevent infection of a cell. Gag proteins have been shown to inhibit the retroviral replication at various steps during the intracellular life cycle. Details on the indicated examples in mouse, cat, sheep and humans are described in the text. Viral RNA is depicted by red wavy lines, viral DNA by blue wavy lines. RT, reverse transcriptase

can mediate resistance to infections by *bona fide* retroviruses. Of note, the genomes of eukaryotes harbor large amounts of ERV sequences; about 8% of the human genome sequence originates from retroviruses (Gifford and Tristem 2003). Most of the human ERVs (HERVs) have invaded the ancestral genome many millions of years ago, and none of the HERV sequences identified in the human genome have any known infectious counterparts still existing; these viruses are most likely extinct except for their endogenous remains. Intensively studied examples of endogenized viral genes are the syncytins which originate from retroviral *env* genes (Lavialle et al. 2013). Specific *env* genes have been independently exapted from different retroviral proviruses at least seventeen times during evolution. Some of these genes are used as syncytins or otherwise functionally related genes which are critical for placentation in mammals and some viviparous lizard species (Cornelis et al. 2017; Imakawa and Nakagawa 2017). Through their immunosuppressive domain (ISD), syncytins likely contribute to the prevention of maternal immune rejection of the fetus via various mechanisms, including the inhibition of leukocytes and the suppression of pro-inflammatory cytokines (Cianciolo et al. 1985; Haraguchi et al. 1995, 1997, 2008).

In addition to syncytins, other retroviral *env* genes have been endogenized that do not exert immunosuppressive, but instead anti-retroviral functions (Fig. 8.1b). For example, the mouse *Friend virus susceptibility 4* (*Fv4*) gene confers resistance of mice to murine leukemia viruses (MuLVs) (Suzuki 1975). Physically, *FV4* is a truncated MuLV-like provirus containing the 3′ portion of the *pol* gene and the entire *env* gene (Ikeda et al. 1985). The *env*-encoded protein binds to the cellular receptor used by MuLV and thereby prevents infection by the exogenous retrovirus, a process referred to as receptor blockade, a variant of SIE. Another captured *env* gene in mice, *resistance to MCF* (*Rmcf*), mediates resistance to mink cell focus-inducing (MCF) viruses and MuLVs, likely also via receptor blockade (Hartley et al. 1983; Brightman et al. 1991; Jung et al. 2002).

In cats, the *refrex-1* gene confers resistance to feline leukemia virus-D (Ito et al. 2013). It is a truncated retroviral *env* gene that contains the putative receptor-binding domain but lacks a C-terminal portion due to a premature stop codon.

Env-mediated interference has also been demonstrated in human cells. The Env protein encoded by a human endogenous retrovirus of the HERV-K(HML-2) family has been shown to interfere with HIV-1 production in vitro (Terry et al. 2017). It is a full-length Env protein that, compared to the consensus, ancestral HERV-K(HML-2) Env, has four mutations that appear to be required for inhibiting HIV-1. Interestingly, HERV-K(HML-2) expression in T cells is activated by HIV-1 infection (Gonzalez-Hernandez et al. 2012), which suggests that expression of Env (and Gag, see Sect. 8.5.2 below) may have evolved as an inducible mechanism of protection against exogenous retroviruses. Another example of an antiviral Env protein in human cells is encoded by a HERV-T provirus (Blanco-Melo et al. 2017). Its expression confers resistance to a reconstructed infectious HERV-T virus (as the virus is extinct) via receptor blockade in vitro. In addition, *Suppressyn*, a truncated *env* gene expressed by a HERV-F element with a known role in placental development, may restrict infection by exogenous retroviruses (Malfavon-Borja and Feschotte 2015).

8.5.2 Endogenous Retrovirus-Mediated Immunity in Eukaryotes: The Gag Protein

Gag is another retroviral gene that has been captured by mammalian hosts for immune defense against exogenous retroviruses (Fig. 8.1b). The best studied *gag*-derived restriction factor is *Friend virus susceptibility 1* (*Fv1*) of mice (Best et al. 1996), which inhibits murine leukemia virus (MuLV) at a stage between entry and proviral integration. The Fv1 protein interacts with the retroviral capsid protein in the preintegration complex of MuLV (Best et al. 1996). *FV1* originates from a MERV-L *gag* gene. In sheep, enJSRV-expressed Gag protein inhibits exogenous JSRV at a late stage of the retroviral life cycle during viral assembly (Palmarini et al. 2004). In human cells, a HERV-K(HML-2) Gag protein inhibits HIV-1 release and reduces infectivity of progeny HIV-1 virions (Monde et al. 2017).

8.5.3 Evolution of Retrovirus-Mediated Immunity in Real Time?

Jaagsiekte sheep retrovirus (JSRV) provides an example of a recent or ongoing endogenization (Armezzani et al. 2014). The youngest identified endogenous elements (enJSRV) were integrated into the sheep genome only 200 years ago, and exogenous, infectious JSRV is still circulating. The sheep genome contains about 27 enJSRV sequences, of which 16 harbor intact *env* genes. enJSRV Env protein likely exerts syncytin-like functions during placentation (Dunlap et al. 2006) and has been shown to prevent infection by exogenous JSRV via receptor blockade (Spencer et al. 2003). Another example of an endogenization in real-time is currently occurring in koalas. Like JSRV, koala retrovirus (KoRV) co-exists in both endogenous and exogenous form (Tarlinton et al. 2006). While endogenous KoRV elements can be identified in the genomes of most koalas, there is substantial inter-individual variation in the integration sites and extensive regional variation, indicative of an ongoing endogenization process. It has been speculated that Env (or other proteins) expressed by endogenous KoRV elements may provide protection against exogenous KoRV infections, analogous to the examples described above (Sarker et al. 2020). In favor of this hypothesis, full-length *env* mRNA appears to be expressed by many endogenous KoRV elements (Tarlinton et al. 2017). However conversely, it has been suggested that koalas show in utero expression of KoRV antigens and those antigens get tolerized as the developing immune system recognizes them as harmless self structures. Consequently, their immune systems may be unable to mount an immune response against exogenous KoRV. In favor of the latter hypothesis, koalas with mostly intact integrated KoRV proviruses are often unable to mount antibody responses against KoRV antigens, even after vaccination, in contrast to animals with less intact KoRV proviruses that can generate antibodies (Tarlinton et al. 2017; Olagoke et al. 2019). Thus, at the current stage integrated and vertically transmitted KoRV elements may either have a beneficial or detrimental effect on the population, and it will take more time for evolutionary selection to resolve this, perhaps by having protection-providing KoRV genes become fixed in the population, as seen for JSRV. It is estimated that KoRV first entered the koala population only between 100 and 200 years ago (Greenwood et al. 2018), whereas the initial infection of the sheep genome with JSRV likely occurred 5–7 million years ago (Armezzani et al. 2014).

8.5.4 Superinfection Exclusion by Other Endogenized Eukaryotic Viruses

Although ERVs constitute the vast majority of endogenous viruses, mammalian genomes also harbor numerous sequences derived from, e.g., *Borna-*, *Filo-*, *Parvo-*, *Circo-*, *Rhabdo-* and *Herpesviridae* (Belyi et al. 2010a, b; Horie et al. 2010; Katzourakis and Gifford 2010; Aswad and Katzourakis 2014). Many integrated (non-reverse transcribing) RNA virus-derived sequences likely originate from reverse transcription and integration via either the replication machinery of retroelements or by nonhomologous recombination (Suzuki et al. 2014). These non-retroviral endogenous viral sequences are frequently referred to as endogenous viral elements (EVEs).

Among these EVEs are the negative sense ssRNA Borna disease virus (BDV) sequences (Belyi et al. 2010b). Interestingly, species that contain genomic BDV sequences (e.g., primates, rats, mice and squirrels) are relatively resistant to infection with exogenous BDVs. In contrast, highly susceptible species like horses, sheep and cattle can develop fatal encephalitis upon BDV infection and do not have detectable BDV sequences in their genomes.

There is also experimental evidence for protection by endogenous BDV sequences. The ground squirrel genome harbors an endogenous bornavirus-like nucleoprotein (itEBLN) sequence with 77% amino acid similarity to circulating infectious BDV (Fujino et al. 2014). The itEBLN RNA binds to the ribonucleoprotein of infectious BDV and is incorporated into virions. This appears to inhibit viral trafficking and cell-to-cell spread (Kim et al. 2020).

Like squirrels, humans usually do not develop Borna disease. Only few anecdotal cases of fatal BDV-induced encephalitis have been reported (Hoffmann et al. 2015). Seven human endogenous bornavirus-like nucleoprotein (hsEBLN) elements are expressed at the RNA level (Sofuku et al. 2015); hsEBLN-2 is also known to be expressed as protein (Ewing et al. 2007). In primate and rodent genomes, EBLNs are enriched within piRNA clusters (Parrish et al. 2015). These EBLNs express functional piRNAs that are antisense to the BDV nucleoprotein mRNA and are expressed in testes. However, whether they mediate inhibition of BDV infection in these or other cells is unknown. Since piRNAs are also known to be expressed in some somatic cell types such as neurons (Lee et al. 2011), EBLN sequences may protect from brain BDV infection. This could at least partially explain why species with endogenous EBLN sequences are relatively resistant to BDV-induced encephalitis. In addition, EBLNs may protect from Borna disease by inducing immune tolerance through in utero EBLN protein expression (Horie 2017). This tolerization of the immune system may limit the possible pathogenicity associated with anti-nucleoprotein immune responses that develop during BDV infection. The BDV nucleoprotein is a major target for cytotoxic T cell responses (Stitz et al. 1993; Planz and Stitz 1999). Thus, tolerance to nucleoprotein may protect against BDV-induced encephalitis which mostly results from immune-mediated inflammation. In addition, EBLN RNAs may act as antisense RNAs to the BDV genome (Horie 2017). It is important to note that EBLN sequences also are found in the genomes of various other species, including whales, birds and lamprey (Kobayashi et al. 2016; Hyndman et al. 2018), where they might also exert antiviral functions.

Aedes mosquitoes are important vectors for human pathogenic flaviviruses such as Dengue and Zika virus. Their genomes contain various endogenous flaviviral sequences (Suzuki et al. 2017). Small RNAs like piRNAs and siRNAs, known to play an important role in antiviral defense in insects (Cullen et al. 2013), are produced from these endogenous viruses and might play a role in antiviral defense (see Sect. 8.5.5).

8.5.5 piRNA-Guided CRISPR-Cas-Like Immunity in Eukaryotes Based on Endogenous Viral Sequences

CRISPR-Cas immunity of prokaryotes will be described in more detail below, but I like to mention it here as there are some interesting similarities between this immune system and piRNA-mediated immunity of eukaryotes. Briefly, CRISPR-Cas is an inheritable immune system that requires three steps. First, fragments of foreign (e.g., viral) DNA or reverse-transcribed RNA are captured and integrated as spacers into specialized genomic regions called CRISPR loci or CRISPR array. Second, the spacers are transcribed and processed into small RNAs (crispr RNAs/crRNAs). Third, the crRNAs guide a nuclease (Cas) to complementary DNA/RNA upon re-exposure of the invader. Cas then inactivates the invading DNA/RNA. It has recently been suggested that the eukaryotic piRNA system may exert analogous functions in some species, especially in insects (Ophinni et al. 2019). Here, foreign viral RNA is reverse-transcribed and preferentially inserted into piRNA clusters (analogous to the CRISPR loci), then transcribed and processed into small RNAs (piRNAs, analogous to crRNAs). These virus-derived piRNAs then guide a nuclease

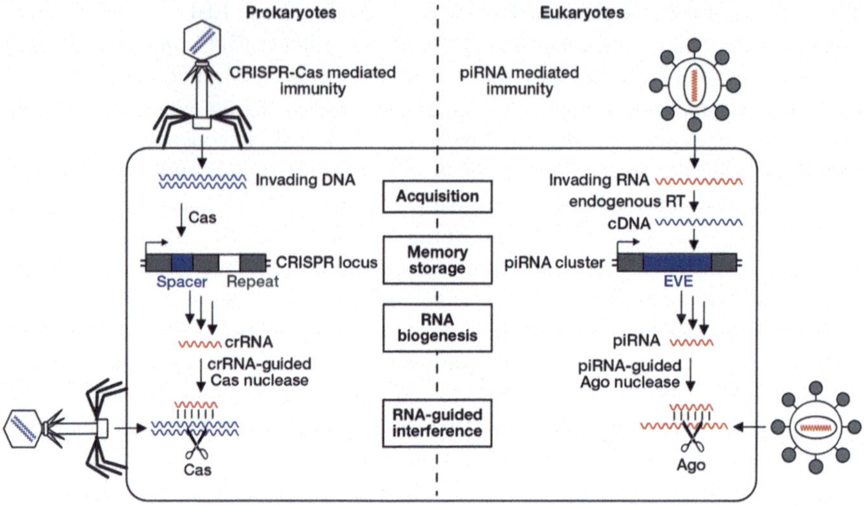

Fig. 8.2 Similarities between CRISPR-Cas immunity in prokaryotes (left) and piRNA-mediated immunity in eukaryotes (right). *Left:* A portion of a phage (top) DNA (or reverse-transcribed RNA in the case of RNA phages) invading the prokaryotic cell is inserted via a Cas protein into a CRISPR locus in the prokaryotic genome. The newly inserted spacer is transcribed along with the CRISPR locus and the transcript is processed into a small RNA (crRNA) that guides another Cas nuclease to an invading DNA (or reverse-transcribed RNA) of another invading phage (bottom) based on sequence complementarity. The Cas nuclease cleaves the invading nucleic acid (indicated by scissors), thus providing immunity against the newly infecting phage. *Right:* In eukaryotes, a similar mechanism is constituted by the piRNA machinery. An invading RNA (of an RNA virus, top) is reverse-transcribed by an endogenous reverse transcriptase and then integrated into a piRNA cluster. The piRNA cluster DNA is transcribed and processed into small RNAs (piRNAs) that guide an Argonaute nuclease to the RNA of a newly infecting RNA virus (right) based on sequence complementarity. The Argonaute protein cleaves the newly invading RNA, thus mediating immunity to the RNA virus. DNA molecules are represented by blue wavy lines, RNA molecules by red wavy lines. Ago, Argonaute; EVE, endogenous viral element; RT, reverse transcriptase. Figure modified from Ophinni et al. (2019)

(Argonaute, analogous to Cas) to the viral RNA, which inactivates the invader (Fig. 8.2).

Such virus-derived piRNAs, specific for *Drosophila* X virus and other RNA viruses, have been first identified in a *D. melanogaster* cell line (Wu et al. 2010). However, to date silencing activity in cells of *D. melanogaster* has only been observed against transposons and endogenous retroviruses (the canonical function of piRNAs), but not against exogenous viruses. In contrast, the link between viral piRNAs and activity against exogenous viruses is stronger in *Aedes* mosquitoes, which show expression of piRNAs derived, for example, from *Toga-*, *Flavi-*, *Bunya-* and *Reoviridae* (all are RNA virus families) in germline cells and somatic tissues. piRNAs are known to be able to guide Argonaute nucleases to complementary RNAs, which induces their degradation. Knockdown of proteins involved in amplification of piRNAs reduced viral piRNA expression and enhanced Dengue virus replication in *Aedes*-derived cells (Miesen et al. 2016). Viral piRNAs have been discovered in other arthropods as well, including whiteflies, and some of these piRNA sequences have been shown to also target DNA viruses (Ophinni et al. 2019). In mammals, the only identified example of endogenous virus-derived piRNAs are those originating from EBLNs, as described in the previous section. It thus appears that the antiviral activity of piRNAs is likely more pronounced in arthropods than in mammals.

8.5.6 Do Endogenous Viruses Render Bats Resistant to Viral Infections?

Bats have been associated with a number of zoonotic viral diseases and constitute an important reservoir for diverse viruses, including members of the families *Flavi-*, *Rhabdo-* and *Bunyaviridae* (Olival et al. 2017). The bat immune system is unique in its ability to tolerate viral infections which are typically lethal to other mammalian species (Banerjee et al. 2020). It has been suggested that endogenous viruses may play a role in the bats' immune tolerance to viruses (Skirmuntt et al. 2020). Bat genomes contain numerous ERVs and sequences originating from *Borna-*, *Filo-*, *Parvoviridae* and others. Interestingly, to date only one exogenous bat-specific retrovirus has been identified in Australian bats, the Hervey pteropid gammaretrovirus (HPG) that is related to KoRV, suggesting that bats may have transmitted this virus to koalas (Hayward et al. 2020). It could be speculated that the apparently very few exogenous bat retroviruses, which stand in contrast to the large diversity of ERVs, are the result of ERV-mediated immunity (Skirmuntt et al. 2020). Interestingly, no HPG-related ERVs have been identified in bat genomes, suggesting that HPG only recently entered bat populations and has either not (yet) been endogenized or is currently in the process of endogenization (Hayward et al. 2020). It remains speculative whether the bat endogenous viruses mediate immunity against their exogenous counterparts. In favor of this hypothesis, the diversity of endogenous viral sequences is higher than in most other mammals (Jebb et al. 2020) and bats are highly immune to many viral infections, including the *Filoviridae* Ebola- and Marburg virus (the bat genome contains endogenous Filovirus elements with intact open reading frames for nucleoprotein and VP35 protein (Skirmuntt et al. 2020)).

8.5.7 Superinfection Exclusion by Endogenized Viruses in Prokaryotes

Superinfection exclusion is not limited to eukaryotes but is also widespread in prokaryotes. Prokaryotic genomes contain up to 10–20% of prophage-derived sequences, and the presence or absence of specific prophage sequences can contribute to prokaryotic interstrain variability (Canchaya et al. 2003). There is often a fitness cost to the host associated with prophage integration (Iranzo et al. 2017). However, prophage sequences can also have beneficial effects, such as increasing virulence, which could extend the ecological range of the bacterium (Canchaya et al. 2004).

One example of prokaryotic SIE is the prophage-encoded Tip protein that suppresses expression of type IV pili on the surface of its host, *Pseudomonas aeruginosa*, with little or no fitness cost to the bacterium (Chung et al. 2014). The type IV pili are common entry receptors for phages. Consequently, Tip expression mediates resistance to various phages (Bondy-Denomy et al. 2016). Indeed, prophage-encoded SIE seems to be relatively broad, as only three prophages make *P. aeruginosa* resistant to at least 30 different phages. Another well-described example for prophage-mediated immunity in bacteria is the sie_{2009} gene expressed by the lactococcal Tuc2009 prophage (McGrath et al. 2002). The Sie_{2009} protein localizes to the bacterial membrane and likely inhibits phage DNA injection. Various other phage-encoded genes that mediate resistance to phage infection have been identified in numerous bacterial hosts (Bondy-Denomy et al. 2016; McGrath et al. 2002); it thus appears that prophage-mediated SIE is a common mechanism of immunity in prokaryotes.

8.6 The Enemy of My Enemy Is My Friend: Harnessing Viruses for Complex Immune Systems

Any organism—unicellular or multicellular—is constantly exposed to a plethora of microorganisms, most benign, some potentially harmful, some opportunistic. Cellular organisms can internalize specific viruses and use them as a weapon against other viruses (or cellular pathogens). Genomic integration allows for the process of endogenization (sometimes referred to as domestication), whereby a given antiviral property of a protein (or any other useful function) may be preserved or even enhanced. Other portions of the integrated viruses may degenerate through accumulation of deleterious mutations over time or by mechanisms of genomic deletion in the absence of positive selection pressure. In the following sections I will highlight three examples of complex immune systems whose evolution involved viruses or virus-like elements. I will start by a relatively simple (yet more complex than SIE described above) example of a virus-based inheritable immune system in the protist *Cafeteria roenbergensis* (a unicellular eukaryote), followed by the more complex adaptive immune system of prokaryotes (CRISPR-Cas), and, lastly, the V(D)J recombination system that diversifies antibodies and T cell receptors in vertebrates, arguably the most complex immune system we know.

8.6.1 *A Small Virus Against a Giant Virus in the Protist* **Cafeteria roenbergensis***: An Adaptive Immune System at the Population Level*

The marine protist *Cafeteria roenbergensis*, a single-celled eukaryote, is infected by a giant virus named *Cafeteria roenbergensis* virus (CroV). The CroV virus is a member of the nucleocytoplasmic large DNA viruses (NCLDV) group with a ~730,000 bp dsDNA genome. After entering the protist host via phagocytosis, CroV replicates in cytoplasmatic viral factories, which are nucleus-like structures in which DNA replication and transcription occur. Viral protein synthesis and virion assembly takes place on the outside of the viral factories (Bell 2020). The infection is fatal to *C. roenbergensis* and, consequently, it has been suggested that CroV may play a role in regulating the protist's population in marine ecosystems (Fischer et al. 2010).

Cafeteria roenbergensis is also host to another, much smaller dsDNA virus termed Mavirus with a ~19,000 bp genome, which can only replicate in the presence of CroV (Fischer and Hackl 2016). Mavirus infection has no known negative consequence for *C. roenbergensis.* As its replication depends on another virus (CroV), Mavirus is designated as a 'virophage' analogous to the bacteria-infecting bacteriophages. The currently known virophages (family *Lavidaviridae*) all have in common their dependence on an NCLDV for replication (Mougari et al. 2019; Fischer 2020). Simultaneous co-infection with Mavirus protects *C. roenbergensis* from CroV-induced lysis by inhibiting the replication of CroV through a yet unknown mechanism. Interestingly, Mavirus can integrate into the protist's genome (forming 'provirophages'), likely due to the presence of a retroviral-like integrase that is packaged into the virion, as well as nuclear localization signals (Born et al. 2018). The provirophages remain transcriptionally silent until they are induced by a CroV superinfection. Notably, activation of the integrated Mavirus genomes does not prevent the lysis of the initial *C. roenbergensis* cell, likely because Mavirus activation occurs only at a late stage of CroV replication. It does, however, induce the massive production of Mavirus particles by the CroV viral factories. Consequently, the lysed cell releases Mavirus particles (along with CroV particles) into the environment. Neighboring *C. roenbergensis* cells are subsequently co-infected with CroV and Mavirus, and Mavirus inhibits CroV replication and cell lysis. This protects the protist's population, at the expense of losing the cell originally infected with CroV by lysis. It is an inheritable immune system with a twist, in which not the direct descendants of the provirophage-carrying cell are protected (as is the case, for example, for CRISPR-Cas immunity described below in Sect. 8.6.2). Instead, immunity is inherited at the population level (via the release of Mavirus particles and 'vaccination' of neighboring cells), and protection from CroV occurs in an altruistic manner. There is also a potential benefit to Mavirus; infection of *C. roenbergensis* and genomic integration may increase the frequency of interaction with the CroV host (Mougari et al. 2019).

It is still not known how wide-spread this type of virophage-mediated immunity is. Recently, it was shown that co-infection with specific virophages rescued *Acanthamoeba castellanii* (an amoebal species) from being lysed by amoeba infecting NCLDVs and reduced the production of NCLDV particles (Mougari et al. 2019, 2020). However, there is currently no evidence for protection mediated by integrated provirophages as is the case for *C. roenbergensis*. In contrast, integrated virophages are found in other organisms such as the unicellular green alga *Bigelowiella natans* (Koonin and Krupovic 2016). There, most of the 38 identified elements were transcriptionally active, and six represented complete provirophages (Blanc et al. 2015). Moreover, polintons, virophage-related transposons, are found in the genomes of various eukaryotes and likely originate from exogenous viruses (Koonin and Krupovic 2018). The provirophages found in the *B. natans* genome and the virophage-like polinton sequences may have been recruited as a defense against yet to be identified (or extinct) NCLDVs. Moreover, a number of novel virophages and their associated NCLDV hosts have been identified recently and become available for experimental testing (Fischer 2020; Xu et al. 2020; Gulino et al. 2020). The identified inheritable immune system of *C. roenbergensis* may therefore just be the tip of an iceberg (Koonin and Krupovic 2016).

8.6.2 CRISPR-Cas Immunity Largely Originates from Viruses and Mobile Genetic Elements

The CRISPR-Cas system is a form of prokaryotic adaptive immunity which utilizes a collection of DNA fragments (or reverse-transcribed RNA) of infecting phages or plasmids that are integrated as spacers into a designated region of the host genome, the CRISPR locus or CRISPR array (Hille et al. 2018). These spacers provide an immunological memory that protects the cell from invaders with similar sequences. Moreover, since the spacer is stably integrated into the CRISPR locus, immunity is passed on to future generations. As such, the spacers bear witness to the types of phages that are or have been infecting a certain prokaryotic species or strain (similar to ERVs that are indicators of past or ongoing retroviral infections in eukaryotes). CRISPR loci are found in about 50% of bacterial and 90% of archaeal genomes, with an average of three and five loci per genome, respectively (Grissa et al. 2007; Hille et al. 2018). The transcribed CRISPR locus RNA (termed pre-crRNA) is processed into smaller crRNAs that guide sequence-specific cleavage of complementary invading nucleic acids by Cas effector nucleases, which resembles the action of viral and transposon-derived piRNAs transcribed from piRNA clusters in eukaryotes (described above in Sect. 8.5.5).

Six types of CRISPR-Cas systems (I through VI) and various subtypes have been described that differ, for example, in the type of nuclease that mediates target cleavage, and whether DNA or RNA invaders are targeted (Koonin and Makarova 2017). The general steps, however, are identical in all systems; (1) adaptation (i.e., spacer acquisition), (2) expression (i.e., crRNA generation) and (3) interference (i.e., target nucleic acid cleavage). Interestingly, a minimum of four different MGEs were involved in CRISPR-Cas evolution (Koonin and Makarova 2017). First (and most importantly), all six systems originate from a transposon called casposon which utilizes Cas1 nuclease for DNA integration (Krupovic et al. 2014). Thus, the crucial invention for prokaryotic adaptive immunity to evolve was the ability to modify the sequence of the host's genome (interestingly, the same ability is at the origin of adaptive immunity of jawed vertebrates, as described in Sect. 8.6.3). Second, Cas2 nuclease and RNase domains of the HEPN family found in several Cas proteins likely originate from toxin-antitoxin modules, which can be regarded as MGEs as they are typically mobilized by plasmids (Koonin and Krupovic 2015; Koonin and Makarova 2017). Third, various Type III systems exapted a reverse transcriptase from an MGE (a mobile group II intron), which allows for spacer acquisition from RNA invaders (e.g., RNA phages). Fourth, the RuvC domains of Type II and V effector Cas nucleases (Cas9 and Cas12, respectively) likely originated from DNA transposons. Functional CRISPR-Cas systems can also be encoded by phages (Seed et al. 2013), suggesting that phages may mediate horizontal gene transfer of this type of immune system. CRISPR-Cas acquired immunity can in theory be transmitted across thousands of microbial generations (Weinberger et al. 2012), although phage evasion by mutation typically occurs within a few generations in coevolution studies (Westra et al. 2019). There is no known mechanism that can discriminate harmful from beneficial invaders (the former could be a lytic bacteriophage, the latter a plasmid conferring antibiotic resistance; both will be equally targeted by CRISPR-Cas). Thus, a CRISPR-Cas locus that provides an evolutionary disadvantage (as it prevents, for example, resistance to antibiotics) might be counter-selected, which might explain the existence of prokaryotes without any CRISPR loci (Westra et al. 2019).

8.6.3 At the Heart of Adaptive Immunity in Jawed Vertebrates Is an Ancient Mobile Genetic Element

In contrast to the prokaryotic adaptive immune system, CRISPR-Cas, immunological memory in multicellular vertebrates is restricted to somatic cells, more specifically, to dedicated cells of the immune system (B and T cells). This memory is therefore not inherited to the next generation. In jawed vertebrates, the diversity of Igs/antibodies and TCRs is generated by a process called V(D)J recombination, in which variable (V), diversity (D) and joining (J) gene segments are recombined. Further antibody diversification is then achieved by somatic hypermutation, whereby random mutations are introduced, and the B cells subjected to a selection process for increased binding affinity to the antigen (Kapitonov and Koonin 2015). In contrast to CRISPR-Cas, elaborate mechanisms have evolved that can differentiate between harmful and harmless, which, for example, renders the adaptive immune system able to mount a response against pathogenic bacteria, but at the same time it tolerates the normal, healthy microbiota of the gastrointestinal tract, the lung, the skin, and other organs. The tolerance to the healthy microbiota is the result of a co-evolution of the hosts and the microbes. Ancestors unable to tolerate the microbes did not survive, similarly those microbes that were not able to evade the immune responses also died. In addition, even though there is no inheritance of the adaptive immunity of jawed vertebrates, there is an evolutionary selection for those individuals that can mount a protective immune response against deadly pathogens.

The ability to produce diversity of antibodies and TCRs in jawed vertebrates developed around 450–500 million years ago (Kapitonov and Koonin 2015). As for CRISPR-Cas, the ability to modify the genome sequence has been the crucial event at the evolutionary origin of adaptive immunity. Both the Rag1 and Rag2 proteins that mediate V(D)J rearrangement by recombining V, D and J segments are encoded by a single genomic locus. They originate from a DNA transposon called *Transib* that today is found in the starfish, oyster and sea urchin genomes, but not anymore in those of jawed vertebrates, where it went extinct (Kapitonov and Koonin 2015). An active *Transib* transposon encoding Rag1 and Rag2-like proteins was recently discovered in the lancelet genome, and its terminal inverted repeats (the sequences flanking the transposon) share similarities with the recombination signal sequences that are recognized by the Rag1/2 complex (Huang et al. 2016). Thus, not only the genes encoding the proteins required for V(D)J recombination but also the recognition sequences for the recombination to occur originate from a transposon.

8.7 Discussion

Viruses have traditionally been regarded mainly as disease-causing agents (hence the name *virus*, Latin for poison). Yet, viruses and their relatives, the MGEs, have also majorly contributed to the evolution of cellular organisms by introducing, mobilizing and amplifying genetic material. The recruitment of sequences from viruses, transposons and other MGEs for pro- and eukaryotic immune systems appears to be strikingly common, as illustrated by the examples presented above. Additional immune systems that have evolved with the involvement of viral or virus-like sequences are discussed briefly below (see also Table 8.1).

One is the prokaryotic restriction-modification (RM) mechanism. The RM systems consist of both a restriction endonuclease that cleaves invading DNA (e.g., of a phage) at a specific sequence motif as well as a methylase that masks that motif in the prokaryotic genome via DNA methylation. The motifs are typically short and are thereby present in many invading DNAs. Thus, RM systems can be regarded as a prokaryotic innate immune system. These RM systems are present in ca. 90% of prokaryotic genomes (Murphy et al. 2013), can be mobilized by phages and frequently co-localize with transposon-derived genes and may be flanked by inverted repeats and target site duplications, with those flanking structures seemingly characteristic of transposons (Naderer et al. 2002; Furuta et al. 2010; Makarova et al.

2011; Takahashi et al. 2011). In addition, transposons can carry functional RM systems (Khan et al. 2010), indicating that these defense systems may have evolutionarily originated from transposons.

Another prokaryotic innate immune system involves Argonaute proteins that act as RNA- or DNA-guided nucleases to cleave invading RNA or DNA (Swarts et al. 2014; Koonin and Krupovic 2015). Like the Argonaute proteins involved in eukaryotic small RNA-guided defense mechanisms, prokaryotic Argonaute proteins share striking structural and functional similarities with the retroviral reverse transcriptase-RNase H proteins (Moelling et al. 2006), suggesting a common evolutionary ancestry.

In both pro- and eukaryotes, antisense transcripts from viral and MGE sequences may act by forming double-stranded RNA complexes with invading RNAs when there is sufficient sequence complementarity, which may induce degradation of the invading RNA (Broecker and Moelling 2019b).

In eukaryotes, interferons are mediators of innate immunity, especially during viral infections. It has been shown that transposons and ERVs have been specifically co-opted by the host to provide enhancers and binding sites for transcription factors for interferon-stimulated genes (Chuong et al. 2016; Ito et al. 2017). Transposons and ERVs have thus significantly shaped the regulation of the interferon response

As the last example, I would like to mention an interesting interaction that happens at the mucosal surfaces of animals (metazoans). Some phages express immunoglobulin-like domains on their surface that bind to mucin glycoproteins expressed in mucosal tissues such as the gastrointestinal tract or lungs (Barr et al. 2013). Thereby, phages are specifically enriched in mucus and protect the underlying epithelium from invading bacteria. The phages also benefit by an increased frequency of interaction with their target bacteria in what can be seen as a symbiotic relationship. Thus, phages enriched in mucosae serve as a non-host derived immune system against bacterial infections.

The fact that genomes of virtually all cellular organisms harbor large numbers of MGEs and viral sequences suggests that yet unknown functionalities will likely be identified in the future. For example, it was recently discovered that prokaryotic defense islands, genomic regions involved in various immune mechanism, are enriched with transposon sequences whose potential functions remain to be determined (Doron et al. 2018; Koonin 2018).

Of note, immune defense is not the only function of endogenized viruses and MGEs. For example, deleting all prophages in *Escherichia coli* results in various fitness deficits, including increased susceptibility to antibiotics and osmotic stress, and causes deficits in growth and biofilm formation (Wang et al. 2010). In eukaryotes, transposons and ERVs do not only modulate the interferon response, but also play roles, amongst others, in cell differentiation, stem cell pluripotency and embryogenesis (Chuong et al. 2016). Thus, viral and virus-like sequences have adopted multifaceted roles, not only for immune defense, and have been a major driving force in the evolution of cellular life.

I would like to end with a comment on the current SARS-Coronavirus-2 pandemic. As viral evolutionist Aris Katzourakis mentioned via Twitter with respect to the analysis of bat genomes for virus-derived sequences (Skirmuntt et al. 2020), there is "no endogenous coronavirus so far" (Tweet by @ArisKatzourakis, May 22, 2020)—even in bats. To my knowledge endogenous coronaviruses have not been reported in any species to date, although there is in vitro evidence that coronavirus RNA can be reverse transcribed and integrated into the genome of human cells (Zhang et al. 2020). However, there seems to be an interplay between SARS-Coronavirus-2 and HERVs in humans. It was shown that various HERV families are upregulated in the lungs of SARS-Coronavirus-2 infected people (Kitsou et al. 2020). However, it remains to be determined whether the upregulated HERVs confer protection, contribute to pathophysiology or if this is simply a by-stander effect.

Acknowledgements I like to express my gratitude to Prof. Karin Moelling (Max Planck Institute for molecular Genetics Berlin) for ongoing support and for stimulating and fostering my interest in the topic of the viral origin of immune systems, among many others. I also like to acknowledge Dr. Günther Witzany who kindly invited me to his exciting symposia *DNA Habitats and its RNA Inhabitants* (2014) and *Evolution—Genetic Novelty/Genomic Variations by RNA Networks and Viruses* (2018) both held in Salzburg, Austria, where many of the ideas described in this chapter have been developed.

References

Anderson KV, Bokla C, Nüsslein-Volhard C (1985a) Establishment of dorsal-ventral polarity in the Drosophila embryo: the induction of polarity by the Toll gene product. Cell 42:791–798

Anderson KV, Jürgens G, Nüsslein-Volhard C (1985b) Establishment of dorsal-ventral polarity in the Drosophila embryo: genetic studies on the role of the Toll gene product. Cell 42:779–789

Armezzani A, Varela M, Spencer TE et al (2014) "Ménage à Trois": the evolutionary interplay between JSRV, enJSRVs and domestic sheep. Viruses 6:4926–4945

Aswad A, Katzourakis A (2014) The first endogenous herpesvirus, identified in the tarsier genome, and novel sequences from primate rhadinoviruses and lymphocryptoviruses. PLoS Genet 10: e1004332

Baccetti B, Benedetto A, Burrini A et al (1994) HIV-particles in spermatozoa of patients with AIDS and their transfer into the oocyte. J Cell Biol 127:903–914

Baccetti B, Benedetto A, Collodel G et al (1998) The debate on the presence of HIV-1 in human gametes. J Reprod Immunol 41:41–67

Bagasra O, Farzadegan H, Sehsamma T et al (1994) Detection of HIV-1 proviral DNA in sperm from HIV-1-infected men. AIDS 8:1669–1674

Banerjee A, Baker ML, Kulcsar K et al (2020) Novel Insights Into Immune Systems of Bats. Front Immunol 11:26

Barboza JM, Medina H, Doria M et al (2004) Use of atomic force microscopy to reveal sperm ultrastructure in HIV-patients on highly active antiretroviral therapy. Arch Androl 50:121–129

Barr JJ, Auro R, Furlan M et al (2013) Bacteriophage adhering to mucus provide a non-host-derived immunity. Proc Natl Acad Sci USA 110:10771–10776

Bayne CJ (2003) Origins and evolutionary relationships between the innate and adaptive arms of immune systems. Integr Comp Biol 43:293–299

Bell PJL (2020) Evidence supporting a viral origin of the eukaryotic nucleus. Virus Res 289:198168

Belshaw R, Pereira V, Katzourakis A et al (2004) Long-term reinfection of the human genome by endogenous retroviruses. Proc Natl Acad Sci USA 101:4894–4899

Belyi VA, Levine AJ, Skalka AM (2010a) Sequence from ancestral single-stranded DNA viruses in vertebrate genomes: the parvoviridae and circoviridae are more than 40 to 50 million years old. J Virol 84:12458–12462

Belyi VA, Levine AJ, Skalka AM (2010b) Unexpected inheritance: multiple integrations of ancient bornavirus and ebolavirus/Marburgvirus sequences in vertebrate genomes. PLoS Pathog 6: e1001030

Berrens RV, Andrews S, Spensberger D et al (2017) An endosiRNA-based repression mechanism counteracts transposon activation during global DNA demethylation in embryonic stem cells. Cell Stem Cell 21:694–703

Best S, Le Tissier P, Towers G et al (1996) Positional cloning of the mouse retrovirus restriction gene Fv1. Nature 382:826–829

Biryukov J, Meyers C (2018) Superinfection exclusion between two high-risk human papillomavirus types during a coinfection. J Virol 92:e01993–e01917

Blanc G, Gallot-Lavallée L, Maumus F (2015) Provirophages in the Bigelowiella genome bear testimony to past encounters with giant viruses. Proc Natl Acad Sci USA 112:E5318–E5326

Blanco-Melo D, Gifford RJ, Bieniasz PD (2017) Co-option of an endogenous retrovirus envelope for host defense in hominid ancestors. eLife 6:e22519

Boehm T, McCurley N, Sutoh Y et al (2012) VLR-based adaptive immunity. Annu Rev Immunol 30:203–220

Bondy-Denomy J, Qian J, Westra ER et al (2016) Prophages mediate defense against phage infection through diverse mechanisms. ISME J 10:2854–2866

Born D, Reuter L, Mersdorf U et al (2018) Capsid protein structure, self-assembly, and processing reveal morphogenesis of the marine virophage mavirus. Proc Natl Acad Sci USA 115:7332–7337

Brightman BK, Li QX, Trepp DJ et al (1991) Differential disease restriction of Moloney and Friend leukemia viruses by the mouse Rcmf gene is governed by the viral long terminal repeat. J Exp Med 174:389–396

Broecker F, Moelling K (2019a) What viruses tell us about evolution and immunity: beyond Darwin? Ann N Y Acad Sci USA 1447:53–68

Broecker F, Moelling K (2019b) Evolution of immune systems from viruses and transposable elements. Front Microbiol 10:51

Burnet FM, Fenner F (1949) The production of antibodies, 2nd edn. Macmillan, Melbourne, Australia

Canchaya C, Proux C, Fournous G et al (2003) Prophage genomics. Microbiol Mol Biol Rev 67:238–276

Canchaya C, Fournous G, Brüssow H (2004) The impact of prophages on bacterial chromosomes. Mol Microbiol 53:9–18

Cardona-Maya W, Velilla P, Montoya CJ et al (2009) Presence of HIV-1 DNA in spermatozoa from HIV-positive patients: changes in the semen parameters. Curr HIV Res 7:418–424

Cardona-Maya W, Velilla PA, Montoya CJ et al (2011) In vitro human immunodeficiency virus and sperm cell interaction mediated by the mannose receptor. J Reprod Immunol 92:1–7

Chu H, Jo Y, Cho WK (2014) Evolution of endogenous non-retroviral genes integrated into plant genomes. Curr Plant Biol 1:55–59

Chung IY, Jang HJ, Bae HW et al (2014) A phage protein that inhibits the bacterial ATPase required for type IV pilus assembly. Proc Natl Acad Sci USA 111:11503–11508

Chuong EB, Elde NC, Feschotte C (2016) Regulatory evolution of innate immunity through co-option of endogenous retroviruses. Science 351:1083–1087

Cianciolo GJ, Copeland TD, Oroszlan S et al (1985) Inhibition of lymphocyte proliferation by a synthetic peptide homologous to retroviral envelope proteins. Science 230:453–455

Citti L, Rainaldi G (2005) Synthetic hammerhead ribozymes as therapeutic tools to control disease genes. Curr Gene Ther 5:11–24

Colson P, Ravaux I, Tamalet C et al (2014) HIV infection en route to endogenization: two cases. Clin Microbiol Infect 20:1280–1288

Cooper MD, Alder MN (2006) The evolution of adaptive immune systems. Cell 124:815–822
Cornelis G, Funk M, Vernochet C et al (2017) An endogenous retroviral envelope syncytin and its cognate receptor identified in the viviparous placental Mabuya lizard. Proc Natl Acad Sci USA 114:E10991–E11000
Cui J, Holmes EC (2012) Endogenous lentiviruses in the ferret genome. J Virol 86:3383–3385
Cullen BR, Cherry S, tenOever BR (2013) Is RNA interference a physiologically relevant innate immune response in mammals? Cell Host Microbe 14:374–378
da Fonseca GC, de Oliveira LFV, de Morais GL et al (2016) Unusual RNA plant virus integration into the soybean genome leads to the production of small RNAs. Plant Sci 246:62–69
de Koning AP, Gu W, Castoe TA et al (2011) Repetitive elements may comprise over two-thirds of the human genome. PLoS Genet 7:e1002384
Diener TO (1989) Circular RNAs: relics of precellular evolution? Proc Natl Acad Sci USA 86:9370–9374
Doron S, Melamed S, Ofir G et al (2018) Systematic discovery of antiphage defense systems in the microbial pangenome. Science 359:eaar4120
Dunlap KA, Palmarini M, Varela M (2006) Endogenous retroviruses regulate periimplantation placental growth and differentiation. Proc Natl Acad Sci USA 103:14390–14395
Enard D, Le C, Gwennap C et al (2016) Viruses are a dominant driver of protein adaptation in mammals. eLife 5:e12469
Escalera-Zamudio M, Greenwood AD (2016) On the classification and evolution of endogenous retrovirus: human endogenous retroviruses may not be 'human' after all. APMIS 124:44–51
Ewing RM, Chu P, Elisma F et al (2007) Large-scale mapping of human protein-protein interactions by mass spectrometry. Mol Syst Biol 3:89
Fischer MG (2020) The virophage family Lavidaviridae. Curr Issues Mol Biol 40:1–24
Fischer MG, Hackl T (2016) Host genome integration and giant virus-induced reactivation of the virophage mavirus. Nature 540:288–291
Fischer MG, Allen MJ, Wilson WH et al (2010) Giant virus with a remarkable complement of genes infects marine zooplankton. Proc Natl Acad Sci USA 107:19508–19513
Flores R, Gago-Zachert S, Serra P et al (2014) Viroids: survivors from the RNA world? Annu Rev Microbiol 68:395–414
Fujino K, Horie M, Honda T et al (2014) Inhibition of Borna disease virus replication by an endogenous bornavirus-like element in the ground squirrel genome. Proc Natl Acad Sci USA 111:13175–13180
Furuta Y, Abe K, Kobayashi I (2010) Genome comparison and context analysis reveals putative mobile forms of restriction-modification systems and related rearrangements. Nucleic Acids Res 38:2428–2443
Ge DT, Zamore PD (2013) Small RNA-directed silencing: the fly finds its inner fission yeast? Curr Biol 23:R318–R320
Gifford R, Tristem M (2003) The evolution, distribution and diversity of endogenous retroviruses. Virus Genes 26:291–315
Gifford RJ, Katzourakis A, Tristem M et al (2008) A transitional endogenous lentivirus from the genome of a basal primate and implications for lentivirus evolution. Proc Natl Acad Sci USA 105:20362–20367
Gilbert C, Maxfield DG, Goodman SM et al (2009) Parallel germline infiltration of a lentivirus in two Malagasy lemurs. PLoS Genet 5:e1000425
Gogvadze E, Buzdin A (2009) Retroelements and their impact on genome evolution and functioning. Cell Mol Life Sci 66:3727–3742
Gonzalez-Hernandez MJ, Swanson MD, Contreras-Galindo R et al (2012) Expression of human endogenous retrovirus type K (HML-2) is activated by the Tat protein of HIV-1. J Virol 86:7790–7805
Greenwood AD, Ishida Y, O'Brien SP et al (2018) Transmission, evolution, and endogenization: lessons learned from recent retroviral invasions. Microbiol Mol Biol Rev 82:e00044–e00017

Grissa I, Vergnaud G, Pourcel C (2007) The CRISPRdb database and tools to display CRISPRs and to generate dictionaries of spacers and repeats. BMC Bioinformatics 8:172
Gulino K, Rahman J, Badri M et al (2020) Initial mapping of the New York City Wastewater Virome. mSystems 5:e00876-19
Han GZ, Worobey M (2012) Endogenous lentiviral elements in the weasel family (Mustelidae). Mol Biol Evol 29:2905–2908
Han GZ, Worobey M (2015) A primitive endogenous lentivirus in a colugo: insights into the early evolution of lentiviruses. Mol Biol Evol 32:211–215
Haraguchi S, Good RA, James-Yarish M et al (1995) Differential modulation of Th1- and Th2-related cytokine mRNA expression by a synthetic peptide homologous to a conserved domain within retroviral envelope protein. Proc Natl Acad Sci USA 92:3611–3615
Haraguchi S, Good RA, Cianciolo GJ et al (1997) Immunosuppressive retroviral peptides: immunopathological implications for immunosuppressive influences of retroviral infections. J Leukoc Biol 61:654–666
Haraguchi S, Good RA, Day-Good NK (2008) A potent immunosuppressive retroviral peptide: cytokine patterns and signaling pathways. Immunol Res 41:46–55
Hartley JW, Yetter RA, Morse HC III (1983) A mouse gene on chromosome 5 that restricts infectivity of mink cell focus-forming recombinant murine leukemia viruses. J Exp Med 158:16–24
Hayward JA, Tachedjian M, Kohl C (2020) Infectious KoRV-related retroviruses circulating in Australian bats. Proc Natl Acad Sci USA 117:9529–9536
Hille F, Richter H, Wong SP et al (2018) The biology of CRISPR-Cas: backward and forward. Cell 172:1239–1259
Hoffmann B, Tappe D, Höper D et al (2015) A variegated squirrel bornavirus associated with fatal human encephalitis. N Engl J Med 373:154–162
Horie M (2017) The biological significance of bornavirus-derived genes in mammals. Curr Opin Virol 25:1–6
Horie M, Honda T, Suzuki Y et al (2010) Endogenous non-retroviral RNA virus elements in mammalian genomes. Nature 463:84–87
Huang S, Tao X, Yuan S et al (2016) Discovery of an active RAG transposon illuminates the origins of V(D)J recombination. Cell 166:102–114
Hyndman TH, Shilton CM, Stenglein MD et al (2018) Divergent bornaviruses from Australian carpet pythons with neurological disease date the origin of extant Bornaviridae prior to the end-Cretaceous extinction. PLoS Pathog 14:e1006881
Ikeda H, Laigret F, Martin MA et al (1985) Characterization of a molecularly cloned retroviral sequence associated with FV-4 resistance. J Virol 55:768–777
Imakawa K, Nakagawa S (2017) The phylogeny of placental evolution through dynamic integrations of retrotransposons. Prog Mol Biol Transl Sci 145:89–109
Iranzo J, Cuesta JA, Manrubia S et al (2017) Disentangling the effects of selection and loss bias on gene dynamics. Proc Natl Acad Sci USA 114:E5616–E5624
Ito J, Watanabe S, Hiratsuka T et al (2013) Refrex-1, a soluble restriction factor against feline endogenous and exogenous retroviruses. J Virol 87:12029–12040
Ito J, Sugimoto R, Nakaoka H et al (2017) Systematic identification and characterization of regulatory elements derived from human endogenous retroviruses. PLoS Genet 13:e1006883
Iwasaki YW, Siomi MC, Siomi H (2015) PIWI-interacting RNA: its biogenesis and functions. Annu Rev Biochem 84:405–433
Jebb D, Huang Z, Pippel M et al (2020) Six reference-quality genomes reveal evolution of bat adaptations. Nature 583:578–584
Jimenez RM, Polanco JA, Lupták A (2015) Chemistry and biology of self-cleaving ribozymes. Trends Biochem Sci 40:648–661
Jones JDG, Vance RE, Dangl JL (2016) Intracellular innate immune surveillance devices in plants and animals. Science 354:aaf6395

Jung YT, Lyu MS, Buckler-White A et al (2002) Characterization of a polytropic murine leukemia virus proviral sequence associated with the virus resistance gene Rmcf of DBA/2 mice. J Virol 76:8218–8224
Kambris Z, Hoffmann JA, Imler JL et al (2002) Tissue and stage-specific expression of the tolls in Drosophila embryos. Gene Expr Patterns 2:311–317
Kapitonov VV, Koonin EV (2015) Evolution of the RAG1-RAG2 locus: both proteins came from the same transposon. Biol Direct 10:20
Katzourakis A, Gifford RJ (2010) Endogenous viral elements in animal genomes. PLoS Genet 6: e1001191
Katzourakis A, Tristem M, Pybus OG et al (2007) Discovery and analysis of the first endogenous lentivirus. Proc Natl Acad Sci USA 104:6261–6265
Khan F, Furuta Y, Kawai M, Kaminska KH, Ishikawa K, Bujnicki JM, Kobayashi I (2010) A putative mobile genetic element carrying a novel type IIF restriction-modification system (PluTI). Nucleic Acids Res 38:3019–3030
Kim KS, Yamamoto Y, Nakaoka S et al (2020) Modeling Borna disease virus in vitro spread reveals the mode of antiviral effect conferred by an endogenous Bornavirus-like element. J Virol 94: e01204–e01220
Kitsou K, Kotanidou A, Paraskevis D et al (2020) Upregulation of human endogenous retroviruses in bronchoalveolar lavage fluid of COVID-19 patients. medRxiv. https://doi.org/10.1101/2020.05.10.20096958
Kobayashi Y, Horie M, Nakano A et al (2016) Exaptation of bornavirus-like nucleoprotein elements in afrotherians. PLoS Pathog 12:e1005785
Koonin EV (2016) Viruses and mobile elements as drivers of evolutionary transitions. Philos Trans R Soc Lond B Biol Sci 371:20150442
Koonin EV (2018) Hunting for treasure chests in microbial defense islands. Mol Cell 70:761–762
Koonin EV, Dolja VV (2013) A virocentric perspective on the evolution of life. Curr Opin Virol 3:546–557
Koonin EV, Dolja VV (2014) Virus world as an evolutionary network of viruses and capsidless selfish elements. Microbiol Mol Biol Rev 78:278–303
Koonin EV, Krupovic M (2015) Evolution of adaptive immunity from transposable elements combined with innate immune systems. Nat Rev Genet 16:184–192
Koonin EV, Krupovic M (2016) Virology: a parasite's parasite saves host's neighbors. Nature 540:2014–2205
Koonin EV, Krupovic M (2018) The depths of virus exaptation. Curr Opin Virol 31:1–8
Koonin EV, Makarova KS (2017) Mobile genetic elements and evolution of CRISPR-Cas systems: all the way there and back. Genome Biol Evol 9:2812–2825
Koonin EV, Wolf YI, Katsnelson MI (2017) Inevitability of the emergence and persistence of genetic parasites caused by evolutionary instability of parasite-free states. Biol Direct 12:31
Kovalskaya N, Hammond RW (2014) Molecular biology of viroid-host interactions and disease control strategies. Plant Sci 228:48–60
Krupovic M, Makarova KS, Forterre P et al (2014) Casposons: a new superfamily of self-synthesizing DNA transposons at the origin of prokaryotic CRISPR-Cas immunity. BMC Biol 12:36
Krupovic M, Dolja VV, Koonin EV (2020) The LUCA and its complex virome. Nat Rev Microbiol 18:661–670
Lange C, Hemmrich G, Klostermeier UC et al (2011) Defining the origins of the NOD-like receptor system at the base of animal evolution. Mol Biol Evol 28:1687–1702
Lavialle C, Cornelis G, Dupressoir A et al (2013) Paleovirology of 'syncytins', retroviral env genes exapted for a role in placentation. Philos Trans R Soc Lond B Biol Sci 368:20120507
Lee EJ, Banerjee S, Zhou H et al (2011) Identification of piRNAs in the central nervous system. RNA 17:1090–1099
Leulier F, Lemaitre B (2008) Toll-like receptors—taking an evolutionary approach. Nat Rev Genet 9:165–178

Mahla RS, Reddy MC, Prasad DV et al (2013) Sweeten PAMPs: role of sugar complexed PAMPs in innate immunity and vaccine biology. Front Immunol 4:248

Makarova KS, Wolf YI, Snir S, Koonin EV (2011) Defense islands in bacterial and archaeal genomes and prediction of novel defense systems. J Bacteriol 193:6039–6056

Malfavon-Borja R, Feschotte C (2015) Fighting fire with fire: endogenous retrovirus envelopes as restriction factors. J Virol 89:4047–4050

McGrath S, Fitzgerald GF, van Sinderen D (2002) Identification and characterization of phage-resistance genes in temperate lactococcal bacteriophages. Mol Microbiol 43:509–520

McKinney HH (1929) Mosaic diseases in the Canary Islands, West Africa and Gibraltar. J Agric Res 39:557–578

Michel N, Allespach I, Venzke S et al (2005) The Nef protein of human immunodeficiency virus establishes superinfection immunity by a dual strategy to down-regulate cell-surface CCR5 and CD4. Curr Biol 15:714–723

Miesen P, Joosten J, van Rij RP (2016) PIWIs Go Viral: arbovirus-derived piRNAs in vector mosquitoes. PLoS Pathog 12:e1006017

Moelling K, Matskevich A, Jung JS (2006) Relationship between retroviral replication and RNA interference machineries. Cold Spring Harb Symp Quant Biol 71:365–368

Monde K, Terasawa H, Nakano Y et al (2017) Molecular mechanisms by which HERV-K Gag interferes with HIV-1 Gag assembly and particle infectivity. Retrovirology 14:27

Mougari S, Sahmi-Bounsiar D, Levasseur A et al (2019) Virophages of giant viruses: an update at eleven. Viruses 11:733

Mougari S, Chelkha N, Sahmi-Bounsiar D et al (2020) A virophage cross-species infection through mutant selection represses giant virus propagation, promoting host cell survival. Commun Biol 3:248

Murphy J, Mahony J, Ainsworth S et al (2013) Bacteriophage orphan DNA methyltransferases: insights from their bacterial origin, function, and occurrence. Appl Environ Microbiol 79:7547–7555

Naderer M, Brust JR, Knowle D, Blumenthal RM (2002) Mobility of a restriction-modification system revealed by its genetic contexts in three hosts. J Bacteriol 184:2411–2419

Nandi S, Chandramohan D, Fioriti L et al (2016) Roles for small noncoding RNAs in silencing of retrotransposons in the mammalian brain. Proc Natl Acad Sci USA 133:12697–12702

Nuovo GJ, Becker J, Simsir A et al (1994) HIV-1 nucleic acids localize to the spermatogonia and their progeny. A study by polymerase chain reaction in situ hybridization. Am J Pathol 144:1142–1148

Obbard DJ, Gordon KHJ, Buck AH et al (2009) The evolution of RNAi as a defence against viruses and transposable elements. Philos Trans R Soc Lond B Biol Sci 364:99–115

Olagoke O, Quigley BL, Eiden MV et al (2019) Antibody response against koala retrovirus (KoRV) in koalas harboring KoRV-A in the presence or absence of KoRV-B. Sci Rep 9:12416

Olival KJ, Hosseini PR, Zambrana-Torrelio C et al (2017) Host and viral traits predict zoonotic spillover from mammals. Nature 546:646–650

Ophinni Y, Palatini U, Hayashi Y et al (2019) piRNA-guided CRISPR-like immunity in eukaryotes. Trends Immunol 40:998–1010

Paez-Espino D, Eloe-Fadrosh EA, Pavlopoulos GA et al (2016) Uncovering Earth's virome. Nature 536:425–430

Palmarini M, Mura M, Spencer TE (2004) Endogenous betaretroviruses of sheep: teaching new lessons in retroviral interference and adaptation. J Gen Virol 85:1–13

Parrish NF, Fujino K, Shiromoto Y et al (2015) piRNAs derived from ancient viral processed pseudogenes are transgenerational sequence specific immune memory in mammals. RNA 21:1691–1703

Planz O, Stitz L (1999) Borna disease virus nucleoprotein (p40) is a major target for CD8+-T-cell-mediated immune response. J Virol 73:1715–1718

Pradeu T, Carosella ED (2006) The self model and the conception of biological identity in immunology. Biol Philos 21:235–252

Sarker N, Fabijan J, Owen H et al (2020) Koala retrovirus viral load and disease burden in distinct northern and southern koala populations. Sci Rep 10:263
Sattler S, Ghadially H, Hofer E (2012) Evolution of the C-type lectin-like receptor genes of the DECTIN-1 cluster in the NK gene complex. ScientificWorldJournal 2012:931386
Seed KD, Lazinski DW, Calderwood SB et al (2013) A bacteriophage encodes its own CRISPR/Cas adaptive response to evade host innate immunity. Nature 494:489–491
Shabalina SA, Koonin EV (2008) Origins and evolution of eukaryotic RNA interference. Trends Ecol Evol 23:578–587
Skirmuntt EC, Escalera-Zamudio M, Teeling EC et al (2020) The potential role of endogenous viral elements in the evolution of bats as reservoirs for zoonotic viruses. Annu Rev Virol 7:103–119
Sofuku K, Parrish NF, Honda T et al (2015) Transcription profiling demonstrates epigenetic control of non-retroviral RNA virus-derived elements in the human genome. Cell Rep 12:1548–1554
Spencer TE, Mura M, Gray CA et al (2003) Receptor usage and fetal expression of ovine endogenous betaretroviruses: implications for coevolution of endogenous and exogenous retroviruses. J Virol 77:749–753
Stitz L, Bilzer T, Richt JA et al (1993) Pathogenesis of Borna disease. Arch Virol Suppl 7:135–151
Suzuki S (1975) FV-4: a new gene affecting the splenomegaly induction by Friend leukemia virus. Jpn J Exp Med 45:473–478
Suzuki Y, Kobayashi Y, Horie M et al (2014) Origin of an endogenous bornavirus-like nucleoprotein element in thirteen-lined ground squirrels. Genes Genet Syst 89:143–148
Suzuki Y, Frangeul L, Dickson LB et al (2017) Uncovering the repertoire of endogenous flaviviral elements in Aedes mosquito genomes. J Virol 91:e00571–e00517
Swarts DC, Makarova K, Wang Y, Nakanishi K, Ketting RF, Koonin EV, Patel DJ, van der Oost J (2014) The evolutionary journey of Argonaute proteins. Nat Struct Mol Biol 21:743–753
Takahashi N, Ohashi S, Sadykov MR, Mizutani-Ui Y, Kobayashi I (2011) IS-linked movement of a restriction-modification system. PLoS One 6:e16554
Tarlinton RE, Meers J, Young PR (2006) Retroviral invasion of the koala genome. Science 442:79–81
Tarlinton RE, Sarker N, Fabijan J et al (2017) Differential and defective expression of Koala Retrovirus reveal complexity of host and virus evolution. bioRxiv. https://doi.org/10.1101/211466
Terry SN, Manganaro L, Cuesta-Dominguez A et al (2017) Expression of HERV-K108 envelope interferes with HIV-1 production. Virology 509:52–59
Travis J (2009) Origins. On the origin of the immune system. Science 324:580–582
Villarreal LP (2011) Viral ancestors of antiviral systems. Viruses 3:1933–1958
Villarreal LP, Witzany G (2013) The DNA habitat and its RNA inhabitants: at the dawn of RNA sociology. Genomics Insights 6:1–12
Wang X, Kim Y, Ma Q, Hong SH, Pokusaeva K, Sturino JM, Wood TK (2010) Cryptic prophages help bacteria cope with adverse environments. Nat Commun 1:147
Weinberger AD, Sun CL, Pluciñski MM et al (2012) Persisting viral sequences shape microbial CRISPR-based immunity. PLoS Comput Biol 8:e1002475
Westra ER, van Houte S, Gandon S et al (2019) The ecology and evolution of microbial CRISPR-Cas adaptive immune systems. Philos Trans R Soc Lond B Biol Sci 374:20190101
Wilkins AS (2010) The enemy within: an epigenetic role of retrotransposons in cancer initiation. Bioeassays 32:856–865
Wolf YI, Silas S, Wang Y et al (2020) Doubling of the known set of RNA viruses by metagenomic analysis of an aquatic virome. Nat Microbiol 5:1262–1270
Wu Q, Luo Y, Lu R et al (2010) Virus discovery by deep sequencing and assembly of virus-derived small silencing RNAs. Proc Natl Acad Sci USA 107:1606–1611

Xu S, Zhou L, Liang X et al (2020) Novel cell-virus-virophage tripartite infection system discovered in the freshwater lake Dishui Lake in Shanghai, China. J Virol 94:e00149–e00120
Zhang L, Richards A, Khalil A et al (2020) SARS-CoV-2 RNA reverse-transcribed and integrated into the human genome. bioRxiv. https://doi.org/10.1101/2020.12.12.422516
Ziebell H, Carr JP (2010) Cross-protection: a century of mystery. Adv Virus Res 76:211–264
Zou J, Chang M, Nie P et al (2009) Origin and evolution of the RIG-I like RNA helicase gene family. BMC Evol Biol 9:85

Chapter 9
Application of Viruses for Gene Therapy and Vaccine Development

Kenneth Lundstrom

Abstract Although viruses have been demonstrated to cause a serious threat to our health, further confirmed by the current coronavirus virus pandemic, engineered viral vectors have been frequently used for gene therapy and vaccine development. A large number of different viral vectors have been subjected to proof-of-concept gene therapy evaluations in animal models for various indications before entering human clinical safety and efficacy studies. Moreover, immunization studies in animals and humans have been conducted with the goal of developing both prophylactic and therapeutic vaccines. Several viral-based gene therapy drugs have been approved for the treatment of various cancers and others have been subjected to clinical trials. Immunization studies have demonstrated protection against lethal challenges with pathogens and for example promising vaccine candidates for Ebola virus have generated favorable results in clinical trials. Several viral vector-based vaccine candidates for the SARS-CoV-2 virus, which causes the disease named COVID-19, have been designed resulting in strong antibody responses in clinical trials. Recently, adenovirus-based COVID-19 vaccines have received Emergency Use Authorization and are currently used for mass vaccinations. In this chapter, the most commonly used viral vectors systems are described, followed by examples of preclinical and clinical evaluation. A special emphasis is dedicated to coronaviruses.

9.1 Introduction

During the years, a large number of viral vectors have been engineered for heterologous gene expression, which has served as the basis for applications in gene therapy and vaccine development (Table 9.1). The spectrum of viral vectors is very broad, including DNA and RNA viruses with single-stranded (ss) and double-stranded (ds) genomes (Lundstrom 2018). It is impossible to recommend a universal viral vector as results have indicated that similar preventive and therapeutic

K. Lundstrom (✉)
PanTherapeutics, Lutry, Switzerland

C. J. Hurst (ed.), *The Biological Role of a Virus*, Advances in Environmental Microbiology 9, https://doi.org/10.1007/978-3-030-85395-2_9

Table 9.1 Examples of gene therapy and vaccine targets

Therapeutic & Vaccine	Vector	Technology	Effect
Gene replacement	AAV, Ad	CRISPR-Cas9	Successful treatment of DMD mice
	Lentivirus	CRISPR-Cas9	Suppression of HSV-1 infection
Gene compensation	MMLV	Ex vivo	Cure of SCID-X1 patients
	AAV	transduction	Therapy of hemophilia with FVIII
	Ad	Systemic delivery	Treatment of OTC
Cytotoxic response	VSV	Cytosine deaminase	Enhanced tumor killing
	GINaTK	HSV-TK	Killing of SW1990 pancreatic cancer cells
	SFV, SIN	HSV-TK-GFP	Killing of glioma cells
	Ad, HIV-1	HSV-TK-GFP	Killing of glioma and renal cancer cells
Oncolytic viral vectors	Ad	ONYX-015	Replication in p53-deficient tumor cells
	Ad	CV706	Replication in prostate cancer cells
	HSV	HSV G207	Tumor-specific tropism in gliomas
Viral vectors with tissue-specific promoters	Ad	AFP promoter	Liver cell-specific expression
	Ad	PSA promoter	Prostate cell-specific expression
	Ad	MUC-1 promoter	Breast cell-specific expression
Tumor suppressor genes	Ad	p53	Prolonged survival in sarcoma patients
	Ad	PTEN	Expression in breast cancer cells
	Ad	Rb	Expression in pituitary tumors
	Ad	MnSOD	50% reduction in tumor growth in hamster
Immunotherapy	Ad	IL-12, CD80	Enhanced CTL response in myeloma cells
	HIV-1	IL12/FasTI	Decreased tumor growth
	HSV-1	IL-12	Enhanced oncolytic activity
	SIN	IL-12, IL-15	Tumor regression in ovarian cancer
	SFV	IL-18	Tumor regression in colon cancer
	ALVAC	IL-2	Significantly reduced tumor recurrence
	NYVAC	IL-2	Significantly reduced tumor recurrence
Apoptosis induction	Ad	TRAIL + ING4	Apoptosis, tumor angiogenesis suppression
	HVS	TRAIL	Apoptosis in colorectal cell lines, disruption of melanoma spheroid cultures
	Ad	Smac	Complete eradication of hepatoma xenografts

(continued)

Table 9.1 (continued)

Therapeutic & Vaccine	Vector	Technology	Effect
Angiogenesis inhibitor genes	HIV-1	PEDF	Neuroprotection of CGCs
	AAV	Endostatin	Tumor regression in hamster
	AAV	Angiostatin	Tumor regression, long-term survival in rats
	EIAV	Endo/Angiostatin	Safe and robust expression in patients
Inhibitor of apoptosis	AAV	XIAP	Protection of SH-SY5Y cells, motor neurons
	AAV	XIAP	Preservation of photoreceptor layer in feline retina
Antigen production	SIN	Her2/neu	Tumor protection, prolonged survival
	Ad	Her2/neu	Therapeutic efficacy in mice
	VEE	TRP-2	Long-term melanoma protection
	VV	TRP-1	Protection against melanoma B16
	VEE	PSA	Immune response, delay of tumor growth
	VEE	PSCA	Tumor prevention
	VEE	STEAP	Tumor prevention, long-term survival
Gene silencing	Ad	TMUV shRNA	Inhibition of TMUV infection in Vero cells
	AAV9	Ad miRNA	Inhibition of Ad infection in Syrian hamster
	RV	TRMP7 siRNA	Induction of apoptosis in RBL-2H3 cells
	HIV-1	HIV-1 shRNA	Shut-down of HIV-1 infection
	VSV	Nodamura B2	Enhanced cytotoxicity, miRNA processing
	SFV	SFV miR124	Reduced neurovirulence
Combination therapy	Ad	HPR + IL-12	Tumor regression, prolonged survival

AAV adeno-associated virus, Ad adenovirus, AFP alpha-fetoprotein, ALVAC canarypox virus, CGCs cerebellar granule cells, CRISPR-Cas-9 clustered regularly interspaced short palindromic repeats/CRISPR-associated nuclease 9, CTL cytotoxic T-cell, DMD Duchenne muscular dystrophy, EIAV equine infectious anemia virus, GINaTK replication-defective retrovirus, HIV-1 human immunodeficiency virus-1, HRP horseradish peroxidase, HSV-tk herpes simplex virus thymidine kinase, HSV-TK-GFP herpes simplex virus thymidine kinase-green fluorescent protein fusion, HVS herpesvirus saimiri, IL interleukin, ING4 inhibitor of growth 4, MnSOD manganese sodium dismutase, NYVAC genetically attenuated vaccinia virus, OTC ornithine transcarbamylase, PEDF pigment epithelium-derived factor, PSA prostate specific antigen, PSCA prostate stem cell antigen, Rb retinoblastoma, RV retrovirus, SFV Semliki Forest virus, SIN Sindbis virus, Smac second mitochondria-derived activator of caspase, STEAP six-transmembrane epithelial antigen of the prostate, TMUV Tembusu virus, TRAIL tumor necrosis factor-related apoptosis-inducing ligand, TRP tyrosinase-related protein, VEE Venezuelan equine encephalitis virus, VSV vesicular stomatitis virus, VV vaccinia virus, XIAP X-linked inhibitor of apoptosis efficacy can be established for different viral vector types for the same indication although differences have also been demonstrated. The therapeutic efficacy also depends on whether the therapeutic effect requires short- or long-term treatment, which relates to the acute or chronic nature of disease. Therefore, it is advantageous to present an overview of the most common viral vector systems as accomplished by this chapter.

9.2 Gene Therapy and Vaccine Targets

Before describing the available viral vector systems, it is appropriate to summarize the target genes applied for gene therapy and vaccine development (Table 9.1). The classic gene therapy approach relates to the compensation of a mutated or malfunctioning gene by a healthy one. This approach has been demonstrated for various indications such as for X-linked severe combined immunodeficiency (SCID-X1) (McCormack and Rabbitts 2004), hemophilia (Nienhuis et al. 2017) and ornithine transcarbamylase (OTC) deficiency (Lehrman 1999). A number of therapeutic genes have been evaluated as listed in Table 9.1. However, with the development of the RNA-guided genome editing tool named clustered regularly interspaced short palindromic repeats/CRISPR-associated nuclease 9 (CRISPR-Cas9), gene replacement for the correction of causal mutations in monogenic disorders became a reality (Xiao-Jie et al. 2015). In one application, CRISPR-Cas9 genome editing based on AAV-delivery was used for the treatment of Duchenne muscular dystrophy (DMD) (Amoasii et al. 2017). The deletion of exon 50 in the dystrophin gene was corrected by systemic delivery of AAV encoding CRISPR/Cas9 editing components, which restored up to 90% of dystrophin expression in skeletal and heart muscles of DMD mice. In addition to AAV, Ad and lentivirus vectors have been utilized for CRISPR-Cas9 delivery (Lino et al. 2018).

A straightforward approach commonly used for cancer therapy is overexpression of either anti-tumor or cytotoxic genes from viral vectors, delivered for initial evaluation to cell cultures before conducting studies in suitable animal models. In the context of cytotoxic genes or suicide genes, aiming at killing tumor cells, the strategy has comprised the introduction of cytotoxic genes by viral vectors into tumor cells, where a non-toxic pro-drug is converted into an activated toxin (Zarogoulidis et al. 2013). One of the most prominent systems is based upon cytosine deaminase (CD) from *Escherichia coli*, which converts the prodrug 5-Flurocytosine (5-FC) into 5-Fluorouracil (5-FU) (Bentires-Alj et al. 2000). Commonly, CD-based therapy has been relying on plasmid-based delivery, but also viral vectors such as vesicular stomatitis virus (VSV) have demonstrated enhanced tumor killing (Porosnicu et al. 2003). Another system based on the herpes simplex virus thymidine kinase gene (HSV-TK) converts ganciclovir (GCV) into GCV monophosphate and further into GCV triphosphate, which is a cytotoxic DNA synthesis inhibitor and cell cycle blocker leading to apoptosis and cell death (Robe et al. 2000). Human SW1990 pancreatic tumor cells treated with a replication-defective retrovirus carrying the HSV-TK demonstrated substantial apoptosis and further provided a bystander effect, where activated GCV diffused to neighbouring bystander cells killing them as well (Wang et al. 2004). Moreover, the HSV-TK approach was evaluated in glioma and renal carcinoma cells using Ad virus, alphavirus and lentivirus vectors (Loimas et al. 2000). Superior cell killing was observed for alphaviruses in glioma cell lines, whereas Ad virus and lentiviruses were more efficient in renal carcinoma cells.

A number of anti-tumor genes have been identified (Liu et al. 2005). In fact, oncolytic viral vectors as such have proven efficacious for cancer therapy. In this context, the oncolytic Ad virus vector ONYX-015 with an E1B-55 kDa-deletion, has been demonstrated to selectively replicate in p53-deficient tumor cells killing them (Reis and Korn 2002). Similarly, the Ad virus vector CV706 selectively replicates in prostate cancer cells (Chen et al. 2001). Moreover, a genetically engineered, conditionally replicating HSV G207 vector showing tumor-specific tropism has been previously evaluated in primates and in three phase I trials in adult patients with recurrent/progressive high-grade gliomas and now is the subject of a trial in children with recurrent or progressive supratentorial malignant tumors (Waters et al. 2017). Another approach has been to engineer tumor-specific promoters to provide selective expression of inserted antitumor genes (see below) in tumor cells (Chiocca 2002). Several tissue-specific promoters such as the alpha-fetoprotein (AFP) promoter for hepatic cancer (Cerghini et al. 1988), the prostate-specific antigen (PSA) promoter for prostate cancer (Pang et al. 1995) and the mucin-1 (MUC-1) promoter for breast cancer (Kurihara et al. 2000) have been introduced into Ad virus vectors.

Several therapeutic (antitumor) genes such as the tumor suppressor genes p53, retinoblastoma (Rb), liver-related putative tumor suppressor (LPTS), phosphatase and tensin homolog (PTEN) and manganese sodium dismutase (MnSOD) have been identified (Liu et al. 2005). For instance, recombinant Ad virus expressing p53 has been subjected to clinical trials in patients with advanced unresectable soft-tissue sarcomas leading to significant improvement of progression-free status and overall survival of patients (Xiao et al. 2018). An Ad vector was engineered for the expression of the PTEN tumor suppressor gene resulting in PTEN expression in 70% of breast cancer cells (Chen et al. 2006). Moreover, Ad vectors applied for the expression of the Rb suppressor oncogene showed successful rescue from pituitary tumors in mice (Bolognani and Goya 2001). Ad vectors have also been evaluated in a hamster cheek pouch model system, where expression of MnSOD resulted in 50% reduction in tumor growth (Lam et al. 2000).

Among apoptosis inducing genes the tumor necrosis factor-related apoptosis-inducing ligand (TRAIL) has been co-expressed with the inhibitor of growth 4 (ING4) from Ad vectors eliciting potent eradicative effects such as apoptosis induction, immune responses and suppression of tumor angiogenesis in an orthotopic human hepatocellular carcinoma mouse model (Zeng et al. 2019; El-Shemi et al. 2018). In another approach, the full-length TRAIL gene expressed from a herpesvirus saimiri (HVS) vector showed considerable apoptosis induction in infected colorectal SW480 cancer cells and disruption of spheroid melanoma Mel888 cultures (Turrell et al. 2012). Moreover, the Ad vector ZD55 overexpressing the fusion protein of TRAIL and the second mitochondria-derived activator of caspase (Smac) showed a broad antitumor effect and completely eradicated xenograft hepatoma tumors in mice (Wang et al. 2012b).

In the context of angiogenesis inhibitor genes, the pigment epithelium-derived factor (PEDF) has been expressed from a lentivirus vector in primary cerebellar granule cells (CGCs), which demonstrated neuroprotection in comparison to mock-infected control cells (Nomura et al. 2001). In another approach, AAV-mediated expression of endostatin was evaluated using an orthotopic metastatic pancreatic cancer model in Syrian golden hamsters (Noro et al. 2004). Intramuscular or intravenous injections of 5×10^{10} AAV particles showed superior antitumor effect in primary pancreatic tumors and in liver metastases. In another approach, AAV-based expression of angiostatin was evaluated in a rat glioma model, which resulted in tumor regression and long-term survival in 40% of treated rats (Ma et al. 2002). Moreover, a lentivirus vector, equine infectious anemia virus (EIAV), has been applied for the expression of endostatin and angiostatin in patients with macular degeneration, providing safe and robust transgene expression for ocular gene therapy (Campochiaro et al. 2017).

The inhibitor of apoptosis (IAP) gene has also been tested for viral-based gene therapy. In this context, the X-linked IAP gene (XIAP) expressed from an Ad vector reduced the percentage of active caspase-3 positive SH-SY5Y neurons, preserved cell density and protected uninfected neighboring cells (Garrity-Moses et al. 2006). Furthermore, inhibition of proapoptotic activity was discovered from studies of motor neurons and dorsal root ganglion cells in a primary E15 model. In another approach, expression of XIAP from an AAV vector was verified in a feline model (Wassmer et al. 2017). It was demonstrated that a significant preservation of the photoreceptor layer in the retina occurred after treatment with AAV-XIAP as compared to AAV-GFP treated animals.

A large number of immunomodulating cytokines have been applied for immunotherapy using viral vectors such as Ad viruses, alphaviruses, retroviruses, lentiviruses, poxviruses, and HSV (Wen et al. 2001; Zajakina et al. 2017; Yang et al. 2018; Jourdier et al. 2003; Todo 2012). For instance, interleukins IL-12, IL-15 and IL-18 have been expressed from alphavirus vectors and subjected to immunization studies in animal models. In this context, tumor regression has been observed for IL-12 and IL-15 in ovarian tumor mouse models (Tseng et al. 2004) and for IL-18 in colon cancer models (Chikkanna-Gowda et al. 2006). Moreover, an Ad virus vector expressing IL-12 and CD80 provoked increase in cytotoxic T-cell responses in human U266 myeloma cells (Wen et al. 2001). It was also demonstrated that lentivirus-based expression of the mIL-12/FasTI fusion protein enhanced killer activation, increased caspase-3 activity, and decreased tumor growth in vitro (Yang et al. 2018). The canarypox ALVAC virus and the genetically attenuated vaccinia virus (VV) NYVAC expressing IL-2 were administered to cats resulting in a significant recurrence of tumors (Jourdier et al. 2003). A conditionally replicating HSV-1 vector expressing IL-12 showed enhanced oncolytic activity (Todo 2012). In a suicide gene therapy combination with immunotherapy Ad virus vectors expressing horseradish peroxidase (HRP) from a human telomerase reverse transcriptase (hTERT) promoter (AdhTERTHRP) and murine interleukin-12 (IL-12) expressed from a CMV promoter (AdCMVmIL-12) showed enhanced tumor inhibition and prolonged survival compared to treatment with either of the Ad virus alone (Xu et al. 2011).

Antigen expression represents an essential part of vaccine development and various viral expression vectors based on Ad viruses, alphaviruses, HSV and Newcastle disease virus (NDV) have been applied (Harrop and Carroll 2006; Lundstrom 2020a). The delivery and expression of full-length antigen-coding genes can provide a broad spectrum of potential antigenic epitopes in a native conformation, which might provide enhanced immunity (Bolhassani et al. 2011). The antigens targeted can be divided into two broad groups, those based on surface structures of infectious agents such as viruses, bacteria, or other pathogens (Lundstrom 2020a) versus tumor-associated antigens (TAAs), commonly associated with different types of tumors (Zajakina et al. 2017).

Among the TAAs expressed from viral vectors, the Her2/neu oncogene has been expressed from alphavirus vectors in mouse breast cancer models, showing tumor prevention, reduction of metastases and prolonged survival of vaccinated animals (Lachman et al. 2001). Moreover, Ad virus-based expression of Her2/neu prevented growth of human A2L2 breast cancer cells, induced high levels of cellular and humoral responses in immunized mice, and showed therapeutic activity in mice injected intravenously with tumor cells prior to immunization (Wang et al. 2005). In the context of melanoma, the tyrosinase-related protein 2 (TPR-2) expressed from an alphavirus vector induced long-term protection in a mouse B16 melanoma model (Avogadri et al. 2010). In another study, it was demonstrated that mice immunized with a vaccinia virus vector expressing TRP-1 were protected against challenges with lethal doses of B16 melanoma (Overwijk et al. 1999). Several prostate cancer TAAs such as prostate specific antigen (PSA) (Riabov et al. 2015) prostate stem cell antigen (PSCA) (Garcia-Hernandez et al. 2008) and six-transmembrane epithelial antigen of the prostate (STEAP) (Garcia-Hernandez et al. 2007) have been expressed from alphavirus vectors resulting in immune responses and tumor prevention.

The phenomenon of RNA interference (RNAi), known as RNA-based gene silencing, although originally discovered in plants and worms (Lee et al. 1993) also occurs frequently in mammals with more than 2300 human micro-RNA (miRNA) sequences having been identified (Alles et al. 2019). Briefly, gene silencing takes place either through mRNA degradation or suppression of translation through the action of small interfering RNAs (siRNAs) (Dana et al. 2017), short hairpin RNAs (shRNAs) (Moore et al. 2010) or miRNAs (Pillai 2005). Delivery of RNAi sequences can be achieved by plasmid DNA or viral vectors. In this context, Ad virus delivery of shRNA demonstrated significant inhibition of Tembusu virus (TUMV) infection in Vero cells (Wang et al. 2017). Moreover, AAV9 vector-based delivery of miRNAs targeting Ad virus resulted in inhibition of Ad virus infections in immunosuppressed Syrian hamsters (Schaar et al. 2017). Additionally, a retrovirus vector has been applied for siRNA targeting of the transient receptor potential melastatin 7 (TRPM7) gene, which resulted in suppression of TRPM7 expression, significantly decreased cellular target survival rates, and increased apoptosis in

RBL-2H3 rat basophil cells (Ng et al. 2012). Related to lentiviruses, a conditionally replicating HIV-1 vector was engineered for the delivery of shRNA to block chronic HIV-1 infections (Westerhout et al. 2006). Rhabdoviruses have also been engineered for RNAi approaches. In this case, the oncolytic VSV expressing the Nodamura virus B2 protein, known to inhibit RNAi-mediated immune responses, showed enhanced replication and cytotoxicity, and altered miRNA processing in cancer cells (Bastin et al. 2018). In the context of alphaviruses, introduction of six tandem neuron-specific miR124 sequences between the nonstructural protein genes nsP3 and nsP4 in the Semliki Forest virus (SFV) genome, generated an attenuated spread in the CNS, providing reduced neurovirulence and demonstrated the potential of using oncolytic alphavirus vectors for virotherapy (Ylösmäki et al. 2013).

9.3 Viral Vectors

There is a large number of viral vectors to choose from for both gene therapy and vaccine development (Table 9.2). Vectors have been engineered for viruses with single-stranded and double-stranded nucleic acid structures and using DNA as well as RNA viruses. Depending on the indication, viral vectors often have been engineered for short-term high-level transgene expression, typically favorable for cancer treatment and vaccine development. As chronic diseases require long-term expression of therapeutic genes, vectors providing chromosomal integration and long-term presence are superior. Below are summarized the features of the most frequently used viral vectors.

9.3.1 Adenoviruses

Adenoviruses (Ad) have been frequently used as both gene therapy and vaccine vectors (Schiedner et al. 1998). The naked (not possessing a membrane envelope) dsDNA Ad has a packaging capacity of 7.5 kb foreign DNA, which typically can generate high levels of transient episomal expression of the gene of interest in a broad range of mammalian host cells. The strong immune response elicited by the original Ad expression vectors triggered the engineering of the so-called gutless second and third generation Ad vectors, which showed significantly reduced immunogenicity due to most of their structural Ad genes having been deleted (Wang et al. 2012b). Due to the deletion of the structural Ad genes, packaging cell lines have been engineered to support large-scale production of Ad particles of GMP-grade for clinical applications (Wei et al. 2017). Ad represented the gold standard in the early days of gene therapy, but suffered an enormous set back, when a patient treated with Ad vectors for OTC deficiency died in 1999 (Lehrman 1999). However, since then substantial progress has been made in relation to viral vector safety and the set-up of clinical procedures, so it is with great satisfaction that current Ad-based vaccine

Table 9.2 Engineered viral vectors

Virus	Genome	Insert size	Features
Adenoviruses			
Ad5 Ad26 ChAd	dsDNA	< 7.5 kb	Broad host cell range, transient expression Initially strong immunogenicity Reduced immunogenicity (gutless Ad) Pre-existing immunity in humans
AAV			
2, 3, 5, 6, 8, 9	ssDNA	< 4 kb	Relatively broad host cell range Limited packaging capacity of foreign DNA > use of dual AAV vectors Chromosomal integration > long-term expression Strong immunogenicity against AAV re-administration > use of different AAV serotypes re-administration
Alphaviruses			
SFV, SIN, VEE, M1	ssRNA + strand	8 kb	Very broad host cell range, risk of neurovirulence Very high transgene expression Low immunogenicity Transient and cytopathogenic vectors not appropriate for chronic diseases but superior for acute diseases and vaccine development Flexibility to use RNA, DNA and viral particles
Flaviviruses			
KUN, WNV, DENV, TBE, YFV	ssRNA + strand	6 kb	Relatively broad host cell range Transient expression Relatively good packaging capacity Established packaging cell lines
Measles virus			
MV-Edm	ssRNA - strand	6 kb	Relatively broad cell host range Transient expression Established reverse genetics and packaging cell lines Oncolytic MV strains available for cancer therapy
Rhabdoviruses			
Rabies VSV	ssRNA - strand	6 kb	Relatively broad host cell range Transient expression High genetic stability Envelope flexibility for pseudotyping Established vaccinia-free packaging cell line
Herpes viruses			
HSV, HSV-1	dsDNA	> 30 kb	Broad host cell range Large insert capacity Latent infection > long-term transgene expression Low cytotoxicity Oncolytic HSV for selective killing of tumor cells

(continued)

Table 9.2 (continued)

Virus	Genome	Insert size	Features
Retroviruses			
MMSV MSCV	ssRNA	8 kb	Generally broad host range, but only dividing cells Random integration into host genome Long-term transgene expression
Lentiviruses			
HIV-1, HIV-2 SIV, FIV, EIAV	ssRNA	8 kb	Broad host range including non-dividing cells Low cytotoxicity Chromosomal integration Long-term expression
Newcastle disease virus			
	ssRNA	4 kb	Limited packaging capacity of foreign genes Specific replication in tumor cells Oncogenic NDV strains with reduced pathogenicity in chicken
Reoviruses			
	dsRNA	ND	Oncolytic viruses targeting and killing cancer cells Probably limited packaging capacity although therapeutic genes are not needed
Poxviruses			
VV Avipox	dsDNA	> 30 kb	Broad host range Excellent packaging capacity Engineering of replication-competent vectors
Picornaviruses			
Coxsackievirus A21, B3 Poliovirus PV-1	ssRNA	6 kb	Cell tropism No chromosomal integration Replication in some non-dividing cells Oncolytic and attenuated strains
Polyoma viruses			
SV40	dsDNA	17.7 kb	Large packaging capacity Broad host range: dividing and non-dividing cells SV40-free VLPs Vero packaging cell line
Chimeric viruses			
SFV-VSV-G	ssRNA	NA	Modified cell tropism, improved biosafety
AAV1-AAV2	ssDNA	NA	Heparin column purified chimeric AAV
VV-MMLV	ssRNA	NA	MMLV genomes expressed from VV-based particles
AAV2-HBoV1-4, AAV2-GBoV	ssDNA	NA	Efficient transduction of pHAE cells and hepatocytes

AAV adeno-associated virus, Ad adenovirus, ChAd Chimpanzee adenovirus, DENV Dengue virus, EIAV equine infectious anemia virus, FIV feline immunodeficiency virus, GBoV gorilla bocavirus, HBoV human bocavirus, HSV herpes simplex virus, KUN Kunjin virus, MMLV Moloney murine leukemia virus, MMSV Moloney murine sarcoma virus, MSCV murine stem cell virus, MV-Edm measles virus Edmonston strain, NA not applicable, ND not determined, NDV Newcastle disease virus, pHAE primary human airway epithelia, RABV rabies virus, SFV Semliki Forest virus, SIN Sindbis virus, SIV simian immunodeficiency virus, SV40 simian virus 40, TBE tick-borne encephalitis virus, VEE Venezuelan equine encephalitis virus, VLPs Virus-like particles, VSV vesicular stomatitis virus, VV vaccinia virus, WNV West Nile virus, YFV yellow fever virus development against Coronavirus disease (COVID-19) has been proven safe showing good vaccine efficacy in human clinical trials (Voysey et al. 2021) receiving Emergency Use Authorization (EUA) for vaccination of humans in the UK and elsewhere. However, recently rare cases of vaccine-induced thrombotic thrombocytopenia (VITT) have been detected after mass vaccinations with Ad-based COVID-19 vaccines (Greinacher et al. 2021; Muir et al. 2021).

9.3.2 Adeno-associated Viruses

Another frequently used virus family, especially for gene therapy, is adeno-associated viruses (AAV). In contrast to Ad, the genome of AAV is ssDNA with a limited packaging capacity of only 4 kb inserts of foreign DNA (Samulski and Muzycka 2014). The advantages of AAV vectors relate to their low pathogenicity and toxicity in vivo. And, due to their integration into the host genome, AAV can provide long-term transgene expression (Park et al. 2008), which has been advantageous in treatment of chronic diseases such as hemophilia (Nienhuis et al. 2017). One major drawback of efficient application of AAV vectors has been the immune response triggered by AAV, which has made gene therapy requiring repeated administration inefficient (Mingozzi and High 2013). One approach to address the AAV-induced immunogenicity comprises the application of different AAV serotypes for each re-administration. Moreover, antibody-binding sites and epitopes have been modified and AAV capsids have been engineered to re-direct tissue tropism and facilitate passage through the blood brain barrier (BBB) (Rabinowitz et al. 2019). Another approach involves the application of exosome-associated AAV (exo-AAV) for liver gene delivery, which allows reduced vector dosage while limiting preexisting AAV capsid immunity (Meliani et al. 2017). The small packaging capacity of AAV (Grieger and Samulski 2005) has also been addressed by engineering of dual AAV vectors either by fragmented, overlapping, trans-splicing or hybrid vectors (McClements and MacLaren 2017).

9.3.3 Alphaviruses

Alphaviruses have an ssRNA genome surrounded by an icosahedral capsid structure and an envelope containing spike proteins (Strauss and Strauss 1994). The unique feature of alphaviruses and other ssRNA viruses is the self-replication of their RNA genome and subgenomic RNA in infected host cells, resulting in approximately 200,000-fold RNA amplification. This feature together with the presence of strong subgenomic promoters has therefore made these self-amplifying RNA (saRNA) viruses to be considered attractive vectors for both gene therapy and vaccine development due to the high levels of heterologous gene expression obtained (Lundstrom 2019). Alphaviruses possess a relatively good packaging capacity of up to 8 kb of foreign gene inserts. The host range is also very broad, although some concern has been associated with the neurovirulence of alphaviruses (Ylösmäki et al. 2013). Generally, replication-deficient alphaviruses have been applied, resulting in transient heterologous gene expression, which is not in line with gene therapy applications for chronic diseases. Moreover, alphaviruses are cytopathogenic by inducing apoptosis of host cells (Li and Stollar 2004). In contrast, high levels of transient expression are advantageous for cancer therapy and vaccine development. Alphaviruses possess a unique flexibility as RNA replicons and DNA plasmids can be used in addition to recombinant alphavirus particles for gene delivery and vaccinations.

9.3.4 Flaviviruses

Flaviviruses are also ssRNA viruses, similar to alphaviruses with the saRNA feature (Pijlman et al. 2006). In the case of Kunjin virus (KUN) RNA-, DNA- and viral particle-based expression systems have been engineered. Other flaviviruses such as West Nile virus (WNV) (Scholle et al. 2004), Dengue virus (DENV) (Pang et al. 2001), tick-borne encephalitis virus (TBE) (Gherke et al. 2003) and yellow fever virus (YFV) (Jones et al. 2005) have been applied as expression vectors. Moreover, packaging cell lines have been engineered for KUN (Khromykh et al. 1998) and TBE (Gherke et al. 2003). Flaviviruses have a relatively good capacity for accommodation of foreign gene sequences and also a relatively broad host range. Similar to alphaviruses, the flavivirus-based expression is transient making them applicable for cancer therapy and vaccine development.

9.3.5 Measles Virus

Although measles viruses (MV) carry an ssRNA genome, in contrast to alphaviruses and flaviviruses the measles virus genome is of negative polarity (Apostolopoulos 2016). The MV expression systems are based on replicating MV rescued from cloned DNA constructs (Radecke et al. 1995). The gene of interest can be introduced into different regions of the MV genome and recombinant MV particles are rescued by reverse genetics in an HEK293 helper cell line (Singh et al. 1999). Syncytia formed from HEK293 helper cells that were transfected with recombinant MV and a plasmid expressing the MV polymerase L gene transferred to Vero cells produce recombinant MV particles after three days. Like other saRNA viral vectors, MV also generates transient expression showing promise in vaccine development (Hu et al. 2016). However, oncolytic MV strains such as the MV-Edmonston strain have also been utilized for cancer therapy (Lange et al. 2013; Reddi et al. 2012).

9.3.6 Rhabdoviruses

Rhabdoviruses are also saRNA viruses with a negative ssRNA polarity (Finke and Conzelmann 2005). Reverse genetics systems have allowed the engineering of efficient vectors for vaccine development and gene therapy, including oncolytic virotherapy. Rhabdoviruses have demonstrated high genetic stability and tolerate changes in the virus envelope. Among rhabdoviruses, both rabies virus (RABV) (Luo et al. 2016) and vesicular stomatitis virus (VSV) (An et al. 2013) have been engineered as expression vectors. Like for MV vectors, rhabdoviruses have been subjected to reverse genetics and efficient recovery of VSV particles was achieved based on vaccinia viruses (An et al. 2013). However, as vaccinia viruses induce strong cytopathogenic effects in transfected cells and may also contaminate recombinant virus stocks, a vaccinia-free packaging system for RABV was engineered in BHK cells (Ito et al. 2003).

9.3.7 Herpes Simplex Viruses

Herpes simplex viruses (HSV) contain a large dsDNA genome and possess an impressive packaging capacity of more than 30 kb of foreign DNA (Holmes et al. 2000). Due to their latent infection mode, HSV vectors can reside for life in neurons and can therefore provide long-term transgene expression (Epstein et al. 2005). Furthermore, deletion of non-essential HSV genes from the expression vector has substantially reduced its cytotoxic effect on host cells (Holmes et al. 2000). Oncolytic HSV-1 vectors have also been engineered for selective killing of human non-small cell lung cancer (NSCLC) cells (Li et al. 2013). Incorporation of four copies of the microRNA-145 target sequences into the $3'$ end untranslated region of the HSV-1 essential viral gene ICP27 resulted in selectively reduced cell proliferation of NSCLC cells.

9.3.8 Retroviruses

Retroviruses are enveloped viruses with an ssRNA genome, which can accommodate up to 8 kb of foreign inserts (Schambach and Morgan 2016). Due to generation of intermediate dsDNA molecules in infected host cells, retroviruses can integrate into the host genome. Because of their random chromosomal integration, problems have arisen as for instance the insertion of the therapeutic gene into the LMO2 oncogene region, which resulted in development of leukemia in patients successfully treated for SCID-X1 (McCormack and Rabbitts 2004; Hacein-Bey-Abina et al. 2008). However, the integration problem has been addressed by designing retrovirus vectors for targeted insertion into the host genome and for large-scale retrovirus production engineering of improved helper cell lines (Hu and Pathak 2000). The major drawback of applying retroviruses is their incapacity to transduce non-dividing cells, which has clearly shifted the focus to using lentivirus vectors instead.

9.3.9 *Lentiviruses*

Lentiviruses belong to the group of ssRNA retroviruses, but in contrast to the latter are capable of transducing both dividing and non-dividing cells (Vigna and Naldini 2000). The packaging capacity of 8 kb is similar to the classic retroviruses, but lentiviruses present a broad host range and low cell cytotoxicity (Kay et al. 2001). In contrast to classic retroviruses, lentiviruses have demonstrated non-random integration, favoring active transcription units (Ciuffi 2008). Moreover, it was demonstrated that the HIV-1 integration site was controlled by the LEDGF/p75 protein and that a fusion protein of the C-terminal HIV integrase-binding region of LEDGF/p75 and the N-terminal chromodomain of heterochromatin protein-1alpha (HP1alpha) could target HIV-1 vector integration and provide a safer lentivirus system for gene therapy (Silvers et al. 2010). In addition to human lentiviruses, simian immunodeficiency virus (SIV) (Nakajima et al. 2000), feline immunodeficiency virus (FIV) (Hartmann 2012) and EIAV (Olsen 1998) have been used for gene therapy applications.

9.3.10 *Newcastle Disease Virus*

The negative sense ssRNA Newcastle disease virus (NDV) has been proposed as an attractive vector for both vaccine development and cancer therapy (Ganar et al. 2014). The packaging capacity of foreign genetic material is limited to approximately 4 kb. NDV has demonstrated specific replication in human tumor cells resulting in their killing (Reichard et al. 1992). This feature has attracted their application as oncolytic agents for both preclinical and clinical studies (Schirrmacher et al. 2001). More recently, a reverse genetics system was applied to engineer an oncolytic NDV vector based on the mesogenic NDV-73 T strain with a modification at the fusion protein (F) cleavage site. It significantly reduced pathogenicity in chicken, but NDV still replicated in and killed human tumor cells (Cheng et al. 2016).

9.3.11 *Reoviruses*

Reoviruses are dsRNA viruses with oncolytic properties, therefore preferentially capable of targeting and killing various types of cancer cells (Clements et al. 2014). In addition to tumor killing, reoviruses invoke immunological stimulation overturning tumor-induced immunosuppression, which promotes antitumor immune responses (Gujar et al. 2010). Generation of a reverse genetics system has allowed the insertion and expression of foreign genes such as the chloramphenicol acetyltransferase (CAT) gene into reovirus vectors (Roner and Joklik 2001).

Moreover, recombinant reovirus systems with tandem repeats and a tetravirus 2A-like element have proven successful for expression of heterologous polyproteins (Demidenko et al. 2013). The size of foreign inserts has not been described in the literature, probably because reovirus strains efficiently kill different types of cancer cells and have not required additional therapeutic genes (Clements et al. 2014).

9.3.12 *Poxviruses*

Poxviruses, which are dsDNA viruses possessing an impressive and exceptional packaging capacity of more than 30 kb of foreign inserts, and particularly vaccinia virus, have been frequently used as delivery vectors for cancer immunotherapy (Kwak et al. 2003). Vectors based upon avipox viruses have been engineered to provide non-replicating vectors that are safer when used in non-avian species (Pastoret and Vanderplasschen 2003). In another approach, tumor-selective replication-competent vaccinia virus (VV) vectors have been engineered (Zeh and Bartlett 2002). As these vectors cause no harm to normal cells, they can be utilized for killing of cancer cells in cancer therapy. Moreover, it was demonstrated that systemic immunity of VV in adult patients did not prevent transfection/infection of tumors in vivo (Mastrangelo and Lattime 2002).

9.3.13 *Picornaviruses*

Among the small ssRNA picornaviruses, coxsackieviruses have been used as vectors in oncology (Bradley et al. 2014). In this context, coxsackievirus A21 (CVA21) (Bradley et al. 2014) and the attenuated coxsackievirus B3 (CVB3) (Kim and Nam 2011) have been used for vaccine development and therapeutic gene delivery. Also, the poliovirus PV-1 has been engineered as an expression vector (Jia et al. 2002). The small genome size and compact viral particle structure have limited the possibilities of genome modifications (Ylä-Pelto et al. 2016). The features of picornaviruses include cell tropism and replication in some non-dividing cells and absence of integration into the host genome (Ylä-Pelto et al. 2016). Due to the global poliovirus eradication campaign, coxsackieviruses are favored as expression vectors.

9.3.14 *Polyoma Viruses*

The simian virus 40 (SV40) has a dsDNA genome of only 5 kb in size (Kimchi-Sarfaty and Gottesman 2004). However, SV40 is capable of incorporating larger DNA inserts than its own genome, even up to a size of 17.7 kb. SV40 shows a broad host range including both dividing and non-dividing cells. SV40-based vectors still contain SV40 sequences, but in vitro packaging can generate virus-like particles

(VLPs) free of any SV40 wild-type DNA sequences. To facilitate SV40 vector production, a Vero cell-based packaging system has been engineered (Toscano et al. 2017). Polyoma viruses and particularly SV40 virus should be considered for gene therapy as they can efficiently deliver anti-viral agents, DNA vaccines, genes for chemoprotection, suicide and anti-angiogenic genes and infect a wide variety of dividing and non-dividing cells.

9.3.15 Chimeric Vectors

Finally, chimeric or hybrid gene therapy vectors are briefly summarized. Chimeric systems have been reported for retroviruses with adeno-alphaviruses, HSV and VV (Falkner and Holzer 2004). The classic approach comprises the generation of pseudotyped vectors, which carry the genetic information of one virus and the envelope from another virus leading to a modified cell tropism and enhanced biosafety. For instance, SFV replicons have been packaged into VLPs expressing the VSV glycoprotein (VSV G) providing modified cell tropism and a higher biosafety level as VSV-G does not share any homology with the SFV genome (Dorange et al. 2004). Moreover, chimeric or hybrid vectors have also been engineered for AAV based on the different properties of AAV1 and AAV2 serotypes (Hauck et al. 2003). Although AAV1 has shown superior expression in muscle and other tissues, AAV2 showed high expression levels in the liver and can be efficiently purified on heparin columns. Chimeric AAV1/AAV2 particles were generated with a mixture of AAV helper plasmids leading to packaging of a mixture of particles with both AAV1 and AAV2 capsid, which could be purified on heparin columns. The chimeric AAV particles generated similar levels of expression in muscle tissue as did AAV1 and equal levels compared to AAV2 in the liver. In another approach, hybrid VV-retrovirus vectors have been engineered by applying defective VV for the expression of functional Moloney murine leukemia virus (MMLV)-based vector genomes (Holzer et al. 2004). The superior VV properties such as high packaging capacity, stability and broad host range make hybrid VV-retroviral vectors promising vehicles for gene therapy applications. Hybrid vectors have also been engineered for AAV and human bocavirus 1 (HBoV1) (Fakhiri et al. 2019). HBoV1, known to cross-package AAV2 genomes can specifically transduce polarized human airway epithelial (pHAE) cells. Moreover, hybrid AAV2-HBoV vectors were engineered for HBoV2-4 and gorilla BoV (GBoV). The hybrid AAV2-HBoV4 and AAV2-GBoV transduced pHAE cells and primary human lung organoids. Moreover, hybrid AAV/BoV vectors transduced primary human hepatocytes, skeletal muscle cells and T cells.

9.4 Preclinical Evaluation

As several examples have already been presented in Sect. 9.2 (Gene Therapy and Vaccine Targets) (Table 9.1) a restricted number of examples of the most prominent preclinical studies are presented in this section (Table 9.3).

Table 9.3 Examples of preclinical gene therapy and vaccine studies

Indication	Vector	Target	Response
Cancer			
Bile duct	MV	SCD	Tumor regression, survival benefit in mice
Breast	HSV HF10	oHSV	Tumor regression, prolonged survival
Canine[a]	VSV	IFN-β, NIS	Safe delivery in dogs with naturally occurring tumors
Colon	HSV-2	oHSV	Significant inhibition of tumor growth
	SFV RNA	LacZ	Protection against tumor challenges in mice
	KUN	GM-CSF	Cure of 50% of treated mice
	CPXV	FUC1	Tumor regression, survival benefit in mice
Glioma	SFV	EGFP	Induced inhibition of tumor growth in mice
	NDV	GFP	Prolonged survival in A549 tumor-bearing mice
	M1	oM1	Suppressed tumor growth, prolonged survival
Liver	MV	SCD	Selective killing of tumor cells
	NDV	IL-2, TRAIL	Induced apoptosis
	NDV	sTRAIL	Superior anti-tumor activity of rNDV-IL-2-TRAIL
Lung	SFV	IL-12	Suppression of carcinoma, no cytotoxicity
	RRV	CD	87% tumor volume regression in rats
Pancreas	Ad	SYE	Prolonged long-term survival of mice
	AdSur	SYE	Oncolysis in PDAC cells
	VSV	MUC1	Complete tumor regression in mice
	Reovirus	oReovirus	Significant reduction of tumor growth in mice
	PANVAC	MUC1/CEA, TRICOM	Suppression tumor growth in xenograft model
	SV40	SST2	MUC1 and CEA CTL responses in mouse models
	M1	oM1	Inhibition of tumor proliferation
	Reovirus	+ anti-PD-1	Selective killing of tumor cells
	CAV21	ICAM-1, DAF	Superior in combination with checkpoint blockade
	CAV21	ICAM-1, DAF + Dox	Tumor regression, elimination of metastases in mice
Melanoma	CPXV	FUC1	Improvement after combination with doxorubicin
	KUN	GM-CSF	Induced inhibition of tumor growth in mice
	NDV	IL-2, IL-15	Cure of 67% of treated mice
	NDV	IL-2. TRAIL	Suppression of tumor growth in mouse model
	CAV21	ICAM-1, DAF	Superior anti-tumor activity of rNDV-IL-2-TRAIL
Plasmacytoma	VSV	IFN-β, NIS	Tumor regression in mice
Prostate	VV	NIS/radioiodide NIS	Dose-dependent tumor regression in mice

(continued)

Table 9.3 (continued)

Indication	Vector	Target	Response
Thyroid	MV-Edm	oMV	Enhanced tumor regression, prolonged survival in mice Enhanced tumor killing
Metabolic			
MPS VII	AAV	GUS	Reversed disease phenotype in liver in mice
OTC	AAV	OTC	High OTC liver levels, prolonged lifespan
FH	AAV	LDLR	Normal serum lipid levels, no severe atherosclerosis
T2D, obesity	AAV	FGF21	Reduced body weight, inflammation, insulin resistance
T2D	MSCV	Insulin	Reversal of diabetes in mice
	MMTV	Ad36 E4orf1	Improved glycemic control in mice on high fat diet
Cardiovascular			
Heart failure	Ad	βARKct	Improved cardiac function in transgenic mice
	pMXs	GMT	Decreased infarct size in mice
	Ad	SERCa2a	Restored heart functions in rat model
	Ad	SERCa2a	Increased coronary blood flow in rat model
	Ad	VEGF	Sustained VEGF expression in dogs
	Ad	HGF	Improved heart function in swine model
Arrhythmia	Ad	Cx43	Prevention of atrial fibrillation in pigs
	Ad	KCNH2-G628S	Elimination of atrial fibrillation in porcine model
Hematological			
Hemophilia A	Ad	FVIII	Long-term expression of FVIII in mice
	HIV	FVIII	Sustained FVIII production, hemostatic correction
	AAV8	FVII	Clinically relevant FVII expression in dogs
Hemophilia B	Ad	FIX	Long-term expression of FIX in mice
	Ad	FIX	Therapeutic levels of FIX in dogs
	AAV8	FIX-Padua	Sustained levels of FIX in mice and dogs
PKD	LV	PKLR	Corrected hematological phenotype in mice
PAD	AAV9	EcSOD	Significant recovery of hind-limb ischemia in mice
Neurological			
PD	AAV	GAD65	Significant improvement in PD symptoms in rats
	AAV	TH, AADC, GCH-I	Spontaneous dopamine production in HEK293 cells
	AAV	GDNF	Regeneration and functional recovery in rats/monkeys
	LV	GDNF	Regeneration and functional recovery in rats/monkeys
	LV	GDNF	Prevention of nigrostriatal degeneration in monkeys

(continued)

Table 9.3 (continued)

Indication	Vector	Target	Response
Chronic pain			
AD	AAV	GAD65	Successful therapy
	AAV2/5	NGF	Neuroprotection in mice for extended time
	AAV	APPsα	Functional rescue of spatial reference memory
	LV	GDNF	Preserved learning and memory
	LV	Klotho	Reduced cognitive deficits and AD-like pathology
	AAV	APOE2	Safe and wide ApoE2 distribution in primate brain
HD	AAV5	HTT miRNA	Prevention of mutant HTT aggregate formation
	AAV5	HTT miRNA	Reduced RTT mRNA and protein levels in minipigs
RTT	AAV	MeCP2	Extended survival in mice
Infections			
Influenza A	Ad5	HA	Full protection in mice and chicken
	VEE	HA	Complete protection in chicken
	SFV	HA, NP	Complete protection in mice
HIV	AAV	HIV Gag	Specific T and B cell responses
	VV	HIV Env-GM-CSF	Superior immune response from fusion in BALB/c mice
	RV	HIV-1 gp160	Solid memory CTL responses
	SFV	HIV-1 Gag/Env/polRT	Antigen-specific T cell responses in mice
EBOV	VEE	EBOV-NP	Protection of mice against EBOV
	VEE	EBOV-GP, -NP	Protection of mice and guinea pigs against EBOV
	VSV	EBOV-GP	Protection of macaques against EBOV
	VSV	EBOV-GP	Protection of non-human primates against EBOV
	VSV	MARV-GP	Protection of non-human primates against MARV
Ophthalmologic			
Glaucoma	AAV	BDNF	Protection of RGCs in glaucoma mouse model
	AAV	MMP-3	Reduced intraocular pressure in mouse eye
XLRS	AAV8	RS1	Improvement of retinal structure and function in mice
EOSRD	AAV2/5	RDH12	Reconstituted retinal reductase activity in mice
Muscular			
DMD	Ad	Δdystrophin	Restoration of dystrophin-associated proteins in vivo
	AAV6	Micro-dystrophin	Reduced skeletal muscle pathology, extended lifespan

(continued)

Table 9.3 (continued)

Indication	Vector	Target	Response
	AAV9	Micro-dystrophin	Whole body skeletal muscle transduction in dogs
OPMD	AAV	PABPN1	Normalization of muscle strength in mice
Lung			
Cystic fibrosis	AAV2	CFTR	CFTR expression in rabbit lung up to 6 months
	AAV2	CFTR	CFTR expression in macaque lung up to 180 days
	HIV	LacZ	β-gal expression in nasal airway epithelium for 92 days
	HIV	CFTR	Recovery of electric function in nasal airways
	HIV-VSV G	LacZ	β-gal expression in airways of marmosets
	SIV-SeV	GFP	Transduction of respiratory epithelium in mouse nose

[a]Naturally occurring tumors in dogs

AADC aromatic L-amino acid decarboxylase, Ad adenovirus, AD Alzheimer's disease, AdSur adenovirus with survivin promoter, APOE2 apolipoprotein E2, APPsα secreted amyloid precursor protein, βARKct β-adrenergic receptor kinase carboxyl terminus, CAV21 Coxsackievirus A21, CD cytosine deaminase, CFTR cystic fibrosis transmembrane conductance regulator, CX43 Connexin 43, DAF decay-accelerating factor, DMD Duchenne muscular dystrophy, EBOV Ebola virus, EcSOD extracellular superoxide dismutase, EORSD Early-onset retinal severe retinal dystrophy, FGF21 fibroblast growth factor 21, FH familial hypercholesterolemia, FIX blood-clothing factor IX FUC1, fusion suicide gene 1, FVII blood-clothing VII, FVIII blood-clothing VIII, GAD65 glutamic acid decarboxylase, GCH-I, GDNF glial cell-derived neurotrophic factor, GFP green fluorescent protein, GTO cyclohydrolase I, GM-CSF granulocyte macrophage-colony stimulating factor, GMT Gat4, Mef2c and Tbx5 genes, GP glycoprotein, GUS β-glucuronidase, HA hemagglutinin, HD Huntington's disease, HGF hepatocyte growth factor, HIV human immunodeficiency virus, HSV herpes simplex virus, HTT huntingtin, ICAM-1 intercellular adhesion molecule 1, IL interleukin, KCNH2-G628S dominant negative mutant of I(Kr) potassium channel alpha-subunit, KUN Kunjin virus, LDLR low-density lipoprotein receptor, LV lentivirus, MARV Marburg virus, MeCP2 methyl CpG-binding protein 2, MMP-3 matrix metalloproteinase-3, MMTV mouse mammary tumor virus, MPS VII mucopolysaccharidosis type VII, MSCV murine stem cell virus, MUC1 mucin 1, MV-Edm measles virus Edmonston strain, NDV Newcastle disease virus, NGF nerve growth factor, NP nucleoprotein, NIS sodium iodide symporter, oHSV oncolytic HSV, oM1 oncolytic M1 alphavirus, oMV oncolytic measles virus, OPMD oculopharyngeal muscular dystrophy, OTC ornithine transcarbamylase, PABPN1 polyA-binding protein nuclear 1, PAD peripheral arterial disease, PANVAC fowl pox and vaccinia virus, PD Parkinson's disease, PDAC pancreatic ductal adenocarcinoma, PKD pyruvate kinase deficiency, PKLR pyruvate kinase L/R, PNET pancreatic neuroendocrine tumor, RDH12 retinol dehydrogenase, RGCs retinal ganglion cells, RRV replicating retrovirus vector, RS1 retinoschisin 1, RTT Rett syndrome, RV rabies virus, SCD super-cytosine deaminase, SeV Sendai virus, SFV Semliki Forest virus, SIV simian immunodeficiency virus, SST2 somatostatin receptor tumor-suppressor 2, SV40 simian virus 40, SYE tumor targeting SYENFSA ligand, T2D type 2 diabetes, TH tyrosine hydroxylase, TRAIL tumor necrosis factor-related apoptosis inducing ligand, TRICOM three costimulatory molecules, VEGF vascular epithelial growth factor, VV vaccinia virus, XLRS X-linked retinoschisis

9.4.1 Cancer

In the context of cancer therapy, oncolytic adenoviruses demonstrated specific targeting of pancreatic cancer cells after introduction of a pancreatic cancer cell-targeting ligand SYENFSA (SYE) resulting in efficient oncolysis of pancreatic ductal adenocarcinoma (PDAC) cells (Nagasato et al. 2017). Moreover, introduction of the survivin promoter resulted in high transduction efficiency of the AdSur-SYE vector in pancreatic neuroendocrine tumors (PNETs) and complete regression of subcutaneous tumors after intratumoral injection of mice (Yamamoto et al. 2017). A chimeric Ad type 5 and type 3 vector expressing the melanoma differentiation associated gene-7 (MDA-7) and interleukin-24 (IL-24) showed selective tumor killing after intratumoral injection, which also spread to distant tumors due to bystander activity (Emdad et al. 2018). Alphaviruses have been applied for cancer therapy due to their capability to generate high-level transient heterologous gene expression and induction of apoptosis in infected cells (Liljeström and Garoff 1991). For instance, administration of SFV particles expressing IL-12 via an implanted cannula generated 87% reduction in RG2 gliomas in rats (Roche et al. 2010). Moreover, local administration of replication-proficient SFV-EGFP particles resulted in prolonged survival of mice implanted with A549 lung tumor xenografts (Määttä et al. 2008). Interestingly, it has also been demonstrated that a single intramuscular injection of 0.1 μg of SFV-LacZ RNA provided protection against tumor challenges in mice implanted with colon tumors (Ying et al. 1999). Moreover, administration of SFV-LacZ RNA to mice with pre-existing tumors generated therapeutic effects and extended survival rates. In another study, the oncolytic alphavirus M1, which can selectively kill zinc-finger antiviral protein (ZAP)-deficient cancer cells, was administered intravenously at a dosage of 3×10^7 pfu of M1 to mice with implanted 4 T1 mammary carcinoma or B16 melanoma cells (Lin et al. 2014). That treatment resulted in high tumor tropism and efficient tumor killing. Related to flaviviruses, KUN vectors expressing the granulocyte macrophage colony-stimulating factor (GM-CSF) were intratumorally injected into mice that had been implanted with subcutaneous CT26 colon tumors (Hoang-Le et al. 2009). The outcome was a cure of more than 50% of treated animals with no tumors detected 18 months after administration. Furthermore, mice with implanted B16-OVA melanoma tumors immunized with KUN-GM-CSF showed significant tumor regression and a cure rate of 67% (Hoang-Le et al. 2009).

Measles viruses, especially the oncolytic MV-Edm, have been subjected to various animal tumor models. For instance, expression of the sodium iodide symporter (NIS) gene from the MV-Edm strain showed enhanced tumor killing in a mouse model for anaplastic thyroid cancer (ATC) (Reddi et al. 2012). MV vectors expressing the yeast-based bifunctional suicide gene encoding cytosine deaminase and uracil phosphoribosyl transferase named super-cytosine deaminase (SCD) showed efficient replication in tumor cells and furthermore resulted in killing of human ovarian cancer cell lines and primary tumors (Hartkopf et al. 2013). Moreover, treatment of mice with hepatocellular carcinoma xenografts showed long-term virus replication in tumor tissue and the suicide therapy induced apoptosis-like cell death (Lampe et al. 2013). Moreover, the MV-SCD was delivered to a human cholangiocarcinoma (huCCT1) xenograft mouse model showing tumor regression and significant survival benefit (Lange et al. 2013). Among rhabdoviruses, the oncolytic VSV vector expressing interferon-β (IFN-β) and NIS showed a dose-dependent tumor regression in mice implanted with syngeneic 5TGM1 plasmacytoma tumors (Zhang et al. 2016). In another dose-escalation study, the VSV- IFN-β-NIS vector was intravenously injected into purposed-bred dogs with naturally occurring tumors (LeBlanc et al. 2013). The maximum tolerated dose (MTD) was 10^{10} $TCID_{50}$ with only mild to moderate adverse events occurring indicating safe systemic delivery. VSV vectors have also been applied for the expression of the human mucin 1 (MUC1) showing a significant reduction in tumor growth in PDAC-bearing mice (Hastie and Grdzelishvili 2012).

In the context of oncolytic HSV vectors, efficient replication in tumor cells has been established leading to induced T cells and natural tumor killer cells, which resulted in significant reduction of tumor growth and prolonged survival of treated mice (Eissa et al. 2017). Moreover, application of an oncolytic HSV-2 vector resulted in significant inhibition of tumor growth (Yang et al. 2016). In the context of retroviruses, the nonlytic amphotropid retroviral replicating vector (RRV) Toca 51 was applied for expression of yeast cytosine deaminase (CD) (Huang et al. 2015). Intravenous or intracranial administration of Toca511 in an orthotopic glioma model resulted in prolonged long-term survival in immune-competent mice. Animals exhibiting pre-existing immune responses to the vector also showed prolonged survival, indicating the feasibility of vector re-administration.

Newcastle disease virus (NDV) possesses oncolytic activity, which has made it attractive for cancer therapy (Niu et al. 2015). For instance, intratumoral administration of both NDV-IL2 and NDV-IL15 showed suppression of tumor growth in a mouse melanoma model. The survival rate was, however, superior for NDV-IL15 treatment compared with NDV-IL2. Moreover, in a tumor re-challenge experiment, the survival rate was 26.6% higher in NDV- IL15 immunized mice than in those receiving NDV-IL2. In attempts to engineer improved NDV vectors, reverse genetics was applied on the oncolytic NDV D90 strain to generate a recombinant NDV vector carrying the GFP gene (Chai et al. 2014). The rescued NDV-GFP vector showed similar suppression of tumor growth and loss of body weight as the parental D90 strain in athymic mice bearing implanted lung tumors. In another study, it was demonstrated that NDV-based expression of TRAIL alone or in combination with IL-2 (rNDV-IL-2-TRAIL) presented superior apoptotic function (Bai et al. 2014). Moreover, the reduction of tumor development in mice treated with rNDV constructs was significantly higher compared to mice treated with the parental virus. Studies in hepatocellular carcinoma and melanoma mouse models showed superior survival of mice receiving rNDV-Il-2-TRAIL compared to mice treated with rNDV, rNDV-IL-2 or rNDV-TRAIL alone. In another approach, the NDV Anhinga strain was engineered to express a soluble form of TRAIL (NDV/Anh-TRAIL) leading to efficient suppression of hepatocellular carcinoma without the presence of any significant cytotoxicity (Wu et al. 2017).

Oncolytic reoviruses are tumor selective and provide efficient tumor killing making them attractive candidates for cancer therapy (Comins et al. 2008). For example, pancreatic cell lines such as Panc1, MIApaca-2, PK1, PK9 and BxPC3 have shown susceptibility to reoviruses and enhanced Ras activity after reovirus infection (Etoh et al. 2003). Moreover, in mouse xenograft models using Panc1 and BxPC3 cell lines intratumoral reovirus administration suppressed tumor growth and local injections demonstrated systemic antitumor responses in bilateral tumor models. An oncolytic reovirus has demonstrated a significantly reduced disease burden and prolonged survival in the syngeneic EMT6 mouse model for breast cancer (Mostafa et al. 2018). Combination therapy of reovirus and the checkpoint inhibitor anti-PD-1 antibody further enhanced the efficacy of breast cancer therapy.

Poxviruses, particularly vaccinia virus, have been applied for various cancer indications such as pancreatic cancer (Al Yaghchi et al. 2015). One approach comprises the PANVAC system, where recombinant vaccinia and fowl pox viruses carry the MUC1 and carcinoembryonic antigen (CEA) genes, respectively (Madan et al. 2007). Moreover, three costimulatory molecules, cluster of differentiation B7.1, intercellular adhesion molecule-1 (ICAM-1) and leukocyte function-associated antigen-3 (LFA-3), collectively named TRICOM elicited CEA and MUC1 CTL responses in preclinical mouse models. In another approach, the light-emitting GLV-2b372 vaccinia virus injected into hepatocellular carcinoma (HCC) xenografts in the flank of athymic nude mice demonstrated a 50% decrease in tumor volumes in comparison to a 400% increase in control mice after 25 days (Ady et al. 2015). Furthermore, the GLV-1 h153 vaccinia virus expressing NIS showed efficient killing of PC3, DU145, LNCaP and WPMY-1 prostate cancer cell lines, which was enhanced by radioiodide treatment (Mansfield et al. 2016). In vivo studies in prostate xenograft and transgenic adenocarcinoma of the mouse prostate (TRAMP) models demonstrated a synergistic effect of VV-NIS and radioiodide treatments related to tumor growth and survival rates compared to administration of either virus-based or radiotherapy alone. In another study, the fusion suicide gene FCU1 was introduced into the cowpox virus (CPXV) vector, which after systemic administration showed accumulation in tumor cells with only low infection and toxicity of normal cells (Ricordel et al. 2017). Intratumoral injection of CPXV-FCU1 resulted in induced inhibition of tumor growth in U-87-MG glioblastoma and LoVo colon cancer models.

Picornaviruses and particularly Coxsackievirus A21 (CAV21) have been used for cancer therapy (Shafren et al. 2014). A single subcutaneous injection of CAV21 expressing ICAM-1 and decay-accelerating factor (DAF) reduced tumor burden and showed efficient tumor regression in melanoma xenograft-bearing non-obese SCID mice. In another study, a single intravenous administration of CAV21-ICAM-1-DAF generated significant regression of pre-established tumors and elimination of metastases in SCID mice with implanted T47D and MDA-MB-231-luc breast tumor xenografts (Skelding et al. 2009). Combination treatment with intraperitoneal doxorubicin hydrochloride further enhanced tumor regression (Skelding et al. 2012).

Finally, SV40 virus should be considered for cancer therapy due to its efficient delivery of suicide and anti-angiogenic genes (Kimchi-Sarfaty and Gottesman

2004). A replication-deficient SV40 vector has been engineered for pancreatic cancer therapy by the introduction of a human telomerase RNA (hTR) tumor-specific promoter and the somatostatin receptor tumor-suppressor 2 (SST2) gene (Cordelier et al. 2007). Intratumoral administration of SV40.hTR-SST2 resulted in inhibition of Capan-1 pancreatic tumor cell progression and proliferation in vivo.

9.4.2 Metabolic Diseases

AAV vectors have been investigated for more than 30 metabolic disease in small animal models demonstrating complete phenotype correction in a substantial proportion of them (Alexander et al. 2008). For instance, AAV vectors expressing β-glucuronidase (GUS) have been applied for gene therapy of the lysosomal storage disease mucopolysaccharidosis type VII (MPS VII) (Watson et al. 1998). It was demonstrated that intramuscular AAV injection generated high local expression of GUS, while intravenous administration resulted in low GUS activity in several tissues reducing glycosaminoglycan levels in the liver to normal levels and reduced storage granules dramatically. Furthermore, it was shown that a single intravenous AAV-GUS injection provided sustained GUS expression, sufficient for reversing the disease phenotype in mice. In another approach, AAV vectors expressing mouse OTC were applied for the correction of metabolic defects in mouse liver (Moscioni et al. 2006). The AAV-OTC treated mice showed high liver OTC activity and a prolonged lifespan compared to control animals. In the context of familial hypercholesterolemia, mutations in the LDL receptor gene causes severe hypercholesterolemia and atherosclerosis. Therapeutic approaches comprise non-human primate AAV-based LDL receptor expression in the liver, which resulted in nearly complete normalization of serum lipid levels and prevention of severe atherosclerosis (Lebherz et al. 2004).

Another potential target for AAV-based gene therapy is type 2 diabetes and obesity targeting fibroblast growth factor 21 (FGF21) as a promising therapeutic agent (Jimenez et al. 2018). Transgenic ob/ob mice or mice subjected to long-term high-fat diet feeding showed substantial reduction in body weight, adipose tissue hypertrophy and inflammation, and insulin resistance for more than a year when treated with AAV-FGF21. In another approach, diabetic mice were subjected to intrahepatic administration of mesenchymal stem cells transduced with the murine stem cell virus (MSCV) retrovirus carrying the human insulin gene (Xu et al. 2007). The procedure resulted in body weight increase and decrease in blood glucose levels. Moreover, increased secretion of insulin into the serum and presence in the liver were detected, and reversal of diabetes was observed for up to 6 weeks. A really interesting finding relates to the anti-hyperglycemic properties of Ad36, more specially its E4orf1 protein (Hegde et al. 2016). A mouse mammary tumor virus (MMLV) retrovirus-based vector was employed to express the Ad36 E4orf1 gene in C57BL/6 mice resulting in significantly improved glucose excursion despite a high fat diet and also enhanced glucose clearance without increased insulin sensitivity, which underscored the insulin-independent effect.

9.4.3 Cardiovascular Diseases

Related to cardiovascular disease, a number of preclinical gene therapy studies have been carried out with various viral vectors (Ishikawa et al. 2018). Cardiovascular gene therapy has targeted adrenergic manipulation, calcium cycling proteins, angiogenesis, cardiac regeneration, cardiac arrhythmias and the etiology of myocardial infarction (Scimia et al. 2014). For instance, due to the negative effect on cardiac functions of down-regulation of β-adrenergic receptors by G protein-coupled receptor kinase-2 (GRK2) activity leading to heart failure, expression of the peptide inhibitor β-adrenergic receptor kinase carboxyl terminus (βARKct) can boost cardiac function and potentially prevent heart failure (Koch et al. 1995). Expression of the βARKct from an Ad virus vector blocked GRK2 activity and resulted in improved cardiac function in transgenic mice (Akhter et al. 1997). In another approach, the pMXs retrovirus vector expressing Gata4, Mef2c and Tbx5 managed to reprogram non-myocytes in the mouse heart to cardiomyocyte-like cells, which decreased infarct size and modestly attenuated cardiac dysfunction (Qian et al. 2012).

In a rat model of heart failure, the decreased sarcoplasmic reticulum Ca^{2+} ATPase (SERCa2a) activity associated with heart failure was restored by Ad virus-based recombinant expression of SERCA2a (Miyamoto et al. 2000). The Ad-SERCA2a treatment restored both systolic and diastolic heart functions to normal levels in aortic constricted rats. Furthermore, Ad-SERCa2a therapy resulted in increased coronary blood flow and reduced cardiomyocyte size in a type 2 diabetic rat model (Sakata et al. 2007). In another approach, an AAV9 vector expressing the inhibitor of protein phosphatase 1 (I-1c) was evaluated in a porcine model of heart failure (Fish et al. 2013). Intracoronary infusion of AAV9-1-1c one month after myocardial infarction prevented deterioration of cardiac function and resulted in decrease in scar size.

Angiogenic peptides such as vascular endothelial growth factor (VEGF) can aid in restoring blood flow in ischemic areas (Zachary and Morgan 2011). Pericardial delivery of an Ad virus vector expressing VEGF (AdCMV.VEGF165) resulted in sustained (8–14 days) pericardial expression of VEGF although it failed to improve myocardial collateral prefusion in mongrel dogs (Lazarous et al. 1999). However, in another approach Ad virus expressing the hepatocyte growth factor (HGF) resulted in improvement of heart function in a postinfarct heart failure model in swine (Yang et al. 2010).

Cardiac arrythmia has also been targeted by virus-based gene therapy (Scimia et al. 2014). For instance, Ad virus-based expression of Connexin 43 (Cx43) increased conduction velocity, prevented atrial fibrillation and reduced susceptibility to tachycardia after myocardial infarction in pigs (Igarashi et al. 2012). Moreover, an Ad virus expressing the dominant negative mutant of the I(Kr) potassium channel alpha subunit (KCNH2-G628S) eliminated atrial fibrillation by prolongation of atrial action potential duration in a swine model (Amit et al. 2010).

9.4.4 Hematological Disorders

Among hematological disorders, hemophilia can be considered as the model target for gene therapy (Nienhuis et al. 2017). Hemophilia, caused by mutations in blood clothing factor VIII (FVIII) (hemophilia A) and factor IX (FIX) (hemophilia B), has been frequently studied and evaluated in clinical trials. Therefore, a limited number of preclinical studies are presented below. For instance, Ad vectors expressing FVIII (Balagué et al. 2000) and FIX have demonstrated long-term expression in preclinical hemophilic mouse models (Dai et al. 1995) and therapeutic levels of FIX have also been obtained in hemophilic dogs (Fang et al. 1995). AAV is the most frequently used vector for hemophilia therapy. In this context, dogs producing less than 1% of normal FVII activity were treated with the liver directed AAV8 serotype expressing FVII, which provided clinically therapeutic levels (15% of normal FVII) (Marcos-Contreras et al. 2016). Moreover, a hyperfunctional FIX, FIX-Padua with the R388L, expressed from AAV8 showed sustained FIX levels in dogs (Crudele et al. 2015). An HIV-based monocyte lineage-restricted, self-activating lentiviral vector (CD68-ET3-LV) expressing FVIII demonstrated safety and efficacy in mouse models (Doering et al. 2018). Administration of CD38-ET3-LV-transduced stem-cell antigen-1 cells to mice with hemophilia A provided sustained production of FVIII and hemostatic correction.

Pyruvate kinase deficiency (PKD), causing hemolytic anemia, was subjected to preclinical gene therapy evaluation in mice by transduction of hematopoietic stem cells with a lentiviral vector expressing the pyruvate kinase L/R (PKLR) gene from the human 3-phosphoglycerate kinase (hPGK) promoter (Garcia-Gomez et al. 2016). Ectopic expression of the R-type specific PK isoform (RKP) normalized the erythroid compartment and corrected the hematological phenotype. Moreover, it was confirmed that the lentiviral chromosomal insertion sites did not generated any genotoxicity in transplanted mice.

In the context of peripheral arterial disease (PAD), murine hind limb ischemia has served as a model (Niiyama et al. 2009). It has been demonstrated that AAV9-based expression of extracellular superoxide dismutase in skeletal muscle, significantly improved recovery from hind-limb ischemia in mice (Saqib et al. 2011).

9.4.5 Neurological Disorders

A large number of preclinical studies have been dedicated to neurological disorders (Sudhakar and Richardson 2019). Mainly two strategies, symptomatic and neurorestorative therapies, have been developed for Parkinson's disease. In the former approach, AAV vectors have been used for expression of glutamic acid decarboxylase (GAD), more precisely GAD65, which significantly improved Parkinson's disease symptoms in rats (Kim et al. 2008). Furthermore, AAV-GAD65 administration provided success in chronic pain models. Related to the neurorestorative approach, co-expression of the dopamine synthetic enzymes tyrosine hydroxylase (TH), aromatic L-amino acid decarboxylase (AADC) and GTP cyclohydroxylase I (GCH I) from three AAV vectors resulted in spontaneous dopamine production in HEK293 cells and might contribute to therapy of Parkinson's disease (Fan et al. 2001). Moreover, both AAV and lentiviral vectors have been employed for the expression of the glial cell line-derived neurotrophic factor (GDNF) showing sustained GDNF delivery for 3–6 months leading to regeneration and significant recovery in both 6-OHDA-lesioned rats and MPTP-lesioned monkeys (Björklund et al. 2000). MPTP (1-methyl-4-phenyl-1,2,3,6-tetrahydropyridine) is a biologically inactive compound which, as a prodrug, leads to the production of MPP+ (1-methyl-4-phenylpyridinium) and thereby affects the substantia nigra of the brain by destroying dopaminergic neurons. That destruction permanently results in Parkinson's disease symptoms. In another study lentiviral-based GDNF injection into the striatum and substantia nigra of rhesus monkeys treated with MPTP reversed functional deficits, completely prevented nigrostriatal degradation, and reversed motor function deficits (Kordower et al. 2000).

In the context of Alzheimer's disease, a hybrid AAV vector with the AAV2 genome and the AAV5 capsid structure was engineered to express the nerve growth factor (NGF) (Wu et al. 2005). It was demonstrated that AAV-mediated NGF delivery provided neuroprotection for an extended period of time. Another approach has relied on overexpression of secreted amyloid precursor protein (APPsα) from AAV vectors to affect Alzheimer's disease (Fol et al. 2016). It was demonstrated that APPsα overexpression resulted in a functional rescue of spatial reference memory and mitigated synaptic and cognitive deficits in mice. Similar to Parkinson's disease therapy, the therapeutic potency of GDNF was evaluated by lentivirus-based delivery in MC65 human neuroblastoma cells and hippocampal astrocytes of 3xTg-AD in vivo (Revilla et al. 2014). Overexpression of GDNF resulted in preserved learning and memory in 3xTg-AD mice. Although recombinant GDNF did not significantly reduce amyloid or tau pathology, it induced upregulation of brain-derived neurotrophic factor (BDNF), which together with GDNF may contribute to the protection of neurons from atrophy and degeneration. In another approach, a lentiviral vector expressing the anti-aging gene Klotho was subjected to intracerebroventricular administration in APP/presenilin-1 transgenic mice (Zeng et al. 2019). Overexpression of Klotho in the brain effectively ameliorated cognitive deficits and Alzheimer's disease-like pathology. As it has been shown that the apolipoprotein E4 (APOE4) allele increases the risk of developing Alzheimer's disease whereas the APOE2 variant can provide protection against late-onset Alzheimer's disease, it has been postulated that overexpression of APOE2 might reverse or prevent progressive neurological damage (Rosenberg et al. 2018). For this reason, an AAV vector expressing APOE2 was administered intracisternally to the CNS of nonhuman primates showing safe and wide distribution of ApoE2.

In the context of the fatal progressive neurogenerative Huntington's disease, caused by a mutation in the huntingtin (HTT) gene, gene therapy approaches have included AAV-based expression of miRNAs targeting HTT transcripts (Miniarikova et al. 2017). AAV5-miHTT showed suppression of mutant HTT mRNA, which almost completely prevented mutant HTT aggregate formation and suppression of DARPP-32-associated neuronal dysfunction. Furthermore, no immune response to AAV5 or therapeutic precursor sequences were observed. Next, the AAV5-miHTT vector was evaluated in the transgenic HD (tgHD) minipig model (Evers et al. 2018). The outcome was significantly reduced human mutant huntingtin mRNA and protein levels in all transduced brain regions.

Other neurodegenerative orders such as the X-chromosome linked Rett Syndrome (RTT) have been investigated by the expression of the transcription regulator methyl CpG-binding protein 2 (MeCP2) delivered by AAV directly to the cerebrospinal fluid (CSF) (Sinnett and Gray 2017). The AAV-MeCP2 administration resulted in extended survival of RTT mice, but also dose-dependent toxicity.

9.4.6 Infectious Diseases

Infectious diseases, especially viral infections, have been targeted in numerous studies for vaccine development (He et al. 2015; Lundstrom 2020a). For instance, Ad5 virus expressing the codon-optimized hemagglutinin (HA) gene for the A/Vietnam/ 1203/2004(H5N1) influenza virus strain provided full protection of immunized mice from challenges with lethal doses of the homologous virus (Gao et al. 2006). Moreover, a single subcutaneous injection generated full protection in chicken. In another study, Venezuelan equine encephalitis virus (VEE), belonging to alphaviruses, was applied for expression of the HA gene from the Hong Kong influenza A virus isolate (A/HK/156/97) (Schultz-Cherry et al. 2000). Immunization of chicken with VEE-HA particles provided protection against influenza A virus challenges. Moreover, partial protection was observed in newborn chicken after inoculation in ovo. In contrast, a single immunization at two weeks of age resulted in complete protection. In a similar manner, Semliki Forest virus (SFV) particles expressing the influenza virus HA and nucleoprotein (NP) genes provided protection against lethal challenges with influenza virus in immunized mice (Fleeton et al. 2001).

Vaccine development against HIV and AIDS has been carried out by using various viral vector systems. For instance, AAV vectors have been engineered for the expression of HIV-1 Gag eliciting specific T and B cell responses (Lin et al. 2008). Moreover, the impact of the presence of AAV-specific neutralizing antibodies has been addressed. It was demonstrated that inhibition of AAV2-directed Gag responses occurred at 10 to 20-fold lower doses of human immunoglobulin than what was required for inhibition of AAV7 and AAV8 vector immunogenicity. In addition, vaccinia virus vectors have been used for HIV vaccine development. For instance, a chimeric antigen consisting of HIV-1 Env and GM-CSF elicited a superior HIV-specific cellular immune response in BALB/c mice compared to Env alone (Rodr Guez et al. 1999). Rabies virus (RV) vectors have also been employed for HIV-1 antigen production (McGettigan et al. 2001). A single inoculation of RV-HIV-1 gp160 elicited a solid and long-lasting memory CTL response in mice, which was not restricted to the homologous HIV-1 gp160 but was also able to cross-kill target cells expressing heterologous HIV-1 gp160. Self-amplifying alphaviruses have also been applied for HIV vaccine development (Lundstrom 2019). For example, SFV particles expressing the HIV-1 Env/Gag/polRT individually or in combination elicited antigen-specific T cell responses in mice (Ajbani et al. 2017).

The dramatic epidemics of Ebola virus disease (EVD) have accelerated vaccine development against Ebola virus (EBOV). For instance, VEE particles expressing EBOV nucleoprotein (NP) were subjected to immunization studies in mice (Wilson and Hart 2001). C57BL/6 mice vaccinated with VEE-EBOV-NP showed protection against challenges with lethal doses of EBOV. In another study, it was demonstrated that a single immunization of BALB/c mice and guinea pigs with VEE-EBOV-GP particles or the combination of VEE-EBOV-GP and -NP particles provided protection against EBOV challenges (Pushko et al. 2000). In contrast, only mice and not guinea pigs were protected after vaccination with VEE-EBOV-NP particles alone. Additionally, macaques immunized with VSV particles expressing EBOV-GP were completely protected against challenges with the West African EBOV-Makona strain (Marzi et al. 2015). In another approach, a single administration of VSV particles expressing the EBOV and Marburg virus (MARV) GPs provided complete protection against three EBOV strains and MARV, respectively, in non-human primates (Geisbert and Feldmann 2011).

9.4.7 Ophthalmological Diseases

Ophthalmological and especially retinal diseases have been frequently targeted for gene therapy. In this context, AAV vectors were employed for the expression of brain-derived neurotrophic factor (BDNF) to evaluate its potential effect in a rat glaucoma model (Martin et al. 2003). Intravitreal injection of AAV-BDNF resulted in retinal ganglion cell protection, which could support the application of neurotrophic glaucoma therapy for lowering intraocular pressure. In another glaucoma study, AAV-based overexpression of matrix metalloproteinase-3 (MMP-3) showed efficient transduction of corneal endothelium, enhanced MMP-3 activity and decrease in intraocular pressure (O'Callaghan et al. 2017). Moreover, the early-age onset macular dystrophy X-linked retinoschisis (XLRS) is caused by loss of the extracellular matrix protein retinoschisin 1 (RS1) (Bush et al. 2016). Administration of an AAV8 vector expressing RS1 to the eye of Rs-1 knockout mice resulted in significant improvement in retinal structure and function. Moreover, early-onset severe retinal dystrophy (EOSRD), a genetically heterogenous group of diseases associated with mutations in the retinol dehydrogenase 12 (RDH12) gene, has been treated with an AAV2/5-RDH12 vector (Feathers et al. 2019). Subretinal injection of Rdh-deficient mice reconstituted retinal reductase activity and decreased susceptibility of light damage associated with Rdh12 deficiency.

9.4.8 Muscular Diseases

Muscular diseases such as Duchenne muscular dystrophy (DMD) have been widely studied as subjects for gene therapy (Chamberlain and Chamberlain 2017). Initially, an Ad virus vector was used for the expression of truncated dystrophin cDNAs, which restored dystrophin-related proteins in adult mouse skeletal muscle in vivo (Yuasa et al. 1998). The large size of the dystrophin gene caused serious problems for gene therapy applications, especially in the context of using AAV with a packaging capacity of only 4 kb. A major improvement was experienced by the demonstration that "micro-dystrophin" cassettes could be engineered (Sakamoto et al. 2002). Moreover, the discovery of the new AAV6, AAV8 and AAV9 serotypes allowed efficient delivery to all striated muscles in adult mice (Gregorevic et al. 2004). For instance, AAV6 expressing micro-dystrophin restored dystrophin expression in respiratory, cardiac and limb musculature in mice, also considerably reducing skeletal muscle pathology and prolonging the lifespan of severely dystrophic mice (Gregorevic et al. 2006). In a dystrophic canine model, intramuscular injection of AAV6-micro-dystrophin showed delivery throughout a group of skeletal muscles (Wang et al. 2012a). Robust micro-dystrophin expression was detected for at least two years. Moreover, another study revealed that a single intravenous injection of AAV9 resulted in whole body skeletal muscle transduction in neonatal dogs (Yue et al. 2008). In contrast, cardiac muscles were barely transduced in dogs.

Oculopharyngeal muscular dystrophy (OPMD) is characterized by the late onset of ptosis, swallowing difficulties, proximal limb weakness and nuclear aggregates in skeletal muscles caused by a trinucleotide repeat expansion in the polyA-binding protein nuclear 1 (PABPN1) gene (Malerba et al. 2017). AAV-based PABPN1 expression in a mouse model of OPMD resulted in substantial reduction of insoluble aggregates, decrease in muscle fibrosis, and normalization of muscle strength. Similar effects were detected in cells derived from OPMD patients.

9.4.9 Lung Diseases

Cystic fibrosis is an inherited disease caused by mutations in the cystic fibrosis transmembrane conductance regulator (CFTR) gene, which causes damage to the lungs, digestive system and other organs (O'Sullivan and Freedman 2009). Gene therapy approaches for cystic fibrosis comprise the use of AAV2 vectors for the expression of the CFTR gene (Flotte et al. 1993). The AAV2-CFTR vector was delivered to one lobe of the rabbit lung via a fiber optic bronchoscope. The recombinant CFTR was detected for up to 6 months in the airway epithelium. Furthermore, a study in rhesus macaques showed vector-specific DNA and recombinant RNA expression for up to 180 days after infection (Conrad et al. 1996).

Lentivirus vectors have also been applied for gene therapy of cystic fibrosis (Marquez Loza et al. 2019). An HIV-based system for in vivo delivery has been developed where mouse nasal epithelium is pre-treated with lysophosphatidylcholine (LPC), which generated significant expression of β-galactosidase from the HIV-LacZ vector (Limberis et al. 2002). Application of the same procedure for the HIV-CFTR vector resulted in partial recovery of electrophysiological functions in the nasal airway epithelium of CF knockout mice for at least 110 days. In another study, a VSV-G pseudotyped HIV-1 based vector expressing the LacZ gene was detected in the conducting airways and in alveolar regions of marmosets but not in liver or spleen (Farrow et al. 2013). Simian immunodeficiency virus (SIV) has also been pseduotyped with envelope proteins from Sendai virus (SeV), which has been demonstrated to efficiently transduce airway epithelial cells (Mitomo et al. 2010). The SIV-SeVGFP vector showed clinically relevant transduction of respiratory epithelial cells in the mouse nose, which persisted for at least 15 months and the vector could be readministered. The SIV vector also showed transduction of human air-liquid interface cultures.

9.5 Clinical Trials

Due to more than 3000 gene therapy clinical trials conducted, of which two thirds have been estimated to involve viral vectors, only a summary can be presented below and in Table 9.4. The vaccine trials based on viral vectors will further increase the number substantially. Treatments of various cancers naturally represent a large part of the conducted clinical trials (Lundstrom 2018). For instance, 17 Japanese patients, 6 with recurrent breast cancer, 3 with recurrent head and neck cancer and 8 with nonresectable pancreatic cancer, were subjected to intratumoral administration of oncolytic HSV HF10 in a phase I dose-escalation clinical trial (Kasuya et al. 2014). No adverse events were registered, and some therapeutic effects were observed based on pathological findings, tumor markers and diagnostic radiography. In another single arm, open-label phase I trial to treat nonresectable locally advanced pancreatic cancer, patients were subjected to intratumoral administration of HSV HF10 (Hirooka et al. 2018). Only adverse events unrelated to HSV HF10 treatment were detected. Partial responses (PR) were observed for 3 patients, stable disease (SD) for 4 patients and progressive disease (PD) for 9 patients. Moreover, preliminary results from a phase II trial with HSV HF10 and ipilimumab (anti-CTLA-4) showed a good safety profile and antitumor activity in patients with nonresectable or metastatic melanoma (Eissa et al. 2017). In an open-label, ascending dose, multicenter phase I trial in patients with recurrent or progressive high-grade glioma (HGG), the Toca 511 replicating retrovirus (RRV) showed an overall survival of 13.6 months, which is statistically superior to the survival in the control group (Cloughesy et al. 2016). The RRV Toca 511 has also been evaluated in phase II/III clinical trials for glioma patients, showing highly encouraging preliminary therapeutic efficacy (Inoko et al. 2018), In the case of MV vectors, a phase I trial with the MV-Edm-Zagreb strain was conducted in patients with cutaneous T-cell lymphoma (Heinzerling et al. 2005). No dose-limiting toxicity was demonstrated, and

Table 9.4 Examples of clinical trials conducted with viral vectors

Indication	Vector/Target	Phase	Outcome
Cancer			
Breast, head & neck, pancreas	HSV HF10	I	No adverse events, some therapeutic potential
Pancreatic	HSV HF10	I	PD and SD in some treated patients
Melanoma	HSV HF10 +	I	Good safety profile, antitumor activity
	PANVAC-VF	III	Significant reduction in tumor size in patients
HGG	Ipilimumab	I	Prolonged survival in glioma patients
	RRV Toca 511	II/III	Preliminary encouraging therapeutic efficacy
CTCL	RRV Toca 511	I	Complete and partial tumor regression
Ovarian	MV-NIS	I/II	Extended overall survival
Glioma	MV-CEA	I	No dose limiting toxicity
Colorectal	MV-CEA	I	Tumor regression in patients
	NDV La Sota IV	III	Prolonged survival
Solid tumors	NDV PV701	I	Progression-free survival for 4–31 months
Hematological			
Sickle-cell anemia	LV- β(A-T87Q)-globin	I/II	Anti-sickling β-globin high for 15 months
Hemophilia A	AAV-FVIII	I/II	8–60% of normal FVIII levels
Hemophilia B	AAV-FIX	I/II	Stable 5% FIX levels for 7 years, reduced spontaneous bleedings in patients
Ophthalmology			
AMD	AAV2-sFLT01	I	Good safety and tolerability
	AAV2-sFLT01	IIa	Maintained/improved vision in patients
LHON	AAV2-ND4	I	Visual acuity improvement in patients
	AAV2-ND4	I	Significant improvements of visual acuity
	AAV2-ND4	I	Improvements in visual acuity
Neurological			
SMA	AAV9-SMN	I	Improved motor function and survival
	AAV9-SMN	I	Improved motor function and prolonged survival
Muscular			
DMD	AAV9-miniDys	I	Patient enrollment in progress
	AAV9-microDys	I/II	Expression for 90 days, study on hold
	AAVrh74-microDys	I/II	Robust expression, functional improvements
	AAV9-miniDys	III	Trial planned for end of 2020
Lung			
Cystic fibrosis	LV-CFTR	Pre	Safety preparation for Phase I
Immunodeficiency			
SCID-X1	γRV-IL2RG	I	Lasting cure for 18 years follow-up, leukemia developed in some treated individuals
	SIN-γRV-IL2RG	I	None of treated patients developed leukemia
ALD	SIN-LV-ABCD1	I	Termination of progressive cerebral demyelination

(continued)

Table 9.4 (continued)

Indication	Vector/Target	Phase	Outcome
ADA-SCID	SIN-γRV-ADA, SIN-LV-ADA	II	Excellent safety and efficacy of treatment
	SIN-LV-ABCD1	II	Excellent safety and efficacy of treatment
Infectious			
CMV	VEE-gB/PP15/IE1 fusion	I	Immune responses against CMV antigens
HIV	VEE-HIV-Gag	I	Only modest immune responses in volunteers
	MRKAd5, ALVAC	I	Limited HIV-1 inhibition
EVD	VSV-ZEBOV (EBOV GP)	III	Protection against EDV
	VSV-ZEBOV (EBOV GP)	III	Protection against EDV

AAV adeno-associated virus, Ad5 adenovirus 5, ADA adenosine deaminase, ALD adrenoleukodystrophy, AMD age-related macular degeneration, CTCL cutaneous T-cell lymphoma, EVD Ebola virus disease, FVIII factor VIII, FIX factor IX, γRV gamma retrovirus, HGG high-grade glioma, HSV HF10 herpes simplex virus HF10 strain, IL2RG interleukin 2 receptor gamma subunit, LHON Leber's hereditary optic neuropathy, LV lentivirus, MV-CEA measles virus expressing carcinoembryonic antigen, MV-NIS measles virus expressing sodium iodide symporter, ND4 NADH dehydrogenase protein subunit 4, PR partial response, RRV replicating retrovirus, SCID-X1 X-linked severe combined immune deficiency, SD stable disease, SMA spinal muscular atrophy, SMN survival of motor neuron gene, VV vaccinia virus

the treatment was well tolerated. Moreover, complete regression of one CTCL tumor was observed in one patient and partial regression was achieved in 4 out 5 treated tumors. In a phase I/II clinical trial MV-CEA was assessed for safety and tolerability in recurrent ovarian cancer patients (Galanis et al. 2010). No dose-limiting toxicity was observed, disease stabilization was achieved in 9 out 9 patients treated with 10^7–10^9 MV-CEA particles, and a median overall survival of 12.15 months was achieved, which is twice the expected overall survival. In another phase I study, MV-CEA has been applied for treatment of recurrent glioblastoma multiforme with doses of 10^5 to 2×10^7 $TCID_{50}$ showing no dose limiting toxicity (Myers et al. 2008) (NCT00390299). In a phase I trial on 11 patients with advanced colorectal cancer and other solid tumors, intravenous administration of poxvirus elicited a potent TH1-mediated immunity against the poxvirus and possibly the cancer (Downs-Canner et al. 2016). A mixed response with resolution of some liver metastasis was observed in one patient and clinical regression of some lesions was detected in another patient with cutaneous melanoma. In the context of NDV, a phase II clinical trial was conducted on 79 patients with solid tumors by administration of 1.2×10^{10}–1.2×10^{11} pfu/m2 of the NDV PV101 strain, which resulted in objective responses and progression-free survival ranging from 4 to 31 months (Pecora et al. 2002). Moreover, in a phase III trial in 335 patients with colorectal cancer NDV immunotherapy showed prolonged survival and short-term improved quality of life (Liang et al. 2003). The poxvirus-based vaccine PANVAC-VF has been subjected to a phase III study in 295 patients with injectable unresectable stage IIIB, IIIC or IV melanoma (Chi et al. 2020). A more than 50% reduction in tumor size was observed in 64% of injected tumors. Moreover, a reduction of more than 50% in tumor size was seen in one-third of un-injected non-visceral tumors and in 15% of visceral tumors.

Hematological diseases have also been targeted for gene therapy applications. In this context, sickle-cell anemia, caused by a homozygous missense mutation in the β-globin gene, has been subjected to therapeutic intervention by lentiviral vector-based introduction of an anti-sickling β-globin gene into autologous hematopoietic stem cells in a phase I/II trial (Ribel et al. 2017; NCT02151526). The level of therapeutic anti-sickling β-globin remained high for 15 months without recurrence of sickle crises. In the context of hemophilia, AAV vectors have been the system of choice (Chapin and Monahan 2018). However, issues such as pre-existing neutralizing antibodies against AAV, elevated liver transaminase levels and immune-related decrease in AAV-based transgene expression have raised concerns about safety and efficacy in hemophilia trials (Mingozzi and High 2013). However, utilization of different AAV serotypes for repeated administration has prevented decrease in transgene expression levels. For this reason, at least 11 gene therapy clinical trials have been conducted in hemophilia patients and at least six phase I/II studies based on liver targeted AAV expression of either FVIII or FIX are in progress (Spencer et al. 2016). One challenge of treating hemophilia A with AAV-based FVIII expression is the FVIII gene size exceeding the packaging capacity of AAV. One solution has been the deletion of the B-domain as it does not affect the hemostatic function of FVIII (Arruda and Samelson-Jones 2015). In this context, a phase I/II clinical trial was initiated and preliminary results from 6 patients receiving a dose of 6×10^{13} AAV-FVIII particles demonstrated 8–60% of normal FVIII levels (Nathwani 2019). One approach to establish clinically relevant FVIII expression has been to bioengineer FVIII constructs showing 10–100-fold higher levels than observed for native human FVIII (Spencer et al. 2016). Related to FIX, it has been demonstrated that AAV can improve FIX deficiency from severe to mild by expression of 3–7% of normal FIX levels (Nathwani et al. 2014). In a phase I/II clinical trial initiated in 2011 stable dose-dependent increase in FIX levels in patients with severe hemophilia B were achieved after a single AAV administration (NCT02576795) (Nathwani 2019). The FIX expression levels remained stable at approximately 5% of normal levels for 7 years providing substantial reduction in spontaneous bleeding in hemophilia B patients.

Spinal muscular atrophy, a genetic disorder characterized by muscle weakness and atrophy leading to infant mortality, is caused by deterioration of motor neurons in the brainstem and spinal cord. In a phase I clinical trial, a one-time administration of an AAV9 vector expressing the survival of motor neuron (SMN) gene provided remarkable improvements in motor function and in survival rates in patients (Pattali et al. 2019). In a phase I trial, 15 patients with SMA received a single intravenous injection of AAV9-SMN, which resulted in prolonged survival, superior achievement of motor milestones and improved motor function (Mendell et al. 2017). Related to the X-linked disorder DMD, a phase I open label, non-randomized, ascending dose study applying the AAV9-mini-dystrophin vector is planned for the evaluation of safety and tolerability (Moorehead et al. 2020) (NCT03362502). In the study, 15 boys in the age group of 4–12 years will be enrolled and they each will receive a single intravenous infusion. The primary endpoints to be evaluated through 12 months include adverse events, dystrophin expression and distribution, and assessment of muscle strength. Additional endpoints include muscle biopsies, biomarkers and functional assessments. Additional AAV-based DMD clinical trials are in progress such as a phase I/II study using AAV9 (NCT03368742), a phase I/II using AAVrh74 (NCT03375164), and a phase III study using AAV9 (NCT04281485). Preliminary results from the phase I/II study with AAVrh74, showing affinity to skeletal and cardiac muscle cells, demonstrated minimal adverse events and good safety in 4 patients (Mendell et al. 2020). The AAVrh74-based expression of micro-dystrophin was robust, the protein localization was correct and functional improvements were seen in the patients.

Related to age-related macular degeneration, VEGF neutralizing proteins have provided proven therapeutic efficacy. In this context, AAV2-based expression of sFLT01 was applied for intravitreous injection in 19 patients, which demonstrated good safety and tolerability at all tested doses (Heier et al. 2017). Next, a phase 2a trial with rAAV-sFLT01 was conducted in 11 patients suffering from age-related macular degeneration (Constable et al. 2016). No treatment-associated serious adverse events were registered. The rAAV-sFLT01 treated patients showed maintained or improved vision in comparison to control individuals.

The optic nerve disorder Leber's hereditary optic neuropathy (LHON) is characterized by rapid visual loss in one eye followed by visual loss in the other eye. This neuropathy is caused by a mutation in the NADH dehydrogenase protein subunit 4 (ND4) gene (Ratican et al. 2018). AAV2-mediated ND4 expression has been subjected to clinical trials, of which the safety and efficacy was evaluated in 9 patients with the G11778A mutation in the ND4 gene (Wan et al. 2016). Injection of AAV2-ND4 into one eye showed significant improvement of visual acuity in 6 out of 9 patients. In another phase I study in 14 LHON patients, intravitreal injection of AAV2-ND4 provided modest but statistically significant improvements of visual acuity (Guy et al. 2017). In a third phase I trial in 15 LHON patients, it was demonstrated that a single injection of rAAV2-ND4 was safe and well tolerated also showing improvement in visual acuity (Vignal et al. 2018).

In preparation for the first-in-man clinical trial for lentivirus-based treatment of cystic fibrosis, a hybrid promoter consisting of a cytosine guanine dinucleotide (CpG)-free CMV enhancer/elongation factor 1 alpha promoter (hCEF) showed highly efficient expression of functional CFTR in murine lung tissue and human air-liquid interface cultures (Alton et al. 2017). The transduction efficacy, lack of toxicity and the chromosomal integration site profile further supported the application of lentivirus vectors for clinical trials on cystic fibrosis patients.

The probably most famous application of viral-based gene therapy relates to SCID X1, the X-linked immunodeficiency in children (Fischer and Hacein-Bey-Abina 2020). The first clinical gene therapy was conducted in 1999 with a defective Moloney γ retrovirus (γRV)-derived vector expressing the interleukin-2 receptor gamma subunit (IL2RG) for ex vivo transduction of $CD34^+$ cells. That first clinical trial in Paris involved 10 SCID-X1 patients (Cavazzana-Calvo et al. 2000). The treatment led to the development of a normal T cell count within 3–6 months in eight patients with lasting clear-cut clinical benefit. After 18 years follow-up, all but one patient is doing well, showing normal growth and not experiencing the opportunistic infections characteristic of SCID-X1 disease. A similar study in London showed sustained clinical benefit in all 10 patients enrolled (Gaspar et al. 2004). Unfortunately, T cell acute lymphoblastic leukemia was discovered in some patients in both studies 2–14 years after treatment due to the integration of the γRV vector into the LMO2 oncogene locus in five patients and in the CCDN2 locus in one patient (Hacein-Bey-Abina et al. 2008; Howe et al. 2008). Because of this setback, self-inactivating γRV and lentivirus vectors have been engineered. A clinical trial with a self-inactivating γRV vector found that none of the 9 treated patients developed leukemia (Hacein-Bey-Abina et al. 2014). Moreover, leukemia has not been discovered in a single of the 44 patients treated with either γRV or lentivirus self-inactivating vectors (Fischer and Hacein-Bey-Abina 2020). In this context, a self-inactivating lentivirus vector carrying the ABCD1 gene coding for an adenosine triphosphate-binding cassette transporter was ex vivo transduced into hematopoietic stem cells (HCTs) for gene therapy of X-linked adrenoleukodystrophy (ALD) patients (Cartier et al. 2009). Re-infusion of HCTs resulted in termination of progressive cerebral demyelination in two patients providing clinical benefits in ALD. Another form of SCID, known as ADA-SCID, is caused by inherited defects in adenosine deaminase (ADA) (Kohn et al. 2019). Today, more than 100 ADA-SCID patients have been treated with either γRV or lentivirus vectors showing excellent safety and efficacy.

Infectious diseases have seen a large number of clinical trials on vaccines so only some examples are given here and in Table 9.4. For instance, 40 CMV seronegative volunteers were subjected to either intratumoral or subcutaneous administration of a VEE alphavirus vector expressing the CMV glycoprotein B (gB) or the PP65/IE1 fusion protein in a randomized, double-blind phase I study (Bernstein et al. 2010). Immune responses against all three CMV antigens were detected in all vaccinated individuals. In another phase I study, healthy HIV-negative individuals received subcutaneously VEE particles expressing a nonmyristoylated form of HIV-Gag in the US and South Africa, which demonstrated good safety and tolerability although only modest local immune and T cell responses were detected (Wecker et al. 2012). Moreover, Ad5 viral vectors have been designed to elicit HIV-1-specific T cells in healthy volunteers (Hayes et al. 2016). Immunizations resulted in CD8 T cell responses although with limited HIV-1 inhibition breadth showing better efficacy in antiretroviral naive HIV-1 infected volunteers naturally controlling viremia. In an open-label, cluster ring vaccination phase III clinical trial 4123 individuals with suspected EVD were subjected to immunization with VSV particles expressing the EBOV GP from the Zaire strain named VSV-ZEBOV (Henao-Restrepo et al. 2015). Another 3528 persons received a delayed vaccination. The immediate vaccination group showed no EVD cases while in the delayed group 16 EVD cases were discovered. In another phase III study in Guinea and Sierra Leone 2119 participants were subjected to immediate immunization with VSV-ZEBOV and for 2041 individuals the vaccination was delayed for 21 days (Henao-Restrepo et al. 2017). Substantial protection against EBOV was also achieved in this study showing no new EVD cases from 10 days after vaccination. VSV-ZEBOV was approved for EVD by the FDA under the brand name Ervebo in December 2019 (Ollmann Saphire 2020).

9.6 COVID-19

For obvious reasons, due to the current pandemic COVID-19 vaccine development has received plenty of attention. In addition to other approaches such as inactivated and live attenuated vaccines, protein subunit and peptide vaccines, and nucleic acid-based vaccines, viral vectors have been applied (Lundstrom 2020b) (Table 9.5). For instance, the simian Ad virus vector ChAdOx1-S expressing the SARS-CoV-2 S protein elicited strong humoral and cellular immune responses in mice and rhesus macaques preventing pneumonia in immunized macaques (Folegatti et al. 2019; van Doremalen et al. 2020). In another Ad5 based preclinical study, expression of SARS-CoV-2 S induced strong immune systemic S-specific antibody and cell-mediated immune responses in immunized mice and rhesus macaques (Feng et al. 2020).

A single intramuscular or intranasal vaccination with Ad5-S-nb2 protected macaques against SARS-CoV-2 challenges. In another approach, a single immunization with Ad26 expressing SARS-CoV-2 S elicited binding and neutralizing antibody responses and protected immunized hamsters against SARS-CoV-2 induced weight loss, pneumonia and death (Tostanoski et al. 2020). Additionally, a single immunization with Ad26-SARS-CoV-2 S induced robust neutralizing antibody responses in rhesus macaques and furthermore provided complete or near-complete protection in bronchoalveolar lavage and naval swabs after SARS-CoV-2 challenges (Mercado et al. 2020).

As lung is a vital organ for SARS-CoV-2 infection, the MVA poxvirus has been proposed for application on mucosal surfaces of the respiratory tract as a candidate for a COVID-19 vaccine (Förster et al. 2020). A non-replicating MVA-based VLP vaccine against COVID-19 has now entered preclinical evaluation (https://www.geovax.com/technology-pipeline/ infectious-diseases). In the case of MV, the full-length SARS-CoV-2 S gene was introduced into two positions in the MV genome (Hörner et al. 2020). The construct providing lower SARS-CoV-2 S protein levels was stable and elicited efficient Th1-biased antibody and T cell responses in mice after two immunizations. VSV vectors expressing the SARS-CoV-2 S protein elicited neutralizing antibodies and provided protection against challenges with SARS-CoV-2 in immunized mice (Case et al. 2020). Moreover, the replication-competent VSVΔG vector, where the VSV G protein was replaced by the SARS-CoV-2 S protein elicited neutralizing antibodies and protected Syrian golden hamsters against SARS-CoV-2 after a single immunization with 5×10^6 pfu of VSVΔG-SARS-CoV-2 S particles (Yahalom-Ronen et al. 2020).

Table 9.5 Viral vector-based COVID-19 vaccines

Viral vector/target	Stage	Response
Adenovirus		
ChAdOx1-S	Pre	Strong immune responses, prevention of pneumonia in primates
ChAdOx1-S	Phase I/II	Humoral and cellular responses in all participants
ChAdOx1-S	Phase III	Trial on hold due to suspect adverse events, but then resumed
ChAdOx1-S	Phase III	Vaccine efficacy 70-90%
ChAdOx1-S	EUA	Vaccine approved for EUA in the UK on January 4, 2021
Ad5-S-nb2	Pre	Strong antibody-specific response, protection against SARS-CoV-2
Ad5-S-nb2	Phase I	Humoral and T cell responses in volunteers
Ad5-S-nb2	Phase II	Robust immune responses
Ad5-S-nb2	Phase III	Good safety and immunogenicity
Ad5-S-nb2	EUA	Vaccine approved for EUA in China and other countries
Ad26.COV2-S	Pre	Antibody responses, protection against SARS-CoV-2 in hamsters
Ad26.COV2-S	Pre	Antibody responses, protection against SARS-CoV-2 in macaques
Ad26.COV2-S	Phase I/II	Neutralizing antibodies in 90% of participants
Ad26.COV2-S	Phase III	Good safety and immunogenicity
Ad26.COV2-S	EUA	Vaccine approved for EUA by the FDA
rAd26-S/rAd5-S	Phase I/II	Good safety, humoral and cellular responses in all participants
rAd26-S/rAd5-S	Phase III	Good tolerability, 91.6% vaccine efficacy
rAd26-S/rAd5-S	EUA	Vaccine approved for EUA in Russia
Poxvirus		
MVA	Pre	Preclinical studies in progress
Measles virus		
MV-SARS-CoV-2 S	Pre	Neutralizing antibody and T cell responses in mice
MV (TMV083)	Phase I	Weak immune responses, study discontinued
Lentivirus		
LV-DCs	Phase I/II	Safety and immunogenicity evaluation of LV-DC + antigen-specific CTLs
Rhabdoviruses		
VSV-SARS-CoV-2 S	Pre	Neutralizing antibodies, protection in mice
VSV-SARS-CoV-2 S	Phase I	Weaker immune response than in COVID-19 patients, study terminated
VSVΔG-SARS-CoV-2 S	Pre	Neutralizing antibodies, protection in hamsters
VSVΔG-SARS-CoV-2 S	Phase I/II	Study in progress

Ad adenovirus, ChAdOx1-S simian adenovirus expressing SARS-CoV-2 S protein, CTLs cytotoxic T lymphocytes, EUA Emergency Use Authorization, LV-DCs lentivirus-transduced dendritic cells, MV measles virus

Based on encouraging results from preclinical evaluation, several clinical trials have been launched (Table 9.5). In this context, in the first-in-human, dose-escalation, non-randomized phase I study, three doses of 5×10^{10}, 1×10^{11} and 1.5×10^{11} Ad5-SARS-CoV-2 S particles were administered to 108 healthy volunteers for safety, tolerability and immunogenicity evaluations (Zhu et al. 2020b) (ChiCTR2000030906). Despite some minor pain reactions, no serious adverse events were recorded. Peak humoral responses against SARS-CoV-2 were detected 28 days after vaccination and rapid SARS-CoV-2-specific T cell responses were registered 14 days post-vaccination. Furthermore, the vaccine candidate was safe and induced robust immune responses in the majority of individuals after a single dose in a phase II clinical trial (Zhu et al. 2020a). The Ad5-SARS-CoV-2 S has been evaluated in phase III trials (NCT04526990, NCT04540419) and has received EUA in China and some other countries. Interim results from a phase I/II study on the Ad26.COV2-S vaccine in 805 healthy volunteers in Belgium and the US elicited neutralizing antibodies in 90% of participants (Sadoff et al. 2021). Moreover, several phase III studies have been conducted (NCT04505722, ISRCTN14722499), which has resulted in EUA by the FDA (Ad26.COV2-S FDA Approval Status. drugs.com/history/ad26-cov2-s.html (accessed on September 2, 2021). Controversially, the Sputnik V vaccine, developed by the Gamaleya Research Institute of Epidemiology and Microbiology in Russia, was approved in Russia before completion of any phase III trial and even publication of preclinical and clinical findings (Callaway 2020). Sputnik V, the rAd26 and rAd5 vector-based vaccine (expression of the S protein, i.e. rAd26-S and rAd5-S) had at the time of approval only been evaluated in 76 volunteers. Finally, weeks after the approval the results from two phase I/II clinical trials were published (Logunov et al. 2020). In phase I, the volunteers received intramuscularly either one dose of rAd26-S or rAd5-S (NCT04436471). In the phase II prime-boost regimen, participants were administered intramuscularly rAd26-S followed by rAd5-S 21 days later (NCT04437875). Most adverse events observed were mild and no serious events were seen. SARS-CoV-2- specific antibodies were detected in all vaccinees. The immunization showed a good safety profile and induced strong humoral and cellular immune responses. Although approved weeks earlier, it is stated in the publication that "further investigation is needed of the effectiveness of this vaccine for prevention of COVID-19" (Logunov et al. 2020). Additionally, interim results from a phase III trial showed good tolerability and 91.6% efficacy of the Sputnik V vaccine (Logunov et al. 2021). Related to the chimpanzee Ad virus vector ChAdOx1, preliminary results on safety, reactogenicity and immunogenicity from a phase I/II trial showed no serious adverse events in healthy volunteers after a single intramuscular injection of 5×10^{10} particles (Folegatti et al. 2020). Moreover, SARS-CoV-2-specific neutralizing antibodies were detected in 32 (91%) of 35 immunized individuals. All participants elicited both humoral and cellular immune responses after a booster dose. These encouraging results enabled a large-scale evaluation of vaccine efficacy in a randomized, double-blind, placebo-controlled multicenter phase III trial in 40,000 adults, which started in August 2020 (NCT04516746). However, in early September it was announced that the phase III trial was put on hold due to suspect adverse events occurring in one participant (Phillips et al. 2020). Following an investigation, the trial resumed. Interim results from the phase III trial (NCT04516746) showed vaccine efficacy of 70.4% after data were combined from two dosing regimens (www.ox.ac.uk/news/2020-11-23-oxford-university-breakthrough-global-covid-19-vaccine). The study demonstrated that that a higher efficacy of 90% was achieved by using a 50% dose for the prime vaccination and a standard dose for the boost immunization. In contrast, standard prime-boost dosing showed 62% efficacy. Moreover, based on interim analysis of a phase III trial, the vaccine showed an acceptable safety profile and efficacious against symptomatic COVID-19 (Voysey et al. 2021). On January 4, 2021, the ChAdOx1-S vaccine was approved in the UK for EUA.

In the context of lentiviruses, a phase I/II clinical trial is in progress in Shenzen China in 100 healthy volunteers using a lentivirus vector containing minigenes of multiple conserved regions of SARS-CoV-2 (NCT04276896). Lentivirus transduced DCs (LV-DC), 5×10^6 cells, will be subcutaneously administered in combination with intravenously injected 1×10^8 antigen-specific CTLs for safety and immunogenicity evaluations. The MV-SARS-CoV-2 S vaccine candidate TMV-083 has also been subjected to a phase I clinical trial. However, early reports of disappointingly weak immune responses in healthy volunteers led to the termination of the trial (NCT04497298). Similarly, a phase I clinical trial on the VSV-SARS-CoV-2 S V590 vaccine was discontinued due to immune responses being weaker than seen in convalescent COVID-19 patients (NCT04569786). The VSVΔG-SARS-CoV-2 vaccine is currently evaluated in a phase I/II clinical trial (NCT04608305).

9.7 Approved Drugs

Although gene therapy has encountered serious setbacks with the death of patients treated with an Ad vector for OTC deficiency, a non-life-threatening disease (Lehrman 1999) and the retrovirus-based therapeutic intervention of SCID-X1 patients resulting in leukemia (McCormack and Rabbitts 2004; Hacein-Bey-Abina et al. 2008; Fischer and Hacein-Bey-Abina 2020), the field has experienced a true renaissance in recent years.

The first gene therapy drug (Gendicine™) based on viral vectors, an oncolytic Ad virus vector expressing the p53 gene, was approved already more than 12 years ago in China (Räty et al. 2008). More than 30,000 patients have been treated for cancers with p53 mutations and head and neck cancers with Gendicine™ showing good safety records and superior response compared to standard therapies, especially in combination with chemotherapy and radiotherapy (Zhang et al. 2018). Moreover, a second-generation oncolytic HSV expressing GM-CSF was approved in the US and Europe for melanoma treatment (Fukuhara et al. 2016; Kaufman et al. 2010). Although success in gene therapy drug development has been achieved, a warning example is illustrated by Glybera™, the AAV-based treatment of the monogeneic inherited disease lipoprotein lipase deficiency (Ylä-Herttuala 2015). Despite approval in Europe, UniQure did not apply for a license renewal for the continued clinical use of Glibera™ because of lack of demand for the rare disease. In the context of viral vector-based vaccines, the EBOV vaccine based on the VSV-ZEBOV vector was approved by the FDA under the brand name Ervebo in December 2019 (Ollmann Saphire 2020). For more information of approved gene therapy-based drugs a list can be found at https://www.fda.gov/vaccines-blood-biologics/cellular-gene-therapy-products/approved-cellular-and-gene-therapy-products.

9.8 Conclusions and Future Aspects

Since the first applications of viral vectors, the progress of gene therapy has been astonishing. Despite the setbacks with the treatment of the non-life-threatening disease OTC with Ad virus (Lehrman 1999) and the retrovirus-based therapy of SCID-X1 leading to leukemia (Fischer and Hacein-Bey-Abina 2020), extensive vector development regarding safety and delivery issues has led to a renaissance in gene therapy. Numerous preclinical studies have showed proof-of-concept for various disease indications applying different viral vector systems. Moreover, clinical trials have confirmed safety, tolerability and efficacy in patients suffering from different ailments. Similarly, viral vector-based vaccine development has shown promising results in both preclinical studies and clinical trials. Moreover, several viral vector-based gene therapy drugs and vaccines have been approved in China, Europe and the US.

Despite these positive developments, there is need for further improvement of gene therapy technologies related to vector engineering from the safety and delivery point of views. It should also be seen as an advantage that a multitude of vector systems are developed as clearly no single system can address all issues related to delivery, targeting, and duration of transgene expression. Obviously, the current COVID-19 pandemic has accelerated vaccine development, and in this context viral vectors also play an important role as has been confirmed by EUA of several Ad virus-based COVID-19 vaccines (Regulatory Approval of COVID-19 Vaccine AstraZeneca—GOV.UK. Available online: www.gov.uk (accessed on September 2, 2021); Ad26.COV2-S FDA Approval Status. drugs.com/history/ad26-cov2-s.html (accessed on September 2, 2021)).

References

Ady JW, Johnsen C, Mojica K et al (2015) Oncolytic gene therapy with recombinant vaccinia virus strain GLV-2b372 efficiently kills hepatocellular carcinoma. Surgery 158(2):331–338. https://doi.org/10.1016/j.surg.2015.03.044

Ajbani SP, Velhal SM, Kadam RB et al (2017) Immunogenicity of virus-like Semliki Forest virus replicon particles expressing Indian HIV-1C gag, env and pol RT genes. Immunol Lett 190:221–232. https://doi.org/10.1016/j.imlet.2017.08.019

Akhter SA, Skaer CA, Kypson AP et al (1997) Restoration of beta-adrenergic signaling in failing cardiac ventricular myocytes via adenoviral-mediated gene transfer. Proc Natl Acad Sci U S A 94(22):12100–12105. https://doi.org/10.1073/pnas.94.22.12100

Al Yaghchi C, Zhang Z, Alusi G et al (2015) Vaccinia virus, a promising new therapeutic agent for pancreatic cancer. Immunotherapy 7(12):1249–1258. https://doi.org/10.2217/imt.15.90

Alexander IE, Cunningham SC, Logan GJ et al (2008) Potential of AAV vectors in the treatment of metabolic disease. Gene Ther 15(11):831–839. https://doi.org/10.1038/gt.2008.64

Alles J, Fehlmann T, Fischer U et al (2019) An estimate of the total number of true human miRNAs. Nucleic Acid Res 47(7):3353–3364. https://doi.org/10.1093/nar/gkz097

Alton EWFW, Beekman JM, Boyd AC et al (2017) Preparation for a first-in-man lentivirus trial in patients with cystic fibrosis. Thorax 72(2):137–147. https://doi.org/10.1136/thoraxjnl-2016-208406. Epub 2016 Nov 16

Amit G, Kikuchi K, Greener ID et al (2010) Selective molecular potassium channel blockade prevents atrial fibrillation. Circulation 121(21):2263–2270. https://doi.org/10.1161/CIRCULATIONAHA.109.911156

Amoasii L, Long C, Li H et al (2017) Single-cut genome editing restores dystrophin expression in a new mouse model of muscular dystrophy. Sci Transl Med 9:418. https://doi.org/10.1126/scitranslmed.aan8081

An H, Kim GN, Kang CY (2013) Genetically modified VSV(NJ) vector is capable of accommodating a large foreign gene insert and allows high level gene expression. Virus Res 171 (1):168–177. https://doi.org/10.1016/j.virusres.2012.11.007

Apostolopoulos V (2016) Vaccine delivery methods into the future. Vaccine 4(2):9. https://doi.org/10.3390/vaccines4020009

Arruda VR, Samelson-Jones BJ (2015) Obstacles and future of gene therapy for hemophilia. Exp Opin Orphan Drugs 3(9):997–1010. https://doi.org/10.1517/21678707.2015.1069179

Avogadri F, Merghoub T, Maughan MF et al (2010) Alphavirus replicon particles expressing TRP-2 provide potent therapeutic effect on melanoma through activation of humoral and cellular immunity. PLoS One 5(9):e12670. https://doi.org/10.1371/journal.pone.0012670

Bai FL, Yu YH, Tian H et al (2014) Genetically engineered Newcastle disease virus expressing interleukin-2 and TNF-related and apoptosis-inducing ligand for cancer therapy. Cancer Ther Biol 15(9):1226–1238. https://doi.org/10.4161/cbt.29686

Balagué C, Zhou J, Dai Y et al (2000) Sustained high-level expression of full-length human factor VIII and a restoration of clotting activity in hemophilic mice using a minimal adenovirus vector. Blood 95(3):820–828

Bastin D, Aitken AS, Pelin A et al (2018) Enhanced susceptibility of cancer cells to oncolytic rhabdo-virotherapy by expression of Nodamura virus protein B2 as a suppressor of RNA interference. J Immunother Cancer 6(1):62. https://doi.org/10.1186/s40425-018-0366-2

Bentires-Alj M, Hellin A-C, Lechanteur C et al (2000) Cytosine deaminase suicide gene therapy for peritoneal carcinomatosis. Cancer Gene Ther 7(1):20–26. https://doi.org/10.1038/sj.cgt.7700093

Bernstein DI, Reap EA, Katen K et al (2010) Randomized, double-blind, phase I trial on an alphavirus replicon vaccine for cytomegalovirus in CMV negative volunteers. Vaccine 28 (2):484–493. https://doi.org/10.1016/j.vaccine.2009.09.135

Björklund A, Kirik D, Rosenblad C et al (2000) Brain Res 886(1–2):82–98. https://doi.org/10.1016/s0006-8993(00)02915-2

Bolhassani A, Safaiyan S, Rafati S (2011) Improvement of different vaccine delivery systems for cancer therapy. Mol Cancer 10:3. https://doi.org/10.1186/1476-4598-10-3

Bolognani F, Goya RG (2001) Gene therapy in the neuroendocrine system: its implementation in experimental models using viral vectors. Neuroendocrinology 73(2):75–83. https://doi.org/10.1159/000054623

Bradley S, Jakes AD, Harrington K et al (2014) Applications of coxsackievirus A21 in oncology. Oncol Virother 3:47–55. https://doi.org/10.2147/OV.S56322

Bush RA, Zeng Y, Colosi P et al (2016) Preclinical dose-escalation study of intravitreal AAV-RS1 gene therapy in a mouse model of X-linked retinoschisis: dose-dependent expression and improved retinal structure and function. Hum Gene Ther 27(5):376–389. https://doi.org/10.1089/hum.2015.142

Callaway E (2020) Russia's fast-track coronavirus vaccine draws outrage over safety. Nature 584 (7821):334–335. https://doi.org/10.1038/d41586-020-02386-2

Campochiaro PA, Lauer AK, Sohn EH et al (2017) Lentiviral vector gene transfer of endostatin/ angiostatin for macular degeneration (GEM) study. Hum Gene Ther 28(1):99–111. https://doi.org/10.1089/hum.2016.117

Cartier N, Hacein-Bey-Abina S, Bartholomae CC et al (2009) Hematopoietic stem cell gene therapy with a lentiviral vector in X-linked adrenoleukodystrophy. Science 326(5954):818–823. https://doi.org/10.1126/science.1171242

Case JB, Rothlauf PW, Chen RE et al (2020) Replication-competent vesicular stomatitis virus vaccine vector protects against SARS-CoV-2-mediated pathogenesis in mice. Cell Host Microbe 28(3):465–474. https://doi.org/10.1016/j.chom.2020.07.018

Cavazzana-Calvo M, Hacein-Bey S, de Saint BG et al (2000) Gene therapy of human severe combined immunodeficiency (SCID)-X1 disease. Science 288:669–672. https://doi.org/10.1126/science.288.5466.669

Cerghini S, Blumenfeld M, Yaniv M (1988) A liver-specific factor essential for albumin transcription differs between differentiated and dedifferentiated rat hepatoma cells. Genes Dev 2 (8):957–974. https://doi.org/10.1101/gad.2.8.957

Chai Z, Zhang P, Fu F et al (2014) Oncolytic therapy of a recombinant Newcastle disease virus D90 strain for lung cancer. Virol J 11:84. https://doi.org/10.1186/1743-422X-11-84

Chamberlain JR, Chamberlain JS (2017) Progress toward gene therapy for Duchenne muscular dystrophy. Mol Ther 25(5):1125–1131. https://doi.org/10.1016/j.ymthe.2017.02.019

Chapin JC, Monahan PE (2018) Gene therapy for hemophilia. BioDrugs 32:9–25

Chen Y, DeWeese T, Dilley J et al (2001) CV706, a prostate cancer-specific adenovirus variant, in combination with radiotherapy produces synergistic antitumor efficacy without increasing toxicity. Cancer Res 61(14):5453–5460

Chen Q, Wang C, Chen D et al (2006) Construction and expression of human PTEN tumor suppressor gene recombinant adenovirus vector. J Huazhong Univ Sci Tehcnolog Med Sci 26 (3):325–328. https://doi.org/10.1007/BF02829565

Cheng X, Wang W, Xu Q et al (2016) Genetic modification of oncolytic Newcastle disease virus for cancer therapy. J Virol 90(11):5343–5352. https://doi.org/10.1128/JVI.00136-16

Chi J, Patel R, Rehman H et al (2020) Recent advances in immunotherapy for pancreatic cancer. J Cancer Metastasis Treat 6:43. https://doi.org/10.20517/2394-4722.2020.90

Chikkanna-Gowda CP, McNally S, Sheahan BJ et al (2006) Inhibition of murine K-BALB and CT26 tumour growth using a Semliki Forest virus vector with enhanced expression of IL-18. Oncol Rep 16(4):713–719. https://doi.org/10.3892/or.16.4.713

Chiocca EA (2002) Oncolytic viruses. Nat Rev Cancer 2(12):938–950. https://doi.org/10.1038/nrc948

Ciuffi A (2008) Mechanisms governing lentivirus integration site selection. Curr Gene Ther 8 (6):419–429. https://doi.org/10.2174/156652308786848021

Clements D, Helson E, Gujar SA et al (2014) Reovirus in cancer therapy: an evidence-based review. Oncol Virother 3:69–82. https://doi.org/10.2147/OV.S51321

Cloughesy TF, Landolfi J, Hogan DJ et al (2016) Phase I trial of vocimagine amiroretrorepvec and 5-fluorocytosine for recurrent high-grade glioma. Sci Transl Med 8:341–375. https://doi.org/10.1126/scitranslmed.aad9784

Comins C, Heinemann L, Harrington K et al (2008) Reovirus: viral therapy for cancer "as nature intended". Clin Oncol (R Coll Radiol) 20(7):548–554. https://doi.org/10.1016/j.clon.2008.04.018

Conrad CK, Allen SS, Afione SA et al (1996) Safety of single-dose administration of an adeno-associated virus (AAV)-CFTR vector in the primate lung. Gene Ther 3(8):658–668
Constable IJ, Pierce CM, Lai C-M et al (2016) Phase 2a randomized clinical trial: safety and post hoc analysis of subretinal rAAV.sFLT-1 for wet age-related macular degeneration. EBioMedicine 14:168–175. https://doi.org/10.1016/j.ebiom.2016.11.016
Cordelier P, Bienvenu C, Lulka H et al (2007) Replication-deficient rSV40 mediate pancreatic gene transfer and long-term inhibition of tumor growth. Cancer Gene Ther 14(1):19–29. https://doi.org/10.1038/sj.cgt.7700987
Crudele JM, Finn JD, Siner JI et al (2015) AAV liver expression of FIX-Padua prevents and eradicates FIX inhibitor without increasing thrombogenicity in hemophilia B dogs and mice. Blood 125(10):1553–1561. https://doi.org/10.1182/blood-2014-07-588194
Dai Y, Schwarz EM, Gu D et al (1995) Cellular and humoral immune responses to adenoviral vectors containing factor IX gene: tolerization of factor IX and vector antigens allows long-term expression. Proc Natl Acad Sci U S A 92(5):1401–1405. https://doi.org/10.1073/pnas.92.5.1401
Dana H, Chalbatani GM, Mahmoodzadeh H et al (2017) Molecular mechanisms and biological functions of siRNA. Int J Biomed Sci 13(2):48–57
Demidenko AA, Blattman JN, Blattman NN et al (2013) Engineering recombinant reovirus with tandem repeats and a tetravirus 2A-like element for exogenous polyprotein expression. Proc Natl Acad Sci U S A 110(20):E1867–E1876. https://doi.org/10.1073/pnas.1220107110
Doering CB, Denning G, Shields JE et al (2018) Preclinical development of a hematopoietic stem and progenitor cell bioengineered factor VIII lentiviral vector gene therapy of hemophilia A. Hum Gene Ther 29(10):1183–1201. https://doi.org/10.1089/hum.2018.137
Dorange F, Piver E, Bru T et al (2004) Vesicular stomatitis virus glycoprotein: a transducing coat for SFV-based RNA vectors. J Gene Med 6:1014–1022
Downs-Canner S, Guo ZS, Ravindranathan R et al (2016) Phase I study of intravenous oncolytic poxvirus (vvDD) in patients with advanced solid cancers. Mol Ther 24(8):1492–1501. https://doi.org/10.1038/mt.2016.101
Eissa IR, Naoe Y, Bustos-Villalobos I et al (2017) Genomic signature of the natural oncolytic herpes simplex virus HF10 and its therapeutic role in preclinical and clinical trials. Front Oncol 7:149. https://doi.org/10.3389/fonc.2017.00149
El-Shemi GA, Ashshi MA, Oh E et al (2018) Efficacy of combining ING4 and TRAIL genes in cancer-targeting gene virotherapy strategy: first evidence in preclinical hepatocellular carcinoma. Gene Ther 25(1):54–65. https://doi.org/10.1038/gt.2017.86
Emdad L, Das SK, Wang XY et al (2018) Cancer terminator viruses (CVT): a better solution for viral-based therapy of cancer. J Cell Physiol 233(8):5684–5695. https://doi.org/10.1002/jcp.26421
Epstein AL, Marconi P, Argnani R et al (2005) HSV-1 derived recombinant and amplicon vectors for gene transfer and gene therapy. Curr Gene Ther 5(5):445–458. https://doi.org/10.2174/156652305774329285
Etoh T, Himenio Y, Matusmoto T et al (2003) Oncolytic viral therapy for human pancreatic cancer cells by reovirus. Clin Cancer Res 9(3):1218–1223
Evers MM, Miniarikova J, Juhas S et al (2018) AAV5-miHTT gene therapy demonstrates broad distribution and strong human mutant huntingtin lowering in Huntington's disease minipig model. Mol Ther 26(9):2163–2177. https://doi.org/10.1016/j.ymthe.2018.06.021
Fakhiri J, Schneider MA, Puschhof J et al (2019) Novel chimeric gene therapy vectors based on adeno-associated virus and four different mammalian bocaviruses. Mol Ther Methods Clin Dev 12:202–222. https://doi.org/10.1016/j.omtm.2019.01.003
Falkner FG, Holzer GW (2004) Vaccinia viral/retroviral chimeric vectors. Curr Gene Ther 4(4):417–426. https://doi.org/10.2174/1566523043346101
Fan D, Shen Y, Kang D et al (2001) Chin Med J 114(12):1276–1279

Fang B, Eisensmith RC, Wang H et al (1995) Gene therapy for hemophilia B: host immunosuppression prolongs the therapeutic effect of adenovirus-mediated factor IX expression. Hum Gene Ther 6(8):1039–1044. https://doi.org/10.1089/hum.1995.6.8-1039

Farrow N, Miller D, Cmielewski P et al (2013) Airway gene transfer in a non-human primate: lentiviral gene expression in marmoset lungs. Sci Rep 3:1287. https://doi.org/10.1038/srep01287

Feathers KL, Jia L, Dayanthi Perera N et al (2019) Development of a gene therapy vector for RDH12-associated retinal dystrophy. Hum Gene Ther 30(11):1325–1335. https://doi.org/10.1089/hum.2019.017

Feng L, Wang Q, Shan C et al (2020) An adenovirus-vectored COVID-19 vaccine confers protection from SARS-CoV-2 challenge in rhesus macaques. Nat Commun 11(1):4207. https://doi.org/10.1038/s41467-020-18077-5

Finke S, Conzelmann KK (2005) Recombinant rhabdoviruses: vectors for vaccine development and gene therapy. Curr Top Microbiol Immunol 292:165–200. https://doi.org/10.1007/3-540-27485-5_8

Fischer A, Hacein-Bey-Abina S (2020) Gene therapy for severe combined immunodeficiencies and beyond. J Exp Med 217(2):e20190607. https://doi.org/10.1084/jem.20190607

Fish KM, Ladage D, Kawase Y et al (2013) AAV9.I-1c delivered via direct coronary infusion in a porcine model of heart failure improves contractility and mitigates adverse remodeling. Clin Heart Fail 6(2):310–317. https://doi.org/10.1161/CIRCHEARTFAILURE.112.971325

Fleeton MN, Chen M, Berglund P et al (2001) Self-replicative RNA vaccines elicit protection against influenza a virus, respiratory syncytial virus, and a tickborne encephalitis virus. J Infect Dis 183(9):1395–1398. https://doi.org/10.1086/319857

Flotte TR, Afione SA, Conrad C et al (1993) Stable in vivo expression of the cystic fibrosis transmembrane conductance regulator with an adeno-associated virus vector. Proc Natl Acad Sci U S A 90(22):10613–10617. https://doi.org/10.1073/pnas.90.22.10613

Fol R, Bradeau J, Ludewig S et al (2016) Viral gene transfer of APPsα rescues synaptic failure in an Alzheimer's disease model. Acta Neuropathol 131(2):247–266. https://doi.org/10.1007/s00401-015-1498-9

Folegatti PM, Bellamy D, Roberts R et al (2019) Safety and immunogenicity of a novel recombinant simian adenovirus ChAdOx2 as a vectored vaccine. Vaccine 7(2):40. https://doi.org/10.3390/vaccines7020040

Folegatti PM, Ewer KJ, Aley PK et al (2020) Safety and immunogenicity of the ChAsOx1 nCoV-19 vaccine against SARS-CoV-2: a preliminary report of a phase 1/2 single-blind, randomised controlled trial. Lancet 396(10249):467–478. https://doi.org/10.1016/S0140-6736(20)31604-4

Förster R, Fleige H, Sutter G (2020) Combating COVID-19: MVA vector vaccines applied to the respiratory tract as promising toward protective immunity in the lung. Front Immunol 11:1959. https://doi.org/10.3389/fimmu.2020.01959

Fukuhara H, Ino Y, Todo T (2016) Oncolytic virus therapy: a new era of cancer treatment at dawn. Cancer Sci 107(10):1373–1379. https://doi.org/10.1111/cas.13027

Galanis E, Hartmann LC, Cliby WA et al (2010) Phase I trial of intraperitoneal administration of an oncolytic measles virus strain engineered to express carcinoembryonic antigen for recurrent ovarian cancer. Cancer Res 70(3):875–882. https://doi.org/10.1158/0008-5472.CAN-09-2762

Ganar K, Das M, Sinha S et al (2014) Newcastle disease virus: current status and our understanding. Virus Res 184:71–81. https://doi.org/10.1016/j.virusres.2014.02.016

Gao W, Soloff AC, Lu X et al (2006) Protection of mice and poultry from lethal H5N1 avian influenza virus through adenovirus-based immunization. J Virol 80(4):1959–1964. https://doi.org/10.1128/JVI.80.4.1959-1964.2006

Garcia-Gomez M, Calabria A, Garcia-Bravo M et al (2016) Safe and efficient gene therapy for pyruvate kinase deficiency. Mol Ther 24(7):1187–1198. https://doi.org/10.1038/mt.2016.87

Garcia-Hernandez ML, Gray A, Hubby B et al (2007) In vivo effects of vaccination with six-transmembrane epithelial antigen of the prostate: a candidate antigen for treating prostate cancer. Cancer Res 67(3):1344–1351. https://doi.org/10.1158/0008-5472.CAN-06-2996

Garcia-Hernandez ML, Gray A, Hubby B et al (2008) Prostate stem cell antigen vaccination induces a long-term protective immune response against prostate cancer in the absence of autoimmunity. Cancer Res 68(3):861–869. https://doi.org/10.1158/0008-5472.CAN-07-0445

Garrity-Moses ME, Teng Q, Krudy C et al (2006) X-linked inhibitor of apoptosis protein gene-based neuroprotection for the peripheral nervous system. Neurosurgery 59(1):172–182. https://doi.org/10.1227/01.NEU.0000219237.69329.B7

Gaspar HB, Parsley KL, Howe S et al (2004) Gene therapy of X-linked severe combined immunodeficiency by use of a pseudotyped gammaretroviral vector. Lancet 364(9452):2181–2187. https://doi.org/10.1016/S0140-6736(04)17590-9

Geisbert TW, Feldmann H (2011) Recombinant vesicular stomatitis virus-based vaccines against Ebola and Marburg infections. J Infect Dis 204(Suppl. S3):S1075–S1081. https://doi.org/10.1093/infdis/jir349

Gherke R, Ecker M, Aberle SW et al (2003) Incorporation of tick-borne encephalitis virus replicons into virus-like particles by a packaging cell line. J Virol 77(16):8924–8933. https://doi.org/10.1128/jvi.77.16.8924-8933.2003

Gregorevic P, Blankinship MJ, Allen JM et al (2004) Systemic delivery of genes to striated muscles using adeno-associated viral vectors. Nat Med 10(8):828–834. https://doi.org/10.1038/nm1085

Gregorevic P, Allen JM, Minami E et al (2006) rAAV6-microdystrophin preserves muscle function and extends lifespan in severely dystrophic mice. Nat Med 12(7):787–789. https://doi.org/10.1038/nm1439

Greinacher A, Thiele T, Warkentin TE et al (2021) Thrombotic thrombocytopenia after ChAdOx1 nCoV-19 vaccination. N Engl J Med. https://doi.org/10.1056/NEJMoa2104840

Grieger C, Samulski RJ (2005) Packaging capacity of adeno-associated virus serotypes: impact of larger genomes on infectivity and postentry steps. J Virol 79(15):9933–9944. https://doi.org/10.1128/JVI.79.15.9933-9944.2005

Gujar SA, Marcato P, Pan D et al (2010) Reovirus virotherapy overrides tumor antigen presentation evasion and promotes protective antitumor immunity. Mol Cancer Ther 9(11):2924–2933. https://doi.org/10.1158/1535-7163.MCT-10-0590

Guy J, Feuer WJ, Davis JL et al (2017) Gene therapy for Leder hereditary optic neuropathy: low- and medium-dose visual results. Ophthalmology 124(11):1621–1634. https://doi.org/10.1016/j.ophtha.2017.05.016

Hacein-Bey-Abina S, Garrigue A, Wang GP et al (2008) Insertional oncogenesis in 4 patients after retrovirus-mediated gene therapy of SCID-X1. J Clin Investig 118(9):3132–3142. https://doi.org/10.1172/JCI35700

Hacein-Bey-Abina S, Pai SY, Gaspar HB et al (2014) A modified γ-retrovirus vector for X-linked severe combined immunodeficiency. N Engl J Med 371(15):1407–1417. https://doi.org/10.1056/NEJMoa1404588

Harrop R, Carroll MW (2006) Viral vectors for cancer immunotherapy. Front Biosci 11:804–817. https://doi.org/10.2741/1838

Hartkopf AD, Bossow S, Lampe J et al (2013) Enhanced killing of ovarian carcinoma using oncolytic measles vaccine virus armed with a yeast cytosine deaminase and uracil phosphoribosyltransferase. Gynecol Oncol 130(2):362–368. https://doi.org/10.1016/j.ygyno.2013.05.004

Hartmann K (2012) Clinical aspects of feline retroviruses: a review. Viruses 4:2684–2710. https://doi.org/10.3390/v4112684

Hastie E, Grdzelishvili VZ (2012) Vesicular stomatitis virus as a flexible platform for oncolytic virotherapy against cancer. J Gen Virol 93(Pt 12):2529–2545. https://doi.org/10.1099/vir.0.046672-0

Hauck B, Chen L, Xiao W (2003) Generation and characterization of recombinant AAV vectors. Mol Ther 7(3):419–425. https://doi.org/10.1016/s1525-0016(03)00012-1

Hayes PJ, Cox JH, Coleman AR et al (2016) Adenovirus-based HIV-1 vaccine candidates tested in efficacy trials elicit $CD8^+$ T cells with limited breadth of HIV-1 inhibition. AIDS 30 (11):1703–1712. https://doi.org/10.1097/QAD.0000000000001122

He B, Zheng B-J, Wang Q et al (2015) Adenovirus-based vaccines against avian-origin H5N1 influenza virus. Microbes Infect 17(2):135–141. https://doi.org/10.1016/j.micinf.2014.11.003

Hegde V, Na H-N, Dubuisson O (2016) An adenovirus-derived protein: a novel candidate for anti-diabetic drug development. Biochemie 121:140–150. https://doi.org/10.1016/j.biochi.2015.12.002

Heier JS, Kherani S, Desai S et al (2017) Intravitreous injection of AAV2-sFLT01 in patients with advanced neovascular age-related macular degeneration: a phase I, open-label trial. Lancet 390 (10089):50–61. https://doi.org/10.1016/S0140-6736(17)30979-0

Heinzerling L, Kunzi V, Oberholzer PA et al (2005) Oncolytic measles virus in cutaneous T-cell lymphomas mounts antitumor immune responses in vivo and targets interferon resistant tumor cells. Blood 106(7):2287–2294. https://doi.org/10.1182/blood-2004-11-4558

Henao-Restrepo AM, Longini IM, Egger M et al (2015) Efficacy and effectiveness of an rVSV-vectored vaccine expressing Ebola surface glycoprotein: interim results from the Guinea ring vaccination cluster-randomised trial. Lancet 386(9996):857–866. https://doi.org/10.1016/S0140-6736(15)61117-5

Henao-Restrepo AM, Camacho A, Longini IM et al (2017) Efficacy and effectiveness of an rVSV-vectored vaccine in preventing Ebola virus disease: final results from the Guinea ring vaccination, open-label, cluster-randomised trial (Ebola Ca Suffit!). Lancet 389(10068):505–518. https://doi.org/10.1016/S0149-6736(16)32621-6

Hirooka Y, Kasuya H, Ishikawa T et al (2018) A phase I clinical trial of EUS-guided intratumoral injection of the oncolytic virus, HF10 for unresectable locally advanced pancreatic cancer. BMC Cancer 18(1):596. https://doi.org/10.1186/s12885-018-4453-z

Hoang-Le D, Smeenk L, Anraku I et al (2009) A Kunjin replicon vector encoding granulocyte macrophage colony-stimulating factor for intra-tumoral gene therapy. Gene Ther 16 (2):190–199. https://doi.org/10.1038/gt.2008.169

Holmes KD, Cassam AK, Chan B et al (2000) A multi-mutant herpes simplex virus vector has minimal cytotoxic effects on the distribution of filamentous actin, alpha-actinin and a glutamate receptor in differentiated PC-12 cells. J Neurovirol 6(1):33–45. https://doi.org/10.3109/13550280009006380

Holzer GW, Mayrhofer JA, Gritschenberger W et al (2004) Poxviral/retroviral chimeric vectors allow cytoplasmic production of transducing defective retroviral particles. Virology 253 (1):107–114. https://doi.org/10.1006/viro.1998.9496

Hörner C, Schürmann C, Auste A et al (2020) A highly immunogenic measles virus-based Th1-biased COVID-19 vaccine. bioRxiv. https://doi.org/10.1101/2020.07.11.198291

Howe SJ, Mansour MR, Schwarzwaelder K et al (2008) Insertional mutagenesis combined with acquired somatic mutations causes leukemogenesis following gene therapy of SCID-X1 patients. J Clin Invest 118(9):3143–3150. https://doi.org/10.1172/JCI35798

Hu WS, Pathak VK (2000) Design of retroviral vectors and helper cells for gene therapy. Pharmacol Rev 52:493–511

Hu HM, Chen HW, Hsiao Y et al (2016) The successful induction of T-cell and antibody responses by a recombinant measles virus-vectored tetravalent dengue vaccine provides partial protection against dengue-2 infection. Hum Vaccin Immunother 12(7):1678–1689. https://doi.org/10.1080/21645515.2016.1143576

Huang TT, Parab S, Burnett R et al (2015) Intravenous administration of retroviral replicating vector, Toca 511, demonstrates efficacy in orthotopic immune-competent mouse glioma model. Hum Gene Ther 26(2):82–93. https://doi.org/10.1089/hum.2014.100

Igarashi T, Finet JE, Takeuchi A et al (2012) Connexin gene transfer preserves conduction velocity and prevents atrial fibrillation. Circulation 125(2):216–225. https://doi.org/10.1161/CIRCULATIONAHA.111.053272

Inoko K, Hiraoka K, Inagaki A et al (2018) Therapeutic activity of retroviral replicating vector-mediated prodrug activator gene therapy for pancreatic cancer. Cancer Gene Ther 25 (7–8):184–195. https://doi.org/10.1038/s41417-018-0020-7

Ishikawa K, Weber T, Hajjar RJ (2018) Human cardiac gene therapy. Circ Res 123(5):601–613. https://doi.org/10.1161/CIRCRESAHA.118.311587

Ito N, Takayama-Ito M, Yamada K et al (2003) Improved recovery of rabies virus from cloned cDNA using a vaccinia virus-free reverse genetics system. Microbiol Immunol 47(8):613–617. https://doi.org/10.1111/j.1348-0421.2003.tb03424.x

Jia Q, Liang F, Ohka S et al (2002) Expression of brain-derived neurotrophic factor in the central nervous system of mice using a poliovirus-based vector. J Neurovirol 8:14–23. https://doi.org/10.1080/13550280231724776

Jimenez V, Jambrina C, Casana E et al (2018) FGF21 gene therapy as treatment for obesity and insulin resistance. EMBO Mol Med 10(8):e8791. https://doi.org/10.15252/emmm.201708791

Jones CT, Patkar CG, Kuhn RJ (2005) Construction and applications of yellow fever virus replicons. Virology 331(2):247–259. https://doi.org/10.1016/j.virol.2004.10.034

Jourdier TM, Moste C, Bonnet MC et al (2003) Local immunotherapy of spontaneous feline fibrosarcomas using recombinant poxvirus expressing interleukin 2 (IL2). Gene Ther 10 (26):2126–2132

Kasuya H, Kodera Y, Nakao A et al (2014) Phase I dose-escalation clinical trial of HF10 oncolytic herpes virus in 17 Japanese patients with advanced cancer. Hepato-Gastroenterology 61 (131):599–605

Kaufman HL, Kim DW, DeRaffele G et al (2010) Local and distant immunity induced by intralesional vaccination with an oncolytic herpes virus encoding GM-CSF in patients with stage IIIc and IV melanoma. Ann Surg Oncol 17(3):718–730. https://doi.org/10.1245/s10434-009-0809-6

Kay MA, Glorioso JC, Naldini L (2001) Viral vectors for gene therapy: the art of turning infectious agents into vehicles of therapeutics. Nat Med 7(1):33–40. https://doi.org/10.1038/83324

Khromykh AA, Varnavski AN, Westaway EG (1998) Encapsidation of the flavivirus Kunjin replicon RNA by using a complementation system providing Kunjin virus structural proteins in trans. J Virol 72(7):5967–5977. https://doi.org/10.1128/JVI.72.7.5967-5977.1998

Kim D-S, Nam J-H (2011) Application of attenuated coxsackievirus B3 as viral vector system for vaccines and gene therapy. Hum Vaccin 7(4):410–416. https://doi.org/10.4161/hv.7.4.14422

Kim J, Yoon YS, Lee H et al (2008) AAV-GAD gene for rat models of neuropathic pain and Parkinson's disease. Acta Neurochir Suppl 101:99–105. https://doi.org/10.1007/978-3-211-78205-7_17

Kimchi-Sarfaty C, Gottesman MM (2004) SV40 pseudovirions as highly efficient vectors for gene transfer and their potential application in cancer therapy. Curr Pharm Biotechnol 5(5):451–458. https://doi.org/10.2174/1389201043376670

Koch WJ, Rockman HA, Samama P et al (1995) Cardiac function in mice overexpressing the beta-adrenergic receptor kinase or a beta ARK inhibitor. Science 268(5215):1350–1353. https://doi.org/10.1126/science.7761854

Kohn DB, Hershfield MS, Puck JM et al (2019) Consensus approach for the management of severe combined immune deficiency caused by adenosine deaminase deficiency. J Allergy Clin Immunol 143(3):852–863. https://doi.org/10.1016/j.jaci.2018.08.024

Kordower JH, Emborg ME, Bloch JH et al (2000) Neurodegeneration prevented by lentiviral vector delivery of GDNF in primate model of Parkinson's disease. Science 290(5492):767–773. https://doi.org/10.1126/science.290.5492.767

Kurihara T, Brough DE, Kovesdi I et al (2000) Selectivity of a replication competent adenovirus for human breast carcinoma cells expressing the MUC1 antigen. J Clin Invest 106:763–771. https://doi.org/10.1172/JCI9180

Kwak H, Honig H, Kaufman HL (2003) Poxviruses as vectors for cancer immunotherapy. Curr Opin Drug Discov Devel 6:161–168

Lachman LB, Rao XM, Kremer RH et al (2001) DNA vaccination against neu reduces breast cancer incidence and metastasis in mice. Cancer Gene Ther 8(4):259–268. https://doi.org/10.1038/sj.cgt.7700300

Lam EW, Hammad HM, Zwacka R et al (2000) Immunolocalization of adenoviral vector-mediated manganese sodium dismutase gene transfer in experimental oral tumors. J Dent Res 79 (6):1410–1417. https://doi.org/10.1177/00220345000790061001

Lampe J, Bossow S, Weiland T et al (2013) An armed oncolytic measles vaccine virus eliminates human hepatoma cells independently of apoptosis. Gene Ther 20(11):1033–1041. https://doi.org/10.1038/gt.2013.28

Lange S, Lampe J, Bossow S et al (2013) A novel armed oncolytic measles vaccine virus for the treatment of cholangiocarcinoma. Hum Gene Ther 24(5):554–564. https://doi.org/10.1089/hum.2012.136

Lazarous DF, Shou M, Stiber JA et al (1999) Adenoviral-mediated gene transfer induces sustained pericardial VEGF expression in dogs: effect on myocardial angiogenesis. Cardiovasc Res 44 (2):294–302. https://doi.org/10.1016/s0008-6363(99)00203-5

Lebherz C, Gao G, Louboutin JP et al (2004) Gene therapy with novel adeno-associated virus vectors substantially diminishes atherosclerosis in a murine model of familial hypercholesterolemia. J Gene Med 6(6):663–672. https://doi.org/10.1002/jgm.554

LeBlanc AK, Naik S, Galvon GD (2013) Et al safety studies on intravenous administration of oncolytic recombinant vesicular stomatitis virus in purpose-bred beagle dogs. Hum Gene Ther Clin Dev 24(4):174–181. https://doi.org/10.1089/humc.2013.165

Lee RC, Feinbaum RL, Ambros V (1993) The C. elegans heterochronic gene lin-4 encodes small RNAs with antisense complimentary to lin-14. Cell 7585:843–854. https://doi.org/10.1016/0092-8674(93)90529-y

Lehrman S (1999) Virus treatment questioned after gene therapy death. Nature 401:517–518

Li M, Stollar V (2004) Alphaviruses and apoptosis. Int Rev Immunol 23(1–2):7–24. https://doi.org/10.1080/08830180490265529

Li JM, Kao KC, Li LF (2013) Micro-RNA-145 regulates oncolytic herpes simplex virus-1 for selective killing of human non-small lung cancer cells. Virol J 10:241. https://doi.org/10.1186/1743-422X-10-241

Liang W, Wang H, Sun TM et al (2003) Application of autologous tumor cell vaccine and NDV vaccine in treatment of tumors of digestive tract. World J Gastroenterol 9(3):495–498. https://doi.org/10.3748/wjg.v9.i3.495

Liljeström P, Garoff H (1991) A new generation of animal cell expression vectors based on the Semliki Forest virus replicon. Biotechnology (NY) 9(12):1356–1361. https://doi.org/10.1038/nbt1291-1356

Limberis M, Anson DS, Fuller M et al (2002) Recovery of airway cystic fibrosis transmembrane conductance regulator function in mice with cystic fibrosis after single-dose lentivirus-mediated gene transfer. Hum Gene Ther 13(16):1961–1970. https://doi.org/10.1089/10430340260355365

Lin J, Calcedo R, Vanderberghe LH et al (2008) Impact of preexisting vector immunity on the efficacy of adeno-associated virus-based HIV-1 gag vaccines. Hum Gene Ther 19(7):663–669. https://doi.org/10.1089/hum.2008.033

Lin Y, Zhang H, Liang J et al (2014) Identification and characterization of alphavirus M1 as a selective oncolytic virus targeting ZAP-defective human cancers. Proc Natl Acad Sci U S A 111 (42):E4504–E4512. https://doi.org/10.1073/pnas.1408759111

Lino CA, Harper JC, Carney JP et al (2018) Delivering CRISPR: a review of the challenges and approaches. Drug Deliv 25(1):1234–1257. https://doi.org/10.1080/10717544.2018.1474964

Liu X-Y, Gu J-F, Shi W-F (2005) Targeting gene-virotherapy for cancer. Acta Biochem Biophys Sin 37(9):581–587. https://doi.org/10.1111/j.1745-7270.2005.00087.x

Logunov DY, Dolzhikova IV, Zubkova OV et al (2020) Safety and immunogenicity of an rAd26 and rAd5 vector-based heterologous prime-boost COVID-19 vaccine in two formulations: two open, non-randomised phase 1/2 studies from Russia. Lancet S0140-6736(20):31866–31863. https://doi.org/10.1016/S0140-6736(20)31866-3

Logunov DY, Dolzhikova IV, Shcheblyakov D et al (2021) Safety and efficacy of an rAd26 and rAd5 vector-based heterologous prime-boost COVID-19 vaccine: an interim analysis of a

randomised controlled phase 3 trial in Russia. Lancet 397(10275):671–681. https://doi.org/10.1016/S0140-6736(21)00234-8
Loimas S, Pasanen T, Gomes A et al (2000) Viral vectors carrying a marker-suicide fusion gene (TK-GFP) as tools for TK/GCV-mediated cancer therapy. Gene Ther Mol Biol 5:147–155
Lundstrom K (2018) Viral vectors in gene therapy. Diseases 6:42. https://doi.org/10.3390/diseases6020042
Lundstrom K (2019) RNA viruses as tools in gene therapy and vaccine development. Genes 10 (3):189. https://doi.org/10.3390/genes10030189
Lundstrom K (2020a) Self-amplifying RNA viruses as RNA vaccines. Int J Mol Sci Int J 2020 (21):5130. https://doi.org/10.3390/ijms21145130
Lundstrom K (2020b) Coronavirus pandemic – therapy and vaccines. Biomedicine 8(5):109. https://doi.org/10.3390/biomedicines8050109
Luo J, Zhao J, Tian Q et al (2016) A recombinant rabies virus carrying GFP between N and P affects viral transcription in vitro. Virus Genes 52(3):379–387. https://doi.org/10.1007/s11262-016-1313-2
Ma H-I, Lin S-Z, Chiang Y-H et al (2002) Intratumoral gene therapy of malignant brain tumor in a rat model with angiostatin delivered by adeno-associated viral (AAV) vector. Gene Ther 9 (1):2–11. https://doi.org/10.1038/sj.gt.3301616
Määttä AM, Mäkinen K, Ketola A et al (2008) Replication competent Semliki Forest virus prolongs survival in experimental lung cancer. Int J Cancer 123(7):1704–1711. https://doi.org/10.1002/ijc.23646
Madan RA, Arlen PM, Gullye JL (2007) PANVAC-VF: Poxviral-based vaccine therapy targeting CEA and MUC1 in carcinoma. Expert Opin Biol Ther 7(4):543–554. https://doi.org/10.1517/14712598.7.4.543
Malerba A, Klein P, Bachtarzi H et al (2017) PABPN1 gene therapy for oculopharyngeal muscular dystrophy. Nat Commun 8:14848. https://doi.org/10.1038/ncomms14848
Mansfield DC, Kyula JN, Rosenfelder N et al (2016) Oncolytic vaccinia virus as a vector for therapeutic sodium iodide symporter gene therapy in prostate cancer. Gene Ther 23(4):357–368. https://doi.org/10.1038/gt.2016.5
Marcos-Contreras OA, Smith SM, Bellinger DA et al (2016) Sustained correction of FVII deficiency in dogs using AAV-mediated expression of zymogen FVII. Blood 127(5):565–571. https://doi.org/10.1182/blood-2015-09-671420
Marquez Loza LI, Yuen EC, McCray PB Jr (2019) Lentiviral vectors for the treatment and prevention of cystic fibrosis lung disease. Genes 10:218. https://doi.org/10.3390/genes10030218
Martin KRG, Quigley HA, Zack DJ et al (2003) Gene therapy with brain-derived neurotrophic factor as a protection: retinal ganglion cells in a rat glaucoma model. Invest Ophthalmol Vis Sci 44(10):4357–4365. https://doi.org/10.1167/iovs.02-1332
Marzi A, Robertson SJ, Haddock E et al (2015) Ebola vaccine. VSV-EBOV rapidly protects macaques against infection with the 2014/2015 Ebola virus outbreak strain. Science 349 (6249):739–742. https://doi.org/10.1126/science.aab3920
Mastrangelo MJ, Lattime EC (2002) Virotherapy clinical trials for regional disease: in situ immune modulation using recombinant poxvirus vectors. Cancer Gene Ther 9(12):1013–1021. https://doi.org/10.1038/sj.cgt.7700538
McClements ME, MacLaren RR (2017) Adeno-associated virus (AAV) dual vector strategies for gene therapy encoding large transgenes. Yale J Biol Med 90(4):611–623. eCollection 2017 Dec
McCormack MP, Rabbitts TH (2004) Activation of the T-cell oncogene LMO2 after gene therapy for X-linked severe combined immunodeficiency. N Engl J Med 350(9):913–922. https://doi.org/10.1056/NEJMra032207
McGettigan JP, Foley HD, Belyakov IM et al (2001) Rabies virus-based vectors expressing human immunodeficiency virus type 1 (HIV-1) envelope protein induce a strong, cross-reactive cytotoxic T-lymphocyte response against envelope proteins from different HIV-1 isolates. J Virol 75(9):4430–4434. https://doi.org/10.1128/JVI.75.9.4430-4434.2001

Meliani A, Boisgerault F, Fitzpatrick Z et al (2017) Enhanced liver gene transfer and evasion of preexisting humoral immunity with exosome-enveloped AAV vectors. Blood Adv 1(23):2019–2031. https://doi.org/10.1182/bloodadvances.2017010181

Mendell JR, Al-Zaidy S, Schell R et al (2017) Single-dose gene-replacement therapy for spinal muscular atrophy. New Engl J Med 377(18):1713–1722. https://doi.org/10.1056/NEJMoa1706198

Mendell JR, Sahenk Z, Lehman K et al (2020) Assessment of systemic delivery of rAAVrh74. MHCK7.micro-dystrophin in children with Duchenne muscular dystrophy. JAMA Neurol 77 (9):1122–1131. https://doi.org/10.1001/jamaneurol.2020.1484

Mercado NNB, Zahn R, Wegmann F et al (2020) Single-shot Ad26 vaccine protects against SARS-CoV-2 in rhesus macaques. Nature 586(7830):583–588. https://doi.org/10.1038/s41586-020-2607-z. Online ahead of print

Mingozzi F, High KA (2013) Immune responses to AAV vectors: overcoming barriers to successful gene therapy. Blood 122(1):23–36. https://doi.org/10.1182/blood-2013-01-306647

Miniarikova J, Zimmer V, Martier R et al (2017) AAV5-miHTT gene therapy demonstrates suppression of huntingtin aggregation and neuronal dysfunction in a rat model of Huntington's disease. Gene Ther 24(10):630–639. https://doi.org/10.1038/gt.2017.71

Mitomo K, Griesenbach U, Inoue M et al (2010) Toward gene therapy for cystic fibrosis using a lentivirus pseudotyped with Sendai virus envelopes. Mol Ther 18(6):1173–1182. https://doi.org/10.1038/mt.2010.13

Miyamoto MI, del Monte F, Schmidt U et al (2000) Adenoviral gene transfer of SERCa2a improves left-ventricular function in aortic-banded rats in transition to heart failure. Proc Natl Acad Sci U S A 97(2):793–798. https://doi.org/10.1073/pnas.97.2.793

Moore CB, Guthrie EH, Huang T-Z et al (2010) Short hairpin RNA (shRNA): design, delivery and assessment of gene knockdown. Methods Mol Biol 629:141–158. https://doi.org/10.1007/978-1-60761-657-3_10

Moorehead T, Yong F, Neelakantan S et al (2020) Safety and tolerability of PF-06939926 in ambulatory boys with Duchenne muscular dystrophy: a phase 1b multicenter, open label, dose ascending study. Mol Ther 28(4S1):272

Moscioni D, Morizono H, McCarter RJ et al (2006) Long-term correction of ammonia metabolism and prolonged survival in ornithine transcarbamylase-deficient mice following liver-directed treatment with adeno-associated viral vectors. Mol Ther 14(1):25–33. https://doi.org/10.1016/j.ymthe.2006.03.009

Mostafa AA, Meyers DE, Thirukkumaran CM et al (2018) Oncolytic Reovirus and immune checkpoint inhibitor as a novel immunotherapeutic strategy for breast Cancer. Cancers 10 (6):205. https://doi.org/10.3390/cancers10060205

Muir K-L, Kallam A, Koepsell SA et al (2021) Thrombotic thrombocytopenia after Ad26.COV.S vaccination. N Engl J Med. https://doi.org/10.1056/NEJMc2105869

Myers R, Harvey M, Kaufmann TJ et al (2008) Toxicology study of repeat intracerebral administration of a measles virus derivative producing carcinoembryonic antigen in rhesus monkeys in support of a phase I/II clinical trial for patients with recurrent gliomas. Hum Gene Ther 19 (7):690–698. https://doi.org/10.1089/hum.2008.035

Nagasato M, Rin Y, Yamamoto Y et al (2017) A tumor-targeting adenovirus with high gene-transduction efficiency for primary pancreatic cancer and ascites cells. Anticancer Res 37 (7):3599–3605. https://doi.org/10.21873/anticanres.11730

Nakajima T, Nakamaru K, Ido E et al (2000) Development of novel simian immunodeficiency virus vectors carrying a dual gene expression system. Hum Gene Ther 11:1863–1874. https://doi.org/10.1089/10430340050129486

Nathwani AC (2019) Gene therapy for hemophilia. Hematology Am Soc Hematol Educ Program 2019(1):1–8. https://doi.org/10.1182/hematology.2019000007

Nathwani AC, Reiss UM, Tuddenham EG et al (2014) Long-term safety and efficacy of factor IX gene therapy in hemophilia B. N Engl J Med 371(21):1994–2004. https://doi.org/10.1056/NEJMoa1407309

Ng N-M, Jiang S-P, Lv Z-Q (2012) Retrovirus-mediated siRNA targeting TRPM7 gene induces apoptosis in RBL-2H3 cells. Eur Rev Med Pharmacol Sci 16(9):1172–1178

Nienhuis AW, Nathwani AC, Davidoff AM (2017) Gene therapy for hemophilia. Mol Ther 25 (5):1163–1167. https://doi.org/10.1016/j.ymthe.2017.03.033

Niiyama H, Huang NF, Rollins MD et al (2009) Murine model of hindlimb ischemia. J Vis Exp 23:1035. https://doi.org/10.3791/1035

Niu Z, Bai F, Sun T et al (2015) Recombinant Newcastle disease virus expressing IL15 demonstrates promising antitumor efficiency in melanoma model. Technol Cancer Res Treat 14 (5):607–615. https://doi.org/10.7785/tcrt.2012.500414

Nomura T, Yabe T, Mochizuki H et al (2001) Survival effects of pigment epithelium-derived factor expressed by a lentiviral vector in rat cerebellar granule cells. Dev Neurosci 23(2):145–152. https://doi.org/10.1159/000048706

Noro T, Miyake K, Suzuki-Miyake N et al (2004) Adeno-associated viral vector-mediated expression of endostatin inhibits tumor growth and metastasis in an orthotopic pancreatic cancer model in hamsters. Cancer Res 64(20):7486–7490. https://doi.org/10.1158/0008-5472.CAN-03-1296

O'Callaghan J, Crosbie DE, Cassidy PS et al (2017) Therapeutic potential of AAV-mediated MMP-3 secretion from corneal endothelium in treating glaucoma. Hum Mol Genet 26 (7):1230–1246. https://doi.org/10.1093/hmg/ddx028

Ollmann Saphire E (2020) A vaccine against Ebola virus. Cell 181(1):6. https://doi.org/10.1016/j.cell.2020.03.011

Olsen JC (1998) Gene transfer vectors derived from equine infectious anemia virus. Gene Ther 5:1481–1487. https://doi.org/10.1038/sj.gt.3300768

O'Sullivan BP, Freedman SD (2009) Cystic fibrosis. Lancet 373(9678):1891–1904. https://doi.org/10.1016/s0140-6736(09)60327-5

Overwijk WW, Lee DS, Surman DR et al (1999) Vaccination with a recombinant vaccinia virus encoding a "self" antigen induces autoimmune vitiligo and tumor cell destruction in mice: requirement for CD4(+) T lymphocytes. Proc Natl Acad Sci U S A 96(6):2982–2987. https://doi.org/10.1073/pnas.96.6.2982

Pang S, Taneja S, Dardashti K et al (1995) Prostate tissue specificity of the prostate-specific antigen promoter isolated from a patient with prostate cancer. Hum Gene Ther 6(11):1417–1126. https://doi.org/10.1089/hum.1995.6.11-1417

Pang X, Zhang M, Dayton AI (2001) Development of dengue virus type 2 replicons capable of prolonged expression in host cells. BMC Microbiol 1:18. https://doi.org/10.1186/1471-2180-1-18

Park K, Kim WJ, Cho YH et al (2008) Cancer gene therapy using adeno-associated virus vectors. Front Biosci 13:2653–2659. https://doi.org/10.2741/2872

Pastoret P-P, Vanderplasschen A (2003) Poxviruses as vaccine vectors. Comp Immunol Microbiol Infect Dis 26(5–6):343–355. https://doi.org/10.1016/S0147-9571(03)00019-5

Pattali R, Mou Y, Li X-J (2019) AAV9 vector: a novel modality in gene therapy for spinal muscular atrophy. Gene Ther 26(7–8):287–295. https://doi.org/10.1038/s41434-019-0085-4

Pecora AL, Rizvi N, Cohen GI et al (2002) Phase I trial of intravenous administration of PV701, an oncolytic virus, in patients with advanced solid cancers. J Clin Oncol 20(9):2251–2266. https://doi.org/10.1200/JCO.2002.08.042

Phillips N, Cyranoski D, Mallapaty S (2020) A leading coronavirus vaccine trial is on hold: scientists react. Nature. https://doi.org/10.1038/d41586-020-02594-w. Online ahead of print

Pijlman GP, Suhrbier A, Khromykh AA (2006) Kunjin virus replicons: An RNA-based, non-cytopathic viral vector system for protein production, vaccine and gene therapy applications. Exp Opin Biol Ther 6(2):134–145. https://doi.org/10.1517/14712598.6.2.135

Pillai R (2005) MicroRNA functions: multiple mechanisms for a tiny RNA. RNA J 11:1753–1756. https://doi.org/10.1261/rna.2248605

Porosnicu M, Mian A, Barber GN (2003) The oncolytic effect of recombinant vesicular stomatitis virus is enhanced by expression of the fusion cytosine deaminase/uracil phosphoribosyltransferase suicide gene. Cancer Res 63(23):8366–8376

Pushko P, Bray M, Ludwig GV et al (2000) Recombinant RNA replicons derived from attenuated Venezuelan equine encephalitis virus protect guinea pigs and mice from Ebola hemorrhagic fever virus. Vaccine 19(1):142–153. https://doi.org/10.1016/S0264-410X(00)00113-4
Qian L, Huang Y, Spencer CI et al (2012) In vivo reprogramming of murine cardiac fibroblasts into induced cardiomyocytes. Nature 485(7400):593–598. https://doi.org/10.1038/nature11044
Rabinowitz J, Chan YK, Samulski RJ (2019) Adeno-associated virus (AAV) versus immune response. Viruses 11(2):102. https://doi.org/10.3390/v11020102
Radecke F, Spielhofer P, Schneider H et al (1995) Rescue of measles viruses from cloned DNA. EMBO J 14(23):5773–5784
Ratican SE, Osborne A, Martin KR (2018) Progress in gene therapy to prevent retinal ganglion cell loss in glaucoma and Leber's hereditary optic neuropathy. Neural Plast 2018:7108948. https://doi.org/10.1155/2018/7108948
Räty JK, Pikkarainen JT, Wirth T et al (2008) Gene therapy: the first approved gene-based medicines, molecular mechanisms and clinical indications. Curr Mol Pharmacol 1(1):13–23. https://doi.org/10.2174/1874467210801010013
Reddi HV, Madde P, McDonough SJ et al (2012) Preclinical efficacy of the oncolytic measles virus expressing the sodium iodide symporter in iodine non-avid anaplastic thyroid cancer: a novel therapeutic agent allowing noninvasive imaging and radioiodine therapy. Cancer Gene Ther 19 (9):659–665. https://doi.org/10.1038/cgt.2012.47
Reichard K, Lorence RM, Cascino CJ et al (1992) Newcastle disease virus selectively kills human tumor cells. J Surg Res 52(5):448–453
Reis S, Korn WM (2002) ONYX-015: mechanisms of action and clinical potential of a replication-selective adenovirus. Br J Cancer 86(1):5–11. https://doi.org/10.1038/sj.bjc.6600006
Revilla S, Ursulet S, Alvarez-Lopez JM et al (2014) Lenti-GDNF gene therapy protects against Alzheimer's disease-like neuropathology in 3xTg-AD mice and MC65 cells. CNS Neurosci Ther 20(11):961–972. https://doi.org/10.1111/cns.12312
Riabov V, Tretyakova I, Alexander RB et al (2015) Anti-tumor effect of the alphavirus-based virus-like particle vector expressing prostatespecific antigen in a HLA-DR transgenic mouse model of prostate cancer. Vaccine 33(41):5386–5395. https://doi.org/10.1016/j.vaccine.2015.08.062
Ribel J-A, Hacein-Bey-Abina S, Paven E et al (2017) Gene therapy in a patient with sickle cell disease. N Engl J Med 376(9):848–855. https://doi.org/10.1056/NEJMoa1609677
Ricordel M, Foloppe J, Pichon C et al (2017) Cowpox virus: a new and armed oncolytic poxvirus. Mol Ther Oncolytics 7:1–11. https://doi.org/10.1016/j.omto.2017.08.003
Robe PA, Princen F, Martin D et al (2000) Pharmacological modulation of the bystander effect in the herpes simplex virus thymidine kinase/ganciclovir gene therapy system: effects of dibutyryl adenosine 3′, 5′-cyclic monophosphate, alpha-glycyrrhetinic acid, and cytosine arabinoside. Biochem Pharmacol 60(2):241–249. https://doi.org/10.1016/s0006-2952(00)00315-4
Roche FP, Sheahan BJ, O'Mara SM et al (2010) Semliki Forest virus-mediated gene therapy of the RG2 rat glioma. Neuropathol Appl Neurobiol 36(7):648–660. https://doi.org/10.1111/j.1365-2990.2010.01110.x
Rodr Guez D, Rodr Guez JR, Llorente M et al (1999) A human immunodeficiency virus type 1 Env-granulocyte-macrophage-colony-stimulating factor fusion protein enhances the cellular immune response to Env in a vaccinia virus-based vaccine. J GenVirol 80(Pt 1):217–223. https://doi.org/10.1099/0022-1317-80-1-217
Roner MR, Joklik WK (2001) Reovirus reverse genetics: incorporation of the CAT gene into the reovirus genome. Proc Natl Acad Sci U S A 98(14):8036–8041. https://doi.org/10.1073/pnas.131203198
Rosenberg JB, Kaplitt MG, De BP et al (2018) AAVrh.10-mediated APOE2 central nervous system gene therapy for APOE4 -associated Alzheimer's disease. Hum Gene Ther Clin Dev 29 (1):24–47. https://doi.org/10.1089/humc.2017.231
Sadoff J, Le Gars M, Shukarev G et al (2021) Interim results of a phase 1-2a trial of Ad26.COV2.S covid-19 vaccine. N Engl J Med 384:1824–1835. https://doi.org/10.1056/NEJMoa2034201

Sakamoto M, Yuasa K, Yoshimura M et al (2002) Micro-dystrophin cDNA ameliorates dystrophic phenotypes when introduced into mdx mice as a transgene. Biochem Biophys Res Comm 293 (4):1265–1272. https://doi.org/10.1016/S0006-291X(02)00362-5

Sakata S, Lebeche D, Sakata Y et al (2007) Transcoronary transfer of SERCa2a increases coronary blood flow and decreases cardiomyocyte size in a type 2 diabetic rat model. Am J Physiol Heart Circ Physiol 292(2):H1204–H1207. https://doi.org/10.1152/ajpheart.00892.2006

Samulski R, Muzycka N (2014) AAV-mediated gene therapy for research and therapeutic purposes. Annu Rev Virol 1(1):427–451. https://doi.org/10.1146/annurev-virology-031413-085355

Saqib A, Prasad K-MR, Katwal AB et al (2011) Adeno-associated virus serotype 9-mediated overexpression of extracellular superoxide dismutase improves recovery from surgical hind-limb ischemia in BALB/c mice. J Vasc Surg 54(3):810–818. https://doi.org/10.1016/j.jvs.2011.03.278

Schaar K, Geisler A, Kraus M et al (2017) Anti-adenoviral artificial MicroRNAs expressed from AAV9 vectors inhibit human adenovirus infection in immunosuppressed Syrian hamsters. Mol Ther Nucleic Acids 8:300–316. https://doi.org/10.1016/j.omtn.2017.07.002

Schambach A, Morgan M (2016) Retroviral vectors for cancer gene therapy. Recent Results Cancer Res 209:17–35. https://doi.org/10.1007/978-3-319-42934-2_2

Schiedner G, Morral N, Parks RS et al (1998) Genomic DNA transfer with a high-capacity adenovirus vector results in improved in vivo gene expression and decreased toxicity. Nat Genet 18(2):180–183. https://doi.org/10.1038/ng0298-180

Schirrmacher V, Griesbach A, Ahlert T (2001) Antitumor effects of Newcastle disease virus in vivo: local versus systemic effects. Int J Oncol 18(5):945–952. https://doi.org/10.3892/ijo.18.5.945

Scholle I, Girard YA, Zhao Q et al (2004) Trans-packaged West Nile virus-like particles: infectious properties in vitro and in infected mosquito vectors. J Virol 78(21):11605–11614. https://doi.org/10.1128/JVI.78.21.11605-11614.2004

Schultz-Cherry S, Dybing JK, Davis NL et al (2000) Influenza virus (a/HK/156/97) hemagglutinin expressed by an alphavirus replicon system protects against lethal infection with Hong Kong-origin H5N1 viruses. Virology 278(1):55–59. https://doi.org/10.1006/viro.2000.0635

Scimia MC, Gumper AM, Koch WJ (2014) Cardiovascular gene therapy for myocardial infarction. Expert Opin Biol Ther 14(2):183–195. https://doi.org/10.1517/14712598.2014.866085

Shafren DR, Au GG, Nguyen T et al (2014) Systemic therapy of malignant human melanoma tumors by a common cold-producing enterovirus, coxsackievirus a21. Clin Cancer Res 10(1 Pt 1):53–60. https://doi.org/10.1158/1078-0432.ccr-0690-3

Silvers RM, Smith JA, Schowalter M et al (2010) Modification of integration site preferences of an HIV-1-based vector by expression of a novel synthetic protein. Hum Gene Ther 21(3):337–349. https://doi.org/10.1089/hum.2009.134

Singh M, Cattaneo R, Billeter MA (1999) A recombinant measles virus expressing hepatitis B surface antigen induces humoral responses in genetically modified mice. J Virol 73 (6):4823–4828. https://doi.org/10.1128/JVI.73.6.4823-4828.1999

Sinnett SE, Gray SJ (2017) Recent endeavors in MECP2 gene transfer for gene therapy of Rett syndrome. Discov Med 24:153–159

Skelding KA, Barry RD, Shafren DR (2009) Systemic targeting of metastatic human breast xenografts by Coxsackievirus A21. Breast Cancer Res Treat 113(1):21–30. https://doi.org/10.1007/s10549-008-9899-2

Skelding KA, Barry RD, Shafren DR (2012) Enhanced oncolysis mediated by Coxsackievirus A21 in combination with doxorubicin hydrochloride. Investig New Drugs 30(2):568–581. https://doi.org/10.1007/s10637-010-9614-0

Spencer HT, Riley BE, Doering CB (2016) State of the art: gene therapy of haemophilia. Haemophilia 22(Suppl. 5):66–71. https://doi.org/10.1111/hae.13011

Strauss JH, Strauss EG (1994) The alphaviruses; gene expression, replication and evolution. Microbiol Rev 58(3):491–562

Sudhakar V, Richardson RM (2019) Gene therapy for neurodegenerative disease. Neurotherapeutics 16(1):166–175. https://doi.org/10.1007/s13311-018-00694-0

Todo T (2012) Active immunotherapy: oncolytic virus therapy using HSV-1. Adv Exp Med Biol 746:178–186. https://doi.org/10.1007/978-1-4614-3146-6_14

Toscano MG, van der Velden J, van der Werf S et al (2017) Generation of a Vero-based packaging cell line to produce SV40 gene delivery vectors for use in clinical gene therapy studies. Mol Ther Methods Clin Dev 6:124–134. https://doi.org/10.1016/j.omtm.2017.06.007

Tostanoski LH, Wegmann F, Martinot AJ et al (2020) Ad26 vaccine protects against SARS-CoV-2 severe clinical disease in hamsters. Nat Med 26(11):1694–1700. https://doi.org/10.1038/s41591-020-1070-6. Online ahead of print

Tseng JC, Hurtado A, Yee H et al (2004) Using Sindbis viral vectors for specific detection and suppression of advanced ovarian cancer in animal models. Cancer Res 64(18):6684–6692. https://doi.org/10.1158/0008-5472.CAN-04-1924

Turrell SJ, Macnab SA, Rose A et al (2012) A herpesvirus saimiri-based vector expressing TRAIL induces cell death in human carcinoma cell lines and multicellular spheroid cultures. Int J Oncol 40(6):2081–2089. https://doi.org/10.3892/ijo.2012.1394

van Doremalen N, Lambe T, Spencer A et al (2020) ChAdOx1 nCov-19 vaccination prevents SARS-CoV-2 pneumonia in rhesus macaques. bioRxiv 586(7830):578–582. https://doi.org/10.1101/2020.05.13.093195

Vigna E, Naldini L (2000) Lentiviral vectors: excellent tools for experimental gene transfer and promising candidates for gene therapy. J Gen Med 2(5):308–316. https://doi.org/10.1002/1521-2254(200009/10)2:5<308::AID-JGM131>3.0.CO;2-3

Vignal C, Uretsky S, Fitoussi S et al (2018) Safety of rAAV2/2-ND4 gene therapy. Ophthalmology 125(6):945–947. https://doi.org/10.1016/j.ophtha.2017.12.036

Voysey M, Costa Clemens SA, Madhi SA et al (2021) Safety and efficacy of the ChAdOx1 nCoV-19 vaccine (AZD1222) against SARS-CoV-2: an interim analysis of four randomized controlled trials in Brazil, South Africa, and the UK. Lancet 397(10269):99–111

Wan X, Pei H, Zhao M-J et al (2016) Efficacy and safety of rAAV2-ND4 treatment for Leber's hereditary optic neuropathy. Sci Rep 6(1):2016

Wang J, Lu X-X, Chen D-Z et al (2004) Herpes simplex virus thymidine kinase and ganciclovir suicide gene therapy for human pancreatic cancer. World J Gastroenterol 10(3):400–403. https://doi.org/10.3748/wjg.v10.i3.400

Wang X, Wang J-P, Rao X-M et al (2005) Prime-boost vaccination with plasmid and adenovirus gene vaccines control Her2/neu$^+$ metastatic breast cancer in mice. Breast Cancer Res 7(5): R580–R588. https://doi.org/10.1186/bcr1199

Wang Z, Storb R, Halbert CL et al (2012a) Successful regional delivery and long-term expression of a dystrophin gene in canine muscular dystrophy: a preclinical model for human therapies. Mol Ther 20(8):1501–1507. https://doi.org/10.1038/mt.2012.111

Wang F, Wang Z, Tian H et al (2012b) Biodistribution and safety assessment of bladder cancer specific oncolytic adenovirus in subcutaneous xenografts tumor model in nude mice. Curr Gene Ther 12(2):67–76

Wang H, Feng Q, Wei L et al (2017) Significant inhibition of Tembusu virus envelope and NS5 gene using an adenovirus-mediated short hairpin RNA delivery system. Infect Genet Evol 54:387–396. https://doi.org/10.1016/j.meegid.2017.08.001

Wassmer SJ, Leonard BC, Coupland SG et al (2017) Overexpression of the X-linked inhibitor of apoptosis protects against retinal degeneration in a feline model of retinal detachment. Hum Gene Ther 6:482–492. https://doi.org/10.1089/hum.2016.161

Waters AM, Johnston JM, Reddy AT et al (2017) Rationale and design of a phase I clinical trial to evaluate HSV G207 alone or with a single radiation dose in children with progressive or recurrent malignant supratentorial brain tumors. Hum Gene Ther Clin Dev 28(1):7–16. https://doi.org/10.1089/humc.2017.002

Watson GL, Sayles JN, Chen C et al (1998) Treatment of lysosomal storage disease in MPS VII mice using a recombinant adeno-associated virus. Gene Ther 5(12):1642–1649. https://doi.org/10.1038/sj.gt.3300775

Wecker M, Gilbert P, Russell N et al (2012) Phase I safety and immunogenicity evaluations of an alphavirus replicon HIV-1 subtype C gag vaccine in healthy HIV-1-uninfected adults. Clin Vaccine Immunol 19(10):1651–1660. https://doi.org/10.1128/CVI.00258-12

Wei Q, Fan J, Liao J et al (2017) Engineering the rapid adenovirus production and amplification (RAPA) cell line to expedite the generation of recombinant adenoviruses. Cell Physiol Biochem 41(6):2383–2398. https://doi.org/10.1159/000475909

Wen XY, Mandelbaum S, Li ZH et al (2001) Cancer Gene Ther 5:361–370. https://doi.org/10.1038/sj.cgt.7700321

Westerhout EM, Vink M, Hasnoot PCJ et al (2006) A conditionally replicating HIV-based vector that stably expresses an antiviral shRNA against HIV-1 replication. Mol Ther 14(2):268–275. https://doi.org/10.1016/j.ymthe.2006.03.018

Wilson JA, Hart MK (2001) Protection from Ebola virus mediated by cytotoxic T-lymphocytes specific for the viral nucleoprotein. J Virol 75(6):2660–2664. https://doi.org/10.1128/JVI.75.6.2660-2664.2001

Wu K, Meyer EM, Bennett JA et al (2005) AAV2/5-mediated NGF gene delivery protects septal cholinergic neurons following axotomy. Brain Res 1061(2):107–113. https://doi.org/10.1016/j.brainres.2005.08.056

Wu Y, He J, Geng J et al (2017) Recombinant Newcastle disease virus expressing human TRAIL as a potential candidate for hepatoma therapy. Eur J Pharmacol 802:85–92. https://doi.org/10.1016/j.ejphar.2017.02.042

Xiao SW, Xu Y-Z, Xiao B-F et al (2018) Recombinant adenovirus-p53 gene therapy for advanced unresectable soft-tissue sarcomas. Hum Gene Ther 29(6):699–707. https://doi.org/10.1089/hum.2017.103

Xiao-Jie L, Hui-Ying X, Zun-Ping K et al (2015) CRISPR-Cas9: a new and promising player in gene therapy. J Med Genet 52(5):289–296. https://doi.org/10.1136/jmedgenet-2014-102968

Xu J, Lu Y, Ding F et al (2007) Reversal of diabetes in mice by intrahepatic injection of bone-derived GFP-murine mesenchymal stem cells infected with the recombinant retrovirus-carrying human insulin gene. World J Surg 31(9):1872–1882. https://doi.org/10.1007/s00268-007-9168-2

Xu Y, Hou J, Liu Z et al (2011) Gene therapy with tumor-specific promoter mediated suicide gene plus IL-12 gene enhanced tumor inhibition and prolonged host survival in a murine model of Lewis lung carcinoma. J Transl Med 9:39. https://doi.org/10.1186/1479-5876-9-39

Yahalom-Ronen Y, Tamir H, Melamed S et al (2020) A single dose of recombinant VSV-ΔG-spike provides protection against SARS-CoV-2 challenge. Nat Commun 11(1):6402. https://doi.org/10.1038/s41467-020-20228-7

Yamamoto Y, Nagasato M, Rin Y et al (2017) Strong antitumor efficacy of a pancreatic tumor-targeting oncolytic adenovirus for neuroendocrine tumors. Cancer Med 6(10):2385–2397. https://doi.org/10.1002/cam4.1185

Yang Z-J, Chen B, Shen Z et al (2010) Improvement of heart function in postinfarct heart failure swine models after hepatocyte growth factor transfer: comparison of low-, medium- and high-dose groups. Mol Biol Rep 37(4):2075–2081. https://doi.org/10.1007/s11033-009-9665-5

Yang H, Peng T, Li J et al (2016) Treatment of colon cancer with oncolytic herpes simplex virus in preclinical models. Gene Ther 23(5):450–459. https://doi.org/10.1038/gt.2016.15

Yang X, Yu X, Wei Y (2018) Lentiviral delivery of novel fusion protein IL12/FasTI for cancer immune/gene therapy. PLoS One 13(7):e0201100. https://doi.org/10.1371/journal.pone.0201100

Ying H, Zaks TZ, Wang R-F et al (1999) Cancer therapy using a self-replicating RNA vaccine. Nat Med 5(7):823–827. https://doi.org/10.1038/10548

Ylä-Herttuala S (2015) Glybera's second act: the curtain rises on the high cost of therapy. Mol Ther 23(2):217–218. https://doi.org/10.1038/mt.2014.248

Ylä-Pelto J, Tripathi L, Suis P (2016) Therapeutic use of native and recombinant enteroviruses. Viruses 3:57. https://doi.org/10.3390/v8030057

Ylösmäki E, Martikainen M, Hinkkanen A et al (2013) Attenuation of Semliki Forest virus neurovirulence by microRNA-mediated detargeting. J Virol 87(1):335–344. https://doi.org/10.1128/JVI.01940-12

Yuasa K, Miaygoe Y, Yamamoto K et al (1998) Effective restoration of dystrophin-associated proteins in vivo by adenovirus-mediated transfer of truncated dystrophin cDNAs. FEBS Lett 425(2):329–336. https://doi.org/10.1016/S0014-5793(98)00251-8

Yue Y, Ghosh A, Long C et al (2008) A single intravenous injection of adeno-associated virus serotype-9 leads to whole body skeletal muscle transduction in dogs. Mol Ther 16(12):1944–1952. https://doi.org/10.1038/mt.2008.207

Zachary I, Morgan RD (2011) Therapeutic angiogenesis for cardiovascular disease: biological context, challenges, prospects. Heart 97(3):181–189. https://doi.org/10.1136/hrt.2009.180414

Zajakina A, Spunde K, Lundstrom K (2017) Application of alphavirus vectors for immunomodulation in cancer therapy. Curr Pharm Des 23(32):4906–4932. https://doi.org/10.2174/1381612823666170622094715

Zarogoulidis P, Darwiche K, Sakkas A et al (2013) Suicide gene therapy for cancer – current strategies. J Genet Syndr Gene Ther 9:4. https://doi.org/10.4172/2157-7412.1000139

Zeh HJ, Bartlett DL (2002) Development of a replication-selective oncolytic proxvirus for the treatment of human cancers. Cancer Gene Ther 9(12):1001–1012. https://doi.org/10.1038/sj.cgt.7700549

Zeng C-Y, Yang T-T, Zhou H-J et al (2019) Lentiviral vector-mediated overexpression of klotho in the brain improves Alzheimer's disease-like pathology and cognitive deficits in mice. Neurobiol Aging 78:18–28. https://doi.org/10.1016/j.neurobiolaging.2019.02.003

Zhang L, Steele MB, Jenks N et al (2016) Safety studies in tumor and non-tumor-bearing mice in support of clinical trials using oncolytic VSV-IFNβ-NIS. Hum Gene Ther Clin Dev 27(3):111–122. https://doi.org/10.1089/humc.2016.061

Zhang WW, Li L, Li D et al (2018) The first approved gene therapy product for cancer ad-p53 (Gendicine): 12 years in the clinic. Hum Gene Ther 29(2):160–179. https://doi.org/10.1089/hum.2017.218

Zhu F-C, Guan X-H, Li Y-H, Huang J-Y, Jiang T, Hou L-H, Li J-X, Yang B-F, Wang L, Wang W-J, Wu S-P, Wang Z, Wu X-H, Xu J-J, Zhang Z, Jia S-Y, Wang B-S, Hu Y, Liu J-J, Zhang J, Qian X-A, Li Q, Pan H-X, Jiang H-D, Deng P, Gou J-B, Wang X-W, Wang X-H, Chen W (2020a) Immunogenicity and safety of a recombinant adenovirus type-5-vectored COVID-19 vaccine in healthy adults aged 18 years or older: a randomised, double-blind, placebo-controlled, phase 2 trial. Lancet 396(10249):479–488

Zhu F-C, Li Y-H, Guan X-H et al (2020b) Safety, tolerability, and immunogenicity of a recombinant adenovirus type-5 vectored COVID-19 vaccine: a dose-escalation, open label, non-randomised, first-in-human trial. Lancet 395(10240):1845–1854. https://doi.org/10.1016/S0140-6736(20)31208-3

Chapter 10
Eukaryotic Virus Interactions with Bacteria: Implications for Pathogenesis and Control

Melissa K. Jones, Erin A. Almand, Anand Soorneedi, and Matthew D. Moore

Abstract As has been highlighted by the recent COVID-19 pandemic, eukaryotic viral infections impose a considerable public health and economic burden. Such viral infection occurs in the presence of other microorganisms like bacteria, and in many cases the composition and presence of these microorganisms has been found to influence numerous factors and outcomes associated with viral infection. Such interactions between eukaryotic viruses and bacteria will be surveyed in this chapter. Specifically, these interactions have potential to enhance or inhibit viral infection in numerous ways through both direct and indirect interactions. Such interactions also occur in numerous tissues throughout the host body, with special focus on the effect of the bacteria in the intestines and lungs. Finally, the chapter will conclude by presenting the latest set of work on the influence of host bacteria on SARS-CoV-2 infection, while identifying areas of needed future research.

10.1 An Introduction to Virus Interactions with Bacteria

Viral infections are dependent on several different factors which can be broadly divided into three categories: host factors, environmental factors, and the microbiota. Host factors include but are not limited to expression of viral receptors, regulation of cellular tropism of the viruses, regulating the immune response to viral infection, and transmission, among others (Neu and Mainou 2020).

Studying the role of host microbiota in virus infections has gained traction in recent years thanks to efforts by various research groups. Briefly, the host microbiota

---

M. K. Jones
Microbiology and Cell Science Department, IFAS, University of Florida, Gainesville, FL, USA

E. A. Almand
United States Air Force Academy, Colorado Springs, CO, USA

A. Soorneedi · M. D. Moore (✉)
Department of Food Science, University of Massachusetts, Amherst, MA, USA
e-mail: mdmoore@umass.edu

C. J. Hurst (ed.), *The Biological Role of a Virus*, Advances in Environmental Microbiology 9, https://doi.org/10.1007/978-3-030-85395-2_10

represents the entirety of microorganisms inhabiting an organism, and plays a major role in viral infection. Much of that role is accomplished through direct interaction, often by interacting either with the virus particles or by producing microbial products which can bind to the virus particles and regulate the viral life cycle (Bradley and Jackson 2008; Wilks and Golovkina 2012). Commenting on the role of components other than bacteria that make up a bulk of the microbiota in regulating viral infections is beyond the scope of this review. This review will focus on the bacterial component of microbiota and its role in regulating viral infection. Though Eukaryotic viruses are not entirely at the mercy of bacteria for their survival, bacteria can regulate several aspects of viral fitness by regulating virion stability, increasing coinfection rates, enhancing viral diversity/recombination, and other functions (Karst 2016). Further, the binding of viruses to bacteria can in some cases impact the bacterial biology.

10.2 Evidence for Direct Interactions Between Viruses and Microbiota

Influenza virus has been shown to bind to several groups of Gram-negative and Gram-positive bacteria which confers a fitness benefit for the bacteria during subsequent binding of those bacteria to host cell receptors. Furthermore, the release of progeny influenza virus particles from host cells can mediate a subsequent enhanced binding of bacterial strains to those host cells (Rowe et al. 2019). This was previously observed for bacterial pathogens including *Streptococcus pneumoniae*, *Morxella cattarhalis* and *Staphylococcus aureus* (Kang and Kang 2021; Spacova et al. 2021). This enhanced binding of bacteria to host cells following attachment to influenza virus is due to neuraminidase activity of the influenza virus particles. Briefly, the neuraminidase can promote cleavage of host sialic acids whose components in turn can provide energy for bacterial growth. Also, the neuraminidase can facilitate bacterial binding by disrupting both host mucus and host glycans (Engevik and Engevik 2021; Wei and Pieters 2021). The activity of neuraminidase could be detrimental to some bacterial species which rely on the host mucus and glycans for efficient binding and entry into respective cells. The association of bacteria with virus could be mutually beneficial in that the binding may help virus particles to remain stable in the extracellular environment during transmission and the bacterial species in return can rely on the virus particles for enhanced adhesion to host cell receptors. These interspecies interactions could also have potential immunological implications because antigen-presenting cells bearing from both pathogen species can elicit a much stronger immune response and is not uncommon (Rowe et al. 2019).

10.2.1 *Respiratory Syncytial Virus-Bacterial Interactions Facilitate* Streptococcus pneumoniae *Infection*

The efficacy of RSV-bound *S. pneumoniae* to infect airway epithelial cells has been shown to increase when compared with unbound bacteria (Smith et al. 2014). Furthermore, flow cytometry studies have unraveled the role of RSV G-protein in mediating the binding of RSV to *S. pneumoniae* (Avadhanula et al. 2007). Mouse studies have also shown that RSV not only helps *S. pneumoniae* bind more efficiently to bronchial epithelial cells but also aids in increasing colonization and invasiveness of the bacterial species.

10.2.2 *Gut Microbiota in the Pathogenesis of Intestinal Viruses*

The human gut microbiome is host to approximately 10^{14} organisms which help maintain homeostasis (Acevedo and Pfeiffer 2021). This is achieved by regulating several aspects of the immune system including immune cell homeostasis and function, mediating T-cell immune response, producing antimicrobial peptides and other products that can regulate inflammation etc. The gut microbiota has been shown to play a major role in regulating infection by the gastroenteritis-producing viruses rotavirus and norovirus (Huang 2020; Segrist and Cherry 2020; Tarris et al. 2021). These effects could either be stimulatory or inhibitory depending on the players involved. Most of the studies on bacteria-virus interactions in the context of the GI-tract were carried out in animal or cell culture models using surrogate viruses and thus have their own shortcomings. A recent study on the role of gut microbiota in rotavirus infection has revealed that bacterial strains belonging to the *Ruminococcus* and *Oxalabacter* genera can interact with the human rotavirus and significantly reduce viral infectivity *in vitro* (Gozalbo-Rovira et al. 2021). The binding of *Ruminococcus gauvreauii* to rotavirus is mediated by histo-blood group antigen (HBGA)-like substances. The data from this study points to the negative correlation between *Ruminococcus* levels and antibody titers against rotavirus and opens new possibilities for antiviral development.

Bacterial derived short chain fatty acids (SCFAs) help fill a gap in our understanding of microbiota-virus interactions. Many SCFAs such as butyrate, propionate and acetate are produced as byproducts of dietary fiber fermentation by bacteria, most notably by anaerobic Gram-positive *Firmicutes* of the *Clostridia* class (Lee et al. 2020). Notably, SCFAs can affect enteric virus infections by (i) altering expression of specific immune response genes and (ii) altering intestinal metabolism in general. Mice treated with streptomycin exhibited reduced shedding of the enteric picornavirus coxsackievirus B3 (species enterovirus B), but not poliovirus (species enterovirus C) pointing to the possibility of distinct microbial dependence among closely related viruses (Acevedo and Pfeiffer 2021). The mechanism behind how

SCFAs can regulate enteric virus infection remains unknown and needs further studies.

Enteric virus infection is also dependent on host bile acid metabolism mediated by the microbiome, as demonstrated by studies on human norovirus. Bile acids can bind to the norovirus capsid protein and facilitate efficient binding to CD3001f, a receptor for murine norovirus (Kong et al. 2021). Bacteria mediated bile acid modification can suppress murine norovirus infection in the proximal intestine indicating the role of bacteria in negatively regulating enteric virus infection (Acevedo and Pfeiffer 2021). Furthermore, bile acids have also been shown to negatively regulate rotavirus infection by downregulating rotavirus-induced lipid synthesis. Taken together, these studies indicate the role of bacteria-modified bile acids in regulating enteric virus infection.

10.3 Circadian Rhythms of the Microbiota: Potential Implications for Viral Infection

Growing evidence about circadian rhythms and their role in regulating the host microbiota and host responses has led to several questions about the possible role of these rhythms in enteric virus infection (Mazzoccoli et al. 2020). Circadian disorders can lead to microbial dysbiosis which in turn could affect enteric virus infection (Borrmann et al. 2020; Maiese 2020; Ray and Reddy 2020; Sultan et al. 2021). While our knowledge about the role of circadian rhythms in regulating host cell immune responses and microbial dysbiosis is limited, it would be interesting to study their role in enteric virus infections as it could help answer some of the most important questions about the role of a host's 24-h cycle in enteric virus infections.

10.4 Role of Viruses in Enhancing Bacterial Superinfections

In the case of Influenza A virus, susceptibility to secondary infections is quite common due to several factors as elucidated by studies in mice models (Paget and Trottein 2019; Rynda-Apple et al. 2015). The multifactorial mechanism involves but is not limited to changes in the respiratory barrier, epithelial cell death and mucin degradation brought about by influenza A infection. These changes render the host cells susceptible to infection by bacteria by exposing new attachment sites (Robinson et al. 2015). Influenza A virus can also disrupt the ciliary mechanism of the lung epithelial cells thereby impairing the bacterial clearing mechanism (Kuek and Lee 2020). The immune response post-influenza A infection is significantly altered as evident by disruption of macrophages and neutrophils leading to a rapid decline in the host's antibacterial defenses. Bacterial superinfection post-influenza infection is mediated by type III interferons, which act by disrupting the nasal microbiome (Barman et al. 2021). While the underlying mechanism of how influenza A infection can lead to subsequent bacterial infections is still under investigation, overwhelming evidence from animal model studies point toward altered nasal barrier functions and dysregulated immune responses. These studies arm us with evidence about active sites that potentially can be targeted during viral infections to prevent subsequent bacterial superinfections.

10.5 Role of Microbiome in Shielding from Viral Infections

Evidence for the role of commensal bacteria (an integral part of the microbiome) in preventing infections by invading viral pathogens is increasing. Commensal bacteria prevent invasion and colonization of viruses by eliciting immune responses that are inhibitory for virus replication. Studies have shown that infant mice primed with *Corynebacterium pseudodipthericum* improved their resistance against RSV infection by modulating the immune response through TLR-3 activation. Similar protective effects were observed when *Lactobacillus rhamnosus* was used to treat infant mice before infection with RSV. The role of commensal bacteria in restricting infection by viral pathogens can be attributed to the various factors expressed by these bacteria. One such factor expressed by *Staphylococcus epidermidis* has been designated an extracellular matrix-binding protein and prevents influenza virus infection by directly attaching to the virus particles and thereby blocking their adhesion to the host cells.

Another study with the enteric picornavirus, encephalomyocarditis virus (EMCV, species cardiovirus A), has shown that the intestinal microbiome plays a prominent role in protecting the host from systemic virus infection by enteric viruses. Depleting the microbiota using an antibiotic cocktail in mice exacerbated mortality, viremia and viral replication in the brain following infection by EMCV. This is due to an impaired innate immune response which otherwise would be mounted against EMCV systemic infection. Upon monocolonization with *Blautia coccoides*, which naturally is an enteric commensal and formerly named *Clostridium coccoides*, the antibiotic-treated mice exhibited alleviated EMCV pathogenesis and restricted viral replication. This antiviral result is achieved by promoting type I interferon response in macrophages, which restricts EMCV infection. That study illustrates the role of specific commensal bacterial species in regulating type I IFN mediated innate immune response upon viral infection (Yang et al. 2021).

10.6 Effects of Virus Binding on Bacteria

Enteric viral infection can have direct implications for the bacteria which comprise the microbiome. These bacteria influence mucin production, alter the permeability of the intestinal barrier through regulation of tight junctions, and modulate innate immune responses (Al-Asmakh and Hedin 2015; Monedero et al. 2018; Schroeder 2019). Therefore, changes in presence or composition of microbiome as the result of viral infection may contribute to the demonstrated bacterial enhancement of enteric viral replication. Viral infection often results in significant disruptions in the commensal microbiota. During human norovirus infection, a significant loss in both microbial diversity and richness are observed with a specific increase in *Proteobacteria* and a decrease in *Bacteroidetes* (Nelson et al. 2012). Similar changes are seen during murine norovirus infection (Hickman et al. 2014). In this murine model of infection, it has also been shown that levels of bacterial species typically considered protective for the host (e.g. Lactobacillus) are reduced during infection (Lee and Ko 2016). When present, these bacteria are hypothesized to increase IFNβ and IFNλ expression, thus, their removal during infection would dampen these critical anti-viral responses leading to enhanced viral replication. Likewise, during rotavirus infection, significant decreases in *Bacteroidetes* and *Firmicutes* with an increase in *Proteobacteria* are also observed (Jang et al. 2019). However, it has been shown that rotavirus infection can not only result in changes at the genus level, but can induce species level changes as well. For example, human rotavirus infection shifts the predominant *Bacteroides spp.* from *B. vulgatus* and *B. stercoris* to *B. fragilis* (Zhang et al. 2020). The implications of this very specific shift are not yet known, however, *B. fragilis* is considered an opportunistic pathogen in the gut and its predominance may lead to increased intestinal barrier permeability. Astrovirus infection has also been shown to decrease the presence of "healthy" commensal bacteria. Studies have shown that *Bifidobacterium* is significantly reduced during human astrovirus infection (Ma et al. 2011). Changes in microbiome diversity are recapitulated in mouse models where astrovirus infection results in a decrease in microbial diversity and a disruption in bacterial composition (Cortez et al. 2019).

In addition to enteric viral infections invoking perturbations in the intestinal microflora, eukaryotic viral infections occurring in other locations of the body have also been demonstrated to result in changes to the gut microbiome. Infections with the retroviruses designated human immunodeficiency viruses (HIV), the orthomyxoviruses that cause influenza, the hepadnavirus hepatitis B (HepB), and the flavivirus hepacivirus C (hepatitis C, HCV) all result in alterations in the composition of the intestinal microbiota with significant reductions in microbial diversity which oftentimes includes an increase in abundance of potentially pathogenic species (Aly et al. 2016; Dang et al. 2012; Groves et al. 2018; Inoue et al. 2018; Li et al. 2014; Wang et al. 2017; Xu et al. 2012; Yildiz et al. 2018). With all of these viruses, changes in the intestinal bacteria not only altered local host responses, but also impacted disease at the local sites of viral replication (Li et al. 2019).

10.7 Enteric Virus-Bacteria Interactions

The ability of commensal bacteria to alter enteric viral infection appears to be ubiquitous given that the microbiota has been shown to play a role in enhancing infection for every viral pathogen studied to date (Jones et al. 2014; Kane et al. 2011; Kuss et al. 2011; Robinson et al. 2014; Robinson and Pfeiffer 2014; Uchiyama et al. 2014). The ability of the intestinal flora to alter infection was first brought to light by two independent research groups studying either the picornavirus species enterovirus C (poliovirus) or the retrovirus mouse mammary tumor virus (MMTV) (Kane et al. 2011; Kuss et al. 2011). These simultaneously published studies not only brought the role of commensal bacteria to the forefront, but they also immediately revealed that bacteria can impact infection through very distinct mechanisms, with enhancement of viral replication at times occurring either via direct interaction between bacteria and viruses, or indirectly through bacterial modulation of the immune response.

10.7.1 Direct Interactions

Norovirus, poliovirus and reovirus are all capable of binding to the bacterial surface (Almand et al. 2017a, b; Berger et al. 2017; Erickson et al. 2018), indicating that these direct interactions may play a role in the bacterial enhancement of viral infection. For poliovirus and MMTV it has been determined that the viruses bind to the bacterial LPS (Kane et al. 2011; Kuss et al. 2011). Human noroviruses are well established to bind to histo-blood group antigens on host cells (Tan et al. 2004). Commensal bacteria can also express HBGA-like glycans on their surface and it is likely that these compounds can mediate norovirus attachment (Miura et al. 2013). However, these viruses can also bind to non-HBGA expressing bacteria, indicating there are other surface molecules that may allow for viral attachment (Almand et al. 2017a, b, c). Reoviruses also associate with commensal bacteria, likely through interactions with LPS and peptidoglycan (Berger et al. 2017). For some viruses, these interactions have been established to play a direct role in enhancement of viral infection. For example, during poliovirus infection, interactions with LPS stabilize the virion and enhance viral attachment to host cells (Robinson et al. 2014). Interactions with LPS also enhance the thermostability of reovirus (Berger et al. 2017). For human noroviruses, interaction with HBGA-like compounds on the bacterial surface can also protect against high heat stress (Li et al. 2015). There is some evidence that viral interaction with bacterial surface components may also enhance viral attachment to target cells, but this is only true for some cell types (Jones et al. 2014). However, a conflicting recent report suggests that binding to bacteria does not enhance virion stability for multiple human norovirus surrogates co-cultured with *E. cloacae* and subjected to bleach and heat treatment (Deng et al. 2019).

10.7.2 Indirect Immune-Mediated Interactions

Enteric virus interactions with commensal bacteria can also have indirect effects on viral infection. One widely described example is the ability of virus-bacterial interaction to suppress host anti-viral immune responses (Baldridge et al. 2015; Grau et al. 2020; Kane et al. 2011; Wilks et al. 2015). Co-presentation of bacterial and viral antigens can lead to the induction of bystander suppression which ultimately dampens anti-viral immune responses and allows for persistence of MMTV (Jude et al. 2003; Kane et al. 2011; Wilks et al. 2015). During murine norovirus (MNV) infection, it appears that commensal bacteria suppress IFN-L responses which allow for viral persistence during chronic infection (Baldridge et al. 2015; Wilen et al. 2018). Type III IFNs also influence regionalization of acute murine norovirus infection through the activities of commensal bacteria. Specifically, transformation of bile acids by commensal bacteria leads to priming of type III IFN response resulting in suppression of norovirus infection in the proximal small intestine (Grau et al. 2020). In addition to altering immune responses, commensal bacteria can indirectly impact viral infection through promotion of viral recombination. Although the precise mechanisms by which this occurs are not yet known, it has been shown that poliovirus interactions with commensal bacteria can lead to increased viral co-infection and promote viral? genetic recombination during viral replication (Erickson et al. 2018). This phenomenon could ultimately increase virus adaptability as well as lead to the emergence of new viral strains.

For example, HIV infection results in a significant reduction in enteric commensal microbial diversity with a simultaneous increase in the abundance of potentially pathogenic species. This reduced microbial richness is accompanied by an increase in *Firmicutes*, and proteobacteria phyla (Sun et al. 2016; Vujkovic-Cvijin et al. 2013). Influenza infection also alters the gut microbiota, although changes to specific bacterial phyla depend on the animal model and viral subtypes used. However, *Firmicutes* are decreased in nearly all the studies. During hepatitis B and hepatitis C infection, composition of the enteric bacterial community is altered and enrichment of the potentially pathogenic bacteria is accompanied by a decrease in beneficial bacteria. There are even some specific bacterial species that can “indirectly maintain barrier permeability by producing metabolites associated with reduced expression of the rotavirus toxin NSP4 (Gonzalez-Ochoa et al. 2017). Rotavirus infection also leads to changes in diversity and composition of the microbiome, however, its impacts on specific bacterial populations differ compared to noroviruses and astroviruses (Chen et al. 2017). Transiently, enteric viral infections can be associated with decreases in enteric bacterial diversity and significant alterations in bacterial composition of the microbiome (Chen et al. 2017; Dinleyici et al. 2018).

Antimicrobial peptides produced by commensal bacteria may also have antiviral properties. The bacteriocin duramycin, that is produced by *Streptomycetes*, has been shown to prevent cellular entry by the flavivirus Zika by blocking the TIM1 co-receptor (Tabata et al. 2016). Subtilosin that is produced by *Bacillus* disrupts late infectious stages of the alphaherpesviruses Human alphaherpesvirus 1 (HSV 1) and Human alphaherpesvirus 2 (HSV 2) (Caignard et al. 2013; Quintana et al. 2014).

10.7.3 Respiratory Virus-Bacteria Interactions

While the gastrointestinal tract provides an ideal, nutrient-rich location for microbes to colonize and multiply, the respiratory tract is generally considered a low-nutrient environment, lined with bacteriostatic compounds, and a wide range of body temperature zones (Dickson et al. 2016, 2017). The upper respiratory tract (URT) is an interconnected pathway that provides the external openings, and thus the exogenous sources of microbes to the respiratory system. The URT contains the anterior nares, nasal and oral cavities, middle ear, sinuses plus the pharynx and top portion of the larynx. The lower respiratory tract (LRT) is comprised of the lower portion of the larynx, trachea and lungs, made up of bronchi, bronchioles and alveoli. The microbial richness is higher in the upper respiratory tract, which contains components of the skin microbiome and species generally isolated to the respiratory tract; however, the lower respiratory tract also contains a distinct bacterial population (Hament et al. 1999; Hanada et al. 2018). Due to their proximity and bidirectional flow, there are a lot of similarities between the microbes that colonize or cause infection in these areas. The lung microbiota, in particular, is influenced by the rest of the respiratory tract, as microbial immigration is its main source of colonization (Dickson and Huffnagle 2015), generally through microaspiration (Venkataraman et al. 2015; Dickson et al. 2016). Microorganisms in this area are also dependent on the rate of elimination from the region, and the rate of reproduction (Dickson and Huffnagle 2015). This bacterial movement may increase during a bacterial infection, as the microbial burden will increase, and is also subject to influence from gastro-intestinal reflux, medications and prolonged periods in a supine position (Dickson et al. 2016). With a lot of bacterial movement occurring between the upper and lower respiratory system, there are lots of opportunities for infections, both bacterial and viral in nature.

Disease in the respiratory tract may follow many different paths. However, typically an acute viral respiratory infection may be followed up by a prolonged and clinically significant bacterial infection (Diavatopoulos et al. 2010; Avadhanula et al. 2006). While there is debate on the timing and scale of interaction during subsequent and potential coinfections, respiratory infections provide an interesting look at the struggle between commensal bacteria, pathogens and the immune system. In general, virus-bacteria interactions in the respiratory system may be broken into two categories: indirect effects on the bacterial contingent of the microbial community and direct effects on its individual component species. There are some viruses that require certain inputs from bacterial neighbors, whereas other viruses cause shifts in the respiratory terrain, affecting the overall diversity of microbes in the area, and skewing the population towards opportunistically pathogenic microbes (Beadling and Slifka 2004; Neu and Mainou 2020). The combination of a viral respiratory infection superimposed by a bacterial infection is the difference between infection of swine by the swine influenza virus, which produces a disease that typically is relatively mild, and the very severe disease known as swine flu. Swine flu occurs when the influenza virus infection is accompanied by infection with the bacterial species *Haemophilus influenzae* (Shope 1931).

For the virus mediated microbiome shifts, these may be due to cellular damage after an infection, accompanied by increased presence of microbial adhesion proteins, and inappropriate immune system function (Vareille et al. 2011; Almand et al. 2017a; Bosch et al. 2013). Epithelial cell damage occurs in a variety of ways during an infection. Many viruses replicate intracellularly, causing damage to the host cell, and in some cases causing cell lysis or apoptosis. This viral cytotoxicity results in a loss of epithelial integrity, (Folkerts et al. 1998) ultimately disrupting the protective barrier and inhibiting the hosts repair mechanisms. The loss of this protective barrier exposes extracellular proteins such as fibronectin, which are often upregulated in the presence of certain viruses, like the picornaviral species rhinovirus A, B, and C, and those upregulated proteins also comprise binding targets for many bacterial species (Bosch et al. 2013). In addition to destroying the cells, and exposing potential binding sites, some viruses, such as the orthomyxoviruses that cause influenza and the paramyxoviruses which cause parainfluenza, make enzymes to aid bacterial adhesion. These viruses create neuraminidase, which cleaves oligosaccharides and exposes bacterial binding sites, creating niches for bacterial colonization (Peltola et al. 2005; Peltola and Mccullers 2004; Tappert et al. 2013). While the enzymatic actions of these viruses directly damaging respiratory epithelium allows the surrounding bacteria to flourish, some respiratory viruses are also capable of causing an immune system malfunction, leading to an ineffective immune response against bacteria in the area, and potentially resulting in a super infection (Vareille et al. 2011). In some instances, the host response feedback loop that should be clearing the microbial invader, instead actually promotes its growth. In the lungs, the production of intra-alveolar catecholamines, part of the alveolar inflammatory response associated with viral infections yields an increase in *Streptococcus pneumoniae* biofilm production, growth and virulence leading to acute cases of pneumonia (Dickson et al. 2016; Kanangat et al. 1999).

The interplay between viruses and bacteria is complex, and each human microbiome is unique. Despite the inherent differences in microbial communities, some pattern shifts and specific pathogen interplay is conserved across infections. In a broad sense, the common respiratory pathogens *Haemophilus influenzae* and *Streptococcus pneumoniae* are isolated more frequently and in larger numbers from patients with viral infections, versus healthy individuals, suggesting at a minimum, there is a link promoting bacterial colonization post viral infection (Avadhanula et al. 2006; Smith et al. 1976).

10.8 The Respiratory System: Microbiome Shifts

Upper respiratory infections constitute the bulk of total respiratory infections, with the vast majority of illness caused by viruses (Kemper Alston and Fahrner 2003). While pneumonia (lower respiratory infection) causes the highest levels of mortality, otitis media is an enormous issue for young children (Bosch et al. 2013). Rhinoviruses are the most prevalent respiratory infections, however parainfluenza virus, respiratory syncytial virus (RSV), coronavirus, adenovirus, and as well the picornaviruses previously known as echoviruses and coxsackieviruses, also are common culprits in terms of causing respiratory infections (Jain et al. 2001; Kemper Alston and Fahrner 2003). Clinically, these infections usually present themselves as the common cold, sinusitis or pharyngitis. When in isolation, these infections have excellent outcomes, many resolving without medical intervention; however, complications may arise when the infection does not stay localized, but rather infects larger portions of the respiratory tract (Jain et al. 2001).

The normal microbiota of the upper respiratory tract depends on the specific physical location, but includes *Corynebacterium, Moraxella, Staphyloccocus* and *Streptococcus* (Depner et al. 2017). Of these common inhabitants, many bacteria are opportunistic pathogens, behaving as commensals until conditions permit additional colonization and disease. In contrast to other body sites, the healthy microbiome of the upper respiratory tract is less diverse than the disease state, making pathogenic shifts and newcomers more obvious. Although the underlying mechanisms may be similar, i.e. increased bacterial adherence to respiratory cells, or impair macrophage function, (Mallia et al. 2012) each of these shifts is specific to the causative viral agent. For influenza, the nasopharynx bacterial community dramatically shifts from the typical *Staphyococcus* and *Streptococcus* dominated microflora to a diversity highlighted by an increased abundance of *Prevotella, Streptobacillus, Porphyromonas, Granuliticatella, Veillonella, Fusobacterium* and *Haemophilus.* Rhinoviruses are the leading cause of upper respiratory tract infections in young children, and this leads to an associated increase in *Streptococcus* and *Haemophilus* (Depner et al. 2017; Rosas-Salazar et al. 2016). While these produce interactions of virus and bacteria that are predominately indirect in nature, they remain distinct, suggesting there are nuances to virus-bacteria co-infections.

Acute Otitis Media (AOM), a classically bacterial infection, has been shown to be poly-microbial in nature, as viruses commonly play a pivotal role in a switch to increased bacterial virulence and pathogenicity. In a study, over half of the children with a viral URT progressed to AOM if either *Streptococcus pneumoniae, Haemophilus influenzae,* or *Moraxella catarrhalis* was present, compared to only 10% of individuals without these pathogens present (Van Den Broek et al. 2019). In this disease association, influenza A virus, coronavirus NL63, and RSV promote inflammation and bacterial adherence. In addition to modifying host immune systems and interfering with antibiotic activity, these viruses also alter mucous properties and prevent the clearance of bacteria in this area (Marom et al. 2012). From a bacterial perspective, there are multiple microbiome profiles associated with this condition, some of which are deemed protective, making individuals less susceptible to infections, while some others are considered riskier, or more prone to infections. The helpful bacterial species include: *Corynebacterium, Dolosigranulum, Propionibacterium, Lactococcus,* and *Staphylococcus.* Conversely, a shift to *Haemophilus, Rothia* and *Actinomyces* improves the likelihood of an infection. Additionally, while these microbes may be part of the normal, healthy flora, an overabundance or large presence of *Moracellaceae, Streptococcaceae* or *Pasteurellaceae,* moves the microbiome from a healthy state to one commonly associated with AOM. To further develop the profile for individuals at risk to contract an AOM infection, when comparing the middle ear fluid, these individuals all contained common pathogens *S. pneumoniae, H. influenzae, Moraxella catarrhalis, Turicella otitidis* and *Staphylococcus auricularis* (Van Den Broek et al. 2019).

Rhinovirus and RSV are common in young children and present similar symptoms clinically. Despite these commonalities, the effect on the microbiome is virus dependent. In infants with only an RSV infection, there is an abundance of *Firmicutes*, particularly *Streptococcus*, with a low abundance of *Proteobacteria*, specifically two commonly implicated pathogens are often present: *Haemophilus* and *Moraxella*. Conversely, rhinovirus trends towards low *Firmicutes* and high *Proteobacteria*, in addition to an increase in *Neisseria* (Allen et al. 2014; Mansbach et al. 2016). This increased pathogenesis be due to the ciliary injury caused by RSV. By promoting localized disruption through ciliostasis, the result is mucus agglomeration, and the mucociliary transport mechanism typically used to remove bacteria becomes ineffective and disorganized (Tristram et al. 1998). Another mechanism, typically observed with rhinovirus infections is macrophage impairment. An increase in viral load additionally suppresses the immune system, limiting the number of antimicrobial peptides released, and promoting bacterial colonization (Mallia et al. 2012). Coupled with these bacterial interactions, both RSV and rhinovirus are commonly linked with another viral co-infection, the parvoviruses of the genus *Bocavirus*. (Hamilos 2015; Lukkarinen et al. 2014; Moesker et al. 2015) While the acute viral presence in the region often may be limited, these infections have lingering effects on the respiratory microbiome. Infants with RSV show a decrease in *Staphylococcus aureus* when compared to their counterparts, with an increase in *Haemophilus influenzae*, a pathogen also linked to the proinflammatory response and the development of childhood asthma (Fraenkel et al. 1995; Rosas-Salazar et al. 2016).

Although not commonly thought of as a respiratory pathogen, the effect of Human Immunodeficiency Virus (HIV) on the oral cavity is well documented (Asai and Nakashima 2018). Opportunistic oral infections may be found in the majority of HIV patients (up to 80%), (Dang et al. 2012) potentially due to a dramatic microbiome remodel in the presence of an HIV infection. The oral microbiome of HIV positive individuals tends to shift away from a commensal population with high levels of *Lactobacillus* and *Streptococcus*, moving towards a pathogenic profile featuring known respiratory nuisances *Prevotella melaninogenica, Rothia mucilaginosa, Veionella parvula, Neisseria* and *Haemophilus*. This trend towards pathogenic species for individuals with an already compromised immune system is troubling, especially given that HIV positive individuals may lose lung function following a bout with pneumonia, a phenomenon not found with the HIV negative community (Asai and Nakashima 2018; Dang et al. 2012; Vidya et al. 2016).

Adenoviruses commonly cause respiratory tract infections, and have been shown to increase the binding of *Streptotoccus pneumoniae,* however in this regard adenoviruses do not affect all *Streptococcus pneumoniae* strains equally. Adenoviruses expose additional pneumococcal binding sites to promote binding of bacterial strains which already exhibit binding capabilities, and thus these bacteria seem to become more adhesive. These cytopathic changes in the respiratory monolayer alter the mucosa, making it more hospitable for colonization. *In vitro* this pattern was observed for human adenovirus types 1, 2, 3, and 5. Human adenovirus type 9, which is not associated with respiratory disease, does not increase bacterial adherence (Hakansson et al. 1994). Adenoviruses also are associated with pediatric acute gastroenteritis (Kumthip et al. 2019).

10.8.1 The Upper Respiratory System: Direct Interactions

In addition to oral infections and lesions, HIV also causes negative effects for periodontitis. In this instance, the same bacterial species commonly implicated in disease, *Porphyromonas gingivalis, Prevotella intermedia,* and *Fusobacterium nucleatum,* may be responsible for triggering HIV infection from latency. These bacteria interact directly with dendritic cells through surface-pattern recognition receptors. The lipopolysaccharides of these Gram-negative bacteria interact with the HIV promoter within these dendritic cells, which can contribute to reactivating the latent virus (Huang et al. 2011).

10.8.2 The Lower Respiratory System: Microbiome Shifts

Lower respiratory infections are a leading cause of morbidity and mortality worldwide, especially amongst young children and the elderly (Troeger et al. 2018). Although many pathogens are capable of creating an infection without additional help, bacterial-virus co-infections are not only common, but may be responsible for increased disease severity. Common members associated with a healthy respiratory microbiome include *Prevotella, Pseudomonas, Fusobacterium, Porphyromonas, Haemophilus, Veillonella,* and *Streptococcus* (Dickson et al. 2017; Morris et al. 2013).

Somewhat surprisingly, members of the *Herpesviridae* family, which include Cytomegalovirus, Epstein-Barr Virus, Herpes Simplex Virus and Varicella-Zoster Virus, are commonly associated with infections in the lower respiratory tract. These viruses are capable of infecting nearly all mucocutaneous sites on the human body, and gain access to the lungs through seeding from other portions of the respiratory tract, like the mucosal layer in the endotracheal tube (Bouza et al. 2011; Chen et al. 2020; Friedrichs et al. 2013; Simoons-Smit et al. 2006). These viruses may also gain entry through reactivation of a latent infection. When herpes simplex virus was recovered from patient bronchoalveolar lavage (BAL) samples, *Pseudomonas aeruginosa* and *Stenotrophomonas maltophilia* were also frequently recovered (Park 2005).

In some instances, such as with influenza, a stable microbiome with *Alloprevotella, Prevotella* and *Bacteroides* is commonly associated with a decreased risk of infection (Hanada et al. 2018). Interestingly, research also shows that when individuals are infected with influenza, the respiratory tract gains diversity, specifically an increased abundance of *Prevotella, Streptobacillus, Porphyromonas, Granuliticatella, Veillonella, Fusobacterium* and *Haemophilus* versus the normal predominant species of *Staphylococcus aureus* and *Streptococcus* (Depner et al. 2017; Rosas-Salazar et al. 2016). This shift in diversity may play a role in the spread of bacteria throughout the LRT. Furthermore, the virus up-regulates bacterial adherence, reduces the activity of neutrophils, and alters the signaling environment, repressing the immune system in the nasopharynx and allowing subsequent colonization within the nasopharynx (Diavatopoulos et al. 2010). *S. pneumoniae* especially is linked with severe clinical outcomes from viral infections of the lower respiratory tract, where cellular changes disrupt cilia on the epithelium, disrupting and causing necrosis in the bronchi, plus lesions in the lungs (Walsh et al. 1961). Although the data is still emerging on the global pandemic coronavirus, SARS-CoV-2, preliminary data suggests this virus also causes a microbiome shift. Conserved across infected patients is an overexpression of *Prevotella* genes. Predictive modeling suggests these proteins support viral growth and replication (Khan and Khan 2020).

10.8.3 The Lower Respiratory System: Direct Interactions

In a healthy lung, the predominant bacteria are *Prevotella*, *Veilonella*, and *Streptococcus*;(Depner et al. 2017) however, the relative abundance of these organisms is greatly impacted by microbial populations within the upper respiratory system, as overgrowths and increased pathogens in the upper respiratory tract will lead to increased microbial growth in the lungs (Hanada et al. 2018). Many illnesses that manifest in the lungs subsequent to viral infections begin with microbial overgrowth in the upper respiratory tract that migrates to the lower respiratory tract (Takase et al. 1999). Of the viruses targeting the lower respiratory tract, Respiratory syncytial virus (RSV) is the most common, especially in children (Thorburn et al. 2006). In contrast to indirect interactions and microbial community adjustments based on the viral infection environment, some respiratory viruses directly interact with their bacterial counterparts.

In the lower respiratory system, paramyxoviruses such as RSV and parainfluenza virus are responsible for increased bacterial adhesion by upregulating bacterial receptors on host cells and promoting proinflammatory cytokine release (Avadhanula et al. 2006). Parainfluenza virus possesses a hemagglutinin-neuraminidase envelope glycoprotein and although the mechanistic role of this enzyme remains uncertain it enhances bacterial binding, including *Streptococcus pneumoniae* (Beadling and Slifka 2004; Hament et al. 1999). A more developed understanding exists regarding the mechanism behind the role of influenza viruses in bacterial adhesion. The influenza viruses have a neuraminidase, which is required for activation of the viral hemagglutinin membrane glycoprotein—a critical step in making the virus infectious—but the influenza neuraminidase may also cleave sialic acid on the surface of respiratory cells, exposing bacterial binding targets. In addition to influenza virus aiding bacterial adhesion, there are three mechanisms that bacterial species such as *Staphylococcus aureus* and *Aerococcus viridans* use to bolster viral infection. These bacteria (1) secrete proteases capable of cleaving hemagglutinin, (2) convert plasminogen to plasmin through kinases that activate some specific influenza strains and (3) promote the release of host proteases by increasing inflammation, further activating influenza (Böttcher-Friebertshäuser et al. 2013; Tashiro et al. 1987; Lee et al. 2019).

There is a strong correlation between human metapneumovirus and lower respiratory infections in otherwise healthy children, specifically bronchiolitis and croup. A potential explanation for this disease burden is direct and indirect interactions between the virus with *S. pneumoniae* during coinfection. The bacterial capsule or pneumolysin of the bacteria penetrates into the respiratory mucosal layer, inhibiting ciliary clearance, and exposing susceptible host cells to viral attack. Once exposed, bacterial cell wall lipopeptides enhance viral binding to these host cells thereby promoting infection and that joint pathogenesis can spread throughout the lower respiratory tract. This increased spread is further enhanced through bacterial evasion of the host innate immune responses, facilitating increased inflammation and activation of the respiratory immune system (Verkaik et al. 2011; Williams et al. 2004).

10.9 SARS-CoV-2 Interaction with Bacteria

The current SARS-CoV-2 pandemic has garnered much attention due to the devastating loss of life and economic impact it has left. At the time of writing, SARS-CoV-2 has claimed more than 3.5 million lives worldwide and continues to claim more lives in developing nations where several factors are contributing towards rapid infection. There is increased interest in understanding the mechanism behind these infections in the context of contributing host factors and the gut microbiome. A recent molecular docking study has shown that Nisin, an antimicrobial peptide produced by lactic acid bacteria and used as a food grade preservative, can compete with the receptor binding domain (RBD) of SARS-CoV-2 in binding to the hACE2 receptor (Bhattacharya et al. 2021). This study underscores the importance of using Nisin as an effective treatment option against SARS-CoV-2. Additional early reports also suggest a potential role of the gut microbiota in SARS-CoV-2 infection and disease outcome (Dhar and Mohanty 2020; He et al. 2020; Kalantar-Zadeh et al. 2020; Moore et al. 2020). SARS-CoV-2 could facilitate dysbiosis of local and distant microbiota by disrupting homeostasis of resident oral bacteria.

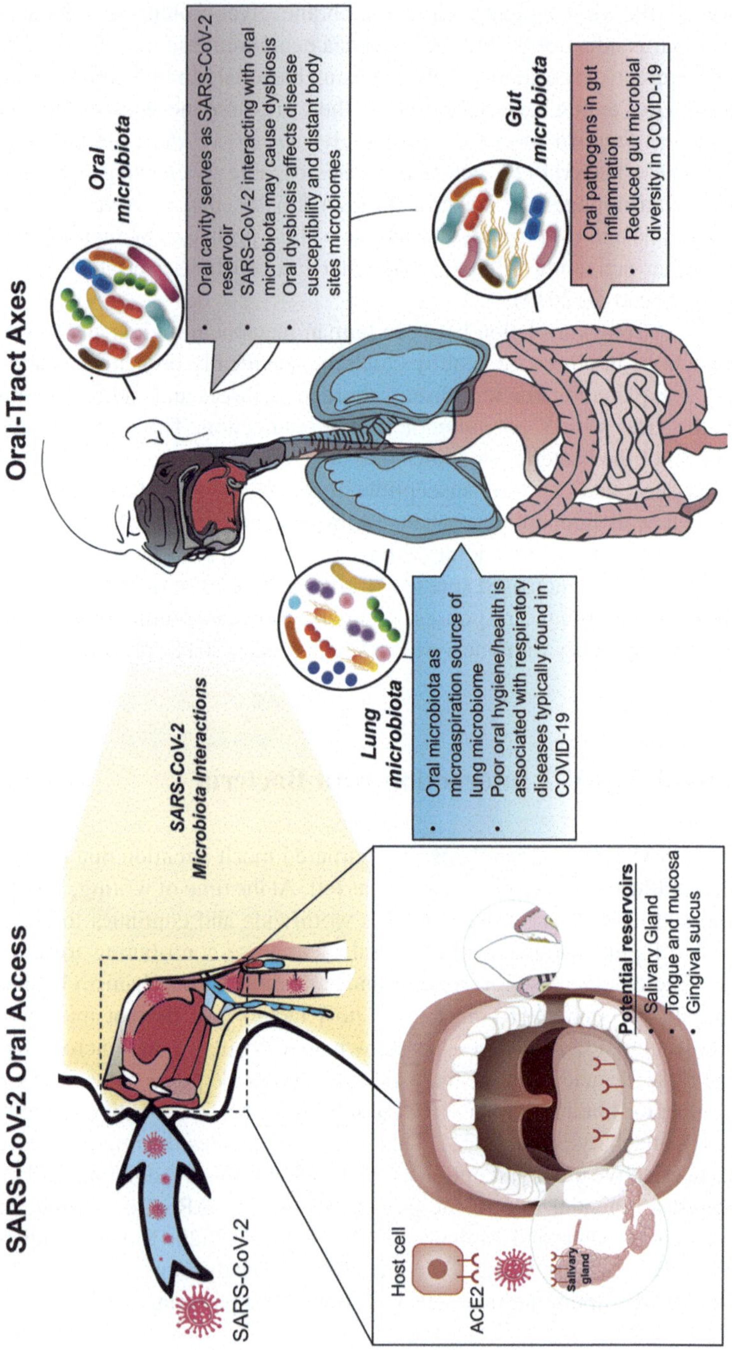

Role of oral microbiota in dysbiosis and susceptibilty of distant microbiomes during SARS-CoV-2 infection (Xiang et al. 2021)

While the oral cavity along with the nasal cavity is one of the primary routes of SARS-CoV-2 entry into the body, much attention has been focused on the nasal route of entry. The oral route of SARS-CoV-2 entry is important as transcriptome studies by Song et al. have shown elevated levels of ACE2 and TMPRS22 expression in the salivary glands than in their lung counterparts (Ren et al. 2021). Recent studies have also pointed to the possibility of contaminated saliva acting as a potential source of infection owing to the fact that salivary glands serve as a reservoir for SARS-CoV-2 (Li et al. 2020). The hallmark symptoms of SARS-CoV-2 infection, loss of smell and taste, can be attributed to the elevated expression of ACE2 in the oral mucosa. However, the role of elevated ACE2 expression in dysregulating chemical perception in non-neural cells needs to be investigated further. The role of SARS-CoV-2 infection in disrupting the local microbial communities cannot be completely ruled out. A study on the nasopharynx microbiota in SARS-CoV-2 positive patients indicated a significant decrease in *Fusobacterium periodonticum* (Nardelli et al. 2021). Further follow up studies employing animal models to better understand the role of SARS-CoV-2 infection in disrupting the local microecological balance and dysbiosis are needed as SARS-CoV-2 infections can predispose patients to superinfections by other opportunistic pathogens. Infection with SARS-CoV-2 can not only influence the microbiota of the oral and nasopharyngeal cavities but can also affect distal organs. Since the respiratory and gastrointestinal systems share the oral cavity as a common route of passage, it is highly likely that disruption in the microbiota of one of these systems could have a cascading effect on the others. This phenomenon is not new as evidenced by development of inflammation in the gut caused by immune response mounted against periodontal pathogens. A few of the questions that remain unanswered about the potential role of SARS-CoV-2 infection are: (i) If SARS-CoV-2 has a significant role in dysbiosis of the oral microbiota then (ii) Is there a correlation between oral microbiota dysbiosis and the outcome of disease severity in SARS-CoV-2 patients and (iii) what might be the impact of medication and other treatment modalities for SARS-CoV-2 oral infection upon symbionts which are critical to microbial health of the respiratory and gastrointestinal systems. By utilizing large scale metagenomics approaches to understand the connection between the human microbiome and SARS-CoV-2 infection, coupled with multi-omics studies, we will gain a better understanding of the interactive role of microbiota on SARS-CoV-2 and vice versa. The data from such studies can help to better design diagnostics, treatment, and prognostic modalities against SARS-CoV-2 infections.

References

Acevedo M, Pfeiffer JK (2021) Microbiota-immune system interactions and enteric virus infection. Elsevier

Al-Asmakh M, Hedin L (2015) Microbiota and the control of blood-tissue barriers. Tissue Barr 3:1–7. https://doi.org/10.1080/21688370.2015.1039691

Allen EK, Koeppel AF, Hendley JO, Turner SD, Winther B, Sale MM (2014) Characterization of the nasopharyngeal microbiota in health and during rhinovirus challenge. Microbiome 2:1–11. https://doi.org/10.1186/2049-2618-2-22

Almand EA, Moore MD, Jaykus L-A (2017a) Norovirus binding to ligands beyond histo-blood group antigens. Front Microbiol 8:2549. https://doi.org/10.3389/fmicb.2017.02549

Almand EA, Moore MD, Jaykus LA (2017b) Virus-bacteria interactions: an emerging topic in human infection. Viruses 9:58–68. https://doi.org/10.3390/v9030058

Almand EA, Moore MD, Outlaw J, Jaykus L-A (2017c) Human norovirus binding to select bacteria representative of the human gut microbiota. PLoS One 12:e0173124. https://doi.org/10.1371/journal.pone.0173124

Aly AM, Adel A, El-Gendy AO, Essam TM, Aziz RK (2016) Gut microbiome alterations in patients with stage 4 hepatitis C. Gut Pathog 8. https://doi.org/10.1186/s13099-016-0124-2

Asai D, Nakashima H (2018) Pathogenic viruses commonly present in the oral cavity and relevant antiviral compounds derived from natural products. Medicines 5:120. https://doi.org/10.3390/medicines5040120

Avadhanula V, Rodriguez CA, DeVincenzo JP, Wang Y, Webby RJ, Ulett GC, Adderson EE (2006) Respiratory viruses augment the adhesion of bacterial pathogens to respiratory epithelium in a viral species- and cell type-dependent manner. J. Virol. 80:1629–1636. https://doi.org/10.1128/jvi.80.4.1629-1636.2006

Avadhanula V, Wang Y, Portner A, Adderson E (2007) Nontypeable *Haemophilus influenzae* and *Streptococcus pneumoniae* bind respiratory syncytial virus glycoprotein. J Med Microbiol 56:1133–1137. https://doi.org/10.1099/jmm.0.47086-0

Baldridge MT, Nice TJ, McCune BT, Yokoyama CC, Kambal A, Wheadon M, Diamond MS, Ivanova Y, Artyomov M, Virgin HW (2015) Commensal microbes and interferon-λ determine persistence of enteric murine norovirus infection, Science (80–). 347:266–269. https://doi.org/10.1126/science.1258025

Barman TK, Racine R, Bonin JL, Califano D, Salmon SL, Metzgeri DW (2021) Sequential targeting of interferon pathways for increased host resistance to bacterial superinfection during influenza. PLoS Pathog.:17. https://doi.org/10.1371/journal.ppat.1009405

Beadling C, Slifka MK (2004) How do viral infections predispose patients to bacterial infections? Curr Opin Infect Dis 17:185–191. https://doi.org/10.1097/01.qco.0000129612.21745.08

Berger AK, Yi H, Kearns DB, Mainou BA (2017) Bacteria and bacterial envelope components enhance mammalian reovirus thermostability. PLoS Pathog. 13. https://doi.org/10.1371/journal.ppat.1006768

Bhattacharya R, Gupta A, Mitra S, Mandal S, Biswas SR (2021) A natural food preservative peptide nisin can interact with the SARS-CoV-2 spike protein receptor human ACE2. Virology 552:107–111

Borrmann H, McKeating JA, Zhuang X (2020) The circadian clock and viral infections. J Biol Rhythms 36:9–22. https://doi.org/10.1177/0748730420967768

Bosch AATM, Biesbroek G, Trzcinski K, Sanders EAM, Bogaert D (2013) Viral and bacterial interactions in the upper respiratory tract. PLoS Pathog 9:e1003057. https://doi.org/10.1371/journal.ppat.1003057

Böttcher-Friebertshäuser E, Klenk H-D, Garten W (2013) Activation of influenza viruses by proteases from host cells and bacteria in the human airway epithelium. Pathog Dis 69:87–100. https://doi.org/10.1111/2049-632X.12053

Bouza E, Giannella M, Torres MV, Catalán Pharmd P, Sánchez-Carrillo Pharmd C, Insa Hernandez Pharmd R, Muñoz P (2011) Herpes simplex virus: a marker of severity in bacterial ventilator-associated pneumonia on behalf of The Gregorio Marañón Task Force for Pneumonia. J Crit Care. https://doi.org/10.1016/j.jcrc.2010.10.008

Bradley J, Jackson JA (2008) Measuring immune system variation to help understand host-pathogen community dynamics. Parasitology 135:807–823

Caignard G, Leiva-Torres GA, Leney-Greene M, Charbonneau B, Dumaine A, Fodil-Cornu N, Pyzik M, Cingolani P, Schwartzentruber J, Dupaul-Chicoine J, Guo H, Saleh M, Veillette A,

Lathrop M, Blanchette M, Majewski J, Pearson A, Vidal SM (2013) Genome-wide mouse mutagenesis reveals cd45-mediated t cell function as critical in protective immunity to HSV-1, PLoS Pathog. 9. https://doi.org/10.1371/journal.ppat.1003637

Chen SY, Tsai CN, Lee YS, Lin CY, Huang KY, Chao HC, Lai MW, Chiu CH (2017) Intestinal microbiome in children with severe and complicated acute viral gastroenteritis. Sci Rep 7. https://doi.org/10.1038/srep46130

Chen C, Feng P, Slots J (2020) Herpesvirus-bacteria synergistic interaction in periodontitis. Periodontology 2000. https://doi.org/10.1111/prd.12311

Cortez V, Margolis E, Schultz-Cherry S (2019) Astrovirus and the microbiome. Curr Opin Virol. https://doi.org/10.1016/j.coviro.2019.05.002

Dang AT, Cotton S, Sankaran-Walters S, Li CS, Lee CYM, Dandekar S, Paster BJ, George MD (2012) Evidence of an increased pathogenic footprint in the lingual microbiome of untreated HIV infected patients. BMC Microbiol. 12:1–10. https://doi.org/10.1186/1471-2180-12-153

Deng W, Almeida G, Gibson KE (2019) Co-culture with *Enterobacter cloacae* does not enhance virus resistance to thermal and chemical treatments. Food Environ Virol 11:238–246. https://doi.org/10.1007/s12560-019-09381-5

Depner M, Ege MJ, Cox MJ, Dwyer S, Walker AW, Birzele LT, Genuneit J, Horak E, Braun-Fahrländer C, Danielewicz H, Maier RM, Moffatt MF, Cookson WO, Heederik D, von Mutius E, Legatzki A (2017) Bacterial microbiota of the upper respiratory tract and childhood asthma. J Allergy Clin Immunol 139:826–834.e13. https://doi.org/10.1016/j.jaci.2016.05.050

Dhar D, Mohanty A (2020) Gut microbiota and Covid-19- possible link and implications. Virus Res. https://doi.org/10.1016/j.virusres.2020.198018

Diavatopoulos DA, Short KR, Price JT, Wilksch JJ, Brown LE, Briles DE, Strugnell RA, Wijburg OL (2010) Influenza A virus facilitates *Streptococcus pneumoniae* transmission and disease. FASEB J. 24:1789–1798. https://doi.org/10.1096/fj.09-146779

Dickson RP, Huffnagle GB (2015) The lung microbiome: new principles for respiratory bacteriology in health and disease. PLoS Pathog. 11:e1004923. https://doi.org/10.1371/journal.ppat.1004923

Dickson RP, Erb-Downward JR, Martinez FJ, Huffnagle GB (2016) The microbiome and the respiratory tract. Annu Rev. Physiol 78:481–504. https://doi.org/10.1146/annurev-physiol-021115-105238

Dickson RP, Erb-Downward JR, Freeman CM, Mccloskey L, Falkowski NR, Huffnagle GB, Curtis JL (2017) Bacterial topography of the healthy human lower respiratory tract. Med Serv. https://doi.org/10.1128/mBio.02287-16

Dinleyici EC, Martínez-Martínez D, Kara A, Karbuz A, Dalgic N, Metin O, Yazar AS, Guven S, Kurugol Z, Turel O, Kucukkoc M, Yasa O, Eren M, Ozen M, Martí JM, Garay CP, Vandenplas Y, Moya A (2018) Time series analysis of the microbiota of children suffering from acute infectious diarrhea and their recovery after treatment. Front Microbiol 9. https://doi.org/10.3389/fmicb.2018.01230

Engevik KA, Engevik MA (2021) Partners in infectious disease: when microbes facilitate enteric viral infections. Gastroenterol Insight. https://doi.org/10.3390/gastroent12010005

Erickson AK, Jesudhasan PR, Mayer MJ, Narbad A, Winter SE, Pfeiffer JK (2018) Bacteria facilitate enteric virus co-infection of mammalian cells and promote genetic recombination. Cell Host Microbe 23:77–88.e5. https://doi.org/10.1016/j.chom.2017.11.007

Folkerts G, Busse WW, Nijkamp FP, Sorkness R, Gern JE (1998) State of the art virus-induced airway hyperresponsiveness and asthma. Am J Respir Crit Care Med 157:1708–1720

Fraenkel DJ, Bardin PG, Sanderson G, Lampe F, Johnston SL, Holgate ST (1995) Lower airways inflammation during rhinovirus colds in normal and in asthmatic subjects. Am J Respir Crit Care Med 151:879–886. https://doi.org/10.1164/ajrccm/151.3_pt_1.879

Friedrichs I, Bingold T, Keppler OT, Pullmann B, Reinheimer C, Berger A (2013) Detection of Herpesvirus EBV DNA in the lower respiratory tract of ICU patients: a marker of infection of the lower respiratory tract? Med Microbiol Immunol 202:431–436. https://doi.org/10.1007/s00430-013-0306-1

Gonzalez-Ochoa G, Flores-Mendoza LK, Icedo-Garcia R, Gomez-Flores R, Tamez-Guerra P (2017) Modulation of rotavirus severe gastroenteritis by the combination of probiotics and prebiotics. Arch Microbiol. https://doi.org/10.1007/s00203-017-1400-3

Gozalbo-Rovira R, Rubio-Del-Campo A, Santiso-Bellón C, Vila-Vicent S, Buesa J, Delgado S, Molinero N, Margolles A, Yebra MJ, Collado MC, Monedero V, Rodríguez-Díaz J, Cabrera Z (2021) Interaction of intestinal bacteria with human rotavirus during infection in children. Int J Mol Sci Artic J Mol Sci. https://doi.org/10.3390/ijms22031010

Grau KR, Zhu S, Peterson ST, Helm EW, Philip D, Phillips M, Hernandez A, Turula H, Frasse P, Graziano VR, Wilen CB, Wobus CE, Baldridge MT, Karst SM (2020) The intestinal regionalization of acute norovirus infection is regulated by the microbiota via bile acid-mediated priming of type III interferon. Nat Microbiol. https://doi.org/10.1038/s41564-019-0602-7

Groves HT, Cuthbertson L, James P, Moffatt MF, Cox MJ, Tregoning JS (2018) Respiratory disease following viral lung infection alters the murine gut microbiota. Front Immunol 9:12. https://doi.org/10.3389/fimmu.2018.00182

Hakansson A, Kidd A, Wadell G, Sabharwal H, Svanborg C (1994) Adenovirus infection enhances in vitro adherence of *Streptococcus pneumoniae*. Infect Immunity 62(7):2707

Hament J-M, Kimpen JL, Fleer A, Wolfs TF (1999) Respiratory viral infection predisposing for bacterial disease: a concise review. FEMS Immunol Med Microbiol 26:189–195. https://doi.org/10.1111/j.1574-695x.1999.tb01389.x

Hamilos DL (2015) Pediatric chronic rhinosinusitis. Am J Rhinol Allergy 29:414–420. https://doi.org/10.2500/ajra.2015.29.4238

Hanada S, Pirzadeh M, Carver KY, Deng JC (2018) Respiratory viral infection-induced microbiome alterations and secondary bacterial pneumonia. Front Immunol. https://doi.org/10.3389/fimmu.2018.02640

He Y, Wang J, Li F, Shi Y (2020) Main clinical features of COVID-19 and potential prognostic and therapeutic value of the microbiota in SARS-CoV-2 infections. Front Microbiol 11:1302. https://doi.org/10.3389/fmicb.2020.01302

Hickman D, Jones MK, Zhu S, Kirkpatrick E, Ostrov DA, Wang X, Ukhanova M, Sun Y (2014) The effect of malnutrition on norovirus infection. MBio 5. https://doi.org/10.1128/mBio.01032-13.Editor

Huang PM (2020) Microbiota and enteric viruses infection. Med Microbiol 3:100006

Huang CB, Alimova YV, Ebersole JL (2011) HIV-1 reactivation in HIV-latently infected dendritic cells by oral microorganisms and LPS. Cell Immunol 268:105–111. https://doi.org/10.1016/j.cellimm.2011.02.003

Inoue T, Nakayama J, Moriya K, Kawaratani H, Momoda R, Ito K, Iio E, Nojiri S, Fujiwara K, Yoneda M, Yoshiji H, Tanaka Y (2018) Gut dysbiosis associated with hepatitis C virus infection. Clin Infect Dis 67:869–877. https://doi.org/10.1093/cid/ciy205

Jain N, Lodha R, Kabra SK (2001) Upper respiratory tract infections. Indian J Pediatr. https://doi.org/10.1007/BF02722930

Jang JY, Kim S, Kwon MS, Lee J, Yu DH, Song RH, Choi HJ, Park J (2019) Rotavirus-mediated alteration of gut microbiota and its correlation with physiological characteristics in neonatal calves. J Microbiol 57:113–121. https://doi.org/10.1007/s12275-019-8549-1

Jones MK, Watanabe M, Zhu S, Graves CL, Keyes LR, Grau KR, Gonzalez-Hernandez MB, Iovine NM, Wobus CE, Vinjé J, Tibbetts SA, Wallet SM, Karst SM (2014) Enteric bacteria promote human and mouse norovirus infection of B cells. Science (80–) 346:755–759. https://doi.org/10.1126/science.1257147

Jude BA, Pobezinskaya Y, Bishop J, Parke S, Medzhitov RM, Chervonsky AV, Golovkina TV (2003) Subversion of the innate immune system by a retrovirus. Nat Immunol 4:573–578. https://doi.org/10.1038/ni926

Kalantar-Zadeh K, Ward SA, Kalantar-Zadeh K, El-Omar EM (2020) Considering the effects of microbiome and diet on SARS-CoV-2 infection: nanotechnology roles. ACS Nano. https://doi.org/10.1021/acsnano.0c03402

Kanangat S, Meduri GU, Tolley EA, Patterson DR, Meduri CU, Pak C, Griffin JP, Bronze MS, Schaberg DR (1999) Effects of cytokines and endotoxin on the intracellular growth of bacteria. Infect Immun 67:2834–2840. https://doi.org/10.1128/iai.67.6.2834-2840.1999

Kane M, Case LK, Kopaskie K, Kozlova A, MacDearmid C, Chervonsky AV, Golovkina TV (2011) Successful transmission of a retrovirus depends on the commensal microbiota. Science (80–) 334:245–249. https://doi.org/10.1126/science.1210718

Kang HM, Kang JH (2021) Effects of nasopharyngeal microbiota in respiratory infections and allergies. Clin Exp Pediatr. https://doi.org/10.3345/cep.2020.01452

Karst SM (2016) The influence of commensal bacteria on infection with enteric viruses. Nat Rev Microbiol 14:197–204

Kemper Alston W, Fahrner KI (2003) Upper respiratory tract infections. Med Manag Infect Dis 68:187–207. https://doi.org/10.5005/jp/books/12168_2

Khan AA, Khan Z (2020) COVID-2019-associated overexpressed Prevotella proteins mediated host-pathogen interactions and their role in coronavirus outbreak. Bioinformatics 36:4065–4069. https://doi.org/10.1093/bioinformatics/btaa285

Kong F, Saif LJ, Wang Q (2021) Roles of bile acids in enteric virus replication. Anim Dis 1. https://doi.org/10.1186/s44149-021-00003-x

Kuek LE, Lee RJ (2020) First contact: the role of respiratory cilia in host-pathogen interactions in the airways. Am J Physiol Lung Cell Mol Physiol. https://doi.org/10.1152/AJPLUNG.00283.2020

Kumthip K, Khamrin P, Ushijima H, Maneekarn N (2019) Enteric and non-enteric adenoviruses associated with acute gastroenteritis in pediatric patients in Thailand, 2011 to 2017. PLoS One 14:e0220263. https://doi.org/10.1371/journal.pone.0220263

Kuss SK, Best GT, Etheredge CA, Pruijssers AJ, Frierson J, Hooper LV, Dermody TS, Pfeiffer JK (2011) Intestinal microbiota promote enteric virus replication and systemic pathogenesis. Science (80–) 334:249–253

Lee H, Ko GP (2016) Antiviral effect of vitamin A on norovirus infection via modulation of the gut microbiome. Sci Rep 6:1–9. https://doi.org/10.1038/srep25835

Lee KH, Gordon A, Shedden K, Kuan G, Ng S, Balmaseda A, Foxman B (2019) The respiratory microbiome and susceptibility to influenza virus infection. PLoS One 14:e0207898. https://doi.org/10.1371/journal.pone.0207898

Lee KH, Song Y, Wu W, Yu K, Zhang G (2020) The gut microbiota, environmental factors, and links to the development of food allergy. Clin Mol Allergy 18:5. https://doi.org/10.1186/s12948-020-00120-x

Li Y, Saxena D, Chen Z, Liu G, Abrams WR, Phelan JA, Norman RG, Fisch GS, Corby PM, Dewhirst F, Paster BJ, Kokaras AS, Malamud D (2014) HIV infection and microbial diversity in saliva. J Clin Microbiol 52:1400–1411. https://doi.org/10.1128/JCM.02954-13

Li D, Breiman A, le Pendu J, Uyttendaele M (2015) Binding to histo-blood group antigen-expressing bacteria protects human norovirus from acute heat stress. Front Microbiol 6:1–8. https://doi.org/10.3389/fmicb.2015.00659

Li N, Ma W-T, Pang M, Fan Q-L, Hua J-L (2019) The commensal microbiota and viral infection: a comprehensive review. Front Immunol 10:1–16. https://doi.org/10.3389/fimmu.2019.01551

Li Y, Ren B, Peng X, Hu T, Li J, Gong T et al (2020) Saliva is a non-negligible factor in the spread of COVID-19. Mol Oral Microbiol. 35:141–145. https://doi.org/10.1111/omi.12289

Lukkarinen H, Söderlund-Venermo M, Vuorinen T, Allander T, Hedman K, Simell O, Ruuskanen O, Jartti T (2014) Human bocavirus 1 may suppress rhinovirus-associated immune response in wheezing children. J Allergy Clin Immunol 133:256–258.e4. https://doi.org/10.1016/j.jaci.2013.10.014

Ma C, Wu X, Nawaz M, Li J, Yu P, Moore JE, Xu J (2011) Molecular characterization of fecal microbiota in patients with viral diarrhea. Curr Microbiol 63:259–266. https://doi.org/10.1007/s00284-011-9972-7

Maiese K (2020) Circadian clock genes: targeting innate immunity for antiviral strategies against COVID-19. Curr Neuorvasc Res 17:531–533

Mallia P, Footitt J, Sotero R, Jepson A, Contoli M, Trujillo-Torralbo M-B, Kebadze T, Aniscenko J, Oleszkiewicz G, Gray K, Message SD, Ito K, Barnes PJ, Adcock IM, Papi A, Stanciu LA, Elkin SL, Kon OM, Johnson M, Johnston SL (2012) Rhinovirus infection induces degradation of antimicrobial peptides and secondary bacterial infection in chronic obstructive pulmonary disease. Am J Respir Crit Care Med 186:1117–1124. https://doi.org/10.1164/rccm.201205-0806OC

Mansbach JM, Hasegawa K, Henke DM, Ajami NJ, Petrosino JF, Shaw CA, Piedra PA, Sullivan AF, Espinola JA, Camargo CA (2016) Respiratory syncytial virus and rhinovirus severe bronchiolitis are associated with distinct nasopharyngeal microbiota. J Allergy Clin Immunol 137:1909–1913.e4. https://doi.org/10.1016/j.jaci.2016.01.036

Marom T, Nokso-Koivisto J, Chonmaitree T (2012) Viral-bacterial interactions in acute otitis media. Curr Allergy Asthma Rep. https://doi.org/10.1007/s11882-012-0303-2

Mazzoccoli G, Vinciguerra M, Carbone A, Relógio A (2020) The circadian clock, the immune system, and viral infections: the intricate relationship between biological time and host-virus interaction. Pathogens 9:83. https://doi.org/10.3390/pathogens9020083

Miura T, Sano D, Suenaga A, Yoshimura T, Fuzawa M, Nakagomi T, Nakagomi O, Okabe S (2013) Histo-blood group antigen-like substances of human enteric bacteria as specific adsorbents for human noroviruses. J Virol 87:9441–9451. https://doi.org/10.1128/JVI.01060-13

Moesker FM, van Kampen JJA, van der Eijk AA, van Rossum AMC, de Hoog M, Schutten M, Smits SL, Bodewes R, Osterhaus ADME, Fraaij PLA (2015) Human bocavirus infection as a cause of severe acute respiratory tract infection in children. Clin Microbiol Infect 21:964.e1–964.e8. https://doi.org/10.1016/j.cmi.2015.06.014

Monedero V, Buesa J, Rodríguez-Díaz J (2018) The interactions between host glycobiology, bacterial microbiota, and viruses in the gut. Viruses. https://doi.org/10.3390/v10020096

Moore MD, Suther C, Zhou Y (2020) Microbiota, viral infection, and the relationship to human diseases. Infect Microbes Dis 3:1–3. https://doi.org/10.1097/im9.0000000000000043

Morris A, Beck JM, Schloss PD, Campbell TB, Crothers K, Curtis JL, Flores SC, Fontenot AP, Ghedin E, Huang L, Jablonski K, Kleerup E, Lynch SV, Sodergren E, Twigg H, Young VB, Bassis CM, Venkataraman A, Schmidt TM, Weinstock GM (2013) Comparison of the respiratory microbiome in healthy nonsmokers and smokers. Am J Respir Crit Care Med 187:1067–1075. https://doi.org/10.1164/rccm.201210-1913OC

Nardelli C, Gentile I, Setaro M, Di Domenico C, Pinchera B, Buonomo AR, Zappulo E, Scotto R, Scaglione GL, Castaldo G, Capoluongo E (2021) Nasopharyngeal microbiome signature in COVID-19 positive patients: can we definitively get a role to Fusobacterium periodonticum? Front Cell Infect Microbiol 11. https://doi.org/10.3389/fcimb.2021.625581

Nelson AM, Walk ST, Taube S, Taniuchi M, Houpt ER, Wobus CE, Young VB (2012) Disruption of the human gut microbiota following norovirus infection. PLoS One 7:e48224. https://doi.org/10.1371/journal.pone.0048224

Neu U, Mainou BA (2020) Virus interactions with bacteria: partners in the infectious dance. PLoS Pathog 16:e1008234. https://doi.org/10.1371/journal.ppat.1008234

Paget C, Trottein F (2019) Mechanisms of bacterial superinfection post-influenza: a role for unconventional T cells. Front Immunol. https://doi.org/10.3389/fimmu.2019.00336

Park DR (2005) The microbiology of ventilator-associated pneumonia. Respir Care 50

Peltola VT, Mccullers JA (2004) Respiratory viruses predisposing to bacterial infections: role of neuraminidase. Pediatr Infect Dis J 23:S87–S97. https://doi.org/10.1097/01.inf.0000108197.81270.35

Peltola VT, Murti KG, McCullers JA (2005) Influenza virus neuraminidase contributes to secondary bacterial pneumonia. J Infect Dis 192:249–257. https://doi.org/10.1086/430954

Quintana VM, Torres NI, Wachsman MB, Sinko PJ, Castilla V, Chikindas M (2014) Antiherpes simplex virus type 2 activity of the antimicrobial peptide subtilosin. J Appl Microbiol 117:1253–1259. https://doi.org/10.1111/jam.12618

Ray S, Reddy A (2020) COVID-19 management in light of the circadian clock. Nat Rev Mol Cell Biol 21:494–495

Ren X, Wen W, Fan X, Hou W, Su B, Cai P, Li J et al (2021) COVID-19 immune features revealed by a large-scale single-cell transcriptome atlas. Cell 184:1895–1913

Robinson CM, Pfeiffer JK (2014) Leaping the norovirus hurdle. Science (80–) 346:700–701. https://doi.org/10.1126/science.aaa0607

Robinson CM, Jesudhasan PR, Pfeiffer JK (2014) Bacterial lipopolysaccharide binding enhances virion stability and promotes environmental fitness of an enteric virus. Cell Host Microbe 15:36–46. https://doi.org/10.1016/j.chom.2013.12.004

Robinson K, Kolls J, Alcorn J (2015) The immunology of influenza virus-associated bacterial pneumonia. Curr Opin Immunol 34:59–67

Rosas-Salazar C, Shilts MH, Tovchigrechko A, Schobel S, Chappell JD, Larkin EK, Shankar J, Yooseph S, Nelson KE, Halpin RA, Moore ML, Anderson LJ, Peebles RS, Das SR, Hartert TV (2016) Differences in the nasopharyngeal microbiome during acute respiratory tract infection with human rhinovirus and respiratory syncytial virus in infancy. J Infect Dis 214:1924–1928. https://doi.org/10.1093/infdis/jiw456

Rowe HM, Meliopoulos VA, Iverson A, Bomme P, Schultz-Cherry S, Rosch JW (2019) Direct interactions with influenza promote bacterial adherence during respiratory infections. Nat Microbiol 4:1328–1336. https://doi.org/10.1038/s41564-019-0447-0

Rynda-Apple A, Robinson K, Alcorn J (2015) Influenza and bacterial superinfection: illuminating the immunologic mechanisms of disease. Infect Immun 83:3764–3770

Schroeder BO (2019) Fight them or feed them: how the intestinal mucus layer manages the gut microbiota. Gastroenterol Rep. https://doi.org/10.1093/gastro/goy052

Segrist E, Cherry S (2020) Using diverse model systems to define intestinal epithelial defenses to enteric viral infections. Cell Host Microbe 27:329–344

Shope RE (1931) Swine influenza: III. filtration experiments and etiology. J Exp Med 54:373–385. https://doi.org/10.1084/jem.54.3.373

Simoons-Smit AM, Kraan EM, Beishuizen A, Strack van Schijndel RJ, Vandenbroucke-Grauls CM (2006) Herpes simplex virus type 1 and respiratory disease in critically-ill patients: real pathogen or innocent bystander? Clin Microbiol Infect. https://doi.org/10.1111/j.1469-0691.2006.01475.x

Smith CB, Golden C, Klauber MR, Kanner R, Renzetti A (1976) Interactions between viruses and bacteria in patients with chronic bronchitis. J Infect Dis 134:552–561

Smith CM, Sandrini S, Datta S, Shafeeq S, Radhakrishnan P, Williams G, Glenn SM, Kuipers OP, Hirst RA, Easton AJ, Andrew PW, O'callaghan C (2014) Respiratory syncytial virus increases the virulence of *Streptococcus pneumoniae* by binding to penicillin binding protein 1a a new paradigm in respiratory infection. Am J Respir Crit Care Med 190:196–207. https://doi.org/10.1164/rccm.201311-2110OC

Spacova I, De Boeck I, Bron PA, Delputte P, Lebeer S (2021) Topical microbial therapeutics against respiratory viral infections. Trends Mol Med. https://doi.org/10.1016/j.molmed.2021.03.009

Sultan A, Ali R, Sultan T, Ali S, Khan NJ, Parganiha A (2021) Circadian clock modulating small molecules repurposing as inhibitors of SARS-CoV-2 Mpro for pharmacological interventions in COVID-19 pandemic. Chronobiol Int. https://doi.org/10.1080/07420528.2021.1903027

Sun Y, Ma Y, Lin P, Tang YW, Yang L, Shen Y, Zhang R, Liu L, Cheng J, Shao J, Qi T, Tang Y, Cai R, Guan L, Luo B, Sun M, Li B, Pei Z, Lu H (2016) Fecal bacterial microbiome diversity in chronic HIV-infected patients in China. Emerg Microbes Infect 5. https://doi.org/10.1038/emi.2016.25

Tabata T, Petitt M, Puerta-Guardo H, Michlmayr D, Wang C, Fang-Hoover J, Harris E, Pereira L (2016) Zika virus targets different primary human placental cells, suggesting two routes for vertical transmission. Cell Host Microbe 20:155–166. https://doi.org/10.1016/j.chom.2016.07.002

Takase H, Nitanai H, Yamamura E, Otani T (1999) Facilitated expansion of pneumococcal colonization from the nose to the lower respiratory tract in mice preinfected with influenza virus. Microbiol Immunol 43:905–907. https://doi.org/10.1111/j.1348-0421.1999.tb01226.x

Tan M, Hegde RS, Jiang X (2004) The P domain of norovirus capsid protein forms dimer and binds to histo-blood group antigen receptors. J Virol 78:6233–6242. https://doi.org/10.1128/JVI.78.12.6233

Tappert MM, Porterfield JZ, Mehta-D'Souza P, Gulati S, Air GM (2013) Quantitative comparison of human parainfluenza virus hemagglutinin-neuraminidase receptor binding and receptor cleavage. J Virol 87:8962–8970. https://doi.org/10.1128/jvi.00739-13

Tarris G, De Rougemont A, Charkaoui M, Michiels C, Martin L, Belliot G (2021) Enteric viruses and inflammatory bowel disease. Viruses 13:104. https://doi.org/10.3390/v13010104

Tashiro M, Ciborowski P, Reinacher M, Pulverer G, Klenk HD, Rott R (1987) Synergistic role of staphylococcal proteases in the induction of influenza virus pathogenicity. Virology 157:421–430. https://doi.org/10.1016/0042-6822(87)90284-4

Thorburn K, Harigopal S, Reddy V, Taylor N, Van Saene HKF (2006) High incidence of pulmonary bacterial co-infection in children with severe respiratory syncytial virus (RSV) bronchiolitis. Thorax 61:611–615. https://doi.org/10.1136/thx.2005.048397

Tristram DA, Hicks W, Hard R (1998) Respiratory syncytial virus and human bronchial epithelium. Arch Otolaryngol Head Neck Surg 124:777–783. https://doi.org/10.1001/archotol.124.7.777

Troeger C, Blacker B, Khalil IA, Rao PC, Cao J, Zimsen SRM, Albertson SB, Deshpande A, Farag T, Abebe Z, Adetifa IMO, Adhikari TB, Akibu M, Al Lami FH, Al-Eyadhy A, Alvis-Guzman N, Amare AT, Amoako YA, Antonio CAT, Aremu O, Asfaw ET, Asgedom SW, Atey TM, Attia EF, Avokpaho EFGA, Ayele HT, Ayuk TB, Balakrishnan K, Barac A, Bassat Q, Behzadifar M, Behzadifar M, Bhaumik S, Bhutta ZA, Bijani A, Brauer M, Brown A, Camargos PAM, Castañeda-Orjuela CA, Colombara D, Conti S, Dadi AF, Dandona L, Dandona R, Do HP, Dubljanin E, Edessa D, Elkout H, Endries AY, Fijabi DO, Foreman KJ, Forouzanfar MH, Fullman N, Garcia-Basteiro AL, Gessner BD, Gething PW, Gupta R, Gupta T, Hailu GB, Hassen HY, Hedayati MT, Heidari M, Hibstu DT, Horita N, Ilesanmi OS, Jakovljevic MB, Jamal AA, Kahsay A, Kasaeian A, Kassa DH, Khader YS, Khan EA, Khan MN, Khang YH, Kim YJ, Kissoon N, Knibbs LD, Kochhar S, Koul PA, Kumar GA, Lodha R, Magdy Abd El Razek H, Malta DC, Mathew JL, Mengistu DT, Mezgebe HB, Mohammad KA, Mohammed MA, Momeniha F, Murthy S, Nguyen CT, Nielsen KR, Ningrum DNA, Nirayo YL, Oren E, Ortiz JR, Postma MJ, Qorbani M, Quansah R, Rai RK, Rana SM, Ranabhat CL, Ray SE, Rezai MS, Ruhago GM, Safiri S, Salomon JA, Sartorius B, Savic M, Sawhney M, She J, Sheikh A, Shiferaw MS, Shigematsu M, Singh JA, Somayaji R, Stanaway JD, Sufiyan MB, Taffere GR, Temsah MH, Thompson MJ, Tobe-Gai R, Topor-Madry R, Tran BX, Tran TT, Tuem KB, Ukwaja KN, Vollset SE, Walson JL, Weldegebreal F, Werdecker A, West TE, Yonemoto N, Zaki MES, Zhou L, Zodpey S, Vos T, Naghavi M, Lim SS, Mokdad AH, Murray CJL, Hay SI, Reiner RC (2018) Estimates of the global, regional, and national morbidity, mortality, and aetiologies of lower respiratory infections in 195 countries, 1990–2016: a systematic analysis for the Global Burden of Disease Study 2016. Lancet Infect Dis 18:1191–1210. https://doi.org/10.1016/S1473-3099(18)30310-4

Uchiyama R, Chassaing B, Zhang B, Gewirtz AT (2014) Antibiotic treatment suppresses rotavirus infection and enhances specific humoral immunity. J Infect Dis 210:171–182. https://doi.org/10.1093/infdis/jiu037

Van Den Broek MFL, De Boeck I, Kiekens F, Boudewyns A, Vanderveken OM, Lebeer S (2019) Translating recent microbiome insights in otitis media into probiotic strategies. Clin Microbiol Rev 32:e00010–e00018. https://doi.org/10.1128/CMR.00010-18

Vareille M, Kieninger E, Edwards MR, Regamey N (2011) The airway epithelium: soldier in the fight against respiratory viruses. Clin Microbiol Rev 24:210–229. https://doi.org/10.1128/CMR.00014-10

Venkataraman A, Bassis CM, Beck JM, Young VB, Curtis JL, Huffnagle GB, Schmidt TM (2015) Application of a neutral community model to assess structuring of the human lung microbiome. MBio 6. https://doi.org/10.1128/mBio.02284-14

Verkaik NJ, Nguyen DT, de Vogel CP, Moll HA, Verbrugh HA, Jaddoe VWV, Hofman A, van Wamel WJB, van den Hoogen BG, Buijs-Offerman RMGB, Ludlow M, de Witte L, Osterhaus

ADME, van Belkum A, de Swart RL (2011) Streptococcus pneumoniae exposure is associated with human metapneumovirus seroconversion and increased susceptibility to in vitro HMPV infection. Clin Microbiol Infect 17:1840–1844. https://doi.org/10.1111/j.1469-0691.2011.03480.x

Vidya KM, Rao UK, Nittayananta W, Liu H, Owotade FJ (2016) Oral mycoses and other opportunistic infections in HIV: therapy and emerging problems - a workshop report. Oral Dis. 22:158–165. https://doi.org/10.1111/odi.12437

Vujkovic-Cvijin I, Dunham RM, Iwai S, Maher MC, Albright RG, Broadhurst MJ, Hernandez RD, Lederman MM, Huang Y, Somsouk M, Deeks SG, Hunt PW, Lynch SV, McCune JM (2013) Dysbiosis of the gut microbiota is associated with HIV disease progression and tryptophan catabolism. Sci Transl Med 5. https://doi.org/10.1126/scitranslmed.3006438

Walsh JJ, Dietlein LF, Low FN, Burch GE, Mogabgab WJ (1961) Bronchotracheal response in human influenza. Type A, Asian strain, as studied by light and electron microscopic examination of bronchoscopic biopsies. Arch Intern Med 108:376–388. https://doi.org/10.1001/archinte.1961.03620090048006

Wang C, Ji D, Chen J, Shao Q, Li B, Liu J, Wu V, Wong A, Wang Y, Zhang X, Lu L, Wong C, Tsang S, Zhang Z, Sun J, Hou J, Chen G, Lau G (2017) Hepatitis due to reactivation of Hepatitis B virus in endemic areas among patients with Hepatitis C treated with direct-acting antiviral agents. Clin Gastroenterol Hepatol 15:132–136. https://doi.org/10.1016/j.cgh.2016.06.023

Wei X, Pieters RJ (2021) Multivalency effects in neuraminidase inhibitor design for influenza virus. Antimicrob Agents Chemother. https://doi.org/10.24820/ark.5550190.p011.368

Wilen CB, Lee S, Hsieh LL, Orchard RC, Desai C, Hykes BL, McAllaster MR, Balce DR, Feehley T, Brestoff JR, Hickey CA, Yokoyama CC, Wang YT, MacDuff DA, Kreamalmayer D, Howitt MR, Neil JA, Cadwell K, Allen PM, Handley SA, van Lookeren Campagne M, Baldridge MT, Virgin HW (2018) Tropism for tuft cells determines immune promotion of norovirus pathogenesis. Science (80–) 360:204–208. https://doi.org/10.1126/science.aar3799

Wilks J, Golovkina T (2012) Influence of microbiota on viral infections. PLoS Pathog 8:e1002681. https://doi.org/10.1371/journal.ppat.1002681

Wilks J, Lien E, Jacobson AN, Fischbach MA, Qureshi N, Chervonsky AV, Golovkina TV (2015) Mammalian lipopolysaccharide receptors incorporated into the retroviral envelope augment virus transmission. Cell Host Microbe 18:456–462. https://doi.org/10.1016/j.chom.2015.09.005

Williams JV, Harris PA, Tollefson SJ, Halburnt-Rush LL, Pingsterhaus JM, Edwards KM, Wright PF, Crowe JE (2004) Human metapneumovirus and lower respiratory tract disease in otherwise healthy infants and children. N Engl J Med 350:443–450. https://doi.org/10.1056/nejmoa025472

Xiang Z, Koo H, Chen Q, Zhou X, Liu Y, Simon-Soro A (2021) Potential implications of SARS-CoV-2 oral infection in the host microbiota. J Oral Microbiol 13:1853451. https://doi.org/10.1080/20002297.2020.1853451

Xu M, Wang B, Fu Y, Chen Y, Yang F, Lu H, Chen Y, Xu J, Li L (2012) Changes of fecal Bifidobacterium species in adult patients with Hepatitis B virus-induced chronic liver disease. Microb Ecol 63:304–313. https://doi.org/10.1007/s00248-011-9925-5

Yang X-L, Wang G, Xie J-Y, Li H, Liu W, Zhu SJ (2021) The intestinal microbiome primes host innate immunity against enteric virus 1 systemic infection through type I interferon 2. mBio 12: e00366-21

Yildiz S, Mazel-Sanchez B, Kandasamy M, Manicassamy B, Schmolke M (2018) Influenza A virus infection impacts systemic microbiota dynamics and causes quantitative enteric dysbiosis. Microbiome 6:1–17. https://doi.org/10.1186/s40168-017-0386-z

Zhang H h, Liu J, Lv Y j, Jiang YL, Pan J x, Zhu Y j, Huang M g, Zhang S k (2020) Changes in intestinal microbiota of type 2 diabetes in mice in response to dietary supplementation with instant tea or matcha. Can J Diabetes 44:44–52. https://doi.org/10.1016/j.jcjd.2019.04.021

The manufacturer's authorised representative in the EU is Springer
Nature Customer Service Centre GmbH, Europaplatz 3, 69115 Heidelberg,
Germany. If you have any concerns regarding our products, please
contact ProductSafety@springernature.com

Printed and bound by CPI Group (UK) Ltd, Croydon, CR0 4YY
30/09/2026
02225703-0003